AF430626

Basic Electronics Engineering

Basic Electronics Engineering

Dr. K. Lal Kishore

Ph.D, MIEEE, FIE, FIETE, MISTE, MISHM.

Director and Professor of Electronics & Communication Engineering,
Jawaharlal Nehru Technological University, Kukatpally,
Hyderabad - 500 085.

BSP **BS Publications**

An unit of **BSP Books Pvt., Ltd.**

4-4-309, Giriraj Lane, Sultan Bazar,
Hyderabad - 500 095
Phone : 040 - 23445605, 23445688

Published by :

BS Publications
An unit of **BSP Books Pvt., Ltd.**

4-4-309, Giriraj Lane, Sultan Bazar,
Hyderabad - 500 095
Phone : 040 - 23445605, 23445688
e-mail : info@bspbooks.net

ISBN : 978-938543-306-1 (HB)

Preface

The field of electronics fascinates many, both educated and not-so-educated. Science reveals the laws of nature, engineering enables designs based on the laws, and Technology puts it to human needs. After industrial revolution, in which mostly Mechanical and Electrical Engineering fields played key role, it is the turn of Electronics Engineering in the last leg of Twentieth century. Subsequently it is Computer Science in the fag end of the Twentieth century and Computer Communications in the beginning of 21st century.

This course is the basic foundation course to understand the principles of Electronic Devices and Basic Circuits. The first book written by this author entitled Electronic Devices and Circuits became popular with students as well as teachers. Though number of books are published in this area, the author felt that there is need for a book which explains clearly the principles and is helpful to students as well as teachers. Though many students of electronic engineering go through this course, still many somehow manage to pass the examination but fail to appreciate the essence of the subject. With sudden expansion of Technical Education in the country all over, there is acute shortage of experienced Teachers to teach this kind of subjects. This prompted the author to write this Book, with the 33 years of Teaching experience and five other books written earlier.

This book enables the teachers as well as students to understand concepts which are not explained in many other books. So conceptual questions with answers are given at the end of every chapter. Some Questions and answers are also given which are generally asked in interviews and competitive exams. Objective type questions and many solved problems are also given to enable better understanding of the subject. This book can also be used for AMIETE, AMIE (Electronics), M.Sc. (Electronics), B.Sc. (Electronics) and Polytechnique students who are studying a course in Electronics.

Suggestions are welcome to further improve the contents and make this book more 'learner friendly'.

- Author

Contents

CHAPTER - 1 | Junction Diode Characteristics

CHAPTER - 2 | Rectifiers, Filters and Regulators

CHAPTER - 3 | Transistor Characteristics

CHAPTER - 4 | Transistor Biasing and Stabilization

Chapter - 5 | Field Effect Transistors (FETs)

Chapter - 6 | Amplifiers and Oscillators

CHAPTER - 7 | Feedback Amplifiers

CHAPTER - 8 | Oscillators

Junction Diode Characteristics

In this Chapter,

- The basic aspects connected with Semiconductor Physics and the terms Effective Mass, Intrinsic, Extrinsic, Semiconductors are introduced.

- Atomic Structure, Configurations, Concept of Hole, Conductivity in *p-type* and *n-type* semiconductors are given. This forms the basis to study Semiconductor Devices in the following chapters.

- After p-type and n-type semiconductors, we shall study semiconductor devices formed using these two types of semiconductors.

- We shall also study the physical phenomena such as conduction, transport mechanism, electrical characteristics, and applications of semiconductor diodes, zener diode, tunnel diode and so on.

I. INTRODUCTION

In this chapter, we shall first study the basic aspects of Atomic Theory, Electronic structure of Silicon and Germanium elements, Energy Band Theory, Semiconductor Device Physics, and then conduction in Semiconductors.

1.1   REVIEW OF SEMICONDUCTOR PHYSICS

1.1.1 ENERGY LEVELS AND ENERGY BANDS

To explain the phenomenon associated with conduction in metals and semiconductors and the emission of electrons from the surface of a metal, we have to assume that atoms have loosely bound electrons which can be removed from it.

Rutherford found that atom consists of a nucleus with electrons rotating around it. The mass of the atom is concentrated in the nucleus. It consist of protons which are positively charged. In hydrogen atom, there is one positively charged nucleus (a proton) and a single electron. The charge on particle is positive and is equal to that of electron. So hydrogen atom is neutral in charge. The proton in the nucleus carries the charge of the atom, so it is immobile. The electron will be moving around it in a closed orbit. The force of attraction between electron and proton follows Coulombs Law. {directly proportional to product of charges and inversely proportional to $(distance)^2$}.

Assume that the orbit of the electron around nucleus is a circle. We want to calculate the radius of this circle, in terms of total energy 'W' of the electron. The force of attraction between the electron and the nucleus is :

$$F \; \alpha \; e^2$$

(Therefore, the nucleus has the proton with electron charge equal to 'e')

$$F \; \alpha \; \frac{1}{r^2}$$

(r is the radius of the orbit)

$$\therefore \qquad F \; \alpha \; \frac{e^2}{r^2}$$

$$F = \frac{e^2}{4\pi \in_o r^2}$$

$$\therefore \qquad F \; \alpha \; \frac{e^2}{r^2}$$

where $\in_o$ is the permitivity of free space. Its value is $\dfrac{10^{-9}}{36\pi}$ F/m.

As the electron is moving around the nucleus in a circular orbit with radius r, and with velocity v, then the force of attraction given by the above expression F should be equal to the

centripetal force $\dfrac{mv^2}{r}$ (according to Newton's Second Law of Motion)

(F = m.a; m is the mass of electron and acceleration, $a = \dfrac{v^2}{r}$)

$\therefore$ The Potential Energy of the electron $= -\dfrac{e^2}{4\pi\varepsilon_0 r^2} \times r = -\dfrac{e^2}{4\pi\varepsilon_0 r}$

(–ve sign is because Potential Energy is by definition work done against the field)

Kinetic Energy $= \frac{1}{2} mv^2$

$\therefore$ Total Energy W possessed by the electron $= \dfrac{1}{2} mv^2 - \dfrac{e^2}{4\pi\varepsilon_0 r}$

But $\qquad mv^2 = \dfrac{e^2}{4\pi\varepsilon_0 r} \qquad$ reduces to

$$W = \dfrac{e^2}{8\pi\varepsilon_0 r} - \dfrac{e^2}{4\pi\varepsilon_0 r} = -\dfrac{e^2}{8\pi\varepsilon_0 r}$$

$\therefore$ Energy possessed by the electron $W = -\dfrac{e^2}{8\pi\varepsilon_0 r}$

W is the energy of the electron. Only for Hydrogen atom W will also be the energy of atom since it has only one electron. The negative sign arises because the Potential Energy of the electron is

$$-\dfrac{e^2}{8\pi\varepsilon_0 r}$$

As radius r increases, Potential Energy decreases. When r is infinity, Potential Energy is zero. Therefore the energy of the electron is negative. If r is $< \infty$, the energy should be less. (Any quantity less than 0 is negative).

The above equation is derived from the classical model of the electron. But according to classical laws of electromagnetism, an accelerated charge must radiate energy. Electron is having charge $= e$. It is moving with velocity v or acceleration $\dfrac{v^2}{r}$ around the nucleus. Therefore this electron should also radiate energy. If the charge is performing oscillations with a frequency 'f', then the frequency of the radiated energy should also be the same. Hence the frequency of the radiated energy from the electron should be equal to the frequency with the electron orbiting round the nucleus.

But if the electron is radiating energy, then its total energy 'W' must decrease by the amount equal to the radiation energy. So 'W' should go on decreasing to satisfy the equation,

$$W = -\dfrac{e^2}{8\pi\varepsilon_0 r}$$

If W decreases, r should also decrease. Since if $-\dfrac{e^2}{8\pi\varepsilon_0 r}$ should decrease, this quantity should become more negative. Therefore, r should decrease. So, the electron should describe smaller and smaller orbit and should finally fall into nucleous. So classical model of atom is not fairly satisfactory.

1.1.2 THE BOHR ATOM

The above difficulty was resolved by Bohr in 1913. He postulated three fundamental laws.

1. *The atom can possess only discrete energies. While in states corresponding to these discrete energy levels, electron does not emit radiation in stationary state.*

2. *When the energy of the electron is changing from W_2 to W_1 then radiation will be emitted. The frequency of radiation is given by*

$$f = \frac{W_2 - W_1}{h}$$

   where 'h' is Plank's Constant.

$$h = 6.626 \times 10^{-34} \text{ J} - \text{sec.}$$

   i.e., when the atom is in stationary state, it does not emit any radiation. When its energy changes from W_2 to W_1 then the atom is said to have moved from one stationary state to the other. The atom remains in the new state corresponding to W_1. Only during transition, will some energy be radiated.

3. *A stationary state is determined by the condition that the angular momentum of the electron in this state must be an integral multiple of $h/2\pi$.*

$$So \quad mvr = \frac{nh}{2\pi}$$

   where n is an integer, other than zero.

1.1.3 EFFECTIVE MASS

An electron mass 'm' when placed in a crystal lattice, responds to applied field as if it were of mass m*. The reason for this is the interaction of the electron even within lattice.

$$E = \text{Kinetic Energy of the Free Electron}$$
$$p = \text{Momentum}$$
$$v = \text{Velocity}$$
$$m = \text{Mass}$$
$$p = mv$$
$$m* = \text{Effective mass of electron}$$

$$E = \frac{p^2}{2m}$$

Electrons in a solid are not free. They move under the combined influence of an external field plus that of a periodic potential of atom cores in the lattice. An electron moving through the lattice can be represented by a wave packet of plane waves grouped around the same value of K which is a wave vector.

Electron velocity falls to zero at each band edge. This is because the electron wave further becomes standing wave at the top and bottom of a band i.e., $v_g = 0$.

Now consider an electronic wave packet moving in a crystal lattice under the influence of an externally applied uniform electric field. If the electron has an instantaneous velocity v_g and moves a distance dx in the direction of an accelerating force F, in time dt, it acquires energy δE where

$$\delta E = F \times \delta x = F \times v_g \times \delta t$$

$$\delta E = \frac{F}{\hbar} \times \frac{\delta E}{\delta k} \times \delta t$$

$$v_g = \frac{\delta E}{\hbar\, \delta k}$$

Within the limit of small increments in K, we can write,

$$\frac{dK}{dt} = \frac{F}{\hbar} \qquad\qquad ....(\ 1.1 \ )$$

But this is not the case for the electron in a solid because the externally applied force is not the only force acting on the electrons. Forces associated with the periodic lattice are also present.

Acceleration of an electronic wave packet in a solid is equal to the rate of change of its velocity.

$$\text{Acceleration of an electron in a solid} = \frac{d v_g}{d t} = \frac{d}{dt}\left(\frac{dE}{dp}\right) = \frac{d^2 E}{(dp)(dt)}$$

$$\therefore \qquad v_g = \frac{dE}{dp}$$

$$\frac{dk}{dt} = \frac{F}{\hbar}$$

$$F = \hbar \times \frac{dk}{dt}$$

But
$$\frac{dk}{dt} = \frac{F}{\hbar} \qquad\qquad \text{from Eq. } (\ 1.1 \ )$$

$$\therefore \qquad \frac{d v_g}{d t} = \left(\frac{dp}{dt}\right)\frac{d^2 E}{dp^2} = \frac{F}{\hbar^2} \cdot \frac{d^2 E}{dp^2}$$

or
$$F = \hbar^2 \times \left(\frac{d^2 E}{dp^2}\right)^{-1} \times \frac{d v_g}{dt}$$

This is of the form, $F = ma$, from Newton's Laws of Motion,

where
$$m^* = \hbar^2 \times \left(\frac{d^2 E}{dp^2}\right)^{-1} \qquad \text{and } a = \frac{dv_g}{dt}$$

$$\therefore \qquad F = m^* \times \frac{dv_g}{dV}$$

where m^* is the effective mass.

If an electric field ε is impressed, the electron will accelerate and its velocity and energy will increase. Hence the electron is said to have positive mass. On the other hand, if an electron is at the upper end of a band, when the field is applied, its energy will increase and its velocity decreases. So the electron is said to have negative mass.

In an atom,

$$\text{Coulombs force of attraction} = \frac{e^2}{4\pi \epsilon_0 r^2}$$

$$\text{Centripetal Force} = \frac{mv^2}{r}$$

Equating these two forces in an atom,

$$\frac{e^2}{4\pi \epsilon_0 r^2} = \frac{mv^2}{r} \qquad v = \text{velocity}; r = \text{radius of Orbit}$$

$$\therefore \qquad \frac{e^2}{4\pi \epsilon_0} = \frac{mv^2}{r} \times r^2 = mv^2r$$

$$\therefore \qquad r = \frac{e^2}{4\pi \epsilon_0 \, mv^2} \qquad \qquad ....(1.2)$$

But $\qquad\qquad mvr = \dfrac{nh}{2\pi} \qquad\qquad h = \text{Plank's Constant}; n = \text{Principle Quantum Number.}$

$$\therefore \qquad v = \frac{nh}{2\pi mr} \qquad\qquad ....(1.3)$$

Substituting the value of v in Equation (1.2),

$$r = \frac{e^2 \times 4\pi^2 m^2 r^2}{4\pi \epsilon_0 .m \times n^2 h^2}$$

$$\therefore \qquad \boxed{r = \frac{n^2 h^2 \epsilon_0}{\pi m e^2}} \qquad\qquad ....(1.4)$$

This is the expression for radii of stable states.

Energy possessed by the atom in the stable state is

$$W = \frac{e^2}{8\pi \epsilon_0 r} \qquad\qquad ....(1.5)$$

Substituting the value of r in Eq. (1.5), then

$$W = - \frac{e^2 \times \pi m e^2}{8\pi\epsilon_0 \, n^2 h^2 \, \epsilon_0}$$

$$W = - \frac{m e^4}{8 h^2 \epsilon_0^2} \cdot \frac{1}{n^2}$$

Thus energy W corresponds to only the coulombs force due to attraction between ground electron (negative charge) and proton (positive charge).

Problem 1.1

Determine the radius of the lowest state of Ground State.

Solution

$$n = 1$$

$$\therefore \quad r = \frac{n^2 h^2 \epsilon_0}{\pi m e^2}$$

Plank's Constant, $\quad h = 6.626 \times 10^{-34}$ J-sec

Permitivity, $\qquad \epsilon_0 = 10^{-9}/36\pi$

Substituting the values and simplifying,

$$\therefore \quad r = 0.58 \text{ A}^\circ$$

1.1.4 ATOMIC ENERGY LEVELS

For different elements, the value of the free electron concentration will be different. By spectroscopic analysis we can determine the energy level of an element at different wavelengths. This is the characteristic of the given element.

The lowest energy state is called the normal level or ground level. Other stationary states are called *excited, radiating, critical* or *resonance* levels.

Generally, the energy of different states is expressed in eV rather than in Joules, and the emitted radiation is expressed by its wavelength λ rather than by its frequency. This is only for convenience since Joule is a larger unit and the energy is small and is in electron volts.

$$F = \frac{W_2 - W_1}{h}$$

$$f = \frac{C}{\lambda}$$

C = Velocity of Light = 3×10^{10} cm / sec.

h = Plank's Constant = 6.626×10^{-34} J - sec.

f = Frequency of Radiation

λ = Wavelength of emitted radiation

W_1 and W_2 are the energy levels in Joules.

$$\boxed{1 \text{ eV} = 1.6 \times 10^{-19} \text{ Joules}}$$

$$\therefore \quad \frac{C}{\lambda} = \frac{(E_2 - E_1) \times 1.6 \times 10^{-19}}{h}$$

$$\frac{3 \times 10^{10}}{\lambda} = \frac{(E_2 - E_1) \times 1.6 \times 10^{-19}}{6.626 \times 10^{-34}}$$

E_1 and E_2 are energy levels in eV.

or
$$\lambda = \frac{3 \times 10^{10} \times 6.626 \times 10^{-34}}{\left(E_2 - E_1\right) \times 1.6 \times 10^{-19}}$$

$$\boxed{\lambda = \frac{12,400}{\left(E_2 - E_1\right)} A^{\circ}}$$

where λ is in Armstrong, $1A^{\circ} = 10^{-8}$ cm $= 10^{-10}$ m, and E_2 and E_1 in eV.

1.1.5 PHOTON NATURE OF LIGHT

An electron can be in the excited state for a small period of 10^{-7} to 10^{-10} sec. Afterwards it returns back to the original state. When such a transition occurs, the electrons will loose energy equal to the difference of energy levels $(E_2 - E_1)$. This loss of energy of the atom results in radiation of light. The frequency of the emitted radiation is given by

$$f = \frac{\left(E_2 - E_1\right)}{h}.$$

According to classical theory it was believed that atoms continuously radiate energy. But this is not true. Radiation of energy in the form of photons takes place only when the transition of electrons will take place from higher energy state to lower energy state, so that ($E_2 - E_1$), is positive. This will not occur if the transition is from lower energy state to higher energy state. When such photon radiation takes place, the number of photons liberated is very large. This is explained with a numercial example given below.

Problem 1.2

For a given 50 W energy vapour lamp. 0.1% of the electric energy supplied to the lamp, appears in the ultraviolet line 2,537 A°. Calculate the number of photons per second, of this wavelength emitted by the lamp.

Solution

$$\lambda = \frac{12,400}{\left(E_2 - E_1\right)}$$

λ = Wavelength of the emitted radiation.

E_1 and E_2 are the energy levels in eV.

$(E_2 - E_1)$ is the energy passed by each photon in eV of wavelength λ. In the given problem, $\lambda = 2,537$ A°. $(E_2 - E_1) = ?$

$$\therefore \qquad (E_2 - E_1) = \frac{12,400}{2,537} = 4.88 \text{ eV/photon}$$

0.1% of 50W energy supplied to the lamp is,

i.e.,
$$\frac{0.1}{100} \times 50 = 0.05 \text{ W} = 0.05 \text{ J/Sec}$$

$$\because \qquad 1W = 1 \text{ J/sec}$$

Converting this into the electron Volts,

$$\frac{0.05 \text{ J/sec}}{1.6 \times 10^{-19} \text{ J/eV}} = 3.12 \times 10^{17} \text{ eV/sec}$$

This is the total energy of all the photons liberated in the λ = Wavelength of 2,537 A°.

$$\therefore \quad \text{Number of photons emitted per sec} = \frac{\text{total energy}}{\text{energy / photon}}$$

$$= \frac{3.12 \times 10^{17} \text{ ev/sec}}{4.88 \text{ ev/photon}}$$

$$= 6.4 \times 10^{16} \text{ photon/sec}$$

The lamp emits 6.4×10^{16} photons / sec of wavelength $\lambda = 2,537 A°$.

1.1.6 IONIZATION POTENTIAL

If the most loosely bound electron (free electron) of an atom is given more and more energy, it moves to stable state (since it is loosely bound, its tendency is to acquire a stable state. Electrons orbiting closer to the nucleus have stable state, and electron orbiting in the outermost shells are loosely bound to the nucleus). But the stable state acquired by the electron is away from the nucleus of the atom. If the energy supplied to the loosely bound electron is enough large, to move it away completely from the influence of the parent nucleus, it becomes detached from it. *The energy required to detach an electron is called Ionization Potential.*

1.1.7 COLLISIONS OF ELECTRONS WITH ATOMS

If a loosely bound electron has to be liberated, energy has to be supplied to it. Consider the case when an electron is accelerated and collides with an atom. If this electron is moving slowly with less energy, and collides with an atom, it gets deflected, i.e., its direction changes. But no considerable change occurs in energy. This is called *Elastic Collision.*

If the electron is having much energy, then this electron transfers its energy to the loosely bound electron of the atom and may remove the electron from the atom itself. So another free electron results. If the bombarding electron is having energy greater than that required to liberate a loosely bound electron from atom, the excess energy will be shared by the bombarding and liberated electrons.

Problem 1.3

Argon resonance radiation, corresponding to an energy of 11.6 eV falls upon sodium vapor. If a photon ionizes an unexcited sodium atom, with what speed is the electron ejected ? The ionization potential of sodium is 5.12 eV.

Solution

Ionization potential is the minimum potential required to liberate an electron from its parent atom. Argon energy is 11.6 eV. Ionization Potential of Na is 5.12 eV.

$$\therefore \quad \text{The energy possessed by the electron which is ejected is}$$

$$11.6 - 5.12 = 6.48 \text{ eV}$$

or Potential, $\qquad V = 6.48$ volts $\qquad$ ($\because$ 1eV energy, potential is 1V)

Its velocity $\qquad v = \sqrt{\dfrac{2eV}{m}} = 5.93 \times 10^5 \sqrt{6.48} - 1.51 \times 10^6$ m/sec.

Problem 1.4

With what speed must an electron be travelling in a sodium vapor lamp in order to excite the yellow line whose wavelength is 5,893 A°.

Solution

$$Ee \geq \frac{12,400}{5,893} \geq 2.11 \text{ eV} \qquad \text{(Corresponding to evergy of 2.11eV,}$$
$$\text{the potential is 2.11 Volts)}$$

$$V = 2.11 \text{ Volts}$$

$$\therefore \text{ Velocity,} \quad v = \sqrt{\frac{2eV}{m}} = 5.93 \times 10^5 \sqrt{2.11} = 8.61 \times 10^5 \text{ m/sec.}$$

Problem 1.5

A radio transmitter radiates 1000 W at a frequency of 10 MHz.

(a) *What is the energy of each radiated quantum in ev ?*

(b) *How many quanta are emitted per second ?*

(c) *How many quanta are emitted in each period of oscillation of the electromagnetic field ?*

Solution

(a) Energy of each radiated quantum $= E = hf$

$$f = 10 \text{ MHz} = 10^7 \text{ Hz, } h = 6.626 \times 10^{-34} \text{ Joules / sec}$$

$$\therefore \qquad E = 6.626 \times 10^{-34} \, 10^7 = 6.626 \times 10^{-27} \text{ Joules / Quantum}$$

$$= \frac{6.626 \times 10^{-27}}{1.6 \times 10^{-19}} = 4.14 \times 10^{-8} \text{ eV / Quantum}$$

(b) $1W = 1$ Joule/sec

$$1000 \text{ W} = 1000 \text{ Joules/sec} = \text{Total Energy}$$

Energy possessed by each quantum $= 6.626 \times 10^{-27}$ Joules/Quantum.

$$\therefore \qquad \text{Total number of quanta per sec, N} = \frac{1000}{6.626 \times 10^{-27}} = 1.5 \times 10^{29}\text{/sec}$$

(c) One cycle $= 10^{-7}$ sec

$$\therefore \qquad \text{Number of quanta emitted per cycle} = 10^{-7} \times 1.51 \times 10^{29}$$

$$= 1.51 \times 10^{22} \text{ per cycle}$$

Problem 1.6

(a) *What is the minimum speed with which an electron must be travelling in order that a collision between it and an unexcited Neon atom may result in ionization of this atom? The Ionization Potential of Neon is 21.5 V.*

(b) *What is the minimum frequency that a photon can have and still be able to cause Photo-Ionization of a Neon atom?*

Solution

(a)   Ionization Potential is 21.5 V

$$\therefore \qquad \text{Velocity} = \sqrt{\frac{2eV}{m}} = 5.93 \times 10^5 \sqrt{21.5} = 2.75 \times 10^6 \text{ m/sec}$$

Wavelength of radiation,

(b)
$$\lambda = \frac{12,400}{E_2 - E_1} = \frac{12,400}{21.5} = 577 \text{ A}^0$$

Frequency of radiation,

$$f = \frac{C}{\lambda} = \frac{3 \times 10^8}{577 \times 10^{-10}} = 5.2 \times 10^{15} \text{ Hz}$$

Problem 1.7

Show that the time for one revolution of the electron in the hydrogen atom in a circular path around the nucleus is

$$T = \frac{m^{1/2} e^2}{4\sqrt{2} \, \epsilon_o \, (-W)^{\frac{3}{2}}}.$$

Solution

$$v = \sqrt{\frac{e^2}{4\pi m \, \epsilon_o \, r}}$$

$$T = \frac{2\pi r}{v} = \frac{2\pi r \left(4\pi m \, \epsilon_o \, r\right)^{1/2}}{e} = \frac{2\pi r^{\frac{3}{2}} \left(4\pi m \, \epsilon_o\right)^{1/2}}{e}$$

But radius,
$$r = \frac{e^2}{4\pi \, \epsilon_o \, W}$$

$$\therefore \qquad T = \frac{m^{1/2} e^2}{4\sqrt{2} \, \epsilon_0 \, (-W)^{3/2}}$$

Problem 1.8

A photon of wavelength 1,400 A^0 is absorbed by cold mercury vapor and two other photons are emitted. If one of these is the 1,850 A^0 line, what is the wavelength λ of the second photon ?

Solution

$$(E_2 - E_1) = \frac{12,400}{\lambda} = \frac{12,400}{1400} = 8.86 \text{ eV}$$

1850^0 A^0 line is from 6.71 eV to 0 eV.

$\therefore$ The second photon must be from 8.86 to 6.71 eV.

So,
$$\Delta E = 2.15 \text{ eV.}$$

$$\lambda = \frac{12,400}{(E_2 - E_1)}$$

$$\lambda = \frac{12,400}{2.15} = 5767 \text{ A}^0.$$

1.1.8 METASTABLE STATES

An atom may be elevated to an excited energy state by absorbing a photon of frequency 'f' and thereby move from the level of energy W_1 to the higher energy level W_2 where $W_2 = W_1 + hf$. But certain states may exist which can be excited by electron bombardment but not by photo excitation (absorbing photons and raising to the excited state). *Such levels are called metastable states*. A transition from a metastable level to a normal state with the emission of radiation has a very low probability of occurence. Transition from higher level to a metastable state are permitted, and several of these will occur.

An electron can be in the metastable state for about 10^{-2} to 10^{-4} sec. This is the mean life of a metastable state. Metastable state has a long lifetime because they cannot come to the normal state by emitting a photon. Then if an atom is in metastable state how will it come to the normal state? It cannot release a photon to come to normal state since this is forbidden. Therefore an atom in the metastable state can come to normal state only by colliding with another molecule and giving up its energy to the other molecule. Another possibility is the atom in the metastable state may receive additional energy by some means and hence may be elevated to a higher energy state from where a transition to normal state can occur.

1.1.9 WAVE PROPERTIES OF MATTER

An atom may absorb a photon of frequency f and move from the energy level W_1 to the higher energy level W_2 where $W_2 = W_1 + hf$.

Since a photon is absorbed by only one atom, the photon acts as if it were concentrated in one point in space. So wave properties can not be attributed to such atoms and they behave like particles.

Therefore according to 'deBroglie' hypothesis, dual character of wave and particle is not limited to radiation alone, but is also exhibited by particles such as electrons, atoms, and molecules. He calculated that a particle of mass 'm' travelling with a velocity **v** has a wavelength λ given by

$$\lambda = \frac{h}{mv} = \frac{h}{p} \qquad (\lambda \text{ is the wavelength of waves consisting of these particles}).$$

where 'p' is the momentum of the particle. Wave properties of moving electrons can be made use to explain Bohr's postulates. A stable orbit is one whose circumference is exactly equal to the wavelength λ or $n\lambda$ where n is an integer other than zero.

Thus $\qquad 2\pi r = n\lambda \qquad$ r = radius of orbit, n = Principle Quantum Number.

But according to DeBroglie,

$$\lambda = \frac{h}{mv}$$

$$\therefore \qquad 2\pi r = \frac{nh}{mv}$$

This equation is identical with Bohr's condition, $mvr = \dfrac{nh}{2\pi}$

1.1.10 SCHRODINGER EQUATION

Schrodinger carried the implications of the wave nature of electrons. A branch of physics called *Wave Mechanics* or *Quantum Mechanics* was developed by him. If deBroglie's concept of wave nature of electrons is correct, then it should be possible to deduce the properties of an electron system from a mathematical relationship called the *Wave Equation* or *Schrodinger Equation*. It is

$$\nabla^2 \phi - \frac{1}{v^2} \cdot \frac{\partial^2 \phi}{\partial t^2} = 0,$$
..........(1.6.1)

where
$$\nabla^2 = \frac{\partial^2}{\partial x^2} + \frac{\partial^2}{\partial y^2} + \frac{\partial^2}{\partial z^2}$$

ϕ can be a component of electric field or displacement or pressure. v is the velocity of the wave, and 't' is the time.

The variable can be eliminated in the equation by assuming a solution.

$$\phi \, (x, y, z, t) = \psi \, (x, y, z) e^{jwt}$$

This represents the position of a particle at 't' in 3-D Motion.

$$\omega = \text{Angular Frequency} = 2 \, \pi f$$

ω is regarded as constant, while differentiating.

ϕ is a function of x, y, z and t.

But $\quad\quad \psi$ is a function of x, y and z only.

To get the independent *Schrodinger Equation*

$$\frac{\partial \phi}{\partial t} = j\omega\psi(x, y, z)e^{j\omega t}$$

$$\frac{\partial^2 \phi}{\partial t^2} = -\omega^2 \psi(x, y, z)e^{j\omega t}$$

But $\quad\quad \omega = 2\pi f$

So, $\quad\quad \omega^2 = 4\pi^2 f^2$

Substituting this in the original *Schrodinger's Equation*,

$$e^{jwt}\left\{ \nabla^2 \psi + \frac{1}{v^2} \times 4\pi^2 f^2 \psi \right\} = 0$$

But $\quad\quad \lambda = \frac{v}{f}; \quad \text{Wavelength} = \frac{\text{Velocity}}{\text{Frequency}}$

$$\therefore \quad\quad \nabla^2 \psi + \frac{4\pi^2}{\lambda^2} \psi = 0$$
..........(1.6.2)

But $\quad\quad \lambda = \frac{h}{mv} = \frac{h}{p} \quad\quad\quad\quad v = \text{Velocity.}$

$\because \quad\quad p = mv$

This is deBrogile's Relationship.

$$\frac{1}{\lambda^2} = \frac{p^2}{h^2}$$

But $\qquad p^2 = m^2 v^2 = 2\,(\text{ Kinetic Energy })m \qquad\qquad \because \text{K.E.} = \tfrac{1}{2}mv^2$

$\qquad\qquad\qquad$ Kinetic Energy = Total Energy (W) − Potential Energy (U)

$\therefore \qquad\qquad p^2 = 2\,(\,W - U\,)\,.\,m$

$\therefore \qquad\qquad \dfrac{p^2}{h^2} = \dfrac{2m}{h^2}\,(\,W - U\,) \qquad\qquad\qquad\qquad\qquad$ ....(1.6.3)

Substituting Eq. (1.7.3) in Eq. (1.7.2) we get,

$$\nabla^2\,\Psi + \frac{8\pi^2 m}{h^2}\,(W - U)\,\Psi = 0.$$

This is the Time Independent Schrodinger Equation Ψ is a function of 't' But this equation as such is not containing the term 't'.

1.1.11   WAVE FUNCTION

Ψ is called as the wave function, which describes the behavior of the particle. Ψ is a quantity whose square gives the probability of finding an electron. $|\Psi|^2(\,dx \times dy \times dz\,)$ is proportional to the probability of finding an electron in volume dx, dy and dz at point P(x, y, z).

Four quantum numbers are required to define the wave function. They are :

1. ***The Principal Quantum Number 'n' :***

   It is an integer 1, 2, 3, …. This number determines the total energy associated with a state. It is same as the quantum number 'n' of Bohr atom.

2. ***The Orbital Angular Momentum Quantum Number l :***

   It takes values 0, 1, 2…l… (n − *l*)

   The magnitude of this angular momentum is $\sqrt{(l)(l+1)} \times \dfrac{h}{2\pi}$

   It indicates the shape of the classical orbit.

3. ***The orbital magnetic number m_l :***

   This will have values 0, ± 1, ±2 …. ± *l*. This number gives the orientation of the classical orbit with respect to an applied magnetic field.

   The magnitude of the Angular Momentum along the direction of magnetic

   $\text{field } = m_\ell\left(\dfrac{h}{2\pi}\right).$

4. ***Electron Spin :***

   It was found is 1925 that in addition to assuming that electron orbits round the nucleus, it is also necessary to assume that electron also spins around itself, in addition to orbiting round the nucleus. This intrinsic electronic angular momentum is called ***Electron Spin***.

   When an electron system is subjected to a magnetic field, the spin axis will orbit itself either parallel or anti-parallel to the direction of the field. The electron

   angular momentum is given by $m_s\left(\dfrac{h}{2\pi}\right)$ where the spin quantum m_s number

   may have values $+\tfrac{1}{2}$ or $-\tfrac{1}{2}$.

1.1.12 ELECTRONIC CONFIGURATION

PAULI'S EXCLUSION PRINCIPLE

No two electrons in an electron system can have the same set of four quantum numbers n, l, m^l and m_s.

Electrons will occupy the lower most quantum state.

ELECTRONIC SHELLS (PRINCIPLE QUANTUM NUMBER)

All the electrons which have the same value of 'n' in an atom are said to belong to the same electron shell. These shells are identified by letters K, L, M, N corresponding to n = 1, 2, 3, 4..... . A shell is subdivided into sub shells corresponding to values of l and identified as s, p, d, f, g, h,. for l = 0, 1, 2, 3... respectively. This is shown in Table 1.1.

Table 1.1

Shell ..... n	K 1	L 2		M 3			N 4			
l ..... subshell	0 s	0 s	1 p	0 s	1 p	2 d	0 s	1 p	2 d	3 f
No. of	2	2	6	2	6	10	2	6	10	14
electron	2	8		18			32			

ELECTRON SHELLS AND SUBSHELLS

Number of Electrons in a sub shell = 2(2l + 1)

$$n = 1 \text{ corresponds to K shell}$$
$$l = 0 \text{ corresponds to s sub shell}$$
$$\therefore \quad l = 0, \ldots, (n - 1) \text{ if } n = 1,$$
$$l = 0 \text{ is the only possibility}$$

Number of electrons in K shell = 2(2l + 1) = 2(0 + 1) = 2 electrons.

$$\therefore \quad \text{K shell will have 2 electrons.}$$

This is written as $1s^2$ pronounced as "*one s two*" (1 corresponds to K shell, n = 1; s is the sub shell corresponds to l = 0 number of electron is 2. Therefore, $1s^2$)

If n = 2,   it is '*l*' shell

If l = 0,   it is 's' sub shell

If l = 1,   it is 'p' sub shell

Number of electron in 's' sub shell (i.e., l = 0) = 2 × (2l + 1) = 2(0 + 1) = 2

Number of electron is 'p' sub shell (i.e., l = 1) = 2[(1 × 2) + 1)] = 6

In l shell there are two sub shells, s and p.

$$\therefore \quad \text{Total number of electron in } l \text{ shell} = 2 + 6 = 8$$
$$\text{This can be represented as } 2s^2\, 2p^6$$

If n = 3,   it is M shell

It has 3 sub shell s, p, d, corresponding to $l = 0, 1, 2$

In 's' sub shell number of electrons $(l = 0) = 2(\therefore l = 0)\ 2(2l + 1)$

In 'p' sub shell number of electrons $(l = 1) = 2(2 + 1) = 6$

In 'd' sub shell number of electrons $(l = 2) = 2(2 \times 2 + 1) = 10$

$\therefore$ Total number of electrons in M shell $= 10 + 6 + 2 = 18$

This can be represented as $3s^2\ 3p^6\ 3d^{10}$

$\therefore$ $1s^2\ 2s^2\ 2p^6\ 3s^2\ 3p^6\ 3d^{10}\ 4s^2\ 4p^6\ 4d^{10}$

Electronic Configuration

Atomic number 'z' gives the number of electrons orbiting round the nucleus. So from the above analysis, electron configuration can be given as $1s^2\,2s^2\,2p^6\,3s^1$. First k shell (n = 1) is to be filled. Then l shell (n = 2) and so on.

In 'k' shell there is one sub shell (s) $(l = 0)$. This has to be filled.

In 'L' shell there are two sub shells and p. First - s and then p are to be filled and so on.

The sum of subscripts $2 + 2 + 6 + 1 = 11$. It is the atomic number represented as Z.

For Carbon, the Atomic Number is 6., i.e., Z = 6.

$\therefore$ The electron configuration is $1s^2\,2s^2\,2p^2$

For the Ge   Z = 32. So the electronic configuration of Germanium is,

$\therefore$ $1s^2\ 2s^2\ 2p^6\ 3s^2\ 3p^6\ 3d^{10}\ 4s^2\ 4p^2$

For the Si   Z = 14. So the electronic configuration of Silicon is,

$\therefore$ $1s^2\ 2s^2\ 2p^6\ 3s^2\ 3p^2$

1.1.13   Types of Electron Emission

Electrons at absolute zero possess energy ranging from 0 to E_F the fermi level. It is the characteristic of the substance. But this energy is not sufficient for electrons to escape from the surface. They must posses energy $E_B = E_F + E_W$ where E_W is the work function in eV.

$$E_B \quad = \quad \text{Barrier's Energy}$$
$$E_F \quad = \quad \text{Fermi Level}$$
$$E_W \quad = \quad \text{Work Function.}$$

Different types of Emission by which electrons can emit are

(1)   *Thermionic Emission*

(2)   *Secondary Emission*

(3)   *Photoelectric Emission*

(4)   *High field Emission.*

1. Thermionic Emission

Suppose, the metal is in the form of a filament and is heated by passing a current through it. As the temperature is increased, the electron energy distribution starts in the metal changes. Some electrons may acquire energy greater than E_B sufficient to escape from the metal.

E_W : *Work function of a metal. It represents the amount of energy that must be given for the electron to be able to escape from the metal.*

It is possible to calculate the number of electrons striking the surface of the metal per second with sufficient energy to be able to surmount the surface barriers and hence escape. Based upon that, the thermionic current is,

$$I_{th} = S \times Ao\ T^2\ e^{-E_w/KT} \qquad\qquad \cdots\cdots\cdots (1.7)$$

It is also written as

$$\frac{I_{th}}{S} = J = AoT^2\ e^{\frac{-E_w}{kT}} = AT^2\ e^{\frac{-B}{T}}$$

where $\qquad A = A_o \quad$ and $\quad B = \dfrac{E_W}{k}$

where $\qquad S \;=\;$ Area of the filament in m^2 (Surface Area)

$\qquad\qquad Ao =\;$ Constant whose dimensions are $A/m^2\ {}^\circ K$

$\qquad\qquad T \;=\;$ Temperature in ${}^\circ K$

$\qquad\qquad K \;=\;$ Boltzman's constant $eV/{}^\circ K$

$\qquad\qquad E_w =\;$ Work function in eV

This equation is called ***Thermionic Emission Current*** or ***Richardson - Dushman Equation***. E_W is also called as latent heat of evaporation of electrons similar to evaporation of molecules from a liquid.

Taking logarithms

$$\log I_{th} = \log (S\ A_o) - \frac{E_w}{kT}\ \log e + \log T^2$$

$$\log I_{th} - 2 \log T = \log S\ Ao - 0.434 \left(\frac{E_W}{kT} \right)$$

$\qquad\qquad\therefore\qquad\qquad \log e = 0.434$

So if a graph is plotted between $(\log I_{th} - 2 \log T)$ V_S $\dfrac{1}{T}$, the result will be a straight line

having a slope $= -\ 0.434 \left(\dfrac{E_W}{kT} \right)$ from which E_W can be determined.

$\qquad\qquad\therefore\qquad\qquad I_{th}$ and T can be determined experimentally.

I_{th} is a very strong function of T. For Tungsten, $E_W = 4.52\ eV$.

CONTACT POTENTIAL

Consider two metals in contact with each other forming junction at C as in Fig 1.1 . The contact difference of potential is defined as the Potential Difference V_{AB} between a point A, just out side metal 1 and a point B just outside metal 2. The reason for the difference of potential is, when two metals are joined, electrons will flow from the metal of lower Work Function, say 1 to the metal of higher Work Function say 2. ($\because\ E_W = E_B - E_F$ electrons of lower work function means E_W is small or E_F is large). Flow of electrons from metal 1 to 2 will continue till metal 2 has acquired sufficient negative charge to repel extra new electrons. E_B value will be almost same for all metals. But E_F differs significantly.

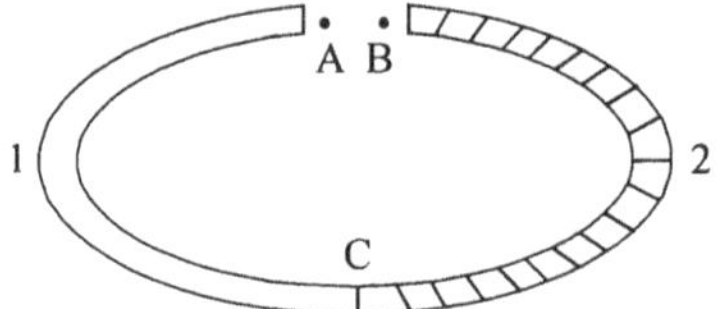

Fig 1.1 Contact Potential

In order that the fermi levels of both the metals are at the same level , the potential energy difference $E_{AB} = E_{W2} - E_{W1}$, that is, *the contact difference of potential energy between two metals is equal to the difference between their work functions*.

If metals 1 and 2 are similar then contact potential is zero. If they are dissimilar, the metal with lower work function becomes positive, since it looses electrons.

ENERGIES OF EMITTED ELECTRONS

At absolute zero, the electrons will have energies ranging from zero to E_F. So the electrons liberated from the metal surface will also have different energies. The minimum energy required is the barrier potential to escape from the metal surface, But the electrons can acquire energy greater than E_B to escape from the surface. This depends upon the initial energy, the electrons are possessing at room temperature.

Consider a case where anode and cathode are plane parallel. Suppose the voltage applied to the cathode is lower, and the anode is indirectly heated. Suppose the minimum energy required to escape from the metal surface is 2 eV. As collector is at a potential less than 2V, electrons will be collected by the collector. If the voltage of cathode is lower, below 2V, then the current also decreases exponentially, but not abruptly. If electrons are being emitted from cathode with 2eV energy, and the voltage is reduced below 2V, then there must be abrupt drop of current to zero. But it doesn't happen This shows that electrons are emitted from surface of the emitter with different velocities. The decrease of current is given by $I = I_{Th}.e^{-Vt}$ where V_t is the retarding potential applied to the collector and V_T is Volt equivalent of temperature.

$$V_T = \frac{kT}{e} = \frac{T}{11,600} \qquad \text{.......... (1.8)}$$

$$T = \text{Temperature in } {}^{o}K$$
$$K = \text{Boltzman's Constant in } J/{}^{o}K$$

SCHOTTKY EFFECT

If a cathode is heated, and anode is given a positive potential, then there will be electron emission due to thermionic emission. There is accelerating field, since anode is at a positive potential. This accelerating field tends to lower the Work Function of the cathode material. It can be shown that under the condition of accelerating field E is

$$I = I_{th} \, e^{+0.44 \, \varepsilon 1/2/T}$$

where I_{th} is the zero field thermionic current and T is cathode temperature in $0 \, {}^{o}K$.

The effect that thermionic current continues to increase as E is increased (even though T is kept constant) is known as *Schottky Effect*.

2. Secondary Emission

This emission results from a material (metal or dielectric) when subjected to electron bombardment.

It depends upon,

1. *The energy of the primary electrons.*
2. *The angle of incidence.*
3. *The type of material.*
4. *The physical condition of surface ; whether surface is smooth or rough.*

Yield or secondary emission ratio S is defined as the ratio of the number of secondary electrons to primary electrons. It is small for pure metals, the value being 1.5 to 2. By contamination or giving a coating of alkali metal on the surface, it can be improved to 10 or 15.

3. Photo Electric Emission

Photo-Electric Emission consists of liberation of electrons by the incidence of light, on certain surfaces. The energy possessed by photons is hf where h is Plank's Constant and f is frequency of incident light. When such a photon impinges upon the metal surface, this energy **hf** gets transferred to the electrons close to the metal surface whose energy is, very near to the barrier potential. Such electrons gain energy, to overcome the barrier potential and escape from the surface of the metal resulting in photo electric emission.

For photoelectric emission to take places, the energy of the photon must at least be equal to the work function of the metal. That is $hf \geq e\phi$ where ϕ is the voltage equivalent Work Function, (i.e. Work Function expressed in Volts). *The minimum frequency that can cause photo-electric emission is called threshold frequency* and is given by

$$f_T = \frac{e\phi}{h}$$

The wavelength corresponding to threshold frequency is called the ***Threshold Wavelength***.

$$\lambda_t = \frac{C}{f_t} = \frac{h_C}{\phi e}$$

If the frequency of radiation is less than f_T, then additional energy appears as kinetic energy of the emitted electron

$$\therefore \quad hf = \phi e + \tfrac{1}{2} m v^2$$

ϕ is the Volt equivalent of Work Function.

$$v = \sqrt{\frac{2eV}{m}}$$

v is the Velocity of electrons in m/sec.

$$\therefore \quad \boxed{hf = = \phi e + eV} \qquad \cdots\cdots\cdots (1.9)$$

Laws of Photo electric Emission

1. *For each photo sensitive material, there is a threshold frequency below which emission does not take place.*

2. *The amount of photo electric emission (current) is proportional to intensity.*

3.   *Photo electric emission is instantaneous. (But the time lag is in nano-sec).*

4.   *Photo electric current in amps/watt of incident light depends upon 'f'.*

4. HIGH FIELD EMISSION

Suppose a cathode is placed inside a very intense electric field, then the Potential Energy is reduced. For fields of the order of 10^9 V/m, the barrier may be as thin as 100 A°. So the electron will travel through the barrier. This emission is called as ***High Field Emission*** or ***Auto Electronic Emission***.

Problem 1.9

Estimate the percentage increase in emission from a tungsten filament when its temperature is raised from 2400 to 2410 °K.

$$A_0 = 60.2 \times 10^4 \text{ A/m}^2/\text{°K}^2$$
$$B = 52,400 \text{ °K}$$

A and B are constants in the equation for current density J

Solution

$$JS_1 = A\,T_1^2 = 1142 \text{ A / m}^2$$
$$JS_2 = AT_2^2\, e^{-B/T} = 1261 \text{ A / m}^2.$$
$$\text{Percentage Increase} = \frac{1261-1142}{1142} \times 100 = 10.35\%$$

Problem 1.10

A photoelectric cell has a cesium cathode. When the cathode is illuminated with light of $\lambda = 5500 \times 10^{-10}$ m, the minimum anode voltage required to inhibit built anode current is 0.55 V. Calculate

(a)   *The work function of cesium*

(b)   *The longest λ for which photo cell can function.*

by applying -0.55V to anode the emitted electrons are repelled. So the current can be inhibited

Solution

(a) $hf = e\phi + eV$ $V = 0.55$ volts ϕ = Work Function (WF) = ?

$$f = \frac{C}{\lambda}$$

$$\therefore \qquad \frac{h.C}{\lambda} = e\,(\,\phi + V\,)$$

Plank's Constant, $h = 6.63 \times 10^{-34}$ J sec.

Charge of Electron, $e = 1.6 \times 10^{-19}$ C

Velocity of Light, $C = 3 \times 10^8$ m/sec

$$= \frac{6.63 \times 10^{-34} \times 3 \times 10^8}{5500 \times 10^{-10}} = 1.6 \times 10^{-19}\,(\phi + 0.55)$$

$$\phi = 1.71 \text{ Volts}$$

(b) *Threshold Wavelength :*

$$\lambda_o = \frac{Ch}{e\phi}$$

$$\therefore \qquad \lambda_o = \frac{12,400}{\phi} = \frac{12,400}{1.71} = 7,250 \ A^o$$

Problem 1.11

If the temperature of a tungsten filament is raised from 2300 to 2320 ^{o}K, by what percentage will the emission change ? To what temperature must the filament be raised in order to double its energy at 2300 ^{o}K. E_W for Tungsten = 4.52 eV. Boltzman's Constant K = 8.62 $\times$ 10^5 eV/^{o}K.

Solution

(a) $\qquad I_{th} = S.A_o \ T^2 \ e^{-Ew/kT}$

Taking Logarithms,

$$ln \ I_{th} - 2 \log T = ln \ S \ A_o - \frac{E_w}{kT}$$

Differentiating, $\qquad \left(\frac{2}{T} + \frac{E_W}{KT^2}\right)\frac{dT}{T^2}$

$$\frac{dI_{th}}{I_{th}} = \left(2 + \frac{E_W}{kT}\right)\frac{dT}{T} = \left(2 + \frac{4.52}{\left(8.62 \times 10^{-5}\right)2310}\right)\frac{20}{2310} = 21.4\%$$

(b) $\qquad I_{th} = S.Ao \ (2300)^2 \ e^{\frac{-4.52}{8.62 \times 10^{-5} \times 2300}}$

$$2I_{th} = S \ Ao \ (T)^2 \ e^{\frac{-4.52}{8.62 \times 10^{-5}}}$$

Ratio of these two equation is

$$2 = \left(\frac{T}{2300}\right)^2 \ e^{-\frac{52,400}{T}+22.8}$$

Taking log to the base 10,

$$\log 2 = 2 \log \left(\frac{T}{2300}\right) - \left(\frac{52,400}{T} + 22.8\right)\log e$$

$$\log 2 = 2 \log \left(\frac{T}{2300}\right) - \frac{52400}{T} \times 0.434 - 22.8 \times 0.434$$

$$9.6 + 2 \log \left(\frac{T}{2300}\right) = \frac{22,800}{T}$$

This is solved by Trail and Error Method to get T = 2370^{0}K

Problem 1.12

In a cyclotron, the magnetic field applied is 1 Tesla. If the ions (electrons) cross the gap between the D shaped discs dees twice in each cycle, determine the frequency of the R.F. voltage. If in each passage through the gap, the potential is increased by 40,000 volts how many passages are required to produce a 2 million volts particle? What is the diameter of the last semicircle?

Solution

$$B = 1 \text{ Tesla} = 1 Wb/m^2$$

Time taken by the particle to describe one circle is $T = \dfrac{35.5\mu\sec}{B}$

$$T = \dfrac{35.5 \times 10^{-6}}{1} \sec,$$

when it describes one circle, the particle passes through the gap twice.

$\therefore$ The frequency of the R.F voltage should be the same.

$\therefore$ $f = \dfrac{1}{T} = \dfrac{1}{35.5 \times 10^{-6}} = 2.82 \times 10^{10}$ Hz

Number of passages required :

In each passage it gains 40,000 volts.

To gain 2 million electron volts $= \dfrac{2 \times 10^6 \text{ V}}{40 \times 10^3 \text{ V}} = 50$

Diameter of last semicircle :

The initial velocity for the 50th revolution is the velocity gained after 49th revolution.

$\therefore$ Accelerating potential $V_0 = 49 \times 40,000 = 1,960$ KV 49th revolution

Radius of the last semicircle $= R = \dfrac{3.37 \times 10^{-6}}{B} \sqrt{V_0}$

$$R = \dfrac{3.37 \times 10^{-6}}{1} \sqrt{1960 \times 10^3} = 47.18 \times 10^{-4} \text{ m}$$
$$\text{Diameter} = 2R = 94.36 \times 10^{-4} \text{ m}$$

Problem 1.13

The radiated power density necessary to maintain an oxide coated filament at 110 $^{\circ}$K is found to be 5.8×10^4 W/m^2. Assume that the heat loss due to conductor is 10% of the radiation loss. Calculate the total emission current and the cathode efficiency η in ma/w.

Take

$$E_W = 1.0 \text{ eV},$$
$$A_0 = 100 \text{ A/m}^2 / {}^{\circ}K^2$$
$$S = 1.8 \text{ cm}^2$$

$$\bar{K} = \text{Boltzman's Constant in eV} / {}^{\circ}K = 8.62 \times 10^{-5} eV / {}^{\circ}K$$

Solution

$$\text{Power Density} = 5.8 \times 10^4 \text{ W/m}^2$$

If 10% is lost by conductor, then the total input power density is $1.1 \times 5.8 \times 10^4 = 6.38$ W/m^2 and the total input power is $6.38 \times 10^4 \times$ Area $(1.8 \times 10^{-4}) = 11.5$ W.

$$I_{th} = S \ Ao \ T^2 \ e^{-Ew/kT}$$

$$= 1.8 \times 10^{-4} \times 100 \times (1100)^2 \ e^{-\frac{1}{8.62 \times 10^{-5} \times 1100}}$$

$$I_{th} = 2.18 \times 10^4 \ e^{-10.55}$$

$$I_{th} = 0.575 \ A$$

Cathode Efficiency, $\eta = \dfrac{0.575}{11.5} = 50$ mA / $^{\circ}$K

1.2 ENERGY BAND STRUCTURES

Insulator is a very poor conductor of electricity. Ex : Diamond

Resistivity, $\rho > 10^9 \ \Omega - cm$

Free electron Concentration, $n \sim 10^7$ electrons/m^3

Energy Gap, $E_G \gg 1eV$

Metal is an excellent conductor of electricity. Ex : Al, Ag, Copper

$$\rho < 10^{-3} \ \Omega - cm$$

Free electron concentration, $n = 10^{28}$ electrons/m^3

Energy Gap, $E_G = 0$

Semiconductor is a substance whose conductivity lies between these two. Ex : C, Si, Ge, GaAs etc.,

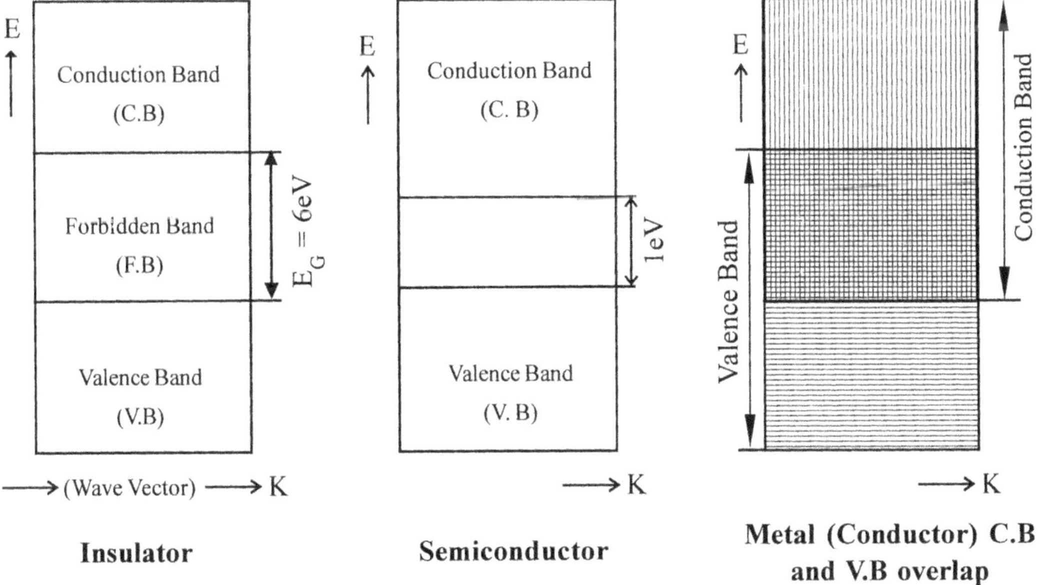

Fig 1.2 Energy band structures.

1.2.1 INSULATOR

For diamond, the value of energy gap Eg is 6eV. This large forbidden band separates the filled valence band region from the vacant conduction band. The energy, which can be supplied to an electron from an applied field, is too small to carry the particle from the filled valence band into the vacant conduction band. Since the electron can not acquire externally applied energy, conduction is impossible and hence diamond is an insulator.

1.2.2 SEMICONDUCTOR

For Semiconducting materials, the value of energy gap Eg will be about 1eV.

Germanium (Ge) has E_G = 0.785 eV, and Silicon is 1.21 eV at 0^oK. Electron can not acquire this much Energy to travel from valence band to conduction band. Hence conduction will not take place. But E_G is a function of temperature T. As T increases, E_G decreases.

For Silicon (Si) E_G decreases at the rate of 3.6×10^{-4} eV/^oK

For Germanium E_G decreases at the rate of 2.23×10^{-4} eV/^oK

$$\therefore \qquad \text{For Si, } E_G = 1.21 - 3.6 \times 10^{-4} \times T$$
$$\text{For Ge, } E_G = 0.785 - 2.23 \times 10^{-4} \times T.$$

The absence of an electron in the semiconductor is represented as hole.

1.2.3 METAL (CONDUCTOR)

In a metal, the valence band may extend into the Conduction Band itself. There is no forbidden band, under the influence of an applied field, the electron will acquire additional energy and move into higher states. Since these mobile electrons constitute a current, this substance is a conductor.

When an electron moves from valence band into conduction band in a metal, the vacancy so created in the valence band can not act as a hole. Since, in the case of metals, the valence electrons are loosely bound to parent atom. When they are also in conduction, the atom can pull another electron to fill its place.

In the energy band diagram, the y – axis is energy. The x–axis is wave vector K, since the energy levels of different electrons are being compared. The Ge has structure of,

$$1s^2 \, 2s^2 \, 2p^6 \, 3s^2 \, 3p^6 \, 3d^{10} \, 4s^2 \, 4p^2$$

The two electrons in the s sub shell and 2 in the p sub shell are the 4 electrons in the outermost 4th shell. The Germanium has a crystalline structure such that these 4 electrons are shared by 4 other Germanium atoms. For insulators the valence shell is completely filled. So there are no free electrons available in the outermost shell. For conductors say Copper, there is one electron in the outermost shell which is loosely bound to the parent atom. Hence the conductivity of Copper is high.

1.3 CONDUCTION IN SEMICONDUCTORS

Conductors will have Resistivity $\rho < 10^{-3}$ Ω–cm; $n = 10^{28}$ electrons/m^3

Insulators will have Resistivity $\rho \geq 10^9$ Ω–cm; $n = 10^7$ electrons/m^3

For semiconductors value lies between these two.

Some of The different types of solid state devices are :

  1. ***Semiconductor Diode*** : *Zener diode, Tunnel diode, Light emitting diode; Varactor Diode*

  2. ***SCR*** : *Silicon Controlled Rectifier*

3.	Transistor (BJT)	:	PNP, NPN
4.	FET	:	Field Effect Transistors
5.	MOSFET	:	Metal Oxide Semiconductor Field Effect Transistors
5.	UJT	:	Unijunction Transistor
6.	Photo Transistors		
7.	IMPATT	:	Impact Ionisation Avalanche Transit Time Device
8.	TRAPATT	:	Trapped Plasma Avalanche Transit Time Device
9.	BARRITT	:	Barrier Injected Avalanche Transit *Time Device*

Advantages of Semiconductor Devices

1. *Smaller in size.*

2. *Requires no cathode heating power (warm up time compared to Vacuum Tubes).*

3. *They operate on low DC power.*

4. *They have long life. (Tubes will pop up frequently).*

Disadvantages

1. *Frequency range of operation is low.*

2. *Smaller power output.*

3. *Low permissible ambient temperature.*

4. *Noise is more (because of recombination between holes and electrons).*

In a metal, the outer or valence electrons of an atom are as much associated with one ion (or parent atom) as with another, so that the electron attachment to any individual atom is almost zero. In other words, band occupied by the valence electrons may not be completely filled and that these are forbidden levels at higher energies. Depending upon the metals at least one and sometimes two or three electrons per atom are free to move throughout the interior of thc metal under the action of applied fields.

Germanium semiconductor has four valence electrons. Each of the valence electrons of a germanium atom is shared by one of its four nearest neighbours. So covalent bonds result. The valence electrons serve to bind one atom to the next also result in the valence electron being tightly bound to the nucleus. Hence in spite of the availability of four valence electrons, the crystal has low conductivity.

In metals the binding force is not strong. So free electrons are easily available.

According to the electron gas theory of a metal, the electrons are in continuous motion. The direction of flight being changed at each collision with the impinging (almost stationary) ions. ***The average distance between collisions is called the mean free path.*** Since the motion is random, on an average there will be as many electrons passing through unit area in the metal in any directions as in the opposite directions, in a given time. Hence the average current is zero, when no electric field is applied.

Now, if a constant electric field 'ε' V/m is applied to the metal, as a result of the electrostatic force, the electrons would be accelerated and the velocity would increase indefinitely with time if the electron will not collides with any other particle. But actually the electron collides with number of ions and it loses energy. Because of these collisions, when the electron loses its energy, its velocity will also decrease. So finally a steady state condition is reached where an infinite value of drift velocity 'v' is attained. This drift velocity is in the direction opposite to that of the electric field and its magnitude is proportional to the electric field E. Thus velocity is proportional to E or v = μE, where μ is called the *mobility of electrons*.

$$v \propto \varepsilon \quad \text{or} \quad v = \mu\varepsilon$$

v is in m/sec. ε is in v/m.

$$\mu = \frac{v}{\varepsilon} = \frac{m \times m}{Sec \times Volts} = m^2/ \text{volts-sec}$$

Mobility is defined as the velocity per unit electric field. The units are m^2/v–sec.

Because of the thermal energy, when no electric field is applied, the motion of electrons is in random nature. When an electric field is applied, steady state drift velocity is superimposed upon the random thermal motion of the electrons. Such a directed flow of electrons constitute a current. If the concentration of free electrons is 'n' (electrons/m^3) the current density is J, (amps/m^2)

$$J = ne \, v \qquad\qquad\qquad .... (1.10)$$

But velocity $\qquad$ $\boxed{v = \mu\varepsilon}$

$\qquad\qquad \therefore \qquad$ $J = ne \, \mu\varepsilon$

But $\qquad\qquad$ ne $\mu = \sigma$ (sigma)

σ is the conductivity of the metal in $(\Omega\text{-m})^{-1}$ or mhos/m.

So $\qquad\qquad$ $\boxed{J = \sigma \times \varepsilon}$ $\qquad\qquad\qquad .... (1.11)$

As temperature increases mobility decreases. This is analogous to large number of people in a small room. The movement of each person is restricted. If less number of persons are there the movement is large or mobility is large. The electrons acquire from the applied field, as a result of collisions is given to the lattice ions. Hence power is dissipated in the metal.

From the above expression for conductivity, σ is proportional to 'n', the number of free electrons. *For a good conductor 'n' is very large, ~10^{28} electrons/m^3. For an insulator n is very small ~10^7 electrons/m^3. For a semiconductor the value of 'n' lies between these two.* The valance electrons in a semiconductor are not free to move like in a metal, but they are trapped between two adjacent ions.

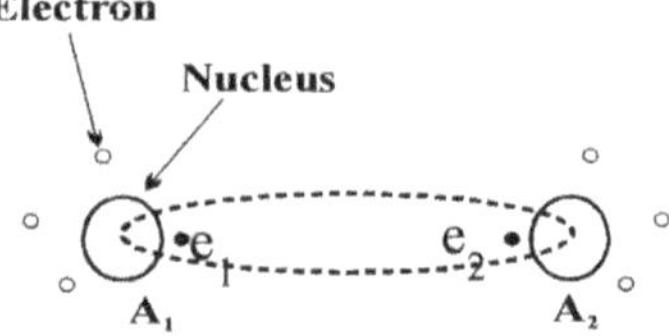

Fig 1.3 Electronic structure.

Germanium has a total of 32 electrons in its atomic structure, arranged in shells. Each atom in a Germanium crystal contribute four valence electrons. The boundry forces between neighbours attains result from, the fact is that each of the valence electrons of a germanium atom is shared by one of its four nearest neighbours.

Electron e_1 of atom A_1 is bound by atom A_2. Electron e_2 of atom A_2 is bond by the atom A_1. So because of this, covalent bonds result. The fact that valence electrons serve to bound atom to the next also results in the valence electrons being bound to the nucleus. Hence inspite of having four valence electrons, the crystal has low conductivity (Fig. 1.3).

At very low temperatures say 0^0 K, the ideal structure is achieved and the semiconductor behaves as an insulator, since no free carriers of electricity are available. However at room temperature, some of the covalent bonds will be broken because of the thermal energy supplied to the crystal, and conduction is made possible. An electron which for the greater period of time forms part of a covalent bond, is shown as being dislodged and so free to wander in a random fashion throughout the crystal.

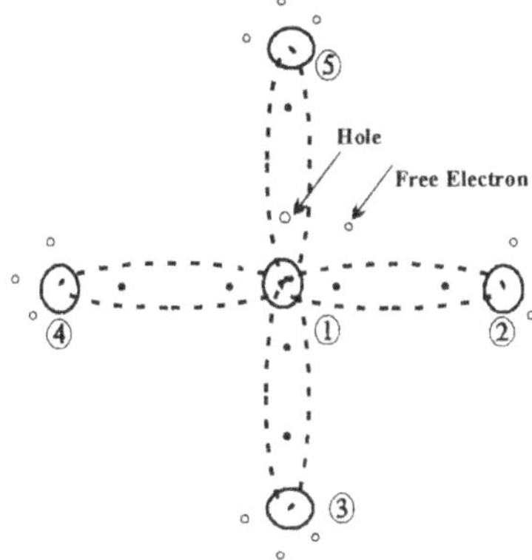

Fig 1.4 Covalent bonds

1, 2, 3, and 4, are Ge atoms with 4 valance electrons (Fig. 1.4). These four valence electrons of each atom are shared by other atoms and so bounded by covalent bonds, except for atom 1. The electron of Ge atom 1 is dislodged from its original position, because of the thermal energy and so the covalent bonds bonding the electron are broken. So this electron is now a free electron to wander anywhere in the material till it collides with some other atom. The absence of the electron in the covalent bond is represented by a hole. This is the concept of hole. The energy required to break different covalent bonds will be different. So at a time at room temperature all the covalent bonds are not broken to create innumerable free electrons.

The energy E_g required to break such a covalent bond is about 0.72 eV for Germanium and 1.1 eV for silicon (at room temperature). A hole can serve as a carrier of electricity. Its significance lies in this characteristics.

The explanation is given below :

When a bond is incomplete so that hole exists, it is relatively easy for a valence electron in a neighbouring atom to leave its covalent bond to fill this hole. An electron moving from a bond to fill a hole leaves a hole in its initial position. Hence the hole effectively moves in the direction opposite to that of the electron. Thus hole in its new position may now be filled by an electron from another covalent bond and the hole will correspondingly move one more step in the direction opposite to the motion of the electron. Here we have a mechanism for the conduction of electricity which does not involve free electrons. Only the electrons are exchanging their position and there by current is flowing. In other words, the current is due to the holes moving in the opposite direction to that of electrons. To explain this further,

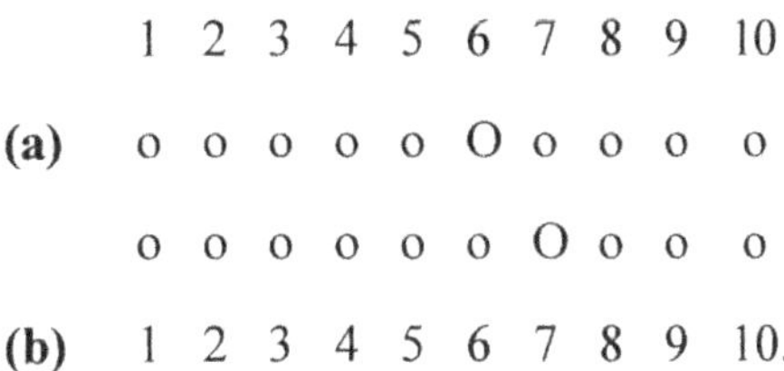

Fig. 1.5 Current flow by the movement of holes

In row (a) there are 10 ions in Fig. 1.5. Except for 6, all the covalent bonds of the ions are intact. The ion 6 has a broken covalent bond or one of its valence electrons got dislodged. So the empty place denotes a hole. Now imagine that an electron from ion 7 moves into the hole at ion 6. Then the configuration is as shown in row (b). Ion 6 in the row is completely filled. There is no broken covalent bond. But ion 7 has a vacancy now, since it has lost one of its valence electrons.

Effectively the hole has moved from ion 6 to ion 7. So the movement of holes is opposite to that of electrons. The hole behaves like a positive charge equal in magnitude to the electron charge.

In a pure semiconductor the number of holes is equal to the number of free electrons. Thermal agitation continues to produce new electron hole pairs where as some other hole electron pairs disappear as a result of recombination.

This is analogous to passengers travelling in a bus. Bus is the semiconducting material. Standing passenger are free electron. When a sitting passenger gets down, a hole is created. This hole is filled by a free electron that is a standing passenger. This process goes on as the bus is moving from stage to stage. If there are many standing passengers, without any vacant seat, it is analogous to n-type semiconductor. If there are many seats vacant without any standing passenger it is like a p-type semiconductor.

So the semiconductros are classified as :

> ***n-type Semiconductor***   :   Free electron concentration 'n' is greater than hole concentration. **n > p**.
>
> ***p-type Semiconductor***   :   Hole concentration 'p' is greater than free electron concentration. **p > n**.
>
> ***Instrinsic Semiconductor***   :   **n = p**.

1.4 CONDUCTIVITY OF AN INTRINSIC SEMICONDUCTOR

When valence electrons are exchanging their positions, we say holes are moving. Current is contributed by these holes current is nothing but rate of flow of charge. Holes are positively charged. So hole movement contributes for flow of current. Because of the positive charge movement, the direction of hole current is same as that of conventional current. Suppose to start with, there are many free electrons, and these will be moving in random directions. A current is constituted by these electrons. So at any instant, the total current density is summation of the current densities due to holes and electrons. The charge of free electrons is negative and its mobility is μ_n. The hole is positive, and its mobility is μ_p. The charge of both holes and electrons are same 'e'. A hole can move from one ion to the nearest where as an electron is free to move anywhere till it collides with another ion or free electron. Electrons and holes move in opposite directions in an electric field E. Though they are of opposite sign, the current due to each ion is in the same direction.

$$\therefore \quad \text{Current Density, } J = \sigma E$$

$$J = (n\,\mu_n + p\mu_p) \times e \times E = \sigma \times E$$

$$n = \text{Magnitude of Free Electron Concentration}$$

$$p = \text{Magnitude of Hole Concentration}$$

$$\sigma = \text{Conductivity } \Omega / \text{cm or Seimens}$$

$$\therefore \quad \sigma = (n\,\mu_n + p\,\mu_p)\, e = ne\mu_n + pe\mu_p$$

A pure or intrinsic semiconductor is one in which $n = p = n_i$ where n_i is intrinsic concentration.

In a pure Germanium at room temperature, there is about one hole-electron pair for every 2×10^9 Germanium atoms. As the temperature increases, covalent bonds are broken and so more free electrons and holes are created. So n_i increases, as the temperature increases, in accordance with the relationship,

$$n_i = A \times T^{\frac{3}{2}} e^{-E_{GO}/2kT} \qquad .... \ (\ 1.12 \ )$$

E_G = Energy Gap in ev.

e = Charge of an Electron or hole

A = Constant for a semiconductor;

for Ge   A = 9.64×10^{21};

for Ge E_G = 0.785 eV at 0 °K

for Si E_G = 1.21 eV at 0 °K

E_G at room temperature,

for Ge = 0.72 eV

for Si = 1.1 eV

T = Temperature in 0 °K

k = Boltzman's Constant.

As n_i increases with T, the conductivity also increases, with temperature for semiconductor. In other words, *the resistivity decreases with temperature for semiconductor*. On the other hand *resistance increases with temperature for metals*. This is because, an increase in temperature for metals results in greater thermal motion of ions and hence decrease in the mean free path of the free electrons. This results in decrease of the mobility of free electrons and so decrease in conductivity or increase in resistivity for metals.

1.5   DONOR TYPE OR n-TYPE SEMICONDUCTORS

Intrinsic or pure semiconductor is of no use since its conductivity is less and it can not be charged much. If a pure semiconductor is doped with impurity it becomes extrinsic. Depending upon impurity doped, the semiconductor may become *n-type, where electrons are the majority carriers or donor type*, since it donates an electron. On the other hand *if the majority carriers are holes, it is p-type or acceptor type semiconductor*, because it accepts an electron to complete the broken covalent bond.

Germanium atom with its electrons arranged in shells will have configuration as

$$1s^2 \ 2s^2 \ 2p^2 \ 3s^2 \ 3p^6 \ 3d^{10} \ 4s^2 \ 4p^2$$

Ge is tetravalent (4). 'Ge' becomes *n-type* if a pentavalent (5), impurity atoms such as Phosphorus (P), or Arsenic are added to it.

The impurity atoms have size of the same order as that of Ge atoms. Because of the energy supplied while doping, the impurity atom dislodges one from its normal position in the crystal lattices takes up that position. But since the concentration of impurity atoms is very small (about 1 atom per million of Ge atoms), the impurity atom is surrounded by Ge atoms. The impurity atom is pentavalent. That is, it has 5 electrons in the outermost orbit (5 valence electrons). Now 4 of these are shared by Ge atoms, surrounding the impurity atom and they form covalent bonds. So one electron of the impurity atom is left free. The energy required to dislodge this fifth electron from its parent impurity atom is very little of the order of 0.01 eV to 0.05 eV. This free electron is in excess to the free electrons that will be generated because of breaking of covalent bonds due to thermal agitation. Since an excess electron is available for each impurity atom, or it can *denote an electron it is called n-type, or donor type semiconductor.*

1.6   ACCEPTOR TYPE OR p - TYPE SEMICONDUCTORS

An intrinsic semiconductor when doped with trivalent (3) impurity atoms like Boron, Gallium Indium, Aluminium etc., becomes p-type or acceptor type.

Because of the energy supplied while doping, the impurity atom dislodges one Ge atom from the crystal lattice. The doping level is low, i.e., there is one impurity atom for one million Ge atoms, the impurity atom is surrounded by Ge atom. Now the three valence electrons of impurity atom are shared by 3 atoms. The fourth Ge atom has no electron to share with the impurity atom. So the covalent bond is not filled or a hole exists. The impurity atom tries to steal one electron from the neighboring Ge atoms and it does so when sufficient energy is supplied to it. So hole moves. There will be a natural tendency in the crystal to form 4 covalent bonds. The impurity atom (and not just 3) since all the other Ge atoms have 4 covalent bonds and the structure of Ge semiconductor is crystalline and symmetrical. The energy required for the impurity atom to steal one Ge electron is 0.01 eV to 0.08 eV. This hole is in excess to the hole created by thermal agitation.

1.7   IONIZATION ENERGY

If intrinsic semiconductor is doped with phosphorus, it becomes *n-type* as Phosphorus is pentavalent. The 4 electrons in the outer orbit of Phosphorus are shared by the 4 Germanium atoms and the fifth electron of Phosphorus in the outer orbit is a free electron. But in order that this electron is completely detached from the parent Phosphorus atom, some energy is to be supplied. This energy required to separate the fifth electron is called *Ionization Energy*. The value of ionization energy for Germanium is 0.012 eV, and in Silicon, it is 0.044 eV. For different impurity materials, these values will be different, in Silicon and Germanium. As this energy is small, at room temperature, we assume that all the impurity atoms are ionized.

1.8   HOLES AND ELECTRONS

In intrinsic semiconductors, $n = p = n_i$. Or the product $n \times p = n_i^2$. In extrinsic semiconductor say n-type semiconductor practically the electron concentration, $\mathbf{n} \gg \mathbf{n_i}$. As a result holes, minority carrier in n type, encounter with free electrons, and this probability is much larger since $n \gg p$. So when a hole encounter a free electron, both electrons and holes recombine and the place of hole is occupied by the free electrons and this probability is much lager since $\mathbf{n} \gg \mathbf{p}$. The result is that both free electron and hole are lost. So the hole density 'p' decreases and also that of electron density 'n' but still $\mathbf{n} \gg \mathbf{n_i}$. This is also true in the case of *p-type* semiconductor $\mathbf{p} \gg \mathbf{n_i}$, n decreases in acceptor type semiconductor, as a result of recombination, 'p' also decreases but $\mathbf{p} \gg \mathbf{n_i}$. It has been observed practically that the net concentration of the electrons and holes follows the realtion $n \times p = n_i^2$. This is an approximate formula but still valid. Though 'p' decreases in *n-type* semiconductor with recombination, '$\mathbf{n}$' also decreases, but $\mathbf{n} \gg \mathbf{n_i}$ and because of breaking of covalent bonds, more free electrons may be created and '$\mathbf{n}$' increases.

In the case of *p-type* semiconductor the concentration of acceptor atoms Na $\gg n_i$. Assuming that all the acceptor atoms are ionized, each acceptor atom contribute at least one hole. So $\mathbf{p} \gg \mathbf{n_i}$. Holes are the majority carriers and the electrons minority carriers. As $\mathbf{p} \gg \mathbf{n_i}$, the current is contributed almost all, by holes only and the current due to electrons is negligibly small.

If impurities of both donor type and acceptor type are simultaneously doped in intrinsic semiconductor, the net result will be, it can be either, *p-type* or *n-type* depending upon their individual concentration. To give a specific example, suppose donor atoms concentration is 100 n_i, and acceptor atoms concentration in 10 n_i. Then $N_D = 0.1\, N_i$. The number of electrons combine contributed by

N_D combine with number of holes contributed by Na. So the net free electrons will be equal to 0.9 N_D = 90 Ni. Such a semiconductor, can be regarded as n type semiconductor. If $N_A = N_D$ the semiconductor remains intrinsic.

1.8.1 INTERSTITIAL ATOMS

Intrinsic or pure semiconductor is practically not available. While doping a semiconductor with impurities, pure Phosphorous, Arsenic, Boron or Aluminium may not be available. These impurities themselves will contain some impurities. Commonly found such undesirable impurities are Lithium, Zinc, Copper, Nickel etc. Sometimes they also act as donor or acceptor atoms. *Such atoms are called as interstitial atoms*, except Copper and Nickle other impurities do not affect much.

1.8.2 EFFECTIVE MASS

When quantum mechanism is used to specify the motion of electrons or holes within a crystal, holes and electrons are treated as imaginary particles with effective masses m_p and m_n respectively. This is valid when the external applied field is smaller than the internal periodic fields produced by the lattice structure.

Most metals and semiconductors are crystalline in structure, i.e., they consist of space array of atoms in a regular tetrahedral or any other fashion. *The regular pattern of atom arrangement is called lattice*. In the case of metals, in each crystal, the atoms are very close to each other. So the valency electron of one atom are as much associated with the other atoms as with the parent atom. In other words, the valance electrons are loosely bound to the parent atom and the valance electrons of one atom are shared by another atom. So every such valence electron has almost zero affinity with any individual atom. Such electrons are free to move within the body of the metal under the influence of applied electric field. *So conductivity of metals is large*.

On the other hand, for a semiconductor also, the valence electrons of one atom are shared by the other atoms. *But these binding forces are very strong*. So the valence electrons are very much less mobile. Hence conductivity is less. As the temperature is increased, the covalent bonds binding the valance electrons are broken and electrons made free to move, resulting in electrical conduction.

Valence Electrons are the outer most electrons orbiting around the nucleus.

Free electrons are those valence electrons which are separated from the parent atom. Since the covalent bonds are broken. Germanium has 4 valence electrons. Number of electrons is equal to number of protons. The atom is neutral when no electric field is applied. In the adjacent figure, the ion is having a charge +4 (circles) with 4 electrons around it. The covalent bonds are shown by lines linking one electron of one atom to the nucleus of other atom (Fig. 1.6).

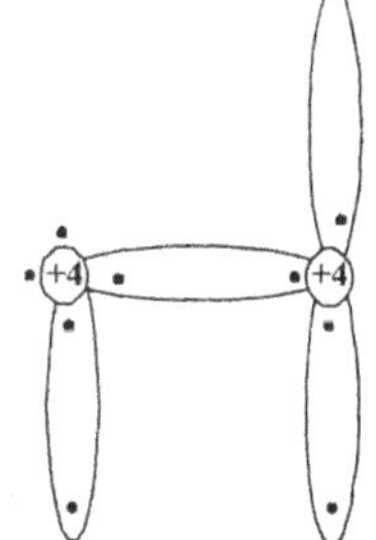

Fig. 1.6 Covalent bonds

1.8.3 HOLES AND EXCESS ELECTRONS

When a covalent bond is broken due to thermal agitation, an electron is released and a hole is created in the structure of that particular atom. The electron so released is called *free electron or excess electron since it is not required to complete any covalent bond in its immediate neighborhood*. Now the ion which has lost electron will seek another new electron to fill the vacancy. So because of the thermal agitation of the crystal lattice, an electron of another ion may come very close to the ion which has lost the electron. The ion which has lost the electron will immediately steal an electron from the closest ion, to fill its vacancy. The holes move from the first ion to the second ion. ***When no electric field is applied, the motion of free electron is random in nature***. But when electric field is applied, all the free electrons are lined up and they move towards the positive electrode. The ***life period of a free electron may be 1μ–sec to 1 millisecond*** after which it is absorbed by another ion.

1.9   MASS ACTION LAW

In an intrinsic Semiconductor number of free electrons $n = n_i$ = No. of holes $p = p_i$

Since the crystal is electrically neutral, $n_i p_i = n_i^2$.

Regardless of individual magnitudes of n and p, the product is always constant,

$\therefore \qquad np = n_i^2$

$$n_i = AT^{\frac{3}{2}}\ e^{\frac{-E_{GO}}{2KT}} \qquad\qquad ....(1.13)$$

This is called ***Mass Action Law***.

1.10 LAW OF ELECTRICAL NEUTRALITY

Let N_D is equal to the concentration of donor atoms in a doped semiconductor. So when these donor atoms donate an electron, it becomes positively charged ion, since it has lost an electron. So positive charge density contributed by them is N_D. If 'p' is the hole density then total positive charge density is $N_D + p$. Similarly if N_A is the concentration of acceptor ions, (say Boron which is trivalent, ion, accepts an electron, so that 4 electrons in the outer shell are shared by the Ge atoms), it becomes negatively charged. So the acceptor ions contribute charge $= (N_A + n)$. Since the Semiconductor is electrically neutral, when no voltage is applied, the magnitude of positive charge density must equal that of negative charge density.

Total positive charge, $N_D + p$ = Total negative charge $(N_A + n)$

$$N_D + p = N_A + n$$

This is known as ***Law of Electrical Neutrality***.

Consider n-type material with acceptor ion density $N_A = 0$. Since it is n-type, number of electrons is >> number of holes.

So 'p' can be neglected in comparison with n.

$\therefore \qquad\qquad n_n \simeq N_D.$ 　　(Since every Donor Atom contributes one free electron.)

In n-type material, the free electron concentration is approximately equal to the density of donor atoms.

In n-type semiconductor the electron density $n_n = N_D$. Subscript n indicates that it is n-type semiconductor.

But $\qquad\qquad n_n \times p_n = n_i^2$

$\therefore \qquad\qquad p_n$ = The hole density in n-type semiconductor $= \dfrac{n_i^2}{N_D}$

$$p_n = \dfrac{n_i^2}{N_D} \qquad\qquad ....(1.14)$$

Similarly Hole concentration in p-type semiconductor,

$$p_p \simeq N_A.$$
$$n_p \times p_p = n_i^2$$

$$\therefore \qquad n_p = \frac{n_i^2}{N_A} \qquad\qquad\qquad .... (\ 1.15\ )$$

Problem 1.14

A specimen of intrinsic Germanium at 300 °K having a concentration of carriers of $2.5 \times 10^{13}/cm^3$ is doped with impurity atoms of one for every million germanium atoms. Assuming that all the impurity atoms are ionised and that the concentration of Ge atoms is $4.4 \times 10^{22} / cm^3$, find the resistivity of doped material. (μ_n for Ge $= 3600$ $cm^2/volt\text{-}sec$).

Solution

The effect of minority carriers (impurity atoms) is negligible. So the conductivity will not appreciably change.

$$\therefore \qquad \text{Conductivity, } \sigma_n = ne\mu_n \ (\ \text{neglecting Hole Concentration}\ )$$

$$n = 4.4 \times 10^{22}/cm^3 \ , \ N_D = \frac{4.4 \times 10^{22}}{10^6} \ /cm^3$$

$$\mu_n = 3600 \ cm^2 / Volt\text{ - }sec$$

$$\therefore \qquad \sigma_n = 4.4 \times 10^{16} \times 3600 \times 1.6 \times 10^{-19}$$

$$= 25.37 \ mhos / cm$$

$$\text{Resistivity, } \quad \rho_n = \frac{1}{25.37} = 0.039 \ \Omega\text{--}cm$$

Problem 1.15

Determine the conductivity (σ) and resistivity (ρ) of pure silicon, at $300°K$ assuming that the concentration of carriers at 300 °K is $1.6 \times 10^{10}/cm^3$ for Si and mobilities as $\mu_n = 1500 \ cm^2/V\text{-}sec$; $\mu_p = 500 \ cm^2/V\text{-}sec$.

Solution

$$\sigma_i = (\ \mu_n + \mu_p\ ) e \ n_i = 1.6 \times 10^{-19} \times 1.6 \times 10^{10} \times (1500 + 500)$$

$$= 5.12 \times 10^{-6} \ \upsilon/cm$$

$$\rho_i = \frac{1}{5.12 \times 10^{-6}} = 195,\ 300 \ \Omega\text{--}cm$$

Problem 1.16

Determine the concentration of free electrons and holes in a sample of Ge at 300 °K which has a concentration of donor atoms equal to 2×10^{14} atoms $/cm^3$ and a concentration of acceptor atoms $= 3 \times 10^{14}$ atom $/cm^3$. Is this p-type or n-type Germanium ?

Solution

$$n \times p = n_i^2$$

$$n_i = AT^{3/2} \ e^{\frac{-E_{go}}{2kT}}$$

$$A = 9.64 \times 10^{14}$$

$$E_G = 0.25 \text{ ev}$$

$$n_i^2 = 6.25 \times 10^{26}/cm^3$$

$$N_A + n = N_D + p$$

$\therefore$ Total negative charge = Total positive charge

or $p - n = N_A - N_D = (3 - 2) \times 10^{14} = 10^{14}$

or $p = n + 10^{14}$

Then $n(n + 10^{14}) = 6.25 \times 10^{26}$

or $n = 5.8 \times 10^{12}$ electrons. $/cm^3$

and $p = n + 10^{14} = 1.06 \times 10^{14}$ holes/cm^3

As $p > n$, this is p-type semiconductor.

Problem 1.17

Find the concentration of holes and electrons in a p type germanium at 300°K, if the conductivity is $100\ \Omega$ - cm. μ_p Mobility of holes in Germanium = $1800\ cm^2$ / V - sec.

Solution

$\because$ It is p-type $p \gg n$.

$\therefore$ $\sigma_p = p\ e\ \mu_p$

$$P_p = \frac{\sigma}{e\mu_p} = \frac{100}{1.6 \times 10^{-19} \times 1800} = 3.47 \times 10^{17} \text{ holes}/cm^3$$

$$n \times p = n_i^2$$

$$n_i = 2.5 \times 10^{13}/cm^3$$

$\therefore$ $$n = \frac{\left(2.5 \times 10^{13}\right)^2}{3.47 \times 10^{17}} = 1.8 \times 10^9 \text{ electrons}/cm^3$$

Problem 1.18

(a)   Find the concentration of holes and electrons in p-type Germanium at 300°K, if $\sigma = 100\ \upsilon$/cm.

(b)   Repeat part (a) for n-type Si, if $\sigma = 0.1\ \upsilon$/cm.

Solution

As it is p-type semiconductor, $p \gg n$.

$\therefore$ $\sigma = pe\mu_p$

$\therefore$ $$p = \frac{\sigma}{e\mu_p} = \frac{100}{1.6 \times 10^{-19} \times 1800} = 3.47 \times 10^{17} \text{ holes}/cm^3$$

$$= 3.47 \times 10^{17} \text{ holes}/cm^3$$

$$n \times p = n_i^2$$

$$n_i = AT^{\frac{3}{2}} e^{-\frac{E_{G_O}}{2kT}} = 2.5 \times 10^{13}/cm^3$$

$$p = 3.47 \times 10^{17}/cm^3$$

$$\therefore \quad n = \frac{n_i^2}{p} = \frac{\left(2.5 \times 10^{13}\right)^2}{3.47 \times 10^{17}} = 1.8 \times 10^9 \text{ electron/m}^3$$

(b) $\sigma = n e \, \mu_n$;

$$n = \frac{0.1}{1300 \times 1.6 \times 10^{-19}} = 4.81 \times 10^{14}/cm^3$$

$$= 4.81 \times 10^{14}/cm^3$$

$$n_i = 1.5 \times 10^{10} \; ;$$

$$p = \frac{n_i^2}{n} = \frac{\left(1.5 \times 10^{10}\right)^2}{4.81 \times 10^{14}} = 4.68 \times 10^{11} \text{ hole/m}^3$$

Problem 1.19

A sample of Ge is doped to the extent of 10^{14} donor atoms/cm^3 and 7×10^{13} acceptor   atoms/cm^3. At room temperature , the resistivity of pure Ge is 60 Ω-cm. If the applied electric field is 2 V/cm, find total conduction current density.

Solution

For intrinsic Semiconductor,

$$n = p = n_i$$

$$\sigma_i = n_i \, e \, (\mu_p + \mu_n)$$

$$= \frac{1}{60} \; \upsilon/cm.$$

μ_p for Ge is 1800 cm^2/V-sec ;

$\mu_n = 3800$ cm^2/V-sec.

$$\therefore \quad n_i = \frac{\sigma_i}{e\left(\mu_p + \mu_n\right)} = \frac{1}{60\left(1.6 \times 10^{-19}\right)\left(3800 + 1800\right)}$$

$$= 1.86 \times 10^{13} \text{ electron/cm}^3$$

$$p \times n = n_i^2 = (1.86 \times 10^{13})^2 \qquad\qquad ....(1)$$

$$N_A + n = N_D + p$$

$$N_A = 7 \times 10^{13}/cm^3$$

$$N_D = 10^{14}/cm^3$$

$$\therefore \quad (p - n) = N_A - N_D = -3 \times 10^{13} \qquad\qquad ....(2)$$

Solving (1) and (2) simultaneously to get p and n,

$$p = 0.88 \times 10^{13}$$

$$n = 3.88 \times 10^{13}$$

$$J = (n\mu_n + p\mu_p).\, e\varepsilon$$

$$= \{ (3.88)(3800) + (0.88)(1800) \} \times 10^{13}.\, e\varepsilon$$

$$= 52.3 \text{ mA/cm}^3$$

Problem 1.20

Determine the concentration of free electrons and holes in a sample of Germanium at $300^{\circ}K$ which has a concentration of donor atoms $= 2 \times 10^{14}$ atoms $/cm^3$ and a concentration of acceptor atoms $= 3 \times 10^{14}$ atoms/cm^3. Is this *p-type* or *n-type* Ge ? In other words, is the conductivity due primarily to holes or electrons ?

Solution

$$n \times p = n_i^2$$

$$N_A + n = P + N_D$$

Solving (a) and (b) to get n and p,

$$p - n = N_A - N_D$$

$$p = \frac{n_i^2}{n}$$

$$\therefore \qquad \frac{n_i^2}{n} - n = (N_A - N_D)$$

or

$$\frac{n_i^2 - n^2}{n} = (N_A - N_D)$$

$$n_i^2 - n^2 = n \times (N_A - N_D)$$

or

$$n^2 + n (N_A - N_D) - n_i^2 = 0$$

This is in the form $ax^2 + bx + c = 0$

$$\therefore \qquad x = -\frac{b}{2} \pm \sqrt{\frac{b^2 - 4ac}{2a}}$$

$$\therefore \qquad n = -\frac{(N_A - N_D)}{2} + \sqrt{\frac{(N_A - N_D)^2 - 4n_i^2}{2}} = 0$$

Negative sign is not taken into consideration since electron or hole concentration cannot be negative.

$$\therefore \qquad n > 0 \text{ and } p > 0.$$

$$N_A = 3 \times 10^{14}/cm^{30}$$

$$N_D = 2 \times 10^{14}/cm^3$$

$$n_i \text{ at } 300^{\circ}K = 2.5 \times 10^{13} / cm^3$$

$$n_i = AT^{\frac{3}{2}}e^{\frac{-E_G}{2KT}}$$

$$E = 0.72 \text{ eV}$$

$$n_i^2 = 6.25 \times 10^{26}/cm^3$$

$$\therefore \qquad n \text{ can be calculated. Similarly p is also calculated.}$$

$$n = 5.8 \times 10^{18}/cm^3$$

$$p = 10.58 \times 10^{19}/m^3$$

as **p > n**, it is *p-type* semiconductor.

Problem 1.21

Calculate the intrinic concentration of Germanium in carries/m^3 at a temperature of 320°K given that ionization energy is 0.75 eV and Boltzman's Constant K = 1.374 × 10^{-23} J/$^\circ$K. Also calculate the intrinsic conductivity given that the mobilities of electrons and holes in pure germanium are 0.36 and 0.17 m^2/ volt-sec respectively.

Solution

$$n_i = AT^{\frac{3}{2}} e^{\frac{-e.E_{GO}}{2KT}}$$

$$\bar{K} = \text{Boltzman's Constant in J}/{}^\circ\text{K}$$
$$K = \text{Boltzman's Constant in eV}/{}^\circ\text{K}$$

$$= 9.64 \times 10^{21} \times (320)^{3/2}.\ e^{\frac{-1.6\times10^{-19}\times0.75}{2\times1.37\times10^{-23}\times320}}$$

$$n_i = 6.85 \times 10^{19} \text{ electrons (or holes)/m}^3.$$

In intrinsic semiconductor, $n = p = n_i$.

$$\therefore \quad \sigma_i = en_i\,(\mu_n + \mu_p)$$
$$= 1.6 \times 10^{-19} \times 6.85 \times 10^{19}\,(0.36 + 0.17) = 5.797 \text{ } \upsilon/m$$

Problem 1.22

Determine the resistivity of intrinsic Germanium at room temperature

Solution

$$T = 300\ {}^\circ\text{K}$$
$$A = 9.64 \times 10^{21}$$
$$E = 0.75 \text{ eV}.$$

$$n_i = AT^{\frac{3}{2}} e^{\frac{-E_{GO}}{2KT}} = 2.5 \times 10^{19} \text{ electrons (or holes)}/m^3.$$
$$\mu_n = 0.36 m^2/V\text{ - sec}$$
$$\mu_p = 0.17\ m^2/V\text{ - sec}$$
$$\sigma_i = en_i(\mu_n + \mu_p)$$
$$= 1.6 \times 10^{-19} \times 2.5 \times 10^{19}\,(0.35 + 0.17) = 2.13 \text{ } \upsilon/m$$

$$\rho = \frac{1}{\sigma} = \frac{1}{2.13} = 0.47 \text{ } \Omega\text{-m}$$

1.11 THE FERMI DIRAC FUNCTION

$$N(E) = \text{Density of states.}$$

i.e., The number of states per ev per cubic meter (number of states/eV/m^3)

The expression for

$$N(E) = \gamma E^{\frac{1}{2}} \qquad\qquad \text{where } \gamma \text{ is a constant.}$$

$$\gamma = \frac{4\pi}{h^3} (2m)^{3/2} (1.6 \times 10^{-19})^{3/2} = 6.82 \times 10^{27}$$

m = Mass of Electron in Kgs

h = Planck's Constant is Joule-secs.

The equation for $f(E)$ is called the ***Fermi Dirac Probability Function***. It specifies the fraction of all states at energy E (eV) occupied under conditions of thermal equilibrium. From Quantum Statistics, it is found that,

$$f(E) = \frac{1}{1 + e^{\frac{(E - E_F)}{kT}}} \qquad \dots\dots (1.16)$$

where   k = Boltzmann Constant, eV/$^\circ$K

T = Temp $^\circ$K

E_F = Fermi Level or Characteristic Energy

The momentum of the electron can be uncertain. Heisenberg postulated that there is always uncertainty in the position and momentum of a particle, and the product of these two uncertainities is of the order of magnitude of Planck's constant 'h'.

If Δ_p is the Uncertainty in the Momentum of a particle, Δ_n is the uncertainty in the position of a particle

$$\Delta_p \times \Delta x \approx h.$$

1.11.1 EFFECTIVE MASS

When an external field is applied to a crystal, the free electron or hole in the crystal responds, as if its mass is different from the true mass. This mass is called the ***Effective Mass*** of the electron or the hole.

By considering this effective mass, it will be possible to remove the quantum features of the problem. This allows us to use Newton's law of motion to determine the effect of external forces on the electrons and holes within the crystal.

1.11.2 FERMI LEVEL

Named after ***Fermi, it is the Energy State, with 50% probability of being filled if no forbidden band exists***. In other words, it is the mass energy level of the electrons, at 0°K.

If E = E_f,

$$f(E) = \tfrac{1}{2} \qquad \text{From Eq.(1.17)}.$$

If a graph is plotted between (E – E_F) and $f(E)$, it is shown in Fig. 1.7

At T = 0°K, if E > E_F then, $f(E) = 0$.

That is, there is no probability of finding an electron having energy > E_F at T = 0^0 K. Since fermi level is the max. energy possessed by the electrons at 0°k. $f(E)$ varies with temperature as shown in Fig. 1.7.

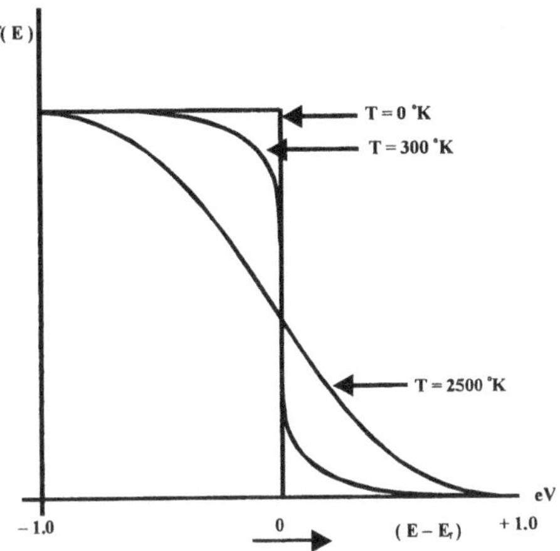

Fig. 1.7 Fermi level variation with temperature.

1.11.3 UNCERTAINTY PRINCIPLE

This was proposed by Heisenberg. The measurement of a physical quantity is characterized in an essential way by lack of precision.

1.11.4 THE INTRINSIC CONCENTRATION

$$f(E) = \frac{1}{1 + e^{(E - E_F)/KT}} \qquad \qquad .... (1.17)$$

$$1 - f(E) = \frac{e^{(E - E_F)/KT}}{1 + e^{(e - E_F)/KT}} \simeq e^{-(E_F - E)/KT}$$

$$e^{(E - E_F)/KT}\left[1 + e^{(E - E_f)/KT}\right]^{-1} \simeq e^{-(E_f - E)/KT}$$

Fermi function for a hole $= 1 - f(E)$.

$$(E_F - E) \gg KT \text{ for } E \leq Ev$$

the number of holes per m^3 in the Valence Band is,

$$p = \int_{-\infty}^{E_V} \gamma (E_V - E)^{1/2} e^{-(E_F - E)/KT} \times dE$$

$$= N_V \times e^{-(E_F - E_V)/KT}$$

where
$$N_V = 2\left(\frac{2\pi m_p \overline{K}T}{h^2}\right)^{\frac{3}{2}}$$

Similarly
$$n = N_e \, e^{-(E_C - E_F)/KT}$$

$$n \times p = N_V \times N_e \, e^{-E_G/KT}$$

Substituting the values of N_C and N_V,

$$n \times p = n_i^2 = AT^3 \, e^{-E_G/KT}$$

1.11.5   CARRIER CONCENTRATIONS IN A SEMICONDUCTOR

$$d\eta \;\; = \;\; N(E) \times f(E) \times dE$$

dn = number of conduction electrons per cubic meter whose energy lies between E and E + dE

$f(E)$ = The probability that a quantum state with energy E is occupied by the electron.

$N(E)$ = Density of States.

$$f(E) \;\; = \;\; \frac{1}{1 + e^{(E-E_F)/KT}}$$

The concentration of electrons in the conduction band is

$$n = \int_{E_C}^{\infty} N(E) \times f(E) \times dE$$

for

$$E \geq E_C$$

$$(E - E_s) \text{ is} >> KT.$$

$$\therefore \qquad f(E) = e^{-(E-E_f)/KT}$$

$$\because \qquad e^{(E-E_F)/KT} >> 1$$

$$\therefore \qquad n = \int_{E_C}^{\infty} \gamma (E - E_C)^{1/2} \times e^{-(E-E_F)KT} \cdot dE$$

Simplifying this integral, we get,

$$n = N_C \times e^{-(E_C - E_F)/KT}$$

where

$$N_C = 2 \left(\frac{2\pi m_n \overline{K} T}{h^2} \right)^{\frac{3}{2}}$$

$$\overline{K} = \text{Boltzman's Constant in J/}^\circ K$$

N_i is constant.

where m_n = effective mass of the electron.

Similarly the number of holes / m^3 in the Valence Band

$$p = N_V \, e^{-(E_F - E_V)/KT}$$

where

$$N_V = 2 \left(\frac{2\pi m_p \overline{K} \times T}{h^2} \right)^{\frac{3}{2}}$$

Fermi Level is the maximum energy level that can be occupied by the electrons at 0 °K. ***Fermi Level*** or characteristic energy represents the energy state with 50% probability of being filled if no forbidden bond exists. If $E = E_F$, then $f(E) = \frac{1}{2}$ for any value of temperature. $f(E)$ is the probability that a quantum state with energy E is occupied by the electron.

1.11.6 Fermi Level in Intrinsic Semiconductor

$$n = p = n_i$$

$$n = N_C \, e^{-(E_C - E_F)/KT}$$

$$p = N_V \, e^{-(E_F - E_V)/KT}$$

$$n = p$$

or

$$N_C \, e^{-(E_C - E_F)/KT} = N_V \, e^{-(E_F - E_V)/KT}$$

Electrons in the valence bond occupy energy levels up to 'E_F'. E_F is defined that way. Then the additional energy that has to bo supplied so that free electron will move from valence band to the conduction band is E_C

$$\frac{N_C}{N_V} = e^{\frac{-(E_F - E_V)}{KT} + \frac{(E_C + E_F)}{KT}}$$

$$= e^{\frac{-2E_F + E_C + E_V}{KT}}$$

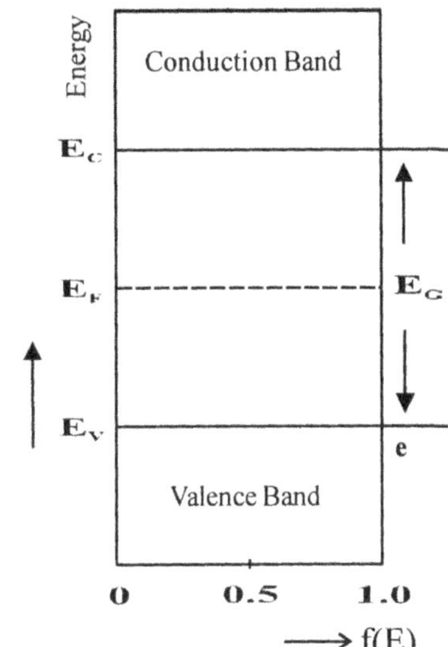

Fig. 1.8 Energy band diagram.

Taking logarithms on both sides,

$$\ln \frac{N_C}{N_V} = \frac{E_C + E_V - 2E_F}{KT}$$

$$\therefore \qquad E_F = \frac{E_C + E_V}{2} - \frac{KT}{2} \ln \frac{N_C}{N_V}$$

$$N_C = 2\left(\frac{2\pi m_n \overline{K}T}{h^2} \right)^{\frac{3}{2}} \qquad \qquad \dots\dots\dots (1.18)$$

$$N_V = 2\left(\frac{2\pi m_p . \overline{K}T}{h^2} \right)^{\frac{3}{2}} \qquad \qquad \dots\dots\dots (1.19)$$

where m_n and m_p are effective masses of holes and electrons. If we assume that $m_n = m_p$, (though not valid),

$$N_C = N_V$$

$$\therefore \qquad \ln \frac{N_C}{N_V} = 0$$

$$\therefore \qquad E_F = \frac{E_C + E_V}{2} \qquad\qquad \dots\dots\dots (1.20)$$

The graphical representation is as shown in Fig. 1.8. ***Fermi Level in Intrinsic Semiconductor lies in the middle of Energy gap E_G.***

Problem 1.23

In p-type Ge at room temperature of 300 °K, for what doping concentration will the fermi level coincide with the edge of the valence bond ? Assume $\mu_p = 0.4$ m.

Solution

$$E_F = E_V$$

when
$$N_A = N_V$$

$\therefore$
$$E_F = E_V + kT \ln . \frac{N_V}{N_A}$$

$\therefore$
$$N_V = 4.82 \times 10^{15} \left(\frac{mp}{m} \right)^{\frac{3}{2}} \times T^{3/2} = 4.82 \times 10^{15}(0.4)^{3/2}(300)^{3/2}$$

$$= 6.33 \times 10^{18}.$$

$\therefore$ Doping concentration $N_A = 6.33 \times 10^{18}$ atoms/cm^3.

Problem 1.24

If the effective mass of an electron is equal to twice the effective mass of a hole, find the distance in electron volts (ev) of fermi level in as intrunsic semiconductor from the centre of the forbidden bond at room temperature.

Solution

For Intrunsic Semiconductor,

$$E_F = \left[\left(\frac{E_C + E_V}{2} \right) - \frac{KT}{2} \ln \left(\frac{N_C}{N_V} \right) \right]$$

If $m_p = m_n$

then $N_C = N_V.$

Hence E_F will be at the centre of the forbidden band. But if $m_p \neq m_n$. E_F will be away from the centre of the forbidden band by

$$\frac{KT}{2} \ln . \frac{N_C}{N_V} = \tfrac{3}{4} \frac{kT}{2} . \ln \frac{m_n}{m_p}$$

$\therefore$
$$N_C = 2 \left(\frac{2\pi m_n \overline{KT}}{n^2} \right)^{3/2}$$

$$N_V = 2 \left(\frac{2\pi m_p \overline{kT}}{n^2} \right)^{3/2}$$

$$= \frac{3}{4} \times 0.026 \, \ln (2)$$

$$= 13.5 \text{ m. eV}$$

1.11.7 FERMI LEVEL IN A DOPED SEMICONDUCTOR

$$\sigma = (\mu_n n + \mu_p p)e,$$

So the electrical characteristics of a semiconductor depends upon '*n*' and '*p*', the concentration of holes and electrons.

The expression for $n = N_C \, e^{-(E_C - E_F)/KT}$ and the expression for $p = N_V \, e^{-(E_F - E_V)/KT}$

These are valid for both intrinsic and extrinsic materials.

The electrons and holes, respond to an external field as if their mass is m*(m* = 0.6m) and not 'm'. So this m* is known as *Effective Mass*.

With impurity concentration, only E_F will change. In the case of intrinsic semiconductors, E_F is in the middle of the energy gap, indicating equal concentration of holes and electrons.

If donor type impurity is added to the intrinsic semiconductor it becomes n-type. So assuming that all the atoms are ionized, each impurity atom contributes at least one free electron. So the first N_D states in the conduction band will be filled. Then it will be more difficult for the electrons to reach Conduction Band, bridging the gap between Covalent Bond and Valence Bond. So the number of electron hole pairs, thermally generated at that temperature will be decreased. *Fermi level is an indiction of the probability of occupancy of the energy states*. Since Because of doping, more energy states in the ConductionBand are filled, the fermi level will move towards the Conduction Band.

EXPRESSION FOR E_G

$$n = N_C \times e^{\frac{-(E_C - E_F)}{KT}}$$

$$p = N_V \cdot e^{\frac{-(E_F - E_V)}{KT}}$$

$$n \times p = N_C \times N_V \times e^{\frac{-(E_C - E_V)}{KT}}$$

But $$(E_C - E_V) = E_G$$

and $$n \times p = n_i^2$$

$\therefore$ $$n_i^2 = N_C \times N_V \times e^{-\frac{E_G}{KT}}$$

$$\frac{n_i^2}{N_C \times N_V} = e^{-\frac{E_G}{KT}}$$

Taking logarithms,

$$\ln\left(\frac{n_i^2}{N_C N_V}\right) = -\frac{E_G}{KT}$$

or $$-\ln\left(\frac{N_C N_V}{n_i^2}\right) = -\frac{E_G}{KT}$$

or $$\boxed{E_G = KT \ln\left(\frac{N_C N_V}{n_i^2}\right)}$$ .... (1.21)

The position of Fermi Level is as shown in Fig. 1.9

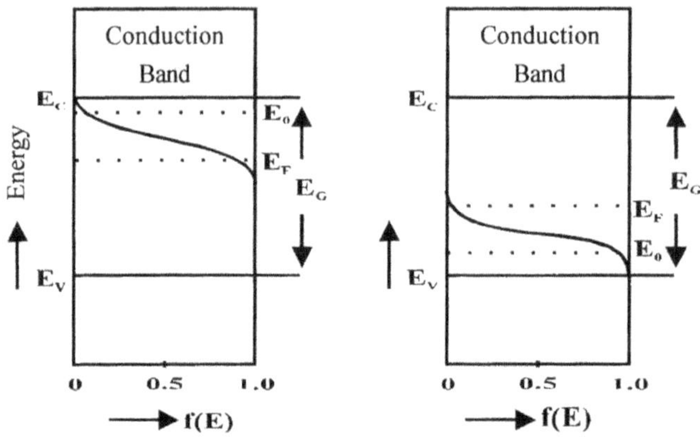

For n-type semiconductor   For p-type semiconductor
Fig. 1.9

Similarly, in the case of *p-type* materials, the Fermi level moves towards the valence band since the number of holes has increased. In the case of *n-type* semiconductor, the number of free electrons has increased. So energy in Covalent Bond has increased. Fermi Level moves to wards conduction band. ***Similarly in p-type semiconductors, fermi Level moves towards Valence Band.*** So it is as shown in Fig. 1.9.

To calculate the exact position of the Fermi Level in n-type Semiconductor:

In *n-type* semiconductor,

$$n \simeq N_D$$

But

$$n = N_C \times e^{-(E_c - E_F)/KT}$$

$\therefore$

$$N_D = N_C \times e^{-(E_c - E_F)/KT}$$

or

$$\frac{N_D}{N_C} = e^{-(E_c - E_F)/KT}$$

Taking logarithms,

$$\ln \frac{N_D}{N_C} = \frac{-(E_c - E_F)}{KT}$$

or

$$KT \times \left\{ \ln \frac{N_D}{N_C} \right\} = -(E_C - E_F)$$

or

$$\boxed{E_F = E_C - KT \times \left\{ \ln \frac{N_C}{N_D} \right\}}$$

.... (1.22)

So Fermi Level E_F is close to Conduction Band E_C in n-type semiconductor.

Similarly for p-type material,

$$p = N_A$$

But $$p = N_V \times e^{-(E_F - E_V)/KT}$$

$$\therefore \quad \frac{N_A}{N_V} = e^{-(E_F - E_V)/KT}$$

Taking Logarithms,

$$\ln \frac{N_A}{N_V} = -\frac{(E_F - E_V)}{KT}$$

$$KT \times \ln \frac{N_A}{N_V} = E_V - E_F$$

or $$\boxed{E_F = E_V + KT \times \ln \frac{N_V}{N_A}} \qquad \dots\dots\dots (1.23)$$

$$\therefore \quad N_A = N_V$$

Fermi Level is close to Valance Band E_V in p-type semiconductor.

Problem 1.25

In n type silicon, the donor concentration is 1 atom per 2×10^8 silicon atoms. Assuming that the effective mass of the electron equals true mass, find the value of temperature at which, the fermi level will coincide with the edge of the conduction band. Concentration of Silicon $= 5 \times 10^{22}$ atom/cm3.

Solution

Donor atom concentration $= $ 1 atom per 2×10^8 Si atom.

Silicon atom concentration $= 5 \times 10^{22}$ atoms/cm^3

$$\therefore \quad N_D = \frac{5 \times 10^{22}}{2 \times 10^8} = 2.5 \times 10^{14}/cm^3.$$

For n-type, Semiconductor,

$$E_F = E_C - KT \ln \left(\frac{N_C}{N_D} \right)$$

If E_F were to coincide with E_C, then

$$N_C = N_D$$
$$N_D = 2.5 \times 10^{14} \ /cm^3.$$

$$N_C = 2 \left\{ \frac{2\pi m_n \overline{K} T}{h^2} \right\}^{\frac{3}{2}}$$

$h = $ Plank's Constant; $\overline{K} = $ Boltzman Constant

m_n the effective mass of electrons to be taken as $= m_E$

$$\therefore \quad N_C = 2 \left\{ \frac{2 \times 3.14 \times 9.1 \times 10^{-31} \times \overline{K} \times T}{h^2} \right\}^{\frac{3}{2}}$$

$$= 4.28 \times 10^{15} \ T^{\frac{3}{2}}$$

$$\therefore \qquad N_C = N_D,$$

$$4.28 \times 10^{15} \ T^{\frac{3}{2}} = 2.5 \times 10^{14}$$

$$\therefore \qquad T = 0.14 \ ^{\circ}K$$

II   TRANSPORT PHENOMENON IN SEMICONDUCTORS

1.12   TOTAL CURRENT IN A SEMICONDUCTOR

1.12.1   DRIFT CURRENT IN AN N-TYPE SEMICONDUCTOR

Within a semiconductor, intrinsic or impure, because of the thermal energy, covalent bonds are broken and electrons and holes move in random directions. These collide with lattices, get deflected and move in a different direction, till they collide with another carrier. Such a random motion is defined as ***mean free path length 'l'***, the distance a carrier travels between collisions and the average time between collisions 't'. The average velocity of motion $v = \dfrac{l}{t}$. Over a period of time which is >> 't', average movement is zero or net current is zero.

But when electric field is applied all the electrons are aligned in a particular direction and move towards the positive electrode and holes in the opposite direction. The resulting current is called ***Drift Current***.

Let the semiconductors be 'n' type. Now using a battery, electric field is applied, under the influence of the electric field, all the free electrons move towards the positive electrode and enter the metal of the positive electrode Fig. 1.10. The donor atoms have thus lost their free electrons. So the donor atoms near the positive electrode pull electrons from the electrode, exactly equal in number to the free electrons which have entered the electrode. So the semiconductors remains electrically neutral. The voltage applied results in voltage drop across semiconductor. If the

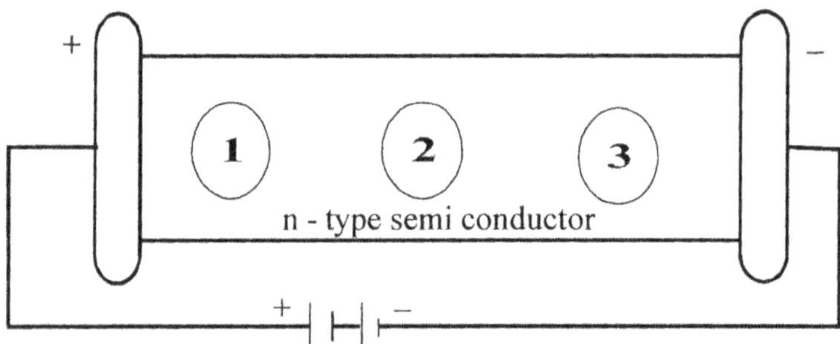

Fig. 1.10 Drift current in n-type semiconductor.

battery is removed, the number of free electrons and holes is same as before the application of field, since semiconductor has taken equal number of electrons from the positive electrode.

When the free electron contributed by phosphorus donor atom goes into the positive electrode, the donor atom loses one electron. Therefore it will pull one more electron from the positive electrode and hence the number of free electrons in the semiconductors remains the same.

Problem 2.26

In p-type Silicon, the acceptor concentration corresponds to 1 atom per 10^8 Silicon atoms. Assume that $m_p = 0.6m$. At room temperature, how far from the edge of the valence band is the Fermi level ? Is E_F above or below E_V? The concentration of Silicon atoms is 5×10^{22} atoms/cm^3.

Solution

Concentration of Silicon atoms $= 5 \times 10^{22}$/cm^{3c}.

Because doping is done at 1 atom per 10^8 Silicon Atoms.

$$N_A = \frac{5 \times 10^{22}}{10^8} = 5 \times 10^{14}/\text{cm}^3$$

$$N_V = 2 \left[\frac{(2\pi m_P KT)^{\frac{3}{2}}}{h^2} \right]$$

K = Boltzman's Constant in eV/°K; $\overline{K}$ = Boltzman constant in J// °K.
h = Planck's Constant $= 6.62 \times 10^{-34}$ J-sec

$\overline{K} = 1.38 \times 10^{-23}$ J/°K

$$N_V = 2 \left(\frac{2\pi m_p \overline{K}T}{h^2} \right)^{\frac{3}{2}} = 4.82 \times 10^{21} \left(\frac{m_p}{m} \right)^{\frac{3}{2}} T^{\frac{3}{2}}$$

$$\left(2 \left(\frac{2\pi.\overline{K}}{h^2} \right)^{\frac{3}{2}} = \frac{4.82 \times 10^{15}}{m^{3/2}} \right)$$

$$= 4.82 \times 10^{15} \times \frac{(m_p)^{\frac{3}{2}}}{m} \times T^{3/2}$$

where $\qquad m_p = 0.6\ m$ (given)

$\qquad T = 300°$K

$\qquad N_V = 4.82 \times 10^{15}(0.6 \times 300)^{3/2} = 1.17 \times 10^{19}$/cm^3

$$E_F - E_V = KT \ln \left(\frac{N_V}{N_A} \right)$$

$$= 0.026 \ln . \left(\frac{1.17 \times 10^{19}}{5 \times 10^{14}} \right)$$

$$= 0.026 \times 10$$

$$= 0.26\ ev$$

or $\qquad E_F - E_V = 0.26\text{ev}$

$\therefore \qquad E_F$ is above E_V.

Problem 2.27

In p-type Ge at room temperature of 300 °K, for what doping concentration will the fermi level coincide with the edge of the valence bond ? Assume $m_p = 0.4\, m$.

Solution

$$E_F = E_V$$

when
$$N_A = N_V$$

$$\therefore \qquad E_F = E_V + kT \ln . \frac{N_V}{N_A}$$

$$\therefore \qquad N_V = 4.82 \times 10^{15}\left(\frac{mp}{m}\right)^{\frac{3}{2}} \times T^{3/2} = 4.82 \times 10^{15}(0.4)^{3/2}(300)^{3/2}$$

$$= 6.33 \times 10^{18}.$$

$$\therefore \qquad \text{Doping concentration } N_A = 6.33 \times 10^{18}\ \text{atoms/cm}^3.$$

Problem 2.28

If the effective mass of an electron is equal to thrice the effective mass of a hole, find the distance in electron volts (ev) of fermi level in as intrunsic semiconductor from the centre of the forbidden bond at room temperature.

Solution

For Intrunsic Semiconductor,

$$E_F = \left[\left(\frac{E_C + E_V}{2}\right) - \frac{KT}{2}\ln\left(\frac{N_C}{N_V}\right)\right]$$

If
$$m_p = m_n$$

then
$$N_C = N_V.$$

Hence E_F will be at the centre of the forbidden band. But if $m_p \neq m_n$. E_F will be away from the centre of the forbidden band by

$$\frac{KT}{2}\ln . \frac{N_C}{N_V} = \tfrac{3}{4}\,\frac{kT}{2}.\ln \frac{m_n}{m_p}$$

$$\therefore \qquad N_C = 2\left(\frac{2\pi m_n \overline{KT}}{n^2}\right)^{3/2}$$

$$N_V = 2\left(\frac{2\pi m_p \overline{kT}}{n^2}\right)^{3/2}$$

$$= \frac{3}{4} \times 0.026\ ln\,(3) \quad = 21.4\ \text{meV}$$

1.12.2 Drift Current in p-Type Semiconductor

The mechanism is the same to as explained above. The holes in the acceptor type semiconductor moves towards the negative electrode and enter into it, pulling out one electron from the negative electrode from the acceptor atoms (Fig. 1.11), the hole has moved away, i.e. it has acquired an electron. So electrical neutrality or of its original condition is disturbed. This results in a electrons from the acceptor atom being

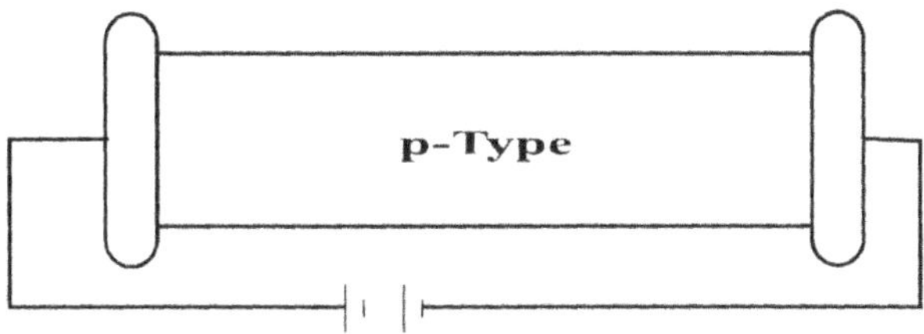

Fig. 1.11 Drift current in p-type semiconductor

pulled away. These free electrons enter the positive electrode. The acceptor atoms having lost one electron steal another electron from the adjoining atom resulting in a new hole. The new holes created thus drift towards negative electrode.

1.12.3 Diffusion Current

This current results due to difference in the concentration gradients of charge carriers. That is, free electrons and holes are not uniformly distributed all over the semiconductor. In one particular area, the number of free electrons may be more, and in some other adjoining region, their number may be less. So the electrons where the concentration gradient is more move from that region to the place where the electrons are lesser in number. This is true with holes also.

Let the concentration of some carriers be as shown in the Fig 1.12. The concentration of carriers is not uniform and varies as shown along the semiconductor length. Area A_1 is a measure of the number of carriers between x_1 and x_2. Area A_2 is a measure of the number of carriers between x_2 and x_3. Area A_1 is greater than Area A_2. Therefore number of carriers in area A_1 is greater than the number of carriers in A_2. Therefore they will move from A_1 to A_2. If these

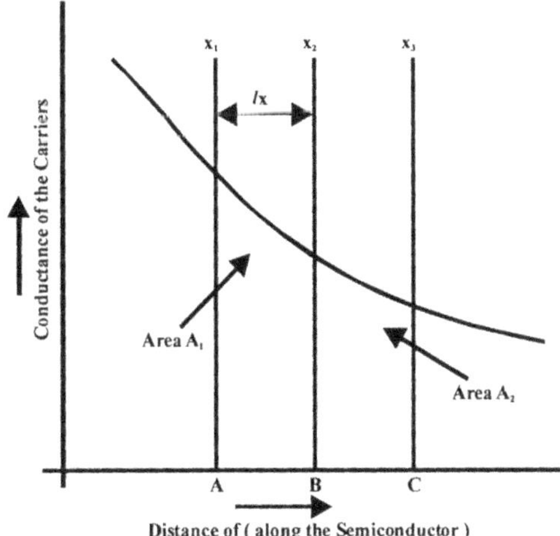

Fig. 1.12 Diffusion current.

charge carriers are free electrons, then electron current directions is from C to A ($\because$ Electrons move from A to C). If these are holes then current direction is from A to C. This current is called diffusion current. It is independent of any applied field. So in semiconductors, at room temperature itself, though no electric field is applied, if the device is connected in a circuit, there will be very small current flowing through, which is called *Diffusion Current*.

Let l_x be the distance travelled by each electron between two successive collisions., i.e. l_x is the mean free path. Let t_x secs be the corresponding time taken.

Because of the thermal agitation, the carriers in areas A_1 and A_2 have equal probability to move from left to right or from right to left. But because the number of carrier in A_1 is greater than in A_2, it results in net diffusion current. During time t_x, half the electrons from A_1 move to A_2 and half the electrons from A_2 move to A_1. The net current is due to the difference of number of electrons from A_1 to A_2

$$= 0.5 \times \left(l_x \frac{dn}{dx} \right) \times dx$$

$$\text{Hence Diffusion Current} = \left((l_x)^2 \times \frac{dn}{dx} \right) \left(\frac{e}{t_x} \right)$$

$$= \tfrac{1}{2} e \times l_x \times v_n \times \frac{dn}{dx}$$

where v_x is the Average Velocity $= \dfrac{l_x}{t_x}$ and e = charge of an electron.

Hence Diffusion Current per Sq. meter of cross section is

$$i_n = e \times D_n \times \frac{dn}{dx}$$
$$D_n = \text{Diffusion current constant in } m^2/\text{sec}$$
$$= \tfrac{1}{2} l_x v_x$$

$\dfrac{dn}{dx}$ is concentration gradient of electrons in Number of carriers / m.

If diffusion current is caused by holes, the equation of diffusion current is

$$i_p = Q \times D_p \times \frac{dp}{dx}$$

For Ge, at room temperature
$$D_n = 93 \times 10^{-4} \ m^2/\text{sec}$$
$$D_p = 44 \times 10^{-4} \ m^2/\text{sec.}$$

Diffusion electrons density J_n is given by the expression

$$J_n = + e \ Dn \times \frac{dn}{dx}$$

positive sign is used since $\dfrac{dn}{dx}$ is negative and direction of current is opposite to the movement of electrons.

where D_n is called Diffusion Constant for electrons. It is in m^2/sec. $\dfrac{dn}{dx}$ is concentration gradient.

D and μ are interrelated. It is given as,

$$\frac{D_n}{\mu_n} = \frac{D_p}{\mu_p} = V_T$$

where $\qquad\qquad V_T = \dfrac{KT}{e} = \dfrac{T}{11,600}$ .... (1.24)

where V_T is volt equivalent. of temperature. At room temperature, $\mu = 39D$.

The thermal energy due to temperature T is expressed as electrical energy in the form of Volts.

1.12.4 TOTAL CURRENT

Both Potential Gradient and Concentration Gradient can exist simultaneously within a Semiconductor. Since in such a case the total current.is the sum of *Drift Current* and *Diffusion Current*.

$$J_p = e\,\mu_p\,p.\,E - e\,D_p \cdot \frac{dp}{dx}$$

Drift Current $\qquad = e\mu_p pE$

Diffusion Current $\;=\; eD_p \dfrac{dp}{dx}$

Similarly the net electron current is

$$Jn = e\,\mu_n\,nE + e\,D_n\,\frac{dn}{dx}$$

Since Diffusion hole current is J_p

$$J_p = -eD_p\,\frac{dp}{dx}$$

p decreases with increase in x. So $\dfrac{dp}{dx}$ is negative. Negative sign is used for J_p, so that, J_p will be positive in the positive x direction. For electrons

$$J_n = +\,e\,.\,D_n\,.\,\frac{dn}{dx}$$

since the electron current is opposite to the directions of conventional current.

There exists a concentration gradient in a semiconductor. On account of this, it results in Diffusion Current. If you consider p-type semiconductor holes are the majority carriers. So the resulting Diffusion Current Density J_p is written as

$$J_p = -\,e \times D_p \times \frac{dp}{dx}$$

where D_p is called diffusion constant for holes. Since p the hole concentration is decreasing with x, $\dfrac{dp}{dx}$ is $-$ve. So $-$ve sign is used in the expression for J_p.

Similarly for electrons also the expression is similar and the slope is $-\dfrac{dn}{dx}$. But since the electron current is opposite to the conventional current,

$$J_n = \left(e.D_n \times \frac{dn}{dx}\right) + e \times D_n \frac{dn}{dx}$$

1.13 EINSTEIN RELATIONSHIP

D, the Diffusion Coefficient and μ are inter related as

$$\boxed{\frac{D_P}{\mu_P} = \frac{D_n}{\mu_n} = V_T}$$

.... (1.25)

where V_T is volt equivalent of temperature.

$$V_T = \frac{KT}{e} = \frac{T}{11,600}$$

$$\overline{K} = 1.6 \times 10^{-19} \text{ J/}^\circ\text{K}$$
$$K = 8.62 \times 10^{-5} \text{ ev/}^\circ\text{K}$$
$$K = 1.381 \times 10^{-23} \text{ J/}^\circ\text{K}$$

$$\mu_p = \frac{D_p}{V_T} = \frac{D_p}{0.026} = 39 D_p$$

.... (1.26)

$$\mu_p = 39 D_p$$
$$\mu_n = 39 D_n$$

or $\qquad\qquad \mu = 39 D$ .... (1.27)

Therefore values of D_p and D_n for Si and Ge can be determined.

Problem 1.29

Determine the values of D_p and D_n for Silicon and Germanium at room temperature.

Solution

For Germanium at room temperature,

$$D_n = \mu_n \times V_T = 3,800 \times 0.026 = 99 \text{ cm}^2/\text{sec}$$
$$D_p = \mu_p \times V_T = 1800 \times 0.026 = 47 \text{ cm}^2/\text{sec}$$

For Silicon, $\qquad D_n = 1300 \times 0.026 = 34 \text{ cm}^2/\text{sec}$

$$D_p = 500 \times 0.026 = 13 \text{ cm}^2/\text{sec}$$

1.14   CONTINUITY EQUATION

Thus Continuity Equation describes how the carrier density in a given elemental volume of crystal varies with time.

If an intrinsic semiconductor is doped with n-type material, electrons are the majority carriers. Electron - hole recombination will be taking place continuously due to thermal agitation. So the concentration of holes and electrons will be changing continuously and this varies with time as well as distance along the semiconductor. We now derive the differential. equation which is based on the fact ***that charge is neither created nor destroyed***. This is called ***Continuity Equation***.

Consider a semiconductor of area A, length dx   (x + dx − x = dx) as shown in Fig. 1.13. Let the average hole concentration be 'p'. Let E_p is a factor of x. that is hole current due to concentration is varying with distance along the semiconductor. Let I_p is the current entering the volume at x at time t, and (I_p + dI_p) is the current leaving the volume at (x + dx) at the same instant of time 't'. So when only I_p colombs is entering, (I_p + dI_p) colombs are leaving. Therefore effectively there is a decrease of (I_p + dI_p − I_p) = dI_p colombs per second within the volume. Or in other words, since more hole current is leaving than what is entering, we can say that more holes are leaving than the no. of holes entering the semiconductor at 'x'.

If dI_p is rate of change of total charge that is

$$dI_p = d\left(\frac{n \times q}{t}\right)$$

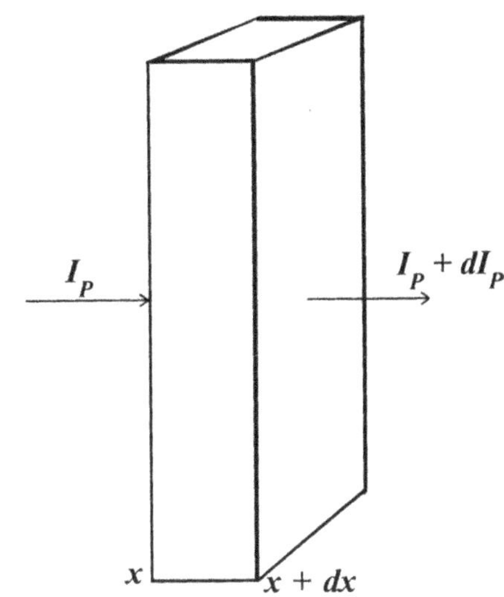

Fig 1.13   Charge flow in semiconductor

$\dfrac{dI_p}{q}$ gives the decrease in the number of holes per second with in the volume A × dx. Decrease in holes per unit volume (hole concentration) per second due to I_p is

$$\frac{dI_p}{A \times dx} \times \frac{1}{q}$$

But
$$\frac{dI_p}{A} = \text{Current Density}$$

$$= \frac{1}{q} \times \frac{dJ_p}{dx}.$$

But because of thermal agitation, more number of holes will be created. If p_0 is the thermal equilibrium concentration of holes,(the steady state value reached after recombination), then, the increase per second, per unit volume due to thermal generation is,

$$g = \frac{p}{\tau_p}$$

Therefore, increase per second per unit volume due to thermal generation,

$$g = \frac{p_0}{\tau_p}$$

But because of recombination of holes and electrons there will be decrease in hole concentration.

The decrease $= \dfrac{p}{\tau_p}$

Charge can be neither created nor destroyed. Because of thermal generation, there is increase in the number of holes. Because of recombination, there is decrease in the number of holes. Because of concentration gradient there is decrease in the number of holes.

So the net increase in hole concentration is the algebraic sum of all the above.

$$\frac{\partial p}{\partial t} = \frac{P_0 - P}{\tau_p} - \frac{1}{q} \times \frac{\partial J_p}{\partial x}$$

Partial derivatives are used since p and Jp are functions of both time t and distance x. $\frac{dp}{dt}$ gives the variation of concentration of carriers with respect to time 't'.

If we consider unit volume of a semiconductor (*n-type*) having a hole density p_n, some holes are lost due to recombination. If p_{no} is equilibrium density, (i.e., density in the equilibrium condition when number of electron = holes).

The recombination rate is given as $\dfrac{P_n - P_{no}}{\tau}$. *The expression for the time rate of change in carriers density is called the Continuity Equation.*

$$\text{Recombination rate } R = \frac{dp}{dt}$$

Life time of holes in n-type semiconductors

$$\tau_p = \frac{\Delta P}{R} = \frac{P_n - P_{no}}{dp/dt}$$

or

$$\frac{dp}{dt} = \left(\frac{P_n - P_{no}}{\tau_p} \right)$$

where p_n is the original concentration of holes in n-type semiconductors and p_{no} is the concentration after holes and electron recombination takes place at the given temperature. In other word p_{no} is the thermal equilibrium minority density. Similarly for a p-type semiconductors, the life time of electrons

$$\tau_n = \frac{n_p - n_{po}}{dx/dt} \; ; \; \frac{dx}{dt} = \frac{n_p - n_{no}}{\tau_n} .$$

1.15 THE HALL EFFECT

*If a metal or semiconductor carrying a current I is placed in a perpendicular magnetic field B, an electric field E is induced in the direction perpendicular to both I and B. This phenomenon is known as the **Hall Effect**. It is used to determine whether a semiconductor is p-type or n-type. By measuring conductivity σ, the mobility μ can be calculated using **Hall Effect**.*

In the Fig. 1.14 current I is in the positive X-direction and B is in the positive Z-direction. So a force will be exerted in the negative Y-direction. If the semiconductor is *n-type*, so that current is carried by electrons, these electrons will be forced downward toward side 1. So side 1 becomes negatively charged with respect to side 2. Hence a potential V_H called the **Hall Voltage** appears between the surface 1 and 2.

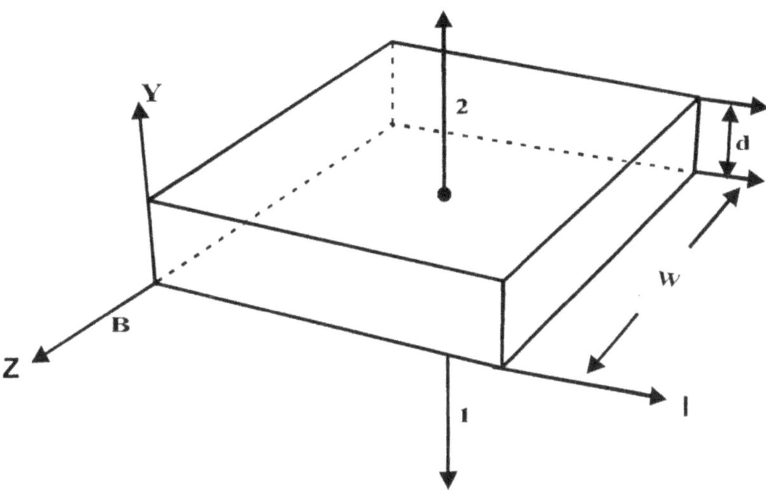

Fig 1.14 Hall effect.

In the equilibrium condition, the force due to electric field intensity 'E', because of Hall effect should be just balanced by the magnetic force or

$$e\varepsilon = B\ ev$$

v = Drift Velocity of carriers in m / sec

B = Magnetic Field Intensity in Tesla (wb/m^2)

or

$$\varepsilon = Bv \qquad \qquad .... (\ a\ )$$

But $\varepsilon = V_H/d$

where V_H = Hall Voltage

d = Thickness of semiconductors.

J = nev or $J = \rho v$

ρ = charge density.

J = Current Density (Amp / m^2)

or $$J = \dfrac{I}{\omega.d}$$

ω = width of the semiconductor; ωd = cross sectional area

I = current

$\therefore$ $$J = \text{Current Density} = \dfrac{I}{\omega d}$$

$$\varepsilon = V_H/d$$

or $$V_H = \varepsilon d$$

But $\qquad \varepsilon = Bv \qquad$ .... From Equation (a)

$\therefore \qquad V_H = B \times v \times d \qquad$ But $\quad v = J/\rho$

$$= \frac{B.J.d}{\rho} \qquad \text{But} \quad J = \frac{I}{\omega d}$$

$$V_H = \frac{B.I.d}{\rho.\omega.d} = \frac{B.I}{\rho\omega}$$

$$\boxed{V_H = \frac{B.I}{\rho\omega}} \qquad\qquad .... (1.28)$$

If the semiconductor is n-type, electrons the majority carriers under the influence of electric field will move towards side 1, side 2 becomes positive and side 1 negative. If on the other hand terminal 1 becomes charged positive then the semiconductor is p-type.

$$\rho = n \times e \ (\text{For n - type semiconductor})$$

or $\qquad \rho = p \times e \ (\text{for p-type semiconductor})$

and $\qquad \rho = \text{Charge density.}$

$$\therefore \qquad V_H = \frac{B.I}{\rho\omega}$$

$$\rho = \frac{B.I}{V_H.\omega}$$

$$\therefore \qquad \boxed{R_H = \frac{V_H.\omega}{BI}} \qquad\qquad .... (1.29)$$

The Hall Coefficient, R_H is defined as $R_H = \dfrac{1}{\rho}$. Units of R_H are m^3 / coulombs

If the conductivity is due primarily to the majority carriers conductivity, $\sigma = ne\mu$ in n-type semiconductors.

$$n.e = \rho = \text{charge density.}$$

$\therefore \qquad \sigma = \rho \times \mu$

But $\qquad \dfrac{1}{\rho} = R_H$

$\therefore \qquad \sigma = \dfrac{1}{R_H} \times \mu$

or $\qquad \mu = R_H \times \sigma = \dfrac{V_H.\omega}{B.I} \times \sigma$

We have assumed that the drift velocity 'v' of all the carriers is same. But actually it will not be so. Due to the thermal agitation they gain energy, their velocity increases and also collision with other atoms increases. So for all particles v will not be the same. Hence a correction has to be made and it has been found that satisfactory results will be obtained if $\dfrac{1}{R_H}$ is taken as $\dfrac{3\pi}{8\rho}$.

$$\therefore \qquad \mu = \left(\dfrac{8\sigma}{3\pi}\right) R_H \qquad\qquad\qquad \text{.......... (1.30)}$$

Multiply R_H by $\dfrac{8}{3\pi}$. Then it becomes **Modified Hall Coefficient.** Thus mobility of carriers (electrons or holes) can be determined experimentally using Hall Effect.

The product Bev is the **Lorentz Force**, because of the applied magnetic field B and the drift velocity v. So the majority carriers in the semiconductors, will tend to move in a direction perpendicular to B. But since there is no electric field applied in that particulars direction, there will develop a Hall voltage or field which just opposes the Lorentz field.

So with the help of Hall Effect, we can experimentally determine

     1.   *The mobility of Electrons or Holes.*

     2.   *Whether a given semiconductor is* p-type *or* n-type *(from the polarity of Hall voltage V_H)*

Problem 1.30

The Hall Effect is used to determine the mobility of holes in a p-type Silicon bar. Assume the bar resistivity is 200,000 Ω-an, the magnetic field $B_z = 0.1$ Wb/m^2 and d = w = 3mm. The measured values of the current and Hall voltage are 10mA and 50 mv respectively. Find μ_p mobility of holes.

Solution

$$B = 0.1 \text{ Wb / m}^2 \text{ (or Tesla)}$$

$$V_H = 50 \text{ mv.}$$

$$I = 10 \text{ mA;}$$

$$\rho = 2 \times 10^5 \ \Omega \text{ - cm ;}$$

$$d = w = 3\text{mm} = 3 \times 10^{-3} \text{ meters}$$

$$\frac{1}{R_H} = \frac{B.I}{V_H \cdot w} = \frac{0.1 \times 10 \times 10^{-3}}{50 \times 10^{-2} \times 3 \times 10^{-3}} = \frac{1}{150} = 0.667.$$

$$\text{Conductivity} = \frac{1}{\rho} = \frac{1}{2 \times 10^5 \times 10^{-2}} = \frac{1}{2000} \text{ mhos / meter.}$$

$$\mu = \sigma \times R_H$$

$$\mu_p = \frac{1}{0.667} \times \frac{1}{2000} = 750 \text{ cm}^2/ \text{ V - sec}$$

III   P-N JUNCTION DIODES

1.16   SEMICONDUCTOR DIODE CHARACTERISTICS

If a junction is formed using *p-type* and *n-type* semiconductors, a diode is realised and it has the properties of a rectifier. In this chapter, the volt ampere characteristics of the diodes, electron-hole currents as a function of distance from the junction and junction capacitances will be studied.

1.16.1   THEORY OF p-n JUNCTION

Take an intrinsic Silicon or Germanium crystal. If donor (*n-type*) impurities are diffused from one point and acceptors impurities from the other, a p-n junction is formed. The donor atoms will donate electrons. So they loose electrons and become positively charged. Similarly, acceptor atoms accept an electron, and become negatively charged. Therefore in the p-n junction on the *p-side*, holes and negative ions are shown and on the *n-side* free electrons and positive ions are shown. To start with, there are only *p-type* carriers to the left of the junction and only *n-type* carriers to the right of the junction. But because of the concentration gradient across the junction, holes are in large number on the left side and they diffuse from left side to right side. Similarly electrons will diffuse to the right side because of concentration gradient.

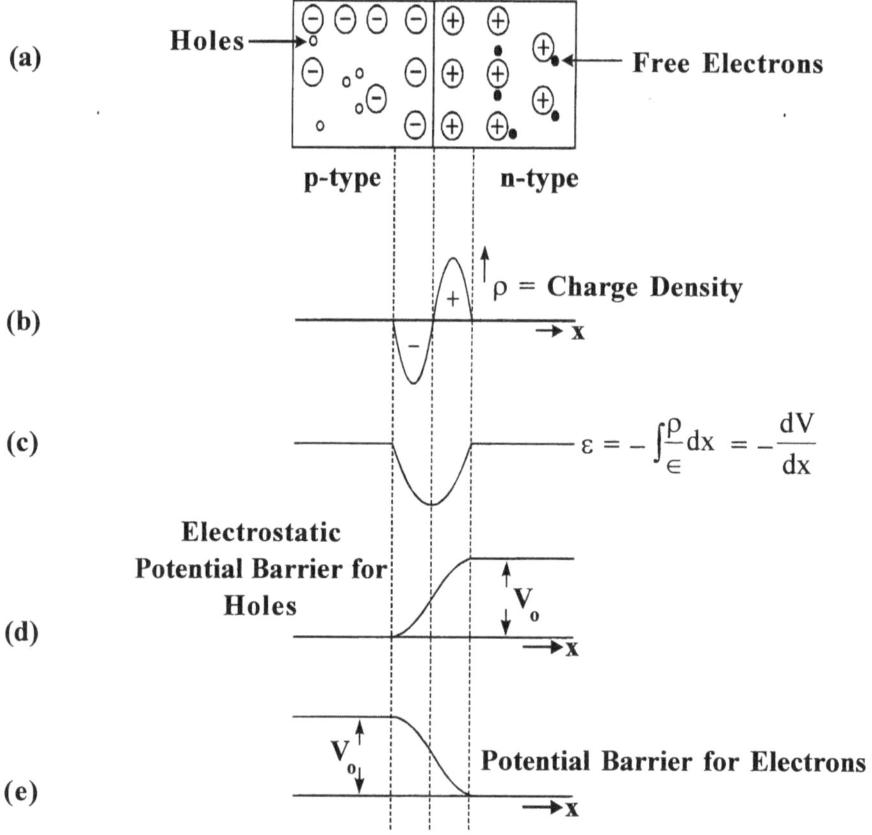

Fig 1.15. Potential distribution in p-n junction diode.

Because of the displacement of these charges, electric field will appear across the junction. Since *p-side* looses holes, negative field exists near the junction towards left. Since *n-side* looses electrons, positive electric field exists on the n side. But at a particular stage the negative field on *p-side* becomes large enough to prevent the flow of electrons from *n-side*. Positive charge on *n-side* becomes large enough to prevent the movement of holes from the *p-side*. The charge distribution is as shown in Fig. 1.15 (b). The charge density far away from the junction is zero, since before all the holes from *p-side* move to *n-side*, the barrier potential is developed. Acceptor atoms near the junction have lost the holes. But for this they would have been electrically neutral. Now these holes have combined with free electrons and disappeared leaving the acceptor atom negative. Donor atoms on *n-side* have lost free electrons. These free electrons have combined with holes and disappeared. So *the region near the junction is depleted of mobile charges. This is called depletion region, space charge region or transition region.* The thickness of this region will be of the order of few microns

$$1 \text{ micron} = 10^{-6} \text{ m} = 10^{-4} \text{cm}.$$

The electric field intensity near the junction is shown in Fig. 1.15 (c). This curve is the integral of the density function ρ. The electro static potential variation in the depletion region is

shown in Fig. 1.15 (d). $\varepsilon = -\int \dfrac{dv}{dx}.$ This variation constitutes a potential energy barrier against further diffusion of holes across the barrier.

When the diode is open circuited, that is not connected in any circuit, the hole current must be zero. Because of the concentration gradient, holes from the *p-side* move towards *n-side*. So, all the holes from *p-side* should move towards *n-side*. This should result in large hole current flowing even when diode is not connected in the circuit. But this will not happen. So to counteract the diffusion current, concentration gradient should be nullified by drift current due to potential barrier. Because of the movement of holes from *p-side* to *n-side*, that region (p-region) becomes negative. A *potential gradient* is set up across the junction such that *drift current flows* in opposite direction to the *diffusion current*. So the net hole current is zero when the diode is open circuited. The potential which exists to cause drift is called *contact potential* or *diffusion potential*. Its magnitude is a few tenths of a volt (0.01V).

1.16.2   p-n JUNCTION AS A DIODE

The p-n Junction shown here forms a semiconductor device called **DIODE**. Its symbol is
A —▶︱—K. **A** is anode. **K** is the cathode. It has two leads or electrodes and hence the name Diode. If the anode is connected to positive voltage terminal of a battery with respect to cathode, it is called **Forward Bias**, (Fig. 1.16 (a)). If the anode is connected to negative voltage terminal of a battery with respect to cathode, it is called **Reverse Bias**, (Fig. 1.16 (b)). The arrow in symbol ($\triangleright$) denotes the direction of current flow when the diode is forward biased

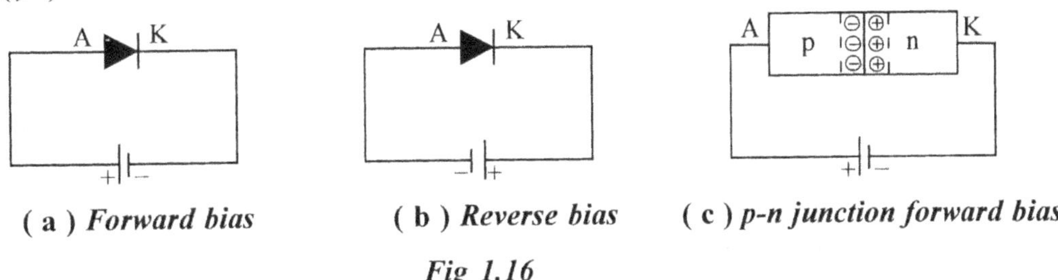

(a) *Forward bias*   (b) *Reverse bias*   (c) *p-n junction forward bias*

Fig 1.16

DENOTESA **1.16.3 OHMIC CONTACT**

In the above circuits, external battery is connected to the diode. But directly external supply cannot be given to a semiconductor. So metal contacts are to be provided for *p-region* and *n-region*. A Metal-Semiconductor Junction is introduced on both sides of p-n junction. So these must be contact potentials across the metal-semiconductor junctions. But this is minimized by fabrication techniques and the contact resistance is almost zero. Such a contact is called ohmic contact. (Fig. 1.17) So the entire voltage appears across the junction of the diode.

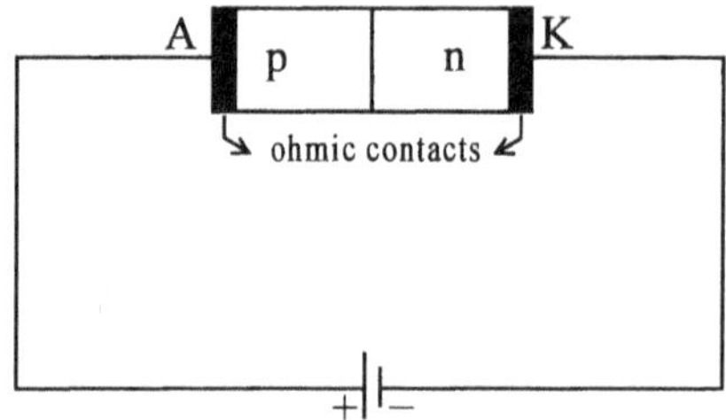

Fig 1.17 Ohmic contacts.

1.17   THE P-N JUNCTION DIODE IN REVERSE BIAS

Because of the battery connected as shown, holes in (Fig. 1.16 (b)) *p-type* and electrons in *n-type* will move away from the junction. As the holes near the junction in *p-region* they will move away from the junction and negative charge spreads towards the left of the junction. Positive charge density spreads towards right. But this process cannot continue indefinitely, because to have continuous flow of holes from right to left, the holes must come from the *n-side*. But *n-side* has few holes. So very less current results. But some electron hole pairs are generated because of thermal agitation. The newly generated holes on the *n-side* will move towards junction. Electrons created on the *p-side* will move towards the junction. So there results some small current called ***Reverse Saturation Current***. It is denoted by I_0. I_0 will increase with the temperature. So the *reverse resistance* or *back resistance decreases with temperature.* I_0 is of the order of a few μA. *The reverse resistance of a diode will be of the order of MΩ. For ideal diode, reverse resistance is* ∞.

The same thing can be explained in a different way. When the diode is open circuited, there exists a barrier potential. If the diode is reverse biased, the barrier potential height increases by a magnitude depending upon the reverse bias voltage. So the flow of holes from *p-side* to *n-side* and electrons from *n-side* to *p-side* is restricted. But this barrier doesn't apply to the minority carriers on the *p-sides* and *n-sides*. The flow of the minority carriers across the junction results in some current.

1.18   THE P-N JUNCTION DIODE IN FORWARD BIAS

When a diode is forward biased, the potential barrier that exists when the diode is open circuited, is reduced. Majority carriers from *p-side* and *n-side* flow across the junction. So a large current results. *For ideal diode, the forward resistance $R_f = 0$.* The forward current I_F, will be of the order of mA (milli-amperes).

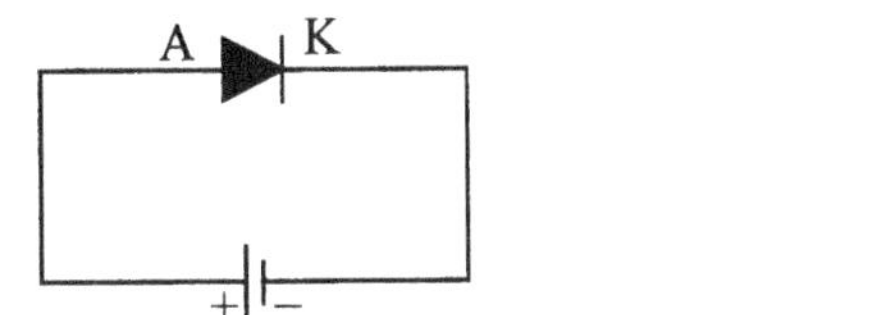 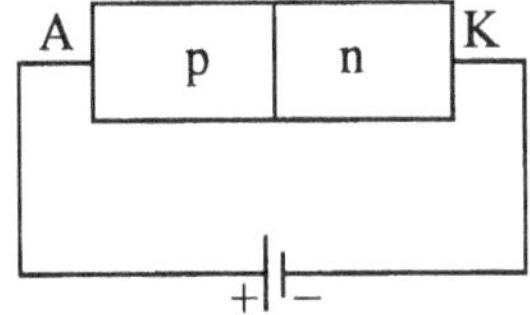

Fig 1.18 Diode in forward bias.

1.18.1   FORWARD CURRENT

If a large forward voltage is applied (Fig. 1.19), the current must increase. If the barrier potential across the junction is made zero, infinite amount of current should flow. But this is not practically possible since the bulk resistance of the crystal and the contact resistance together will limit the current. We may see in the other sections that when the diode is conducting, the voltage across it remains constant at V_γ *cut in voltage*. If the applied voltage is too large junction   breakdown will occur.

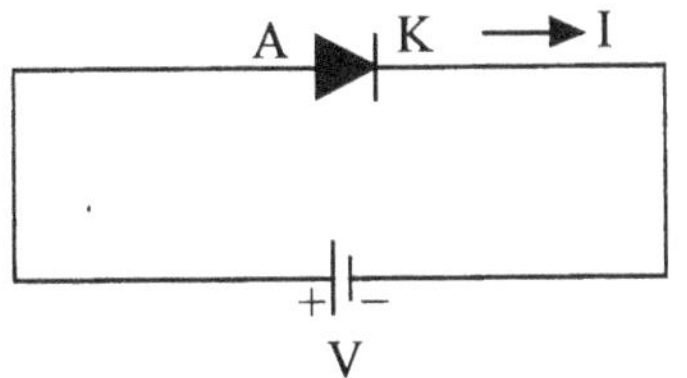 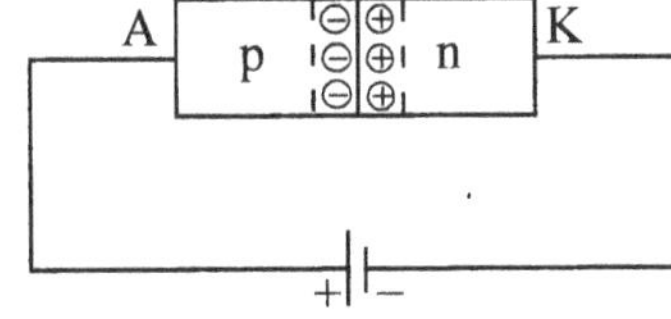

Fig 1.19   Forward biasing.

1.19   BAND STRUCTURE OF AN OPEN CIRCUIT p-n JUNCTION

When *p-type* and *n-type* semiconductors are brought into intimate contact *p-n junction* is formed. Then the fermilevel must be constant throughout the specimen. If it is not so, electrons on one side will have higher energy than on the other side. So the transfer of energy from higher energy electrons to lower energy electrons will take place till fermilevel on both sides comes to the same level. But we have already seen that in *n-type* semiconductors, E_F is close to conduction band E_{cn} and it is close to valence band edge E_{vp} on *p-side*. So the conduction band edge of *n-type* semiconductor cannot be at the same level as that of *p-type* semiconductor Hence, as shown in Fig. 1.20 the energy band diagram for a *p-n junction* is where a shift in energy levels E_0 is indicated.

$$E_G \quad = \quad \text{Energy gap in eV}$$
$$E_F \quad = \quad \text{Fermi energy level}$$
$$E_O \quad = \quad \text{Contact difference of potential}$$
$$E_{cn} \quad = \quad \text{Conduction Band energy level on the } n\text{-side.}$$
$$E_{cp} \quad = \quad \text{Conduction Band energy level on the } p\text{-side.}$$
$$E_{vn} \quad = \quad \text{Valence Band energy level on the } n\text{-side.}$$
$$E_{vp} \quad = \quad \text{Valence Band energy level on the } p\text{-side.}$$

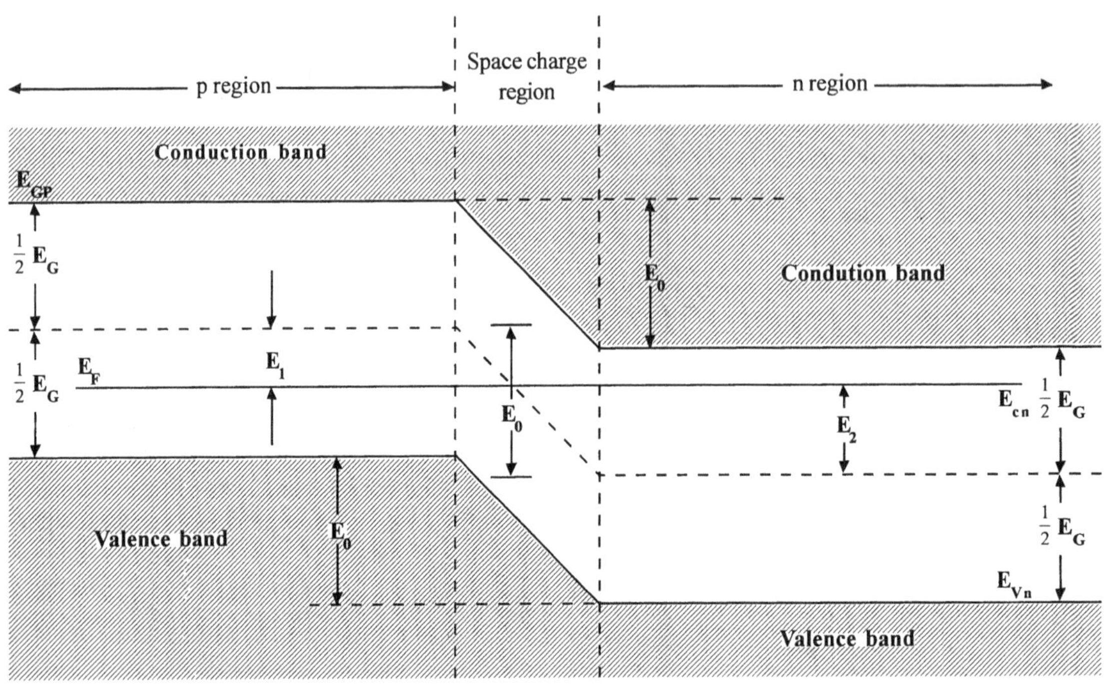

Fig 1.20 Band structure of open circuited diode.

If a central line $\dfrac{E_G}{2}$ is taken, the shift in energy levels is the difference between the two

central lines $\dfrac{E_G}{2}$ of the two semiconductors.

$$E_0 = E_{cp} - E_{cn} = E_{vp} - E_{vn}.$$

$$E_1 = \frac{E_G}{2} - (E_F - E_{vp})$$

$$E_2 = \frac{E_G}{2} - (E_{cn} - E_F)$$

$$E_1 + E_2 = E_G - E_F + E_{vp} - E_{cn} + E_F$$

But $\qquad E_G = E_{cp} - E_{vp}$

$\therefore \qquad E_1 + E_2 = E_{cp} - E_{vp} - E_{cn} + E_{vp}.$

But $\qquad E_{cp} - E_{cn} = E_O$

$$E_1 + E_2 = E_O.$$

This energy E_0 represents the potential energy barrier for electrons. The contact difference of potential

$$E_F - E_{vp} = \tfrac{1}{2}\,(E_G) - E_1 = \left(\frac{E_G}{2} \right) - E_1 \qquad\qquad \dots\dots\dots(1)$$

and
$$E_{cn} - E_F = \tfrac{1}{2}\,(E_G) - E_2 = \left(\frac{E_G}{2}\right) - E_2 \qquad \ldots\ldots\ldots(2)$$

Adding (1) and (2),

$$(E_F - E_{vp}) + (E_{cn} - E_F) = E_G - E_1 - E_2$$

or
$$(E_1 + E_2) = E_G - (E_{cn} - E_F) - (E_F - E_{vp})$$
$$(E_1 + E_2) = E_O$$

$$(E_{cn} - E_F) = KT \ln\left(\frac{N_C}{N_D}\right)$$

$N_D \qquad = \;$ Donor Atom Concentration N_o/m^3.

$N_A \qquad = \;$ Acceptor Atom Concentration N_o/m^3.

$$(E_F - E_{vp}) = KT \ln\left(\frac{N_V}{N_A}\right)$$

$$E_G = KT \ln\left(\frac{N_C.N_V}{n_i^2}\right)$$

$$\therefore \qquad E_0 = KT\left[\ln\left(\frac{N_C.N_V}{n_i^2}\right) - \ln\left(\frac{N_C}{N_D}\right) - \ln\left(\frac{N_V}{N_A}\right)\right]$$

$$E_0 = KT \ln\left(\frac{N_C.N_V}{n_i^2} \times \frac{N_D}{N_C} \times \frac{N_A}{N_V}\right) = KT \ln\left(\frac{N_A.N_D}{n_i^2}\right) \qquad \ldots\ldots\ldots (1.31)$$

The energy is expressed in electron volts eV.

K is Botlzman's Constant in eV / °K $= 8.62 \times 10^{-5}$ eV / °K

Therefore, E_0 is in eV and V_0 is the contact difference potential in volts V_0 is numerically equal to E_0. In the case of *n-type* semiconductors, $n_n = N_D$. (Subscript 'n' indicates electron concentration in *n-type* semiconductor)

$$n_i^2 = n_n \times p_n = N_D \times p_n$$
$$\because \qquad n_n = N_D$$

$n_i \;=\;$ Intrinsic Concentration

$n_p \;=\;$ Electron Concentration in *p-type* semiconductor

$n_n \;=\;$ Electron Concentration in *n-type* semiconductor

$p_p \;=\;$ Hole Concentration in *p-type* semiconductor

$p_n \;=\;$ Hole Concentration in *n-type* semiconductor

$$p_n = \frac{n_i^2}{N_D} \qquad \text{and} \qquad N_D = \frac{n_i^2}{p_n}$$

$$n_p = \frac{n_i^2}{N_A} \qquad \text{and} \qquad N_A = \frac{n_i^2}{n_p}$$

$$n_i^2 = n_n \times p_n$$

Substituting all these value in

$$E_0 = KT \ln\left(\frac{N_A . N_D}{n_i^2}\right)$$

$$E_0 = KT \ln\left(\frac{n_i^2}{p_n} \times \frac{n_i^2}{n_p} \times \frac{1}{n_i^2}\right)$$

$$= KT \ln\left(\frac{n_i^2}{p_n . n_p}\right)$$

$$= KT \ln\left(\frac{n_n . p_n}{p_n . n_p}\right) = KT \ln\left(\frac{n_n}{n_p}\right)$$

$$E_0 = KT \ln\left(\frac{n_{n0}}{n_{p0}}\right) = KT \ln\left(\frac{p_{pn}}{p_{n0}}\right)$$

Taking reasonable values of $n_{n0} = 10^{16} / cm^3$

$$n_{p0} = 10^4 / cm^3$$
$$KT = 0.026 \ eV$$

$$E_0 = 0.026 \times \ln\frac{10^{16}}{10^4} = 0.718 \ eV.$$

1.20   THE CURRENT COMPONENTS IN A P-N JUNCTION DIODE

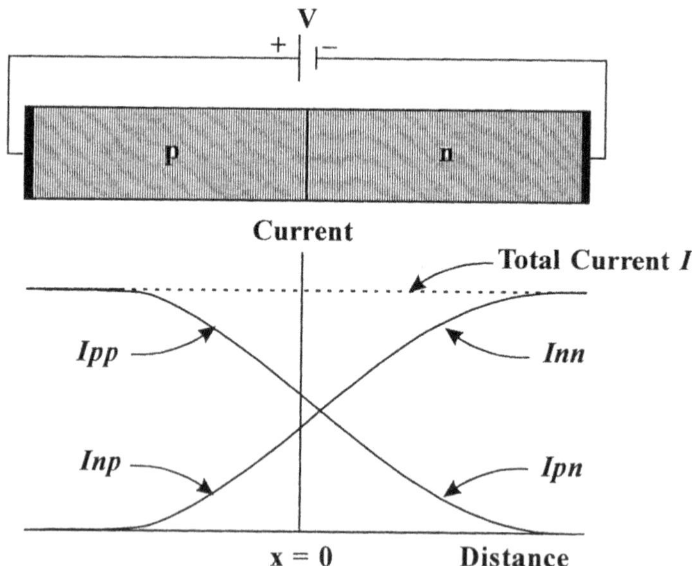

Fig 1.21 Current components in a p-n junction.

When a forward bias is applied to the diode, holes are injected into the *n-side* and electrons to the *p-side*. The number of this injected carriers decreases exponentially with distance from the junction. Since the diffusion current of minority carriers is proportional to the number of carriers, the minority carriers current decreases exponentially, with distance. There are two minority currents, one due to electrons in the *p-region Inp*, and due to holes in the *n-region Ipn*. As these currents vary with distance, they are represented as *Ipn(x)*.

Electrons crossing from *n* to *p* will constitute current in the same direction as holes crossing from *p* to *n*. Therefore, the total current at the junction where x = 0 is

$$I = Ipn(0) + Inp(0)$$

The total current remains the same. The decrease in *Ipn* is compensated by increase in *Inp* on the *p-side*.

Now deep into the *p-region* (where x is large) the current is because of the electric field (since bias is applied) and it is drift current *Ipp* of holes. As the holes approach the junction, some of them recombine with electrons crossing the junction from *n* to *p*. So *Ipp* decreases near the junction and is just equal in magnitude to the diffusion current *Inp*. What remains of *Ipp* at the junction enters the *n-side* and becomes hole diffusion current *Ipn* in the *n-region*. Since holes are minority carriers in the *n-side*, *Ipn* is small and as hole concentration decreases in the *n-region*, *Ipn* also exponentially decreases with distance.

In a forward biased *p-n junction* diode, at the edge of the diode on *p-side*, the current is hole current (majority carriers are holes). This current decreases at the junction as the junction approaches and at a point away from the junction, on the *n-side*, hole current is practically zero. But at the other edge of the diode, on the *n-side*, the current is electron current since electrons are the majority carriers. Thus in a *p-n junction* diode, the current enters as hole current and leaves as electron current.

1.21 LAW OF THE JUNCTION

$$p_{po} = \text{Thermal Equilibrium Hole Concentration on } p\text{-side}$$

$$p_{no} = \text{Thermal equilibrium hole concentration on n side}$$

$$p_{po} = p_{no}\, e^{V_0/V_T} \qquad \qquad ....(1)$$

where V_O is the Electrostatics Barrier Potential that exists on both sides of the junction. But the thermal equilibrium hole concentration on the *p-side*

$$p_{po} = p_n(0)\, e^{(V_0-V)/V_T} \qquad \qquad ....(2)$$

where $\qquad p_n(0) = $ Hole concentration on *n-side* near the junction

$\qquad \qquad V = $ Applied forward bias voltage.

This relationship is called Boltzman's Relationship.

$\therefore \qquad$ Equating (1) and (2).

$$p_n(0)\, e^{(V_0-V)/V_T} = p_{no} \times e^{V_0/V_T}$$

or $\qquad\qquad p_n(0) = p_{no} \times e^{\frac{V_0}{V_T} - \frac{V_0}{V_T} + \frac{V}{V_T}}$

$$p_n(0) = p_{no} \times e^{V/V_T}$$

Therefore, the total hole concentration in 'n' region at the junction varies with applied forward bias voltage V as given by the above expression.

This is called the **Law of the Junction**.

$$p_n(0) = p_n(0) - p_{no}$$

$$= p_{no} \, e^{V/V_T} - p_{no}$$

$$\boxed{p_n(0) = p_{no}\left(e^{V/V_T} - 1\right)}$$ .... (1.32)

IV. DIODE EQUATION

1.22 DIODE CURRENT EQUATION

The hole current in the *n-side Ipn(x)* is given as

$$Ipn(x) = \frac{Ae \times Dp}{L_p} \, p_n(0) \, e^{-x/L_p}$$

But

$$p_n(0) = p_{no}\left(e^{V/V_T} - 1\right)$$

$$Ipn(0) = \frac{Ae \times Dp}{L_p} \times p_{no}\left(e^{V/V_T} - 1\right)$$

$$D_p = \text{Diffusion coefficient of holes}$$

$$D_n = \text{Diffusion coefficient of electrons.}$$

$$\therefore \qquad e^{-x/L_p} \text{ at } x = 0 \text{ is } 1.$$

Similarly the electron current due to the diffusion of electrons from *n-side* to *p-side* is obtained from the above equation itself, by interchanging n and p.

$$\therefore \qquad Inp(0) = \frac{Ae \times Dn}{L_n} \times n_{po}\left(e^{V/V_T} - 1\right)$$

The total diode current is the sum of *Ipn (0)* and *Inp(0)*

or

$$\boxed{I = I_0\left(e^{V/V_T} - 1\right)}$$ .... (1.33)

where

$$I_0 = \frac{AeDp}{Lp} \times p_{no} + \frac{AeDn}{L_n} \times n_{po}$$

In this analysis we have neglected charge generation and recombination. Only the current that results as a result of the diffusion of the carriers owing to the applied voltage is considered.

Reverse Saturation Current

$$I = I_0 \times \left(e^{V/V_T} - 1\right)$$

This is the expression for current I when the diode is forward biased. If the diode is reverse biased, V is replaced by $-V$. V_T value at room temperature is ~ 26 mV. If the reverse bias voltage is very large, $e^{\frac{-V}{V_T}}$ is very small. So it can be neglected.

$$\therefore \qquad I = -I_0$$

I_0 will have a small value and I_0 is called the ***Reverse Saturation Current***.

$$I_0 = \frac{AeD_p p_{no}}{L_p} + \frac{AeD_n n_{po}}{L_n}.$$

In *n-type* semiconductor,
$$n_n = N_D$$

But $\qquad n_n \times p_n = n_i^2 \qquad \therefore p_n = \frac{n_i^2}{N_D}$

In *p-type* semiconductor,

$$p_p = N_A \qquad \therefore n_p = \frac{n_i^2}{N_A}$$

Substituting these values in the expression for I_0,

$$I_0 = Ae \left(\frac{D_p}{L_p N_D} + \frac{D_n}{L_n \cdot N_A} \right) \times n_i^2$$

where $\qquad n_i^2 = A_0 T^3 e^{-E_G/KT}$

E_G is in electron volts $= e.V_G$, where V_G is in Volts.

$$\therefore \qquad n_i^2 = A_0 T^3 e^{-\frac{E_{GO}}{KT}}$$
$$E_{GO} = V_G \cdot e$$

But $\qquad \dfrac{KT}{e} = $ Volt equivalent of Temperature V_T.

For Germanium, D_p and D_n decrease with temperature and n_i^2 increases with T. Therefore, temperature dependance of I_0 can be written as,

$$I_0 = K_1 T^2 e^{-V_G/V_T}$$

For Germanium, the current due to thermal generation of carriers and recombination can be neglected. But for Silicon it cannot be neglected. So the expression for current is modified as

$$I = I_0 \left(e^{\frac{V}{\eta V_T}} - 1 \right)$$

where n = 2 for small currents and n = 1 forM large currents.

V. V-I CHARACTERISTICS

1.23 VOLT-AMPERE CHARACTERISTICS OF A P-N JUNCTION DIODE

The general expression for current in the *p-n junction* diode is given by

$$I = I_0 \left(e^{\frac{V}{\eta V_T}} - 1 \right).$$

$\eta = 1$ for Germanium and 1 or 2 for Silicon. For Silicon, I will be less than that for Germanium.
$$V_T = 26 \text{ mV}.$$

If V is much larger than V_T, 1 can be neglected. So I increases exponentially with forward bias voltage V. In the case of reverse bias, if the reverse voltage $-V \gg V_T$, then e^{-V/V_T} can be neglected and so reverse current is $-I_0$ and remains constant independent of V. So the characteristics are as shown in Fig. 1.22 and not like theoretical characteristics. The difference is that the practical characteristics are plotted at different scales. If plotted to the same scale, (reverse and forward) they may be similar to the theoretical curves. Another point is, in deriving the equations the breakdown mechanism is not considered. As V increases **Avalanche multiplication** sets in. So the actual current is more than the theoretical current.

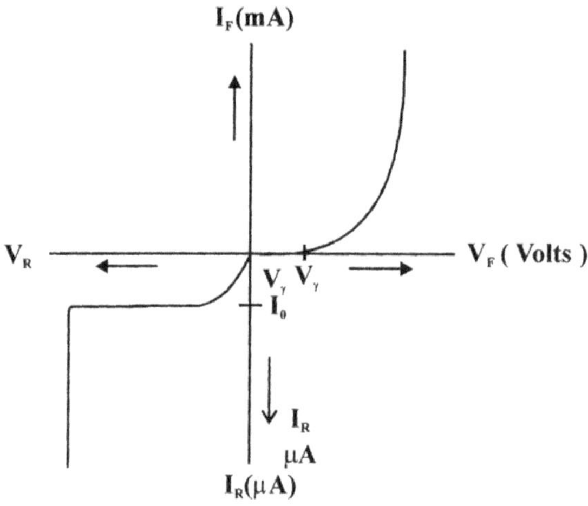

Fig 1.22 V-I Characteristics of p-n junction diode.

CUT IN VOLTAGE V_γ

In the case of Silicon and Germanium, diodes there is a *Cut In* or *Threshold* or *Off Set* or *Break Point Voltage*, below which the current is negligible. It's magnitude is 0.2V for Germanium and 0.6V for Silicon (Fig. 1.23).

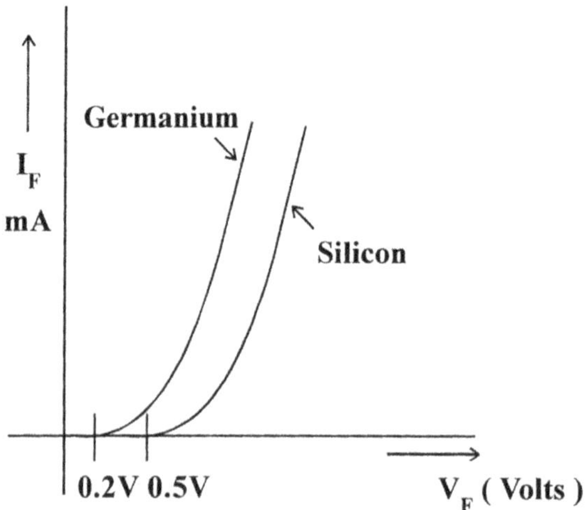

Fig 1.23 Forward characteristics of a diode.

VI. DIODE RESISTANCE

The static resistance (R) of a diode is defined as the ratio of $\dfrac{V}{I}$ of the diode. Static resistance varies widely with V and I. The dynamic resistance or incremental resistance is defined as the reciprocal of the slope of the Volt-Ampere Characteristic $\dfrac{dV}{dI}$. This is also not a constant but depends upon V and I.

VII. EFFECT OF TEMPERATURE ON V-I CHARACTERISTICS

1.24 TEMPERATURE DEPENDANCE OF P-N JUNCTION DIODE CHARACTERISTICS

The expression for reverse saturation current I_0

$$I_0 = Ae\left(\frac{D_p}{L_p N_D} + \frac{D_n}{L_n \cdot N_A}\right) \times n_i^2$$

$$n_i^2 \; \alpha \; T^3$$

$$\therefore \quad n_i^2 = A_0 T^3 \, e^{-E_{G0}/K_T}$$

D_p decreases with temperature.

$$\therefore \quad I_0 \; \alpha \; T^2$$

or $\qquad I_0 = KT^m \, e^{-V_{G0}/\eta V_T}$

where V_G is the energy gap in volts. (E_G in eV)

For Germanium, $\quad \eta = 1, m = 2$

For Silicon, $\qquad \eta = 2, m = 1.5$

$$I_0 = KT^m \, e^{-V_{G0}/\eta V_T}$$

Taking ln, (Natural Logarithms) on both sides,

$$\ln I_0 = \ln(K) + m \times ln\,(T) \; \frac{-V_{G0}}{\eta V_T}$$

Differentiating with respect to Temperature,

$$\frac{1}{I_0} \times \frac{dI_0}{dT} = 0 + \frac{m}{T} - \left(-\frac{V_{G0}}{\eta V_T} \times \frac{1}{T}\right)$$

$$\frac{1}{I_0} \times \frac{dI_0}{dT} = \frac{m}{T} + \frac{V_{G0}}{\eta T V_T}$$

$\dfrac{m}{T}$ value is negligible

$$\therefore \qquad \boxed{\frac{1}{I_0} \times \frac{dI_0}{dT} \simeq \frac{V_{G0}}{\eta T V_T}}$$

Experimentally it is found that reverse saturation current increases $\sim$ 7% / °C for both Silicon and Germanium or for every 10°C rise in temperature, I_0 gets doubled. The reverse saturation current increases if expanded during the increasing portion.

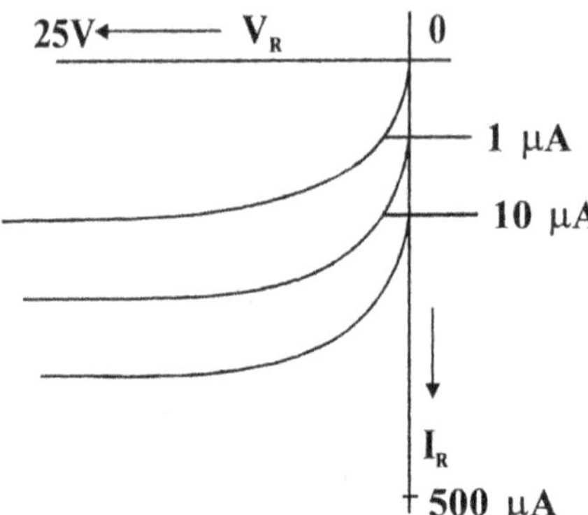

Fig 1.24 Reverse characteristics of a p-n junction diode.

Reverse Saturation Current increases 7% /°C rise in temperature for both Silicon and Germanium. For a rise of 1°C in temperature, the new value of I_0 is,

$$I_0 = \left(1 + \frac{7}{100}\right) I_0$$
$$= 1.07 \, I_0.$$

For another degree rise in temperature, the increase is 7% of $(1.07 \, I_0)$.

Therefore for 10 °C rise in temperature, the increase is $(1.07)^{10} = 2$.

Thus for every 10 °C rise in temp I_0 for Silicon and Germanium gets doubled.

Ideal Diode characteristics

For a practical diode, forward resistance R_f will be less (few Ω) and reverse resistance R_r will be high (KΩ or HΩ). cut-in voltage $V_v = 0.1v$ for Ge and 0.5_v for S$_i$.

But for ideal diode, $R_f = 0$; $R_r = \infty$ and $V_v = 0$.

Junction capacitance values are also zero.

VIII IDEAL DIODE CAPACITANCE

For a practical diode forward resistance R_f will be les (few Ω) and reuse reactance will be high (KΩ or M Ω). Cut = in voltage Vv = 0.1v for ...... and 0.5 for sc.

But for ideal diode, $R_f = 0$; $R_r = 0$; $R_r = ¥$ and Vv =0 Junction capacitance values are also zero.

1.25   SPACE CHARGE OR TRANSITION CAPACITANCE C_T

When a reverse bias is applied to a *p-n junction* diode, electrons from the *p-side* will move to the *n-side* and vice versa. When electrons cross the junction into the *n-region*, and hole away from the junction, negative charge is developed on the *p-side* and similarly positive charge on the *n-side*. Before reverse bias is applied, because of concentration gradient, there is some space charge region. Its thickness increases with reverse bias. So space charge Q increases as reverse bias voltage increases.

But $$C = \frac{Q}{V}$$

Therefore, Incremental Capacitance,

$$C_T = \left|\frac{dQ}{dV}\right|$$

where |dQ| is the magnitude of charge increase due to voltage dV. It is to be noted that there is negative charge on the *p-side* and positive charge on *n-side*. But we must consider only its magnitude.

$$\text{Current } I = \frac{dQ}{dt}$$

Therefore, if the voltage dV is changing in time dt, then a current will result, given by

$$I = C_T \times \frac{dV}{dt}$$

This current exists for A.C. only. For D.C, Voltage is not changing with time. For D.C. capacitance is open circuit.

The knowledge of C_T is important in considering diode as a circuit element. C_T *is called Transition region capacitance or space charge capacitance or barrier capacitance, or depletion region capacitance.* This capacitance is not constant but depends upon the reverse bias voltage V. If the diode is forward biased, since space charge $\approx$ 0,(this doesn't exist). It will be negligible. C_T is of the order of 50 pf etc.

ALLOY JUNCTION

Indium is trivalent. If this is placed against *n-type* Germanium, and heated to a high temperature, indium diffuses into the Germanium crystal, a pn junction will be formed and for such a junction there will be abrupt change from acceptor ions on one side to donor ions on the other side. Such a junction is called **Alloy Junction** or **Diffusion Junction**. In the figure, the acceptor ion concentration N_A and donor atom concentration, N_D is shown in Fig. 1.25. There is sudden change in concentration levels. To satisfy the condition of charge neutrality,

$$e \times N_A \times W_P = e \times N_D \times W_n.$$

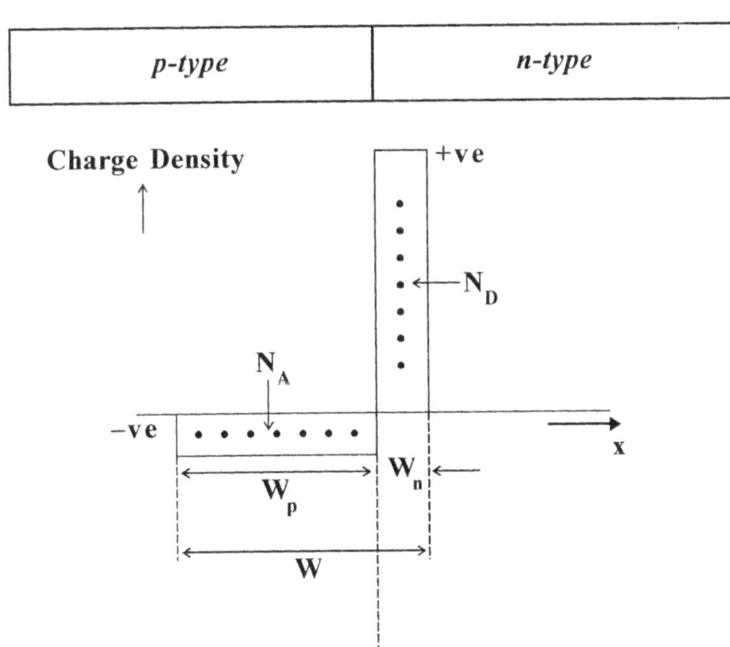

Fig 1.25 Abrupt p-n junction.

If $N_A \ll N_D$, then $W_p \gg W_n$. In practice the width of the region, W_n will be very small. So it can be neglected. So we can assume that the entire barrier potential appears across the *p-region* just near the junction.

Poisson's Equation gives the relation between the charge density and potential.

It is

$$\frac{d^2V}{dx^2} = \frac{eN_A}{\in}$$

where $\in$ is the permitivity of the semiconductor

$$\frac{dV}{dx} = \frac{eN_A}{\in} x$$

$$V = \frac{e.N_A}{\in} \times \frac{x^2}{2}$$

At $x = w_p$, $V = V_B$ the barrier potential. $W_p = W$.

$$\therefore \quad \boxed{V_B = \frac{e.N_A}{2 \in} \times W^2} \qquad \dots\dots\dots (\,1.34\,)$$

The value of W depends upon the applied reverse bias V. If V_0 is the contact potential, $V_B = V_0 - V$ where V is the reverse bias voltage with negative sign.

So as V increases, W also increases and V_B increases

$$W \; \alpha \; \sqrt{V_B}$$

If A is the area of the junction, the charge in the width W is

$$Q = e \times N_A \times W \times A$$

where

$$W \times A = \text{Volume}$$

$$e \times N_A = \text{Total charge density}$$

$$C_T = \left|\frac{dQ}{dV}\right| = e \times N_A \times A \times \left|\frac{dW}{dV}\right|$$

$$Q = \sqrt{e \times N_A} \sqrt{2 \in V_B \times A}$$

But

$$V_B = \frac{e.N_A}{2 \in} \times W^2$$

or

$$W = \sqrt{\frac{2 \in V_B}{e \times N_A}}$$

$$\frac{dW}{dV} = \frac{1}{2} \sqrt{\frac{2 \in}{e \times N_A}} \; V_B^{-1/2}$$

But

$$\sqrt{V_B} = W \sqrt{\frac{e \times N_A}{2 \in}}$$

$$\frac{dW}{dV} = \frac{1}{2}\sqrt{\frac{2\epsilon}{e \times N_A}} \times \frac{\sqrt{2\epsilon}}{\sqrt{e \times N_A \times W}}$$

$$= \frac{\epsilon}{e \times N_A} \times \frac{1}{W}$$

$$= \frac{\epsilon}{e \times N_A \times W}$$

$$\therefore \quad \boxed{C_T = \frac{\epsilon \times A}{W}} \qquad \dots\dots\dots (\ 1.35\ )$$

This expression is similar to that of the Parallel Plate Capacitor.

1.26 DIFFUSION CAPACITANCE, C_D

When a *p-n junction* diode is forward biased, the junction capacitance will be much larger than the transition capacitance C_T. When the diode is forward biased, the barrier potential is reduced. Holes from *p-side* are injected into the *n-side* and electrons into the *p-side*. Holes which are the minority carriers in the *n-side* are injected into the *n-side* from the *p-region*. The concentration of holes in the *n-side* decrease exponentially from the junction. So we can say that a positive charge is injected into the *n-side* from *p-side*. This injected charge is proportional to the applied forward bias voltage 'V'. So the rate of change of injected charge 'Q' with voltage 'V' is called the Diffusion Capacitance C_D. Because of C_D total capacitance will be much larger than C_T in the case of forward bias, (C_D is few mF (2mF.)) C_T value will be a few pico farads.

DERIVATION FOR C_D

Assume that, the *p-side* is heavily doped compared to *n-side*. When the diode is forward biased, the holes that are injected into the *n-side* are much larger than the electrons injected into the *p-side*. So we can say that the total diode current is mainly due to holes only. So the excess charge due to minority carriers will exist only on *n-side*. The total charge Q is equal to the area under the curve multiplied by the charge of electrons and the cross sectional area A of the diode. $P_n(0)$ is the Concentration of holes/cm^3. Area is in cm^2, x in cm,

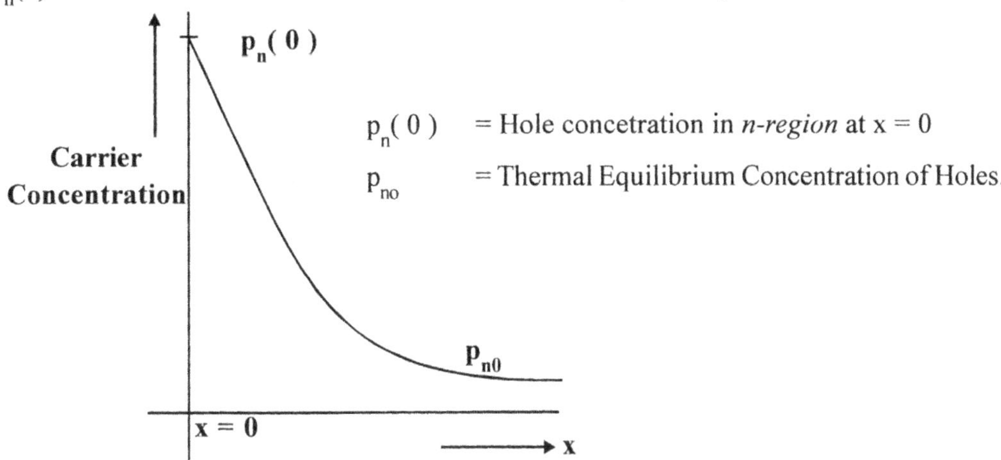

Fig 1.26 Carrier concentration variation.

$$\therefore \qquad Q = \int_{0}^{\infty} Ae P_n(0) e^{-x/Lp}\, dx$$

$$= Ae\, P_n(0) \int_{0}^{\infty} e^{-x/Lp}\, .dx$$

$$Q = AeP_n(0)\,[\,-Lp\,[0\text{-}1]\,]$$

$$= AeL_p P_n(0)$$

$$C_D = \frac{dQ}{dv} = AeL_p \times \frac{dp_n(0)}{dv} \qquad\qquad ....(1)$$

We know that

$$Ipn(x) = +\frac{e \times A \times Dp \times p_n(0) \times e^{-x/Lp}}{Lp}$$

or $\qquad\qquad$
$$Ipn(0) = I = \frac{AeD_p p_n(0)}{L_p}$$

$$P_n(0) = L_p \times I / AeD_p$$

$$\frac{dp_n(0)}{dv} = \frac{L_p}{AeD_p} \times \frac{dI}{dv}$$

$$= \frac{L_p}{AeD_p} \times g \qquad\qquad ....(2)$$

where g is the conductance of the diode.

Substitute equation (2) in (1).

$$\therefore \qquad C_D = Ae \times L_p \times \left(\frac{L_p}{AeD_p} \right) \times g$$

$$= \frac{L_p^2}{D_p} \times g$$

The lifetime for holes $\tau_p = \tau$ is given by the eq.

$$\tau = \frac{L_p^2}{D_p}$$

where $\qquad\qquad D_p$ = Diffusion coefficient for holes.

$\qquad\qquad\qquad\quad L_p$ = Diffusion length for holes.

$\qquad\qquad\qquad\quad D_p$ = cm^2/sec

$\therefore \qquad\qquad C_D = \tau \times g.$

But diode resistance

$$r \simeq \frac{\eta.V_T}{I}$$

where

$$\eta = 1 \text{ for Germanium}$$
$$\eta = 2 \text{ for Silicon}$$

$\therefore$
$$g = \frac{I}{\eta.V_T}$$

$\therefore$
$$C_D = \frac{I.\tau}{\eta.V_T} \qquad\qquad \dots\dots\dots (3)$$

C_D is proportional to I. In the above analysis we have assumed that the current is due to holes only. So it can be represented as C_{DP}. If the current due to electron is also to be considered then we get corresponding value of C_{Dn}. The total diffusion capacitance $= C_{DP} + C_{Dn}$. Its value will be *around 20µF*.

$$C_D = \tau \times g$$

or
$$r \times C_D = \tau$$

$r \times C_D$ is called the time constant of the given diode. It is of importance in circuit applications. Its value ranges from nano-secs to hundreds of micro-seconds.

Charge control description of a diode :

$$Q = A \times e \times L_p \times p_n(0)$$

$$I = \frac{AeD_p p_n(0)}{L_p}$$

$\therefore$
$$\frac{Q}{L_p} = A \times e \times p_n(0)$$

$$I = Q \times D_p/L_p^2$$

But
$$L_p^2/Dp = \tau$$

$\therefore$
$$I = \frac{Q}{\tau}$$

1.27 DIODE SWITCHING TIMES

When the bias of a diode is changed from forward to reverse or viceversa. the current takes definite time to reach a steady state value.

1.27.1   FORWARD RECOVERY TIME (T_{FR})

Suppose a voltage of 5V is being applied to the diode. Time taken by the diode to reach from 10% to the 90% of the applied voltage is called as the forward recovery time t_{fr}. But usually this is very small and so is not of much importance. This is shown in Fig. 1.26.

1.27.2   DIODE REVERSE RECOVERY TIME (t_{rr})

When a diode is forward biased, holes are injected into the 'n' side. The variation of concentration of holes and electrons on *n-side* and *p-side* is as shown in Fig. 1.27. P_{no} is the thermal equilibrium concentration of holes on *n-side*. P_n is the total concentration of holes on 'n'-side.

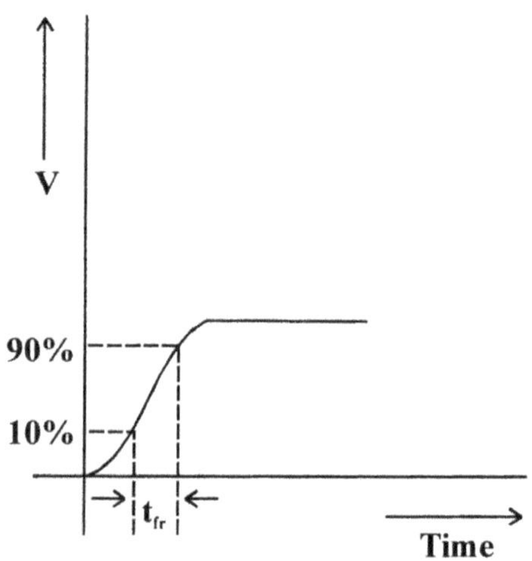
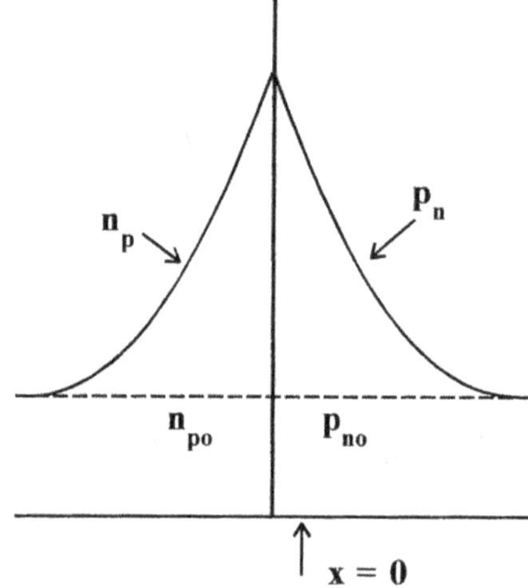

Fig 1.27 Rise time.   *Fig 1.28 Carrier concentration in reverse bias.*

1.27.3 STORAGE AND TRANSITION TIMES

Suppose the input given to the circuit is as shown in Fig. 1.29 (i). During $0 - t_1$, the diode is forward biased. Forward resistance of the diode is small compared to R_L. Since all the voltage drop is across R_L itself, the voltage drop across the diode is small.

$$\therefore \qquad I_F = \frac{V_F}{R_L}$$

This is shown in Fig 1.29 (ii) and voltage across the diode in Fig 1.29 (iii) up to t_1. Now when the forward voltage is suddenly reversed, at $t = t_1$, because of the *Reverse Recovery Time*, the diode current will not fall suddenly. Instead, it reverses its direction

and $\approx \dfrac{V_R}{R_L}$ (R_L is small compared to reverse resistance of the diode). At $t = t_2$, the equilibrium

level of the carrier density at the junction takes place. So the voltage across the diode falls slightly but not reverses and increases to V_R after time t_3. The current also decreases and reaches a value = reverse saturation current I_0 .

The interval time $(t_1 - t_2)$ for the stored minority charge to become zero is called *Storage Time* t_s. The time $(t_2 - t_3)$ when the diode has normally recovered and the reverse current reaches I_0 value is called *Transition Time* t_t. These values range from few milli-seconds to a few micro-seconds.

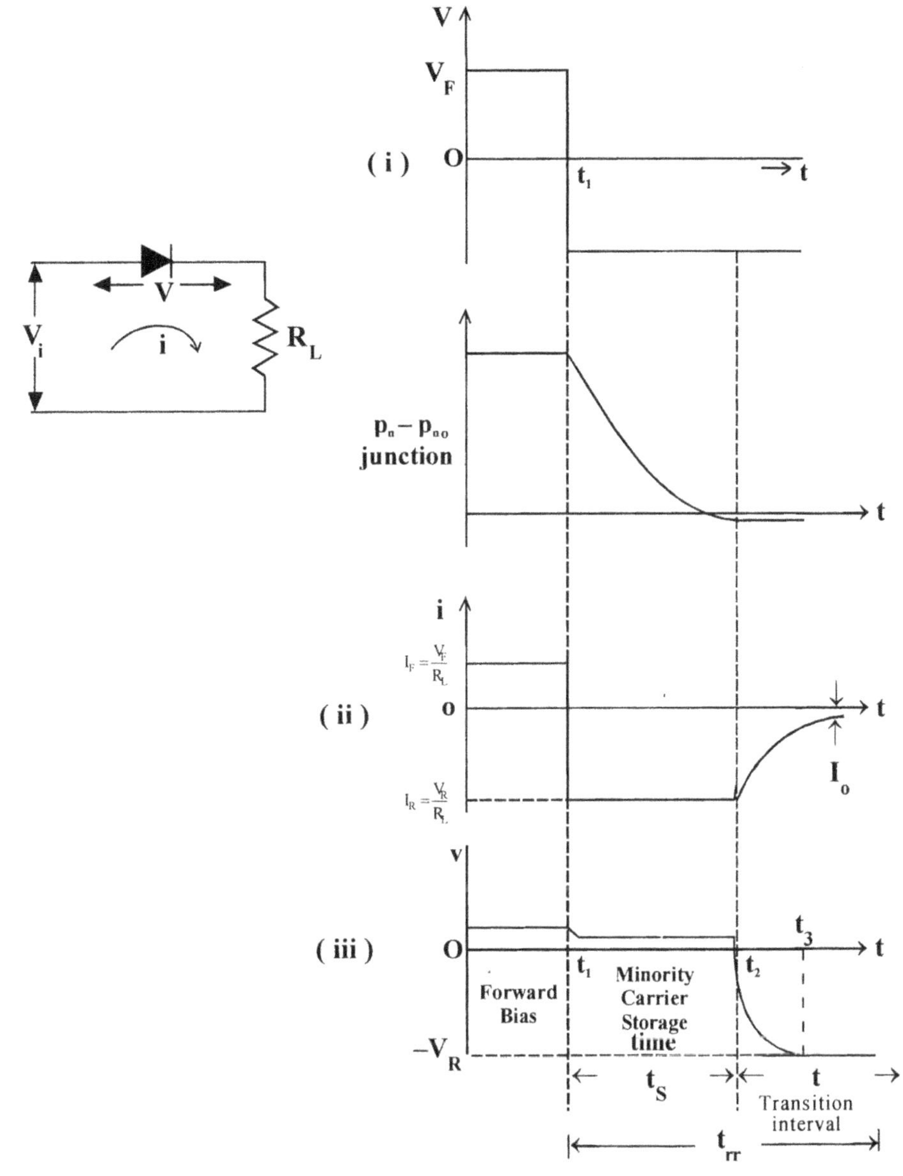

Fig 1.29 Storage and Transition times.

Problem 1.31

(a) For what voltage will the reverse current in p-n junction Germanium diode reaches 90% of its saturation value at room temperature ?

(b) If the reverse saturation current is 10 μA, calculate the forward currents for a voltage of 0.2, and 0.3V respectively.

Solution

(a) $V_T = \dfrac{T}{11,600} = 0.026V$ at room temperature. Or it is 26mV.

V_T is volt equivalent of temperature. T is in °K.

$$V_T = \frac{KT}{e}$$

It is the Thermal Energy expressed in equivalent electrical units of Volts.

$$I = I_0\left(e^{\frac{V}{\eta V_T}} - 1\right)$$

$\eta = 1$, for Germanium Diode.

$$0.9\,I_0 = I_0\left(e^{\frac{V}{0.026}} - 1\right)$$

or $V = -0.017V$

(b)   For $V = 0.2$, $I = 10\left(e^{200/26} - 1\right) = 21.85$ mA

        $\because$ $0.2V = 200mV$

                  For $V = 0.3$, $I = 10\left(e^{11.52} - 1\right) = 1.01A$

Problem 1.32

Find the value of (i) D.C resistance and a.c resistance of a Germanium junction diode, if the temperature is 25°C and $I_0 = 20\mu A$ with an applied voltage of 0.1 V

Solution

$$I = I_0\left(e^{V/\eta V_T} - 1\right)$$

$$T = 273 + 25 = 298°K$$

For Germanium, $\eta = 1$

$$V_T = \frac{T}{11,600} = \frac{(273+25)}{11,600} = 0.026V$$

$$I = 20 \times 10^{-6}\left(e^{\frac{0.1}{V_T}} - 1\right) = 0.916 \text{ mA}$$

$$r_{DC} = \frac{V}{I} = \frac{0.1}{0.196 \times 10^{-3}} = 109.2\,\Omega$$

$$r_{AC} = \frac{dV}{dI} = \frac{\Delta V}{\Delta I} = ?$$

$$r_{AC} = \frac{1}{g_{ac}} = \frac{dI}{dv}$$

$$\frac{dI}{dv} = \frac{I_0 \left(e^{\frac{V}{V_T}} \right)}{V_T} = 38.1 \times 10^{-3}\ \mho$$

$$r_{AC} = 26.3\ \Omega$$

Problem 1.33

Find the width of the depletion layer in a germanium junction diode which has the following specifications Area A = 0.001 cm^2, σ_n = 1mhos / cm, σ_p = 100 mhos / cm, μ_n = 3800 cm^2/sec μ_p = 1800 cm^2/sec.

Solution

Permitivity of Germanium,

$$\in\ = 16 \times 8.85 \times 10^{-14}\ F/cm$$
$$n_i^2 = 6.25 \times 10^{26}$$
$$T\ = 300°K.$$

Applied reverse bias voltage = 1V.

$$W = \sqrt{\frac{2\in.V_B}{e.N_A}}$$

In this formula, in the denominators, N_A is used since in the expression we have assumed that *p-side* of the p-n junction is heavily doped. If *n-side* is heavily doped, it would be N_D.

V_B = Total barrier potential = Applied reverse bias voltage + the contact difference of potential (V_o).

$$V_B = V_R + V$$

So first V_0 should be calculated.

$$V_0 = KT\ \ln\ .\ \frac{n_n.p_p}{n_i^2}$$

$$KT = 0.026\ eV$$

$$N_D = n_n = \frac{\sigma_n}{e\mu_n} = \frac{1}{1.6 \times 10^{-19} \times 3800} = 1.64 \times 10^{15}/cm^3$$

$$N_A = p_p = \frac{\sigma_p}{e\mu_p} = \frac{100}{1.6 \times 10^{-19} \times 1800} = 3.5 \times 10^{17}/cm^3$$

$$\therefore \qquad V_0 = 0.026\ \ln\ .\ \frac{3.5 \times 10^{17} \times 1.64 \times 10^{15}}{6.25 \times 10^{26}} = 0.357V$$

$$W = \sqrt{\frac{2 \times 16 \times 8.85 \times 10^{-14} \times (0.35 + 1)}{1.6 \times 10^{-19} \times 3.5 \times 10^{17}}} = 0.083 \times 10^{-4} cm$$

Problem 1.34

Calculate the dynamic forward and reverse resistance of a p - n junction diode, when the applied voltage is 0.25V for Germanium Diode. $I_0 = 1\,\mu A$ and $T = 300^\circ K$.

Solution

$$I_0 = 1\,\mu A$$
$$T = 300^\circ K$$
$$V_f = 0.25V$$
$$V_r = 0.25V$$
$$I = I_0\left(e^{V/V_T} - 1\right)$$

For Germanium, $\eta = 1$

Dynamic Forward Resistance :

V is positive

$$\frac{1}{r_f} = \frac{dI}{dV} = \frac{I_0}{V_T}\, e^{\frac{V}{V_T}} - 0$$

$$= \frac{1 \times 10^{-6}}{0.026} \cdot e^{\frac{0.25}{0.026}}$$

$$\frac{1}{r_f} = 0.578 \text{ mhos}$$

$$r_f = 1.734\ \Omega$$

Dynamic Reverse Resistance :

$$I = I_0\left(e^{\frac{-V}{V_T}} - 1\right)$$

$$\frac{1}{r_r} = \frac{dI}{dV} = \cdot \frac{I_0}{V_T} \cdot e^{-V/V_T}$$

$$= \frac{1 \times 10^{-6}}{0.026}\ e^{\frac{-0.25}{0.026}}$$

$$\frac{1}{r_r} = 2.57 \times 10^{-9} \text{ mhos}$$

$$r_r = \frac{1}{2.57 \times 10^{-9}} = 389.7\ M\Omega$$

In practice r_r is much smaller due to surface leakage current.

SUMMARY

♦ Energy possessed by an electron rotating in an orbit with radius r;

$$E = -\frac{e^2}{8\pi \in_0 r}$$

♦ Expression for the radius of orbit, $r = \dfrac{n^2 h^2 \in_0}{\pi m e^2}$

♦ Expression for wavelength of emitted radiation, $\lambda = \dfrac{12,400}{(E_2 - E_1)}$

♦ Types of Electronic Emissions from the surface.
    1. Thermionic Emission   2. Secondary Emission
    3. Photo electric Emission   4. High Field Emission

♦ Expression for Threshold Frequency f_T in photoelectric emission,

$$f_T = \frac{e\phi}{h}$$

♦ Energy Gap E_G decreases with temperature in semiconductors.

♦ Mobility μ decreases with temperature. Units are cm^2/v-sec.

♦ $n\,p = n_i^2; \; n_i = AT^{3/2}\, e^{-\frac{eE_G}{2KT}}$

♦ Law of Electrical Neutrality $N_A + p = N_D + n;$

♦ Fermi Level lies close to E_C in *n-type* semiconductor

♦ Fermi Level lies close to E_V in *p-type* semiconductor

♦ $\sigma = ne\mu_n + pe\mu_p$

♦ Hall Voltage, $V_H = \dfrac{BI}{\rho\omega}$

♦ Expression for current through a *p-n junction*.

$$\text{Diode is } I = I_O \left(e^{\frac{V}{\eta V_T}} - 1 \right)$$

♦ Forward Resistance R_f of a diode will be small of the order of few Ω. Reverse Resistance R_γ will be very high, of the order of $M\Omega$

♦ Cut in voltage V_v for Germanium diode is 0.1 V and for Silicon diode it is 0.5 V at room temperature. It decreases with increase in temperature.

♦ I_O the reverse saturation current of a diode will be of the order of a few μA. It increases with temperature. I_O gets doubled for every 10°C rise in temperature.

♦ When the diode is reverse biased, transition capacitance C_T results when the diode is forward biased, diffusion capacitance C_D is exhibited.

◆ The three types of breakdown mechanisms in semiconductor diodes are

    1. Avalanche Breakdown

    2. Zener Breakdown

    3. Thermal Breakdown.

◆ Tunnel diode exhibits negative resistance characteristics

◆ Varactor diodes are operated in reverse bias and their junction capacitance varies with the voltage.

◆ Zener diode is also operated in reverse bias for voltage regulation.

OBJECTIVE TYPE QUESTIONS

1. Free electron concentration in semiconductors is of the order of ....

2. Insulators will have resistivity of the order of ....

3. Expression for J in terms of E and σ is ....

4. Expression for J in the case of a semiconductor with concentrations n and p is ....

5. Value of E_G at room temperature for Silicon is ....

6. According to Law of Mass Action in semiconductors, ...

7. Intrinsic concentration depends on temperature T as ....

8. The equation governing Law of Electrical Neutrality, is ....

9. Cutin Voltages .... with temperature

10. Depletion region width varies with reverse bias voltages as ....

11. In *p-type* semiconductor, Fermi Level lies close to ....

12. In *n-type* semiconductor, Fermi Level lies close to ....

13. The rate at with I_o changes with temperature in Silicon diode is ....

14. Value of Volt equivalent of Temperature at $25^{o}C$ is ....

15. Einstein's relationship in semiconductors is ....

16. A very poor conductor of electricity is called ....

17. When donor impurities are added allowable energy levels are introduced a little .... the .... band.

18. Under thermal equilibrium, the product of the free negative and positive concentration is a constant independent of the amount of .... doping and this relationship, called the mass action low and is given by np = ...

19. The unneutralized ions in the neighbourhood of the junction are referred to as .... charges and this region depleted of mobile charges is called .... charge region.

20. General expression for E_o, the contact difference of potential in an open circuited *p-n junction* in terms of N_C, N_A and n_i is ....

21. Typical value of E_o = ...eV

22. The equation governing the law of the junction is ....

23. The expression for the current in a forward biased diode is ....

24. The value of cut in voltage in the case of Germanium diode .... and Silicon Diode .... at room temperature.

25. Expression for V_B the barrier potential in terms of depletion region width W is V_B = ...

26. Expression for I_o in terms of temperature T and V_T is ....

27. Zener breakdown mechanism needs relatively .... electric field compared to Avalanche Breakdown.

28. Expression for the diffusion capacitance C_D in terms of L_P and D_P is ....

29. Expression for Volt equivalent of temperature V_T in terms of temperature T is V_T = ...

30. The salient feature of a Tunnel diode is, it exhibits .... characteristics.

ESSAY TYPE QUESTIONS

1. Explain the concept of 'hole'. How *n-type* and *p-type* semiconductors are formed? Explain.

2. Derive the expression for E_G in the case of intrinsic semiconductor.

3. Derive the expression for E_G in the case of of *p-type* and *n-type* semiconductors.

4. With the help of necessary equations, Explain the terms Drift Current and Diffusion Current.

5. Explain about Hall Effect. Derive the expression for Hall Voltage. What are the applications of Hall Effect?

6. Distinguish between Thermistors and Sensistors.

7. Derive the expression for contact difference of potential V_o in an open circuited p-n junction.

8. Draw the forward and reverse characteristics of a p-n junction diode and explain them qualitatively.

9. Derive the expression for Transistor Capacitance C_T in the case of an abrupt p-n junction.

10. Compare Avalanche, Zener and Thermal Breakdown Mechanisms.

11. Derive the expression for E_o in the case of open circuited *p-n junction* diode.

12. Qualitatively explain the forward and reverse characteristic of *p-n junction* diode.

13. Derive the expression Transistor Capacitance C_T in the case of *p-n junction* diode.

14.   How junction capacitances come into existance in *p-n junction* diode.

15.   Write notes on Varactor Diode.

16.   Distinguish between Avalanche, Zener and Thermal Breakdown Mechanisms

17.   Derive the expression for the diffusion capacitance C_D in the case of *p-n junction* diode.

MULTIPLE CHOICE QUESTIONS

1. The force of electron 'F' between nucleus and electron with charge 'e' and radius 'r' is, proportional to, F α

(a) $\dfrac{e^2}{r^2}$ (b) $\dfrac{r^2}{e^2}$ (c) $e^2 r^2$ (d) $\dfrac{e}{r}$

2. The energy possesed by the electron, W orbitting round the nucleus is,

(a) $\dfrac{-e}{8\pi\,\varepsilon_0 r}$ (b) $\dfrac{-e^2}{8\pi\,\varepsilon_0 r}$ (c) $\dfrac{-e.r}{8\pi\,\varepsilon_0}$ (d) $\dfrac{-e^2}{8\pi\,\varepsilon_0 r^2}$

3. The expression for the Kinetic Energy E of free electron in terms of momentum of the electron 'p' and mass of the electron m is,

(a) $\dfrac{p}{2\,m}$ (b) $\dfrac{p}{2\,m^2}$ (c) $\dfrac{p^2}{2\,m^2}$ (d) $\dfrac{p^2}{2\,m}$

4. The expression for radius of stable state 'e' r, is

(a) $\dfrac{n^2 h^2 \varepsilon_0}{\pi\,m e^2}$ (b) $\dfrac{n\,h\,\varepsilon_0}{\pi^2\,m^2 e^2}$ (c) $\dfrac{n^2\,h\,\varepsilon_0{}^2}{\pi\,m\,e^2}$ (d) $\dfrac{n^2\,h^2\,\varepsilon_0{}^2}{\pi\,m\,e^2}$

5. The value of the radius of the lowest state or ground state is, given, h = 6.626 $\times$ 10^{-34} J - Secs $\varepsilon_0 = 10^{-9}/36\pi$

(a)   0.28 A$^\circ$ (b)   0.98 A$^\circ$ (c)   0.18 A$^\circ$ (d)   0.58 A$^\circ$

6. The energy required to detach an electron from its parent atom is called,

(a)   Ionization potential   (b)   Eletric potential

(c)   Kinetic energy   (d)   Threshold potential

7. Electron collision without transfer of energy in collision is called,

(a)   Null collision   (b)   Impact collision

(c)   Elastic collision   (d)   Stiff collision

8. Wave mechanics in electron theory is also known as

(a)   Eienstein theory   (b)   Quantum mechanics

(c)   Bohr mechanics   (d)   Classical theory

9. The expression for threshold frequency f_T to cause photoelectric emission is, with usual notation is,

   (a) $\dfrac{e\phi^2}{h}$
   (b) $\dfrac{e\phi}{h}$
   (c) $\dfrac{eh}{\phi}$
   (d) $\dfrac{e^2\phi^2}{h^2}$

10. The conductivity of a good conductor is typically

    (a) 10^{-3} Ω/cm
    (b) 10^{3} Ω/cm
    (c) 10^{3} ℧/cm
    (d) 10 ℧/m

11. The free electron concentration of a good conductor is of the order of

    (a) 10^{10} electrons/m3
    (b) 10^{28} electron/m^3
    (c) 10^{2} electrons/cm^3
    (d) 10^{10} electrons/m^3

12. Fermi level in Intrinsic semiconductor lies

    (a) close to conduction band
    (b) close to valence band
    (c) In the middle
    (d) None of the these

13. The potential which exists in a p-n junction to cause drift of charge carriers is called

    (a) contact potential
    (b) diffusion potential
    (c) ionisation potential
    (d) threshold potential

14. The equation governing the law of the junction is

    (a) $pn\,(o) = (e^{V/VT})\,n_{no}$
    (b) $n_{po} = P_{no} + n_i$

    (c) $pn\,(o) = P_{no}\left(e^{\frac{V}{VT}} - 1\right)$
    (d) $n_{po} = pno\left(e^{\frac{V}{VT}} - 1\right)$

15. The rate of increase of reverse saturation current for Germanium diode is,

    (a) 5%
    (b) 4%
    (c) 1%
    (d) 7%

16. Cut in voltage V_γ of a silicon diode is also called as

    (a) break over potential
    (b) breakdown potential
    (c) critical potential
    (d) offset voltage

17. A diode which is formed by using lightly doped GaAs or silicon with metal is called

    (a) Zener diode
    (b) Schottky diode
    (c) Varactor diode
    (d) tunnel diode

18. The symbol shown ⊸⊳⊢ is that of a

    (a) LED
    (b) Schottky diode
    (c) Varactor diode
    (d) Gunn diode

CONCEPTUAL QUESTIONS (Interview Questions)

1. When it is said that an electron moves from valence band to conduction band, will there be physical movement of the electron ?

Ans : No, The energy band diagram represents the energy levels of the electrons. When an electron "moves" from valence band to conduction band, it only means that the energy possessed by the electron has increased from Ev electron volts to Ec+ electron volts. The energy diagram is only Pictorial representation of the energy possessed by the electrons. The energy levels of different electrons will be different.

2. What does the arrow in the symbol of a P-n junction diode (—▷—) indicate?

Ans : The arrow mark $\underset{(p)}{\overset{A}{\longrightarrow}}\hspace{-0.5em}\triangleright\hspace{-0.3em}\underset{(n)}{\overset{K}{\vdash}}$ indicates the direction of current flow when the P - n juction is forward biased.

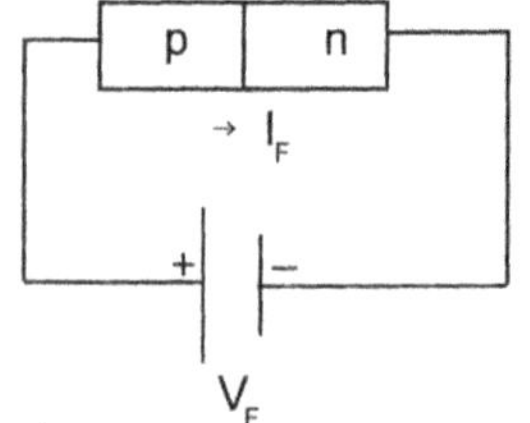

Direction of I_F is the same as that of the direction of arrow.

3. When a Semiconductor device 'breaks down'; what does it mean?

Ans : It does not mean physical breakdown. The shape of the device remains the same. But the device looses its Semiconductor properties. The device may become a short or open circuited.

4. Why tunnel diode device exhibits negative resistance characteristic?

Ans : The current flowing through a device is proportional to the number of free electrons. The higher the number, the more the current. If current through a device 'I' decreases as the voltage across it is being increased, the device exhibits negative resistance.

    In the case of a tunnel diode, there are two conditions to be satisfied for tunnelling to take place.

(i) There must be empty energy states on the P-side corresponding to filled energy states on the n-side.

(ii) The barrier width at the junction must be very narrow ($\sim 3A^{\circ}$), corresponding to doping concentration of 10^{19} carries/cm^3.

    When forward voltage is applied to the tuneel diode, because of the energy gained from the electric field, electrons with higher energy levels tunnel through the barrier, to occupy the equivalent empty energy states on P-side. As the forward voltage is further increased, electrons with lesser energy also acquire sufficient energy from the increased electric field and will be able to tunnel through the barrier. But as the empty energy levels on the P-side get filled, after reaching the peak current I_P, even though forward voltage is being increased, the number of electrons reaching P-side decreases and hence tunnelling current decreases. So the device exhibits negative resistance region, as current is decreasing with increasing voltage. When the voltage is further increased, tunnelling phenomenon ceases, as the empty energy states on P - side would have been filled and normal forward current of the diode only flows.

5. What is the difference between Avalanche and zener breakdown mechanisms?

   Ans : In Zener breakdown mechanism, due to sudden increase in the electric field strength ε, when it reaches a critical value, covalent bonds break and due to this there will be sudden increase in the number of free electrons and hence, the reverse current which was small till that time increases suddenly. But in Avalanche breakdown the field strength is not that high. A free electron with lot of energy collides with another electron, transfers its energy and makes it a free electron. This in turn generates some more free electrons. Thus there is multiplication of carriers. This will occur usually in a forward biased device. Avalanche breakdown precedes Zener breakdown, for comparision, though both cannot occur in the same device at a time.

6. How for a low voltage like 6.8 V, in the case of Fz 6.8 Zener doide zener breakdown occurs, which requires high electric field strength?

   Ans : Electric field strength $\varepsilon = \dfrac{V}{d}$. Even if V is small, if 'd' is very small, ε can be very high. This is what happens in a Zener diode. Due to heavy doping depletion, region width near the junction becomes very small. So, when the voltage V_z reaches the required value, electric field 'ε' reaches the critical value, covalent bonds will be broken and Zener breakdown occurs.

Specifications for Silicon Diode

Sl.No.	Parameter	Symbol	Typical Value	Units
1.	Reverse breakdown voltage	V_{br}	75	V
2.	Static Reverse Current	I_R	5	μA
3.	Static Forward voltage (for silicon)	V_F	0.5	V
4.	Total Capacitance	C_T	2	pf
5.	Reverse Recovery Time	t_{rr}	4	ns
6.	Continuous Power Dissipation	P	500	mw
7.	Max. For Ward Current	I_F	50	mA

7. why semiconductor materials any are used for electronic devices? Why not metals or insulators?

   Ans: current control can be effectively done using semiconductor materials rather than with conducting or Insulating materials for a small change in Input, measeable change in output can be obtained. This in not the case with conductors or Insulator

Some type numbers of Diodes

1.	0A79	:	$0 \rightarrow$ Semiconductor Device.
			$A \rightarrow$ Denotes Germanium device.
2.	BY 127	:	$B \rightarrow$ Denotes Silicon Device.
			$Y \rightarrow$ Rectifying Diode.
			Number 127 is owing a type no and has significance.
3.	1 N 4153	:	$N \rightarrow$ Bipolar Device.
			$1 \rightarrow$ Single Polar function.
4.	BY 100 :800V, 1 A Silicon Diode.		

Specifications of a junction diode

Sl.No.	Parameter	Symbol	Typical Value	Units
1.	Working Inverse voltage	WIV	80	V
2.	Average Rectifield current	I_U	100	A
3.	Continious forward current	I_F	300	mA
4.	Peak repetetive forward current	i_f	400	mA
5.	Forward Voltage	V_f	0.6	V
6.	Reverse current	I_R	500	nA
7.	Breakdown voltage	BV	100	V

Specifications of a Germanium Tunnel Diode IN 2939

Sl.No.	Parameter	Symbol	Typical Value	Units
1.	Forward current	I_F	5	mA
2.	Reverse current	I_R	10	mA
3.	Peak current	I_p	1	mA
4.	Valley current	I_v	0.1	mA
5.	Peak voltage	V_P	50	V
6.	Valley voltage	V_V	30	V
7.	Ratio of I_p to I_V	I_P/I_V	10	-

Rectifiers, Filters and Regulators

2

In this Chapter,

- Circuit applications of p-n junction diode device namely Half Wave Rectifier (HWR), Full Wave Rectifier (FWR) and Bridge Rectifier circuits, for rectification applications are described.

- Capacitance, Inductor, L-section and π-section filter circuits are explained.

- Zener voltage regulator circuits, series and shunt voltage regulator circuits are also explained.

2.1   RECTIFIERS

The electronic circuits require a D.C. source of power. For transistor A.C. amplifier circuit for biasing, D.C supply is required. The input signal can be A.C. and so the output signal will be amplified A.C. signal. But without biasing with D.C. supply, the circuit will not work. So more or less all electronic A.C. instruments require D.C. power. To get this, D.C. batteries can be used. But they will get dried quickly and replacing them every time is a costly affair. Hence it is economical to convert A.C. power into D.C. Such circuits, their efficiency (η) etc., will be discussed in this Chapter.

Rectifier is a circuit which offers low resistance to the current in one direction and high resistance in the opposite direction.

Rectifier converts sinusoidal signal to unidirectional flow and not pure D.C.

Filter converts unidirectional flow into pure D.C.

If the input to the rectifier is a pure sinusoidal wave, the average value of such a wave is zero, since the positive half cycle and negative half cycle are exactly equal.

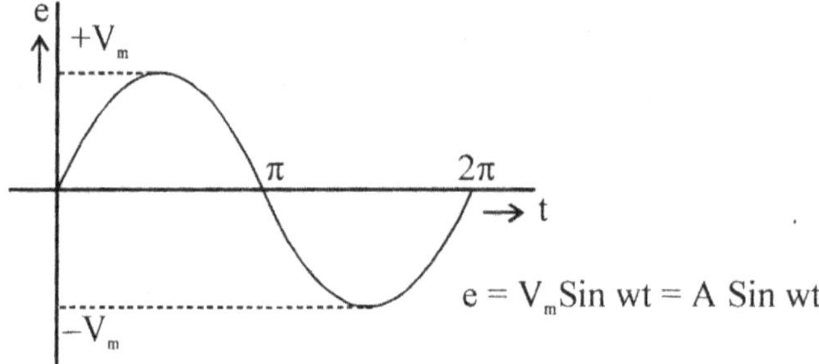

Fig 2.1 AC input wave form.

$$T_{av} = \frac{1}{2\pi} \int_{0}^{2\pi} A \, Sin(\omega t) \, dt$$

2.2   HALF-WAVE RECTIFIER

If this signal is given to the rectifier circuit, say Half Wave Rectifier Circuit, the output will be as shown in Fig. 2.2.

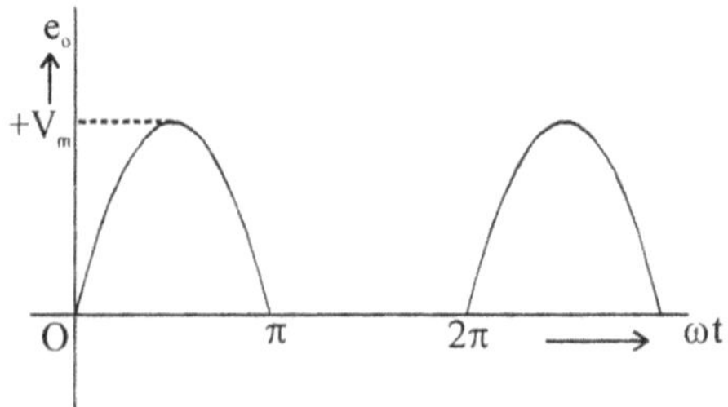

Fig 2.2 Half wave rectified output (unidirectional flow).

Now the average value of this waveform is *not zero,* since there is no negative half. Hence a rectifier circuit converts A.C. Signal with zero average value to a unidirectional wave form with non zero average value.

The rectifying devices are semiconductor diodes for low voltage signals and vacuum diodes for high voltage circuit. A basic circuit for rectification is as shown in Fig. 2.3.

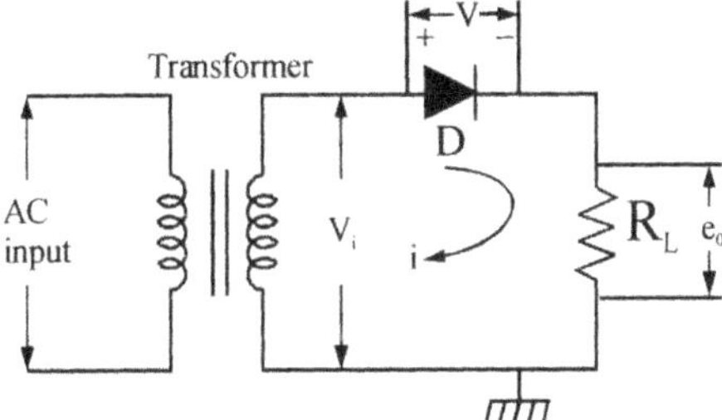

Fig 2.3 Halfwave Rectifier (HWR) circuit.

A.C input is normally the A.C. main supply. Since the voltage is 230V, and such a high voltage cannot be applied to the semiconductor diode, step down transformer should be used. If large D.C. voltage is required vacuum tubes should be used. Output voltage is taken across the load resistor R_L. Since the peak value of A.C. signal is much larger than V_γ, we neglect V_γ, for analysis.

2.2.1 MAXIMUM OR PEAK CURRENT

The output current waveform for half wave rectification is shown in Fig. 2.4.

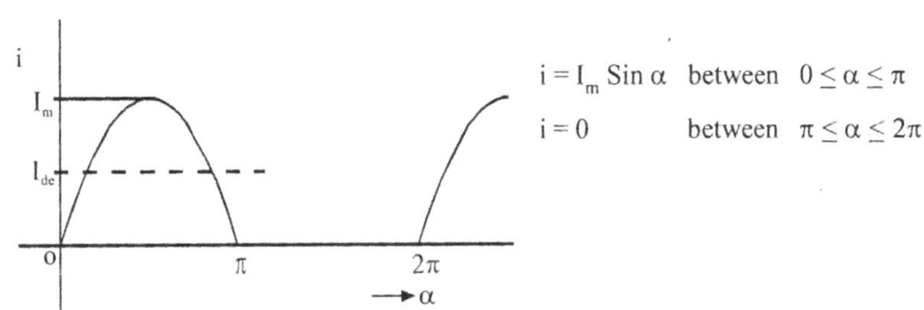

Fig 2.4 Half wave rectified output.

$$\therefore \qquad i = I_m \sin \alpha \qquad \alpha = \omega t \qquad 0 \le \alpha \le \pi$$
$$i = 0 \qquad\qquad\qquad \pi \le \alpha \le 2\pi$$

$$I_m = \frac{V_m}{R_f + R_L}$$

R_f is the forward resistance of the diode. R_L is the load resistance

2.2.2 READING OF DC AMMETER

If a D.C. Ammeter is connected in the rectifier output circuit, what reading will it indicate? Is it the peak value or will the needle oscillate from O to maximum and then to O and so on, or will it indicate average value? The meter is so constructed that it reads the average value.

$$\text{By definition, average value} = \frac{\text{Area of the curve}}{\text{Base}}$$

$\therefore$ For Half wave rectified output, base value is 2π for one cycle,

$$I_{DC} = \frac{1}{2\pi} \int_0^\pi I_m Sin\alpha \, d\alpha = \frac{I_m}{2\pi} \cdot \left| -Cos\alpha \right|_0^\pi$$

$$I_{DC} = \frac{I_m}{2\pi}[1+1] = \frac{I_m}{\pi}$$

Upper limit is only π, because the signal is zero from π to 2π. The complete cycle is from 0 to 2π.

2.2.3 READING OF A.C AMMETER

An A.C. ammeter is constructed such that the needle deflection indicates the effective or RMS current passing through it.

Effective or RMS value of an A.C. quantity is the equivalent D.C. value which produces the same heating effect as the alternating component. If some A.C. current is passed through a resistor, during positive and negative half cycles, also because of the current, the resistor gets heated, or there is some equivalent power dissipation. What is the value of D.C. which produces the same heating effects as the A.C. quantity? The magnitude of this equivalent D.C. is called the RMS or Effective Value of A.C.

$$\text{By definition} \quad I_{rms} = \sqrt{\frac{1}{2\pi} \int_0^{2\pi} (I_m Sin\alpha)^2 \, d\alpha} = \sqrt{\frac{I_m^2}{2\pi} \int_0^{2\pi} Sin^2\alpha \cdot d\alpha}$$

$$= \frac{I_m}{\sqrt{2\pi}} \sqrt{\int_0^{2\pi} \left\{ \frac{1 - Cos2\alpha}{2} \right\} d\alpha} = \frac{I_m}{\sqrt{2\pi}} \sqrt{\left. \frac{\alpha}{2} \right|_0^{2\pi} + \left. \frac{Sin2\alpha}{4} \right|_0^{2\pi}}$$

$$\therefore \quad \text{RMS value of a sine wave} = \frac{I_m}{\sqrt{2\pi}} \times \sqrt{\frac{2\pi}{2} + 0}$$

$$= \frac{I_m \sqrt{\pi}}{\sqrt{2}\sqrt{\pi}} = \frac{I_m}{\sqrt{2}} = 0.707 \, I_m$$

$$\text{Form Factor} = \frac{\text{RMS Value}}{\text{Average Value}}$$

$$\text{For a sine wave,} \quad I_{average} = 0.636 I_m = 2 \times \frac{1}{2\pi} \int_0^\pi (I_m \, Sin\alpha) \, d\alpha$$

$$I_{rms} = \frac{I_m}{\sqrt{2}} = 0.707 I_m$$

$$\text{Form factor} = \frac{\left(\dfrac{I_m}{\sqrt{2}} \right)}{\left(\dfrac{I_m}{\pi} \right)} = \frac{0.707 I_m}{0.636 I_m} = 1.11$$

2.2.4 PEAK INVERSE VOLTAGE

For the circuit shown, the input is A.C. signal. Now during the positive half cycle, the diode conducts. The forward resistance of the diode R_f will be small.

$$\therefore \quad \text{The voltage across the diode } v = i \times R_f$$

$$i = I_m \, Sin \, \alpha \qquad\qquad 0 \le \alpha \le \pi.$$

$$\therefore \quad v = I_m \, R_f \, Sin \, \alpha \qquad\qquad 0 \le \alpha \le \pi.$$

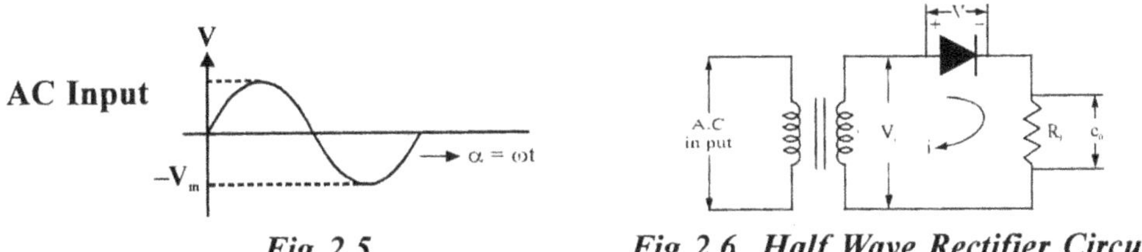

Fig 2.5   *Fig 2.6 Half Wave Rectifier Circuit*

Since R_f is small, the voltage across the diode V during positive half cycle will be small, and the waveform is as shown. But during the negative half cycle, the diode will not conduct. Therefore, the current i through the circuit is zero. So the voltage across the diode is not zero but the voltage of the secondary of the transformer V_i will appear across the diode ($\because$ effectively the diode is across the secondary of the transformer.)

$$\therefore \qquad v = V_m \text{ Sin } \alpha \qquad\qquad \pi \leq \alpha \leq 2\pi.$$

The waveform across the diode is Fig. 2.7.

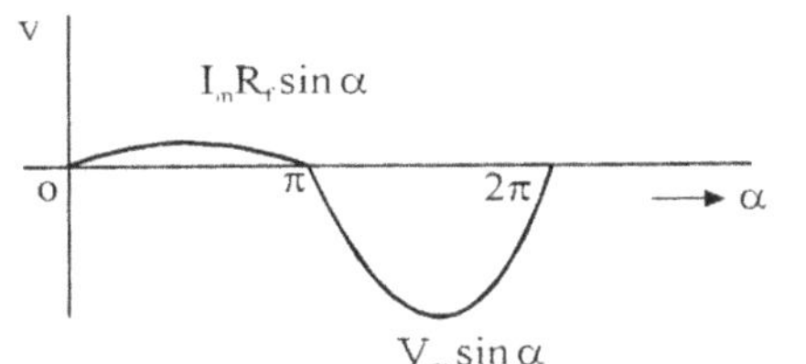

Fig 2.7 Voltage Waveform across the Diode

So the D.C. Voltage that is read by a D.C. Voltmeter is the average value.

$$V_{DC} = \frac{1}{2\pi} \left[\int_0^\pi I_m R_f \cdot \text{Sin}\alpha \, d\alpha + \int_\pi^{2\pi} V_m \text{Sin}\alpha \, d\alpha \right]$$

$$= \frac{1}{2\pi} \left[(2)I_m R_f - (2)I_m (R_f + R_L) \right]$$

$$\because \qquad V_m = I_m (R_f + R_c)$$

$$\therefore \qquad \boxed{V_{DC} = \frac{-I_m R_L}{\pi}} \qquad\qquad\qquad \dots\dots\dots (2.1)$$

If we connect a CRO across the diode, this is the waveform that we see is as shown in Fig. 2.8. So in the above circuits the diode is being subjected to a maximum voltage of V_m. It occurs when the diode is not conducting. Hence it is called the ***Peak Inverse Voltage (PIV)***

2.2.5 REGULATION

The variation of D.C output voltage as a function of D.C load current is called '*regulation*'.

$$\% \text{ Regulation} = \frac{V_{\text{No Load Voltage}} - V_{\text{Full Load}}}{V_{\text{Full Load}}} \times 100\%$$

For an ideal power supply output voltage is independent of the load or the output in voltage remains constant even if the load current varies, like in Zener diode near breakdown. Therefore, *regulation is zero or it should be low for a given circuit.*

EXPRESSION FOR V_{DC} THE OUTPUT DC VOLTAGE

For half wave rectifier circuit (Fig. 2.8), I_{DC} the average value is :

$$I_{DC} = \frac{1}{2\pi} \int_0^{2\pi} i.d\alpha$$

$$= \frac{1}{2\pi} \int_0^{\pi} I_m \sin\alpha.d\alpha = \frac{I_m}{\pi}$$

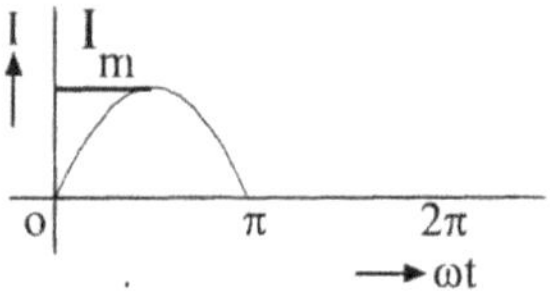

Fig 2.8 HWR current output

$$I_{DC} = \frac{I_m}{\pi} \qquad \qquad .... (2.2)$$

But $\qquad I_m = \dfrac{V_m}{R_f + R_L}$ $\qquad\qquad$ $R_L = $ Load Resistance

$\qquad\qquad\qquad\qquad\qquad\qquad\qquad\qquad$ $R_f = $ Forward Resistance of the Diode.

$$V_{DC} = I_{DC}\, R_L$$

$$\therefore \qquad I_{DC} = \frac{V_m/\pi}{R_f + R_L}$$

But $\qquad I_{DC} = \dfrac{V_m}{\pi(R_f + R_L)}$

$$\therefore \qquad V_{DC} = \frac{V_m.R_L}{\pi(R_f + R_L)}$$

Adding and Subtracting R_f,

$$V_{DC} = \frac{V_m \cdot (R_L + R_f - R_f)}{\pi(R_f + R_L)} = \frac{V_m}{\pi} - \frac{V_m \cdot R_f}{\pi(R_f + R_L)}$$

$$I_{DC} = \frac{V_m}{\pi(R_f + R_L)}$$

I_{DC} is determined by R_L. Hence V_{DC} depends upon R_L.

$$\boxed{V_{DC} = \frac{V_m}{\pi} - I_{DC} \times R_f} \qquad \dots\dots\dots(\,2.3\,)$$

This expression indicates that V_{DC} is $\dfrac{V_m}{\pi}$ at *no load or when the load current is zero*, and it decreases with increase in I_{DC} linearly since R_f is more or less constant for a given diode. The larger the value of R_f, the greater is the decrease in V_{DC} with I_{DC}. But, the series resistance of the secondary winding of the transformer R_s should also be considered.

For a given circuit of half wave rectifier, if a graph is plotted between V_{DC} and I_{DC} the slope of the curve gives $(\,R_f + R_s\,)$ where R_f is the forward resistance of diode and R_s the series resistance of secondary of transformer.

Voltage Regulation

REGULATION FOR HWR

The regulation indicates how the DC voltage varies as a function of DC load current. In general, the percentage of regulation for ideal power supply is zero. The percentage of regulation is defined as

$$\% \text{ Regulation} = \frac{V_{NL} - V_{FL}}{V_{FL}} \times 100$$

As we know that,

$$V_{DC} = \frac{V_m}{\pi} - I_{DC} \cdot R_f$$

$$V_{NL} = \frac{V_m}{\pi} \quad \text{and} \quad V_{FL} = \frac{V_m}{\pi} - I_{DC} \cdot R_f$$

$$\therefore \% \text{ Regulation} = \frac{\dfrac{V_m}{\pi} - \dfrac{V_m}{\pi} + I_{DC} \cdot R_f}{\dfrac{V_m}{\pi} - I_{DC} \cdot R_f} \times 100$$

$$= \frac{I_{DC} \cdot R_f}{\dfrac{V_m}{\pi} - I_{DC} \cdot R_f} = \frac{1}{\dfrac{V_m}{\pi\,(I_{DC} \cdot R_f)} - 1} \times 100$$

$$= \frac{1}{\dfrac{V_m}{\pi} \times \dfrac{\pi}{V_m} \dfrac{(R_f + R_L)}{R_f} - 1} \times 100 \left[\because I_{DC} = \frac{V_m}{\pi(R_f + R_L)}\right]$$

$$\boxed{\% \text{ Regulation} = \frac{R_f}{R_L} \times 100}$$

Suppose for a given rectifier circuit, the specifications are 15V and 100 mA, i.e., the no load voltage is 15V and max load current that can be drawn is 100 mA. If the value of $(R_f + R_s) = 25\ \Omega$ then the percentage regulation of the circuit is :

$$\text{No Load Voltage} = 15V$$

$$\text{Drop across Diode} = I_m\ (R_f + R_s)$$

$$R_S = \text{Transformer Secondary Resistance}$$

$$\text{Max. Voltage with load} = 15V - (I_m \times (R_f + R_s))$$

$$= 15 - 100\ \text{mA} \times 25\Omega$$

$$= 15 - 2.5\ \text{volts} = 12.5V$$

$$\therefore \qquad \text{Percentage Regulation} = \frac{15 - 12.5}{12.5} \times 100 = \frac{2.5}{12.5} \times 100 \simeq 20\%.$$

2.2.6   RIPPLE FACTOR

The purpose of a rectifier circuit is to convert A.C. to D.C. But the simple circuit shown before will not achieve this. Rectifier converts A.C. to unidirectional flow and not D.C. So filters are used to get pure D.C. Filters convert unidirectional flow into D.C. Ripple factor is a measure of the fluctuating.components present in rectifier circuits.

$$\text{Ripple factor, } \gamma = \frac{\text{RMS Value of alternating components of the waveform}}{\text{Average Value of the waveform}}$$

$$\gamma = \frac{I'_{rms}}{I_{DC}} = \frac{V'_{rms}}{V_{DC}}$$

I'_{rms} and V'_{rms} denote the value of the A.C components of the current and voltage in the output respectively. While determining the ripple factor of a given rectifier system experimentally, a voltmeter or ammeter which can respond to high frequencies (greater than power supply frequency 50 Hz) should be used and a capacitor should be connected in series with the input meter in order to block the D.C. component. Ripple factor should be small. (Total current $i = I_m$ Sin ωt according to Fourier Series, only A.C. is sum of D.C. and harmonics).

We shall now derive the expression for the ripple factor. The instantaneous current is given by $i' = i - I_{DC}$.

i is the total current. In a half wave rectifier, some D.C components are also present. Hence A.C component is,

$$i' = (i - I_{DC}) [(\text{Total current} - I_{DC})]$$

$$\text{RMS value is } I'_{rms} = \sqrt{\frac{1}{2\pi} \int_0^{2\pi} (i - I_{DC})^2\, d\alpha}$$

$$\therefore \qquad I'_{rms} = \sqrt{\frac{1}{2\pi}\int_{0}^{2\pi}\left(i^2 - 2I_{DC}.i + I_{DC}^{\,2}\right)d\alpha}$$

Now, $\qquad \dfrac{1}{2\pi}\int_{0}^{2\pi}i^2 d\alpha$ = Square of the rms value of a sine wave by definition.

$$= (I_{rms})^2$$

$$\frac{1}{2\pi}\int_{0}^{2\pi}i.d\alpha = \text{The average value or D.C value } I_{DC}.$$

I_{DC} is constant. So taking this term outside,

$$\frac{I_{DC}^{\,2}}{2\pi}\int_{0}^{2\pi}d\alpha = \frac{I_{dc}}{2\pi}\left[2\pi\right]$$

$$\therefore \qquad I'_{rms} = \sqrt{\left(I_{rms}\right)^2 - 2I_{DC}^{\,2} + I_{DC}^{\,2}}$$

The rms ripple current is

$$I'_{rms} = \sqrt{\left(I_{rms}\right)^2 - I_{DC}^{\,2}}$$

Ripple factor, $\qquad \gamma = \dfrac{I'_{rms}}{I_{dc}} = \dfrac{\sqrt{\left(I_{rms}\right)^2 - I_{DC}^{\,2}}}{I_{DC}}$

$$\gamma = \sqrt{\left(\frac{I_{rms}}{I_{DC}}\right)^2 - 1}. \qquad\qquad \dots\dots\dots (\,2.4\,)$$

This is independent of the current waveshape and is not restricted to a half wave configuration. If a capacitor is used to block D.C. and then I_{rms} or V_{rms} is measured,

Then, $\qquad \gamma = \dfrac{I_{rms}}{I_{DC}}$

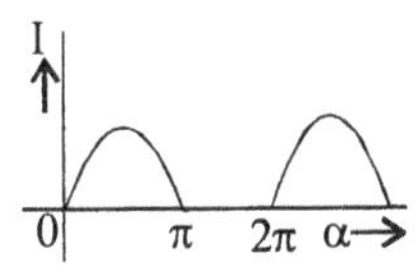

$$\therefore \qquad I'_{rms} = \sqrt{I_{rms}^{\,2} - I_{DC}}$$

and $\qquad I_{DC} = 0$ (blocked by Capacitor) output waveform

For Half Wave Rectifier Circuit (HWR),

$$I = I_m \sin\alpha$$

$$I_{av} = I_{DC} = \frac{1}{2\pi}\int_{0}^{\pi}I_m \sin\alpha.d\alpha = \frac{I_m}{\pi}$$

$$I_{rms} = \sqrt{\frac{1}{2\pi} \int_0^\pi I_m^2 Sin^2(\alpha)\, d\alpha} = \frac{I_m}{2}$$

$$I_{DC} = \frac{I_m}{\pi} ; \quad V_{DC} = \frac{V_m}{\pi} \quad \text{Peak Inverse Voltage (PIV) } = V_m$$

Ripple Factor $\quad \gamma = \dfrac{\text{RMS value of ripple current}}{\text{Average value of the current}} = \dfrac{I'_{rms}}{I_{DC}}$

Total current $\quad I = I_{DC} + I'(\text{ripple})$

$I'(\text{ripple}) = (I - I_{DC})$

$$I'_{rms} = \sqrt{\frac{1}{2\pi} \int_0^{2\pi} \left(I - I_{DC}\right)^2 d\alpha} = \sqrt{\frac{1}{2\pi} \int_0^{2\pi} \left(I^2 - 2I.I_{DC} + I_{DC}^2\right).d\alpha}$$

$$= \sqrt{\left(I_{rms}\right)^2 - 2I_{DC}^2 + I_{DC}^2}$$

$\therefore \qquad I'_{rms} = \sqrt{\left(I_{rms}\right)^2 - I_{DC}^2}$

$$\gamma = \frac{I'_{rms}}{I_{DC}} = \sqrt{\left(\frac{I_{rms}}{I_{DC}}\right)^2 - 1}$$

for Half Wave Rectifier, $\gamma = \sqrt{\left(\dfrac{I_m \times \pi}{2 \times I_m}\right)^2 - 1} = \sqrt{\left(\dfrac{\pi}{2}\right)^2 - 1} = 1.21$

2.2.7 RATIO OF RECTIFICATION (η) : $\dfrac{P_{DC}}{P_{AC}}$

$$\text{Ratio of Rectification} = \frac{\text{DC power delivered to the load}}{\text{AC input power from transfer secondary}}$$

$$P_{DC} = I_{DC}^2 \times R_L$$

But $\qquad I_{DC}$ for HWR $= \dfrac{I_m}{\pi}$

$\therefore \qquad P_{DC} = \left(\dfrac{I_m}{\pi}\right)^2 \times R_L$

P_{AC} is what a Wattmeter would indicate if placed in the HWR circuit with its voltage terminal connected across the secondary of the transformer.

From $\pi - 2\pi$ the diode is not conducting. Hence $I_{ac} = 0$.

$$P_{ac} = (I_{rms})^2 (R_f + R_L)$$

Q from $\pi - 2\pi$ the AC is not being converted to D.C. So this power is wasted, as heat across diode and transformer.

$$\therefore \qquad I_{rms}(\pi - 2\pi) \text{ portion of AC is} = \sqrt{\frac{1}{2\pi} \int_{\pi}^{2\pi} I_m^2 \sin^2 \alpha \, d\alpha} = \frac{I_m}{2}$$

I_{rms} for HWR is $I_m/2$. During negative half cycle, diode is not conducting and current I in the loop is zero even though Voltage V is present.

$$\therefore \qquad P_{AC} = \left(\frac{I_m}{2}\right)^2 (R_f + R_L) = \left(\frac{I_m}{2}\right)^2 \times R_L$$

$$\therefore \qquad \frac{P_{DC}}{P_{AC}} = \frac{I_m^2 . R_L \times 4}{\pi^2 I_m^2 \times R_L} = \frac{4}{\pi^2} = 0.406$$

$$\therefore \qquad \text{Ratio of rectification for HWR} = 0.406$$

Considering ideal diode. If we consider R_f also, the expression $= \dfrac{0.406}{1 + \left(\dfrac{R_f}{R_L}\right)}$

The AC input power is not converted to D.C. Only part of it is converted to D.C and is dissipated in the load. The balance of power is also dissipated in the load itself as AC power. We have to consider the rating of the secondary of the transformer. In ratio of rectification we have considered only the A.C output as the secondary of the transformer.

2.2.8 TRANSFORMER UTILIZATION FACTOR

$$\text{Transformer Utilization Factor (TUF)} = \frac{\text{D.C power delivered to the load}}{\text{AC rating of transformer secondary}}$$

$$= \frac{P_{DC}}{P_{AC} \text{ rated}}$$

This term TUF is not ratio of rectification, because all the rated current of the secondary is not being drawn by the circuit.

$$P_{ac} = I_{DC}^2 . R_L - \left(\frac{I_m}{\pi}\right)^2 . R_L$$

$$P_{ac} = V_{rms} \times I_{rms}$$

$$V_{rms} = \frac{V_m}{\sqrt{2}} = \text{Rated voltage of (Secondary Transformer)}$$

$$I_{rms} = I_m/2$$

$$P_{ac} = V_{rms} \times I_{rms}$$

But $\qquad V_m = I_m (R_f + R_L) \simeq I_m R_L \ (R_f \text{ is small})$

$$\text{TUF} = \frac{I_m^2}{\pi^2} \times R_L \bigg/ \frac{I_m R_L}{\sqrt{2}} \times \frac{I_m}{2} = \frac{2\sqrt{2}}{\pi^2} = 0.287$$

Transformer Utilization Factor for Half Wave Rectifier is 0.287

2.2.9   DISADVANTAGES OF HALF WAVE RECTIFIER

1. High Ripple Factor (1.21)   **2.** Low ratio of rectification (0.406)
3. Low TUF (0.287)   **4.** D.C saturation of transformer.

For half wave configuration,

$$I_{rms} = \sqrt{\frac{1}{2\pi}\int_0^\pi I_m^2 \sin^2 \alpha.d\alpha} \quad = \frac{I_m}{2}$$

Because during negative half cycle the diode will not conduct hence i = 0 in the loop from
$\pi - 2\pi$. So integration is from $0 - \pi$ only. Even though V is present, i = 0 for one half cycle.
Therefore, Power is zero (V × I = P = 0, since, I = 0)

$$I_{DC} = \frac{1}{2\pi}\int_0^\pi I_m \sin \alpha.d\alpha \quad = \frac{I_m}{2\pi}\left|-\cos\alpha\right|_0^\pi = \frac{I_m}{2\pi}\left|1+1\right| = \frac{I_m}{\pi}$$

∴   Ripple Factor for Half Wave Rectification $\dfrac{I_{rms}}{I_{DC}} = \dfrac{I_m/2}{I_m/\pi} = \dfrac{\pi}{2}$

$$\gamma = \sqrt{(1.57)^2 - 1} = 1.21$$

∵ $\dfrac{\pi}{2} = 1.57$

∴ $\gamma > 1$

So the ripple voltage exceeds D.C voltage. Hence HWR is a poor circuit for converting AC to DC.

2.2.10   POWER SUPPLY SPECIFICATIONS

The input characteristics which must be specified for a power supply are :

1. *The required output D.C voltage*   **2.** *Regulation*
3. *Average and peak currents in each diode*   **4.** *Peak inverse voltage (PIV)*
5. *Ripple factor.*

2.3   FULL WAVE RECTIFIER (FWR)

Since half wave rectifier circuit has poor ripple factor, for ripple voltage is greater than DC voltage,
it cannot be used. So now analyze a full wave rectifier circuit.

The circuit is as shown in Fig. 2.9.

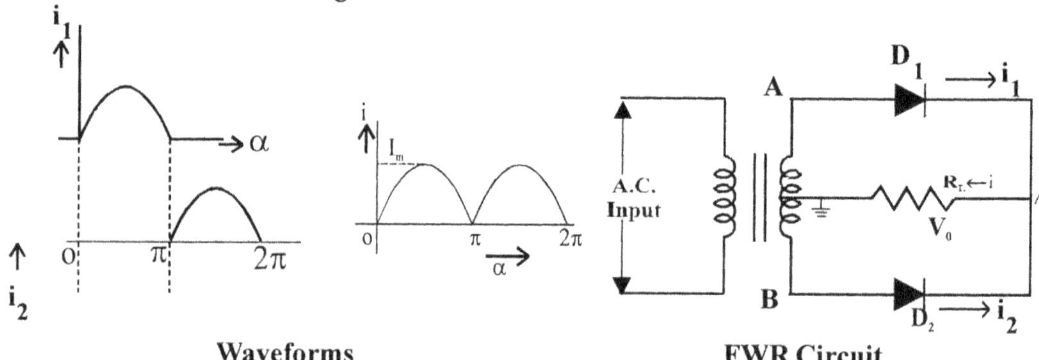

Waveforms   **FWR Circuit**

Fig 2.9 FWR

During the +ve half cycle, D_1 conducts and the current through D_2 is zero

During the – ve half cycle, D_2 conducts and the current through D_1 is zero

A centre tapped transformer is used.

The total current i flows through the load resistor R_L and the output voltage V_0 is taken across R_L.

For half wave rectifier, circuit, $I_{DC} = \dfrac{1}{2\pi} \displaystyle\int_0^{2\pi} I_m \sin\alpha\, d\alpha = \dfrac{I_m}{\pi}$

For full wave rectifier, circuit, I_{DC} = Twice that of half wave rectifier circuit

$\therefore \qquad I_{DC} = 2\, I_m/\pi$

For half wave rectifier circuit , $I_{rms} = \sqrt{\dfrac{1}{2\pi} \displaystyle\int_0^{\pi} I_m^2 \sin\alpha\, d\alpha} = \dfrac{I_m}{2}$

For full wave rectifier circuit, $I_{rms} = \sqrt{2 \times \dfrac{1}{2\pi} \displaystyle\int_0^{\pi} I_m^2 \sin^2\alpha\, d\alpha}$

$$I_{rms} = \sqrt{2} \times \dfrac{I_m}{2} = \dfrac{I_m}{\sqrt{2}}$$

$$I_m = \dfrac{V_m}{R_f + R_L}$$

R_f is the forward resistance of each diode.

A centre tapped transformer is essential to get full wave rectification. So there is a phase shift of 180^0, because of centre tapping. So D_1 is forward biased during the input cycle of 0 to π, D_2 is forward biased during the period π to 2π since the input to D_2 has a phase shift of 180^0 compared to the input to D_1. So positive half cycle for D_2 starts at a full wave rectifier circuit, while the D.C. current starts through the load resistance R_L is twice that of the Half Wave Rectifier Circuit.

Therefore, for Half Wave Rectifier Circuit,

$$I_{DC} = \dfrac{I_m}{\pi}$$

For Full Wave Rectifier Circuit,

$$I_{DC} = \dfrac{2I_m}{\pi}$$

Hence, Ripple Factor is improved.

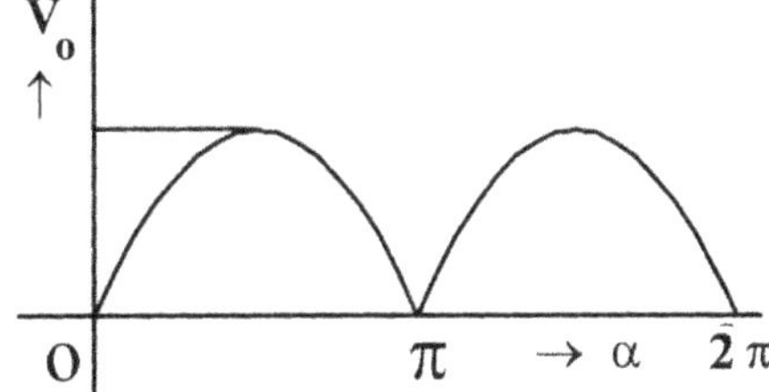

Fig 2.10 FWR voltage output.

2.3.1 RIPPLE FACTOR

$$\dfrac{I'_{rms}}{I_{DC}} = \dfrac{I_m/\sqrt{2}}{2I_m/\pi} = \dfrac{\pi}{2\sqrt{2}} = 1.11$$

$$I'_{rms} = \sqrt{I_{rms}^2 - I_{DC}^2} \; ; \quad \gamma = \dfrac{\sqrt{I_{rms}^2 - I_{DC}^2}}{I_{DC}}$$

Therefore, $\gamma = 1.21$ for HWR, and it is is 0.482 for FWR.

2.3.2 REGULATION FOR FWR

The regulation indicates how D.C voltage varies as a function of load current. The percentage of regulation is defined as

$$\% \text{ Regulation} = \frac{V_{NL} - V_{FL}}{V_{FL}} \times 100$$

The variation of DC voltage as a function of load current is as follows

$$V_{DC} = I_{DC} R_L$$

$$= \frac{2I_m}{\pi} \cdot R_L \quad \left[\because I_m = \frac{V_m}{R_f + R_L} \right]$$

$$= \frac{2V_m}{\pi(R_f + R_L)} \cdot R_L \quad \left[\because I_m = \frac{V_m}{R_f + R_L} \right]$$

$$= \frac{2V_m}{\pi} \left[1 - \frac{R_f}{R_f + R_L} \right]$$

$$= \frac{2V_m}{\pi} - \frac{2V_m}{\pi(R_f + R_L)} \times R_f$$

$$\boxed{V_{DC} = \frac{2V_m}{\pi} - I_{DC} \cdot R_f}$$

$$V_{NL} = \frac{V_m}{\pi}, \qquad V_{FL} = \frac{V_m}{\pi} - I_{DC} \cdot R_f$$

$$\% \text{ Regulation} = \frac{\dfrac{2V_m}{\pi} - \dfrac{2V_m}{\pi} + I_{DC} \cdot R_f}{\dfrac{2V_m}{\pi} - I_{DC} \cdot R_f} \times 100$$

$$= \frac{I_{DC} R_f}{\dfrac{2V_m}{\pi} - I_{DC} . R_f} \times 100$$

$$= \frac{I_{DC} R_f}{I_{DC} R_f \left(\dfrac{2V_m}{\pi \, I_{DC} . R_f} \right) - 1} \times 100$$

$$= \frac{1}{\left(\dfrac{2V_m}{\pi} \times \dfrac{(R_f + R_L)}{\dfrac{2V_m}{\pi} . R_f} - 1 \right)} \times 100$$

$$\boxed{\% \text{ Regulation} = \frac{R_f}{R_L} \times 100}$$

2.3.3 Ratio of Rectification

$$\frac{\text{D.C. power delivered to Load}}{\text{A.C. input power from transformer secondary}}$$

$$P_{DC} = I_{DC}^2 \times R_L \text{ for FWR,}$$

$$I_{DC} = \frac{2I_m}{\pi} = \frac{4.I_m^2}{\pi^2} R_L.$$

P_{AC} is what a Wattmeter would indicate if placed in the FWR Circuit with its voltage terminals connected across the transformer secondary.

$$P_{ac} = (I_{rms})^2 . (R_F + R_L); \ I_{rms} \text{ for FWR} = \frac{I_m}{\sqrt{2}}$$

$$= \left(\frac{I_m}{\sqrt{2}}\right)^2 .(R_F + R_L).$$

$$= \frac{I_m^2}{2} (R_F + R_L).$$

If we assume that R_f is the forward resistance of the diode is very small, compared to R_L.

$$R_f + R_L \cong R_L.$$

$$\therefore \quad P_{AC} = \frac{I_m^2}{2} (R_L).$$

$$\text{Ratio of Rectification} = \frac{P_{DC}}{P_{AC}} = \frac{4 I_m^2 R_L \times 2}{\pi^2 I_m^2 \times R_L} = \frac{8}{\pi^2} = 0.812$$

For HWR it is 0.406. Therefore, for FWR the ratio of rectification is twice that of HWR.

2.3.4 Transformer Utilization Factor : TUF

In fullwave rectifier using center-tapped transformer, the secondary current flows through each half separately in every half cycle. While the primary of transformer carries current continuously. Hence TUF is calculated for primary and secondary windings separately and then the average TUF is determined.

$$\text{Secondary TUF} = \frac{\text{DC power delevered to load}}{\text{AC rating of transformer secondary}}$$

$$= \frac{(I_{DC})^2 \ R_L}{V_{rms} \ I_{rms}} = \frac{\left(\frac{2 I_m}{\pi}\right)^2 . R_L}{\frac{V_m}{\sqrt{2}} \times \frac{I_m}{\sqrt{2}}}$$

$$= \frac{4I_m^2}{\pi^2} \times \frac{2R_L}{I_m^2 (R_f + R_L)} = \frac{8}{\pi^2} \times \frac{1}{1 + \frac{R_f}{R_L}}$$

if $R_f \ll R_L$, then secondary TUF = 0.812.

The primary of the transformer in feeding two HWR's separately. These two HWR's work independently of each other but feed a common load. Therefore,

$$\text{TUF for primary} = 2 \times \text{TUF of HWR} = 2 \times \frac{0.287}{1 + \dfrac{R_f}{R_L}}$$

if $R_f \ll R_L$, then TUF for primary = 0.574

The average TUF for FWR using center tapped transformer

$$= \frac{\text{TUF of primary} + \text{TUF of secondary}}{2} = \frac{0.812 + 0.574}{2} = 0.693$$

Therefore, the transformer is utilized 69.3% in FWR using center tapped transformer.

2.3.5 PEAK INVERSE VOLTAGE (PIV) FOR FULL WAVE RECTIFIER

With reference to the FWR circuit, during the -ve half cycle, D_1 is not conducting and D_2 is conducting. Hence maximum voltage across R_L is V_m, since voltage is also present between A and O of the transformer, the total voltage across $D_1 = V_m + V_m = 2V_m$ (Fig. 2.9).

2.3.6 D.C. SATURATION

In a FWR, the D.C. currents I_1 and I_2 flowing through the diodes D_1 and D_2 are in opposite direction and hence cancel each other. So there is no problem of D.C. current flowing through the core of the transformer and causing saturation of the magnetic flux in the core.

PIV is $2\,V_m$ for each diode in FWR circuit because, when D_1 is conducting, the drop across it is zero. Voltage delivered to R_L is V_m. D_2 is across R_L. During the same half cycle, D_2 is not conducting. Therefore, peak voltage across it is V_m from the second half of the transformer. The total voltage across D_2 is $V_m + V_m = 2V_m$.

2.4   BRIDGE RECTIFIERS

The circuit is shown in Fig. 2.11. During the positive half cycle, D_1 and D_2 are forward biased. D_3 and D_4 are open. So current will flow through D_1 first and then through R_L and then through D_2 back to the ground. During the –ve half cycle D_4 and D_3 are forward biased and they conduct. The current flows from D_3 through R_L to D_4. Hence the direction of current is the same. So we get full wave rectified output.

In Bridge rectifier circuit, there is no need for centre tapped transformer. So the transformer secondary line to line voltage should be one half of that, used for the FWR circuit, employing two diodes.

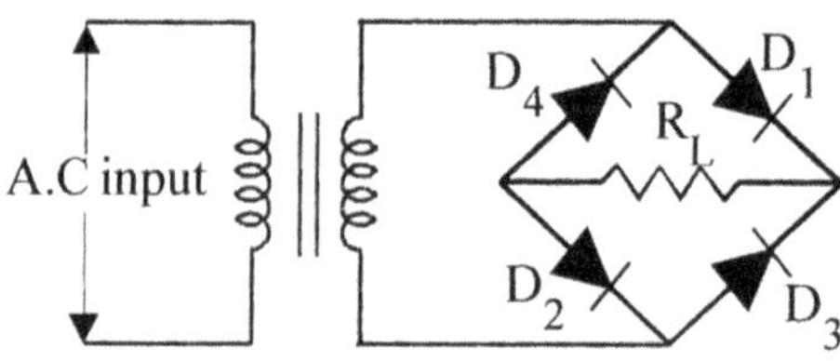

Fig 2.11   Bridge rectifier circuit.

$$I_{DC} = \frac{2I_m}{\pi}$$

$$\text{where} \qquad I_m = \frac{V_m}{R_S + 2R_f + R_L}$$

$$V_{DC} = \frac{2V_m}{\pi} - I_{DC}(R_S + 2R_F)$$

$$R_S = \text{Resistance of Transformer Secondary.}$$

$$R_f = \text{Forward Resistance of the Diode.}$$

$$R_L = \text{Load Resistance.}$$

$2R_f$ should be used since two diodes in series are conducting at the same time. The ripple factor and ratio of rectification are the same as for Full Wave Rectifier.

2.4.1 ADVANTAGES OF BRIDGE RECTIFIER

1. *The peak inverse voltage (PIV) across each diode is V_m and not $2V_m$ as in the case of FWR. Hence the Voltage rating of the diodes can be less.*
2. *Centre tapped transformer is not required.*
3. *There is no D.C. Current flowing through the transformer since there is no centre tapping and the return path is to the ground. So the transformer utilization factor is high.*

2.4.2 DISADVANTAGES

1. *Four diodes are to be used.*
2. *There is some voltage drop across each diode and so output voltage will be slightly less compared to FWR. But these factors are minor compared to the advantages.*

Bridge rectifiers are available in a package with all the 4 diodes incorporated in one unit. It will have two terminals for A.C. Input and two terminals for DC output. Selenium rectifiers are also available as a package.

2.5 COMPARISON OF RECTIFIER CIRCUITS

	HWR	FWR	Bridge Rectifier Circuit
Peak Inverse Voltage	V_m	$2V_m$	V_m
No Load Voltage	$\dfrac{Vm}{\pi}$	$\dfrac{2Vm}{\pi}$	$\dfrac{2V_m}{\pi}$
Ripple factor	1.21	0.482	0.482
Number of Diodes required	1	2	4
Ratio of Rectification $\dfrac{P_{dc}}{P_{AC}}$	0.406	0.812	0.812
$TUF = \dfrac{\text{DC Power delivered to the load}}{\text{AC ratings of transformer secondary}}$	0.287	0.693	0.812

2.6   VOLTAGE DOUBLER CIRCUIT

2.6.1 HALF WAVE VOLTAGE DOUBLER CIRCUIT

The circuit is shown in Fig. 2.12 and the wave forms are shown in Fig. 2.13. During the +ve half cycle, D_1 is forward biased. So C_1 gets charged to V_m. During the –ve half cycle, D_1 is reverse biased and the PIV across it V_m. Therefore, the total voltage during –ve half cycle V_m across C_1 +Vm of –ve half cycle = $2V_m$. So, D_2 is forward biased during the –ve half cycle and C_2 gets charged to $2V_m$. This is the no-load voltage. The actual voltage depends upon the value of R_L connected. In A.C. Voltage measurements, the A.C. input is given to a voltage doubler and filter circuits. The output of D.C. meter ($2V_m$) is proportional to A.C. Input. The meter calibrated in terms of the rms value of the A.C. Input.

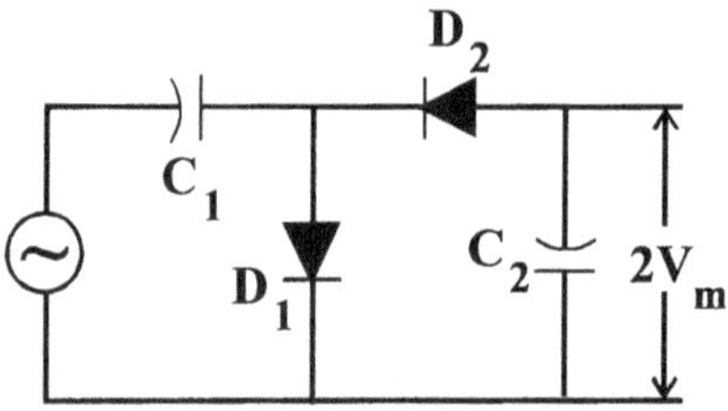

Fig 2.12   Halfwave voltage doubler circuit.

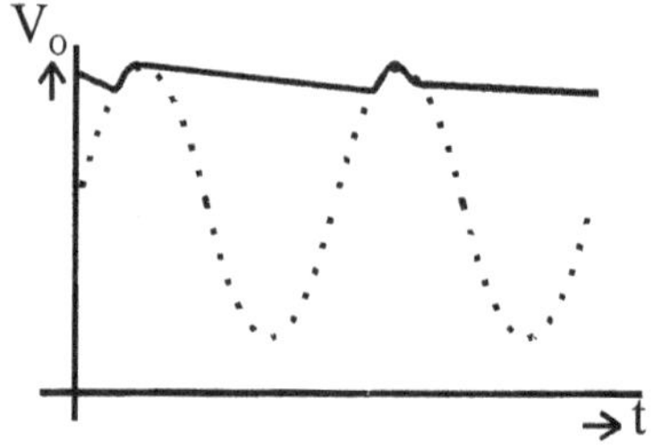

Fig 2.13   Output waveform.

2.6.2 FULL WAVE VOLTAGE DOUBLER CIRCUIT

The circuit is shown in Fig. 2.14. Wave forms are shown in Fig. 2.15. During +ve half cycle, D_1 conducts C_1 gets charged to V_m. During –ve half cycle, C_2 charges through D_2 and to V_m. R_L is across the series combination of C_1 and C_2. Therefore, the total output voltage $V_0 = 2V_m$. Here the PIV across each diode is only V_m and the capacitors are charged directly from the input and not C_2 will not get charged through C_1 During the time period 0 to 2π we have two cycles. Current flows from 0 to π and π to 2π also. So it is FW Voltage Doubler Circuit.

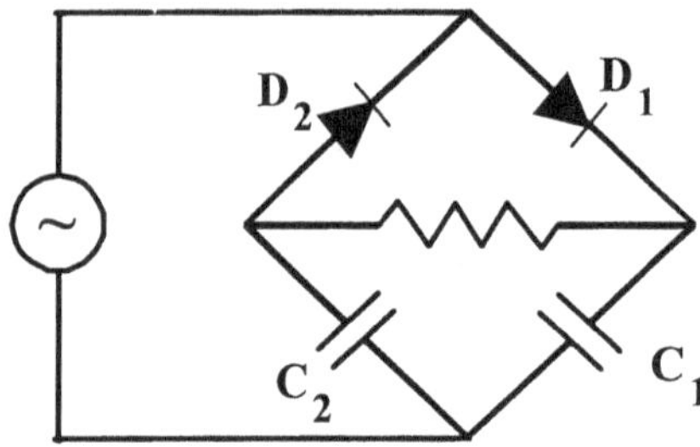

Fig 2.14   Fullwave voltage doubler circuit.

Fig 2.15   Output waveforms.

FILTER CIRCUITS FOR POWER SUPPLIES

2.7   INDUCTOR FILTER CIRCUITS

FILTERS

A power supply must provide ripple free source of power from an A.C. line But the output of a rectifier circuit contains ripple components in addition to a D.C. term. It is necessary to include a filter between the rectifier and the load in order to eliminate these ripple components. Ripple components are high frequency A.C. Signals in the D.C output of the rectifier. These are not desirable, so they must be filtered. So filter circuits are used.

Flux linkages per ampere of current $L = \dfrac{N\phi}{I}$. The ability of a component to develop induced voltage when alternating current is flowing through the element is the property of the inductor. Types of Inductors are :

       1. *Iron Cored*   **2.** *Air Cored*

An inductor opposes any change of current in the circuit. So any sudden change that might occur in a circuit without an inductor are smoothed out with the presence of an inductor. In the case of AC, there is change in the magnitude of current with time.

Inductor is a short circuit for DC and offers some impedance for A.C. ($X_L = j\omega L$). So it can be used as a filter. AC voltage is dropped across the inductors, where as D.C. passes through it. Therefore, A.C is minimized in the output.

Inductor filter is used with FWR circuit. Therefore, HWR gives 121% ripple and using filter circuit for such high ripple factor has no meaning . FWR gives 48% ripple and by using filter circuit we can improve it. According to Fourier Analysis, current, I in HWR Circuit is

$$i = \frac{I_m}{\pi} + \frac{I_m}{2}\,\text{Sin}\,\omega t - \frac{2I_m}{\pi}\,\frac{\text{Cos}\,4\omega t}{3}$$

Current in the case of FWR is $i \cong \dfrac{2I_m}{\pi} - \dfrac{4I_m}{3\pi}\cos 2\omega t - \dfrac{4I_m \cos 4\omega t}{15\pi}$.

The reactance offered by inductor to higher order frequencies like $4\omega t$ can be neglected. So the output current is

$$i \cong \frac{2I_m}{\pi} - \frac{4I_m}{3\pi}\cos 2\omega t$$

the fundamental harmonic ω is eliminated.

The circuit for FWR with inductor filter is as shown in Fig. 2.16.

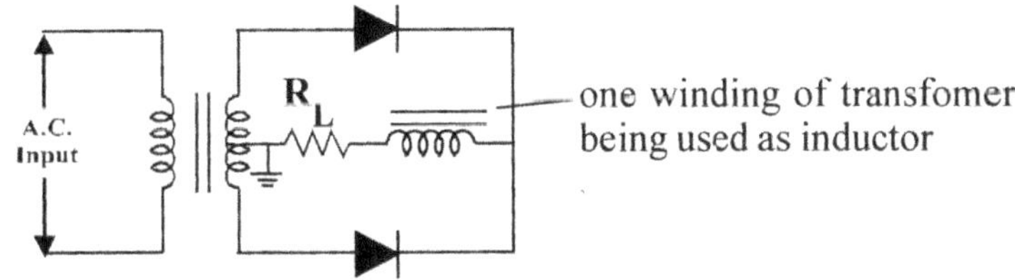

Fig 2.16 Inductor filter circuit.

One winding of transformer can be used as 'L'. For simplification, diode and choke (inductor) resistances can be neglected, compared with R_L. The D.C. component of the current is

$$I_m = \frac{V_m}{R_L}$$

Impedance due to L and R_L in Series $|Z| = \sqrt{R_L^{\,2} + (2\omega L)^2}$

For second Harmonic, the frequency is 2ω.

Therefore, A.C. Component of current $I_m = V_m / \sqrt{R_L{}^2 + 4\omega^2 L^2}$. So substituting these values in the expression, for current,

$$I = \frac{2I_m}{\pi} - \frac{4I_m}{3\pi} \cos 2\omega t$$

by inductor to higher order frequencies like $4\omega t$ etc., we get,

$$i = \frac{2V_m}{\pi R_L} - \frac{4V_m}{3\pi \sqrt{R_L{}^2 + 4\omega L^2}} \cos(2\omega t - \theta)$$

where θ is the angle by which the load current lags behind the voltage and is given by

$$\theta = \operatorname{Tan}^{-1} \frac{2\omega L}{R_L}.$$

2.7.1   RIPPLE FACTOR

$$\gamma = I_{r \cdot rms}/I_{D.C.} = I'_{rms}/I_{DC} \quad I_{DC} = 2V_m/\pi R_L$$

$$I_{rms} = \frac{I_m}{\sqrt{2}} = \frac{4V_m}{3\sqrt{2\pi}\sqrt{R_L^2 + 4\omega^2 L^2}}$$

$\therefore$ Ripple factor $= \dfrac{4V_m \times \pi R_L}{3\sqrt{2\pi}\sqrt{R_L^2 + 4\omega^2 L^2} \times 2V_m}$

$$= \frac{2}{3\sqrt{2}} \times \frac{1}{\sqrt{1 + \dfrac{4\omega^2 L^2}{R_L{}^2}}}$$

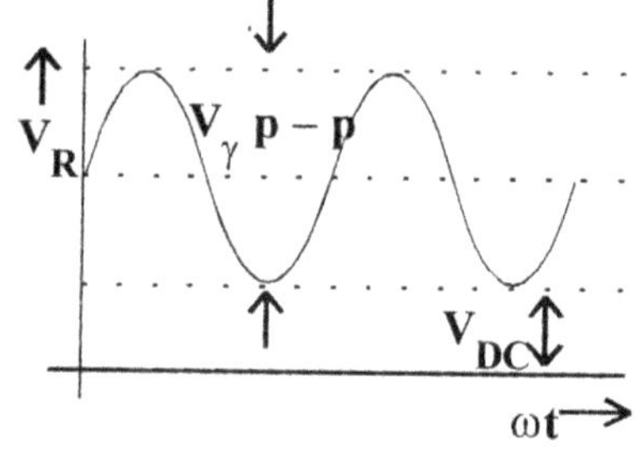

If $\dfrac{4\omega^2 L^2}{R_L{}^2} \gg 1$, then

$$\text{Ripple Factor } r = \frac{R_L \times 2}{3\sqrt{2} \times 2\omega L}$$

Fig 2.17 Ripple Voltage V_V

$\therefore$ $\boxed{\text{Ripple factor} = \dfrac{R_L}{3\sqrt{2}\,\omega L}}$ .... (2.5)

Therefore, for higher values of L, the ripple factor is low. If R_L is large, then also γ is high. Hence inductor filter should be used where the value of R_L is low. Suppose the output wave form from FWR supply is as shown in Fig. 2.17, then,

$V_{\gamma p - p}$ is the peak to peak value of the ripple voltage. Suppose

$$V_{DC} = 300V, \text{ and } V_{\gamma p - p} \text{ is } 10V.$$

then V_R maximum $= \dfrac{10}{2} = 5V$. Therefore, $V_\gamma rms = \dfrac{V_\gamma \max}{\sqrt{2}} = \dfrac{5}{\sqrt{2}} = 3.54V$

$\therefore$ Ripple Factor $\gamma = \dfrac{V_r rms}{V_x} = \dfrac{3.54}{300} = 0.0118.$

$$\% \text{ ripple} = \text{Ripple factor} \times 100\,\% = 1.18$$

2.7.2 REGULATION

$$V_{DC} = I_{DC}.R_L; \ I_{DC} \text{ for F W R is } \frac{2I_m}{\pi}.$$

$$V_{DC} = \frac{2I_m R_L}{\pi} = \frac{2V_m}{\pi}$$

$$I_m.R_L = V_m.$$

Therefore, V_{DC} is constant irrespective of R_L. But this is true if L is ideal. In practice

$$V_{DC} = \frac{2V_m}{\pi} - I_{DC}.R_f$$

where R_f is the resistance of diode.

An inductor stores magnetic energy when the current flowing through it is greater than the average value and releases this energy when the current is less than the average value.

Another formula that is used for Inductor Filter for

$$\boxed{\text{Ripple Factor} = 0.236 \ \frac{R_C + R_L}{wL}} \qquad \dots\dots (\ 2.6\ )$$

where R_L is load resistance R_c is the series resistance of the inductors. This is the same as the

above formula $\dfrac{R_L}{3\sqrt{2}wL}$

$$\therefore \qquad \frac{1}{3\sqrt{2}} = 0.236.$$

Problem 2.1

A FWR is used to supply power to a 2000Ω Load. Choke Inductors of 20 H inductance and capacitors of 16μf are available. Compute the ripple factor using **1**. One Inductor filter **2**. One capacitor filter **3**. Single L type filter.

Solution

1. One Inductor Filter :

$$I = \frac{2I_m}{\pi} - \frac{4I_m}{3\pi} \cos(2\omega t - \phi) \text{ where } \phi = \text{Tan}^{-1}\left(\frac{2\omega L}{R}\right)$$

$$I_{DC} = \frac{2V_m}{\pi R_L}$$

$$V = \frac{I \text{ ripple}}{I_{DC}} = \frac{R_L}{3\sqrt{2}\omega_L} = \frac{100}{3 \times 1.414 \times 2} = 0.074$$

2. Capacitor filter :

The ripple voltage for a capacitor filter is of Triangular waveform approximately. The rms

value for Triangular wave is $\dfrac{E_d}{2\sqrt{3}}$ or $\dfrac{V_{rms}}{2\sqrt{3}}$.

$$\therefore \qquad E_{rms} = \frac{E_d}{2\sqrt{3}}$$

Suppose the discharging of the capacitor will cut one for one full half cycle .then the change last by the capacitor

$$Q = \omega t = I_{DC} \times \frac{1}{2f}.$$

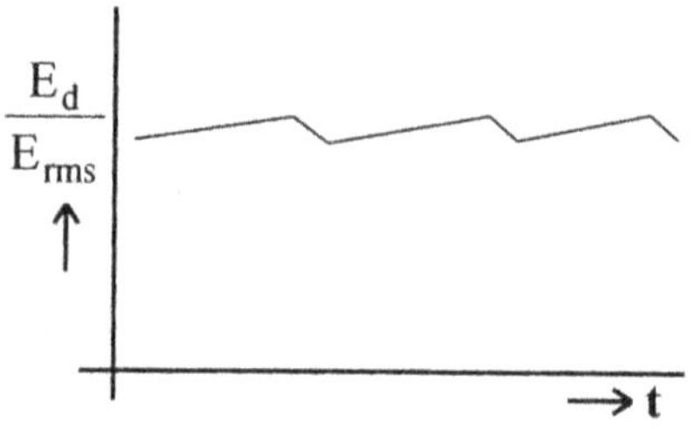

$\therefore$ $\quad$ voltage $= \dfrac{Q}{c}$

$\therefore$ $\quad E_d = \dfrac{I_{DC}}{2fc}$

Fig 2.18 V_0 of Capacitor filter.

$$\gamma = \frac{I_{DC}}{2fc \times 2\sqrt{3} \times I_{DC}} = \frac{1}{4fc\sqrt{3}R_L} = \frac{1}{4 \times \sqrt{3} \times 50 \times 16 \times 10^{-6} \times 2000} = 0.09$$

3. L Type factor :

$$\gamma = \frac{\sqrt{2}}{3} \times \frac{1}{4\omega^2 LC} = 0.0037$$

Problem 2.2

A diode whose internal resistance is 20 Ω is to supply power to a 1000Ω load from a 110V ((rms) source of supply. Calculate **(a)** The peak load current. **(b)** The DC load current **(c)** AC Load Current **(d)** The DC diode voltage. **(e)** The total input power to the circuit. **(f)** % regulation from no load to the given load.

Solution

Since only one diode is being used, it is for HWR Circuit.

(a) $\qquad I_m = \dfrac{V_m}{R_f + R_L} = \dfrac{110\sqrt{2}}{1020} = 152.5\text{mA.}$

(b) $\qquad I_{DC} = \dfrac{I_m}{\pi} = \dfrac{152.5}{\pi} = 48.5\text{mA.}$

(c) $\qquad I_{rms} = \dfrac{I_m}{2} = \dfrac{152.5}{2} = 76.2\text{mA.}$

(d) $\qquad V_{dc} = \dfrac{-I_m.R_L}{\pi} = -48.5 \times 1 = -48.5\text{V.}$

(e) $\qquad P_i = I_{rms} \times (R_f + R_L) = 5.92\Omega.$

(f) $\qquad$ % regulation $= \dfrac{V_{NL} - V_{FL}}{V_{FL}} \times 100\% = \dfrac{\dfrac{V_m}{\pi} - I_{DC}.R_L}{I_{DC}.R_L} = 2.06\%$

Problem 2.3

Show that the maximum DC output power $P_{DC} = V_{DC} I_{DC}$ in a half wave single phase circuit occurs when the load resistance equals the diode resistance R_f.

Solution

$$P_{DC} = I_{DC}^2 \cdot R_L = \frac{V_m^2 \cdot R_L}{\pi^2 (R_F + R_L)}$$

when R_L is very large, $\quad V_{DC} \cong \frac{V_m}{\pi}; \quad I_{DC} = \frac{V_{DC}}{R_L} = \frac{V_m}{\pi \cdot R_L}$

for P_{DC} to be maximum, $\quad \dfrac{dp_{DC}}{dR_L} = 0.$

or

$$\frac{V_m^2}{\pi^2}\left(\frac{(R_F + R_L)^2 - (R_F + R_L).R_L}{(R + R_L)^4} \right) = 0.$$

$$R_F + R_L = 2R_L \text{ or } R_L = R_F.$$

Therefore, P_{DC} is maximum when R_L is equal to R_F.

Problem 2.4

A 1mA meter whose resistance is 10Ω is calibrated to read rms volts when used in a bridge circuit with semiconductor diodes. The effective resistance of each element may be considered to be zero in the forward direction and ∞ in the reverse direction. The sinusoidal input voltage is applied in series with a 5 - KΩ resistance. What is the full scale reading of this meter?

Solution

$$I_{DC} = \frac{2I_m}{\pi} \text{ for bridge rectifier circuit.}$$

$$I_m = \frac{V_m}{R_L};$$

$$V_m = \sqrt{2}V_{rms}$$

$\therefore \qquad I_{DC} = \dfrac{2\sqrt{2} \times V_{rms}}{\pi R_L};$

$$R_L = 5K\Omega + 10\ \Omega = (5000 + 10)\Omega = 5010\ \Omega$$

$$1mA = \frac{2\sqrt{2} \times V_{rms}}{\pi \times 5010.}$$

$\therefore \qquad V_{rms} = \dfrac{1 \times 10^{-3} \times \pi \times 5010}{2\sqrt{2}} = 5.56V.$

2.8 CAPACITOR FILTER

Here X_c should be smaller than R_L. Because, current should pass through C and C should get charged. If C value is very small, X_C will be large and hence current flows through R only (Fig. 2.19) and no filtering action takes place. During +ve half cycle for a H W R circuit, with C filter, C gets charged when the diode is conducting and gets discharged (when the diode is not conducting) through R_L. When the input voltage $e = E_m \sin \omega t$ is greater than the capacitor voltage, C gets charged. When the input voltage is less than that of the capacitor voltage, C will discharged through R_L. The stored energy in the capacitor maintains the load voltage at a high value for a long period. The diode conducts only for a short interval of high current. The waveforms are as shown in Fig. 2.20. Capacitor opposes sudden fluctuations in voltage across it. So the ripple voltage is minimised.

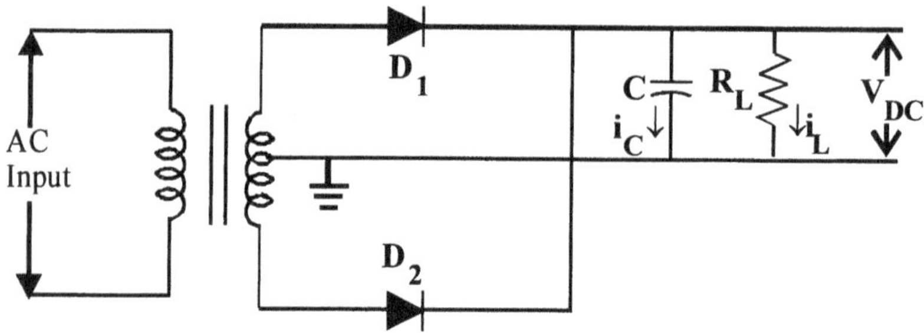

Fig 2.19 FWR Circuit with C filter.

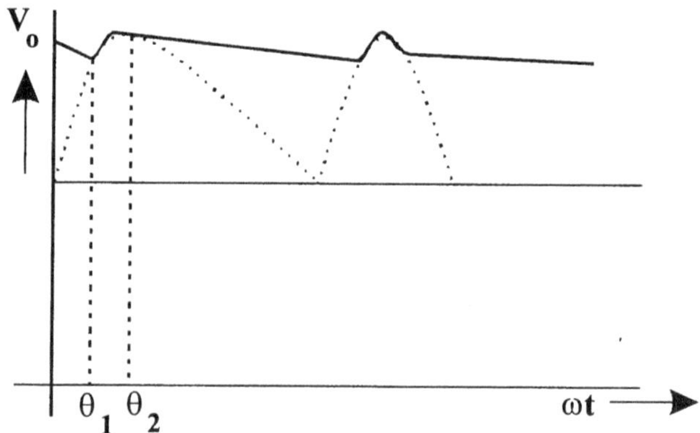

Fig 2.20 Output of capacitor filter circuit.

So the voltage across R_L is more or less constant and hence ripple is reduced, as shown in the graph, the voltage V_o is closer to DC wave form. At θ_1 'C' gets charged and at θ_2 C starts discharging.

The point at which diode starts conducting and C gets charged is known as **cut in point** θ_1. The point at which the diode stops conducting or the capacitor gets discharged is known as *Cut Out Point* θ_2.

C will not start charging at π itself. Therefore, even though the second diode is forward biased, e_c is almost $\simeq V_m$. So the first diode conducts from θ_1 to θ_2 and the second diode conducts from $\pi + \theta_1$ to $\pi + \theta_2$. Diode conducts only for a short of time when input current is high and charges C to the peak voltage.

Considering one diode, $i_B = i_C + i_R$ (Fig. 2.19)

$$i_C = C.\frac{dV}{dt} = C.\frac{de_c}{d_t}.$$

$$i_R = \frac{e_c}{R} \qquad \text{where } e_c \text{ is the voltage to which c gets charged.}$$

$$i_B = C.\frac{de_c}{d_t} + \frac{e_c}{R}.$$

But $\qquad e_c = E_m \, Sin \, \omega t.$

Therefore, at that point, diode conducts and C gets charged.

$$\theta_1 = \omega t_1.$$

$$\therefore \qquad \frac{de_c}{dt} = \omega E_m \cdot Cos \, \omega t.$$

General expression for

$$i_B = \omega C \, E_m \, \cos \omega t + \frac{E_m}{R} \, Sin \, \omega t \qquad \theta_1 < \omega t < \theta_2.$$

By some mathematical manipulation, i_b can be written as

$$i_B = \frac{E_m}{R} \sqrt{1 + \omega^2 R^2 C^2} \, Sin(wt + \phi)$$

where $\qquad \phi = Tan^{-1}(\omega RC)$

At an ω instant θ_2, when the capacitor fully discharges, $i_B = 0$

$$\therefore \qquad \omega C \, E_m \, \cos \theta_2 + \frac{E_m}{R} Sin\theta_2 = 0.$$

or $\qquad \theta_2 = tan^{-1} (-\Omega RC)$

Thus the value of conduction angle depends upon the values of R and C, and also the value of i_B. During the non conducting interval, the capacitor discharges into the load R supplying load current.

$$- i_C = i_R.$$

The circuit equation is $-C \times \dfrac{de_c}{d_t} = \dfrac{e_c}{R}$

or $\qquad e_C = A \, e^{-t/R_L}$

At $\quad \omega t = \theta_2 \qquad e_C = E_m \, sin \, \theta_2.$

The final expression for E_{DC} seems a little complicated. So a simplified expression which is widely used is $E_{DC} = (E_m - \dfrac{E_R}{2} \cdot)$, where E_R is the peak to peak ripple voltage. The expression for Ripple Factor,

$$\boxed{\gamma = \frac{E'_{AC}}{E_{DC}} = \frac{\pi + (\theta_1 - \theta_2)}{2\sqrt{3}\omega RC}.}$$

where $(\theta_1 - \theta_2)$ is the angle of conduction θ_C. For large ωRC (typical value 100), $\theta_c = 14.6^\circ$. The larger the value of R, the smaller the γ.

2.8.1 RIPPLE FACTOR FOR C FILTER

Initially, the capacitor charges to the peak value V_m when the diode D_1 is conducting. When the capacitor voltage is equal to the input voltage, the diode stops conducting or the current through the diode D_1 is zero. Now the capacitor starts discharging through R_L depending upon the time constant $C.R_L$. Therefore, R_L value is large. Rate of charging is different from rate of discharging. When the voltage across the capacitor falls below the input voltage and when the diode D_2 is forward biased, the capacitor will again charge to the peak value Vm and the current through D_2 becomes zero when $V_c = V_{in}$. Thus the diodes D_1 and D_2 conduct for a very short period θ_1 to θ_2 and $\pi + \theta_1$ to $\pi + \theta_2$ respectively (Fig. 2.21).

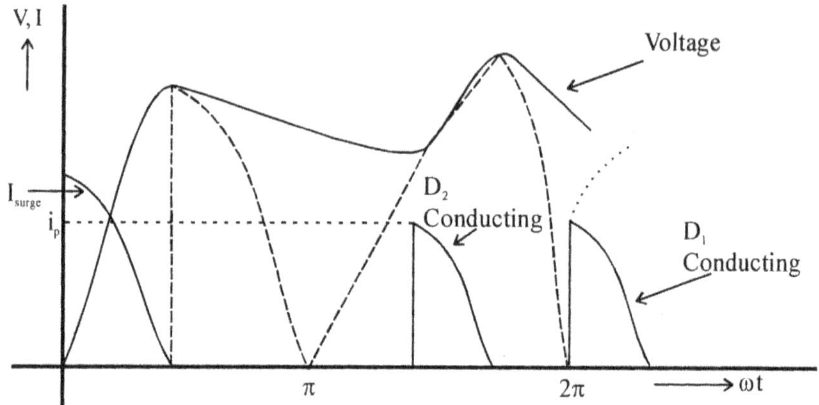

Fig 2.21 Output waveforms of a capacitor filter.

T_2 = Time period of the discharging of the capacitor

T_1 = Time period of the charging of the capacitor

The amount of charge lost by the capacitor when it is discharging = $I_{DC} \cdot T_2$. Because, I_{DC} is the average value of the capacitor discharge current. This charge is replaced during a short interval T_1 during which the voltage across the capacitor changes by an amount = peak to peak voltage of the ripple $V'_{r_{p-p}}$.

$$Q = V \times C.$$

$$\therefore \quad Q \text{ charge } = V'_{p-p} \times C$$

$$Q \text{ charge } = Q \text{ Discharge.}$$

$$\therefore \quad V'_{p-p} \times C = I_{DC} \times T_2.$$

$$\text{or} \quad V'_{p-p} = \frac{I_{DC} \times T_2}{C}$$

The output waveform can be assumed to be a triangular wave $T_2 \gg T_1 (T_1 + T_2) = T/2$ when the diode is conducting (0 to π). ($T_1 + T_2$ corresponds to one half cycle).

$$\therefore \quad T_2 = \frac{T}{2} = \frac{1}{2f} \therefore T_2 = \frac{1}{2f}.$$

$$\therefore \quad V' \text{ p-p} = \frac{I_{DC} \times 1}{C2f} \therefore T_2 = \frac{1}{2f}. \quad \text{for the triangular wave, form factor} = \frac{1}{\sqrt{3}}.$$

$$\therefore \quad V'_{rms} = \frac{V'_{p-p}}{2\sqrt{3}}$$

$$I_{DC} = \frac{V_{DC}}{R_L}.$$

$$V' \text{ rms} = \frac{V_{DC}}{4\sqrt{3}fCR_L}.$$

Ripple Factor $\qquad \gamma = \dfrac{v'_{rms}}{V_{DC}} = \dfrac{1}{4\sqrt{3}fCR_L}$.

Therefore, Ripple factor may be decreased by increasing C or R_L or both.

Expression for V_{DC} :

$$V_{DC} = V_m - \frac{V'p-p}{2} \; ; \quad V'\text{P-P} = \frac{I_{DC}}{2fc}.I_{DC} = \frac{V_{DC}}{R_L}$$

$$V_{DC} = V_m - \frac{V_{DC}}{4fC.R_L}$$

or $\qquad V_{DC}\left(1 + \dfrac{1}{4fcR_L}\right) = V_m$

$\therefore \qquad V_{DC} = V_m\left(\dfrac{4fR_LC}{1+4fR_LC}\right)$

2.9   LC FILTER

In an inductor filter, ripple decreases with increase in R_L. In a capacitor filter, ripple increases with increase in R_L. A combination of these two into a choke *input or L-Section Filter* should then make the ripple independent of load resistance.

If L is small, the capacitor will be charged to V_m, the diodes will be cut off allowing a short pulse of current. As L is increases, the pulses of current are smoothed and made to flow for a larger period, but at reduced amplitude. But for a critical value of inductance L_c, either one diode or the other will be always conducting, with the result that the input voltage V and I to the filter are full wave rectified sine waves.

The graph of DC output voltage for LC Filter is as shown in Fig. 2.22.

$$V_{DC} = \frac{2V_m}{\pi}$$

for Full Wave Rectifier Circuit. considering ideal elements, conduction angle is increased when inductor is placed because there is some drop across L. So C cannot charge to V_m. Therefore, the diode will be forward biased for a longer period.

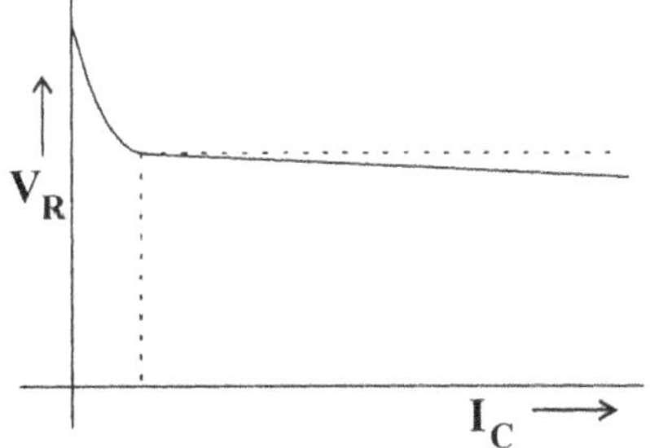

Fig 2.22   LC filter characteristic.

The ripple circuit which passes through L, is not allowed to develop much ripple voltage across R_L, if X_c at ripple frequency is small compared to R_L. Because the current will pass through C only. Since X_c is small and Capacitor will get charged to a constant voltage. So V_o across R_L will not vary or ripple will not be there. Since for a properly designed LC Filter,

$$X_c \ll R_L$$

and $\qquad X_L \gg X_c \qquad$ at $\omega = 2\pi f$

X_L should be greater than X_c because, all the AC should be dropped across X_L itself so that AC Voltage across C is nil and hence ripple is low.

2.9.1   RIPPLE FACTOR IN LC FILTER

AC Component of current through L is determined by X_L.

$$X_L = 2\omega L$$

RMS value of ripple current for Full Wave Rectifier with L Filter is,

$$I'\,rms = \frac{4V_m}{3\pi\sqrt{2}X_L} = \text{RMS Value of Ripple Current for L Filter}$$

$$I'\,rms = \frac{2}{3\sqrt{2}X_L} = \frac{2V_m}{\pi}$$

$$\therefore \qquad I'\,rms = \frac{2}{3\sqrt{2}X_L}\cdot V_{DC} = \frac{\sqrt{2}V_{DC}}{3X_L}$$

The ripple voltage in the output is developed by the ripple current flowing through X_C.

$$\therefore \qquad V'\,rms = I'\,rms\,.\,X_C = \frac{\sqrt{2}}{3}\cdot\frac{X_C}{X_L}\cdot V_{DC}$$

$$\text{Ripple factor} = \frac{V'rms}{V_{DC}}$$

$$\gamma = \frac{\sqrt{2}.X_C}{3X_L}$$

$$X_C = \frac{1}{j2\omega fc}\,.\,X_L = jL\omega L = 2\,\pi fL$$

$$\boxed{\gamma = \frac{\sqrt{2}}{3}\cdot\frac{1}{2\omega C}\cdot\frac{1}{2\omega L}}$$

$X_L = 2wL$ since ripple is being considered at a frequency, twice the line frequency.

If $\qquad\boxed{f = 60\text{ Hz,}}\quad \gamma = \dfrac{0.83}{LC}$

2.9.2 BLEEDER RESISTANCE

For the LC filter, the graph between I_{DC} and V_{DC} is as shown in Fig. 2.23. For light loads, i.e., when the load current is small, the capacitor gets charged to the peak value and so the no load voltage is $2V_m/\pi$. As the load resistance is decreased, I_{DC} increases so the drop across other elements V_{iz} diodes and choke increases and so the average voltage across the capacitor will be less than the peak value of $2V_m/\pi$. The out put voltage remains constant beyond a certain point I_B and so the regulation will be good. The voltage V_{DC} remains constant even if I_{DC} increases, because, the capacitor gets charged every time to a value just below the peak voltage, even though the drop across diode and choke increases. So for currents above I_B, the filter acts more like an inductor filter than C filter and so the regulation is good. Therefore, for LC filters, the load is chosen such that, the $I_{DC} \geq I_B$. The corresponding resistance is known as **Bleeder Resistance** (R_B). So Bleeder Resistance is the value for which $I_{DC} \geq I_B$ and good regulation is obtained, and conduction angle is 180^0. When $R_L = R_B$, the conduction angle $= 180^0$, and the current is continuous. So just as we have determined the critical inductance L_C, when $R_L = R_B$, the Bleeder Resistance, I_{DC} = the negative peak of the second harmonic term.

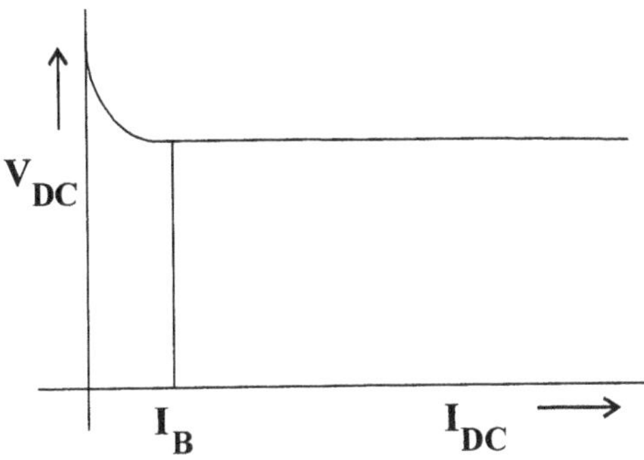

Fig 2.23 LC filter characteristic.

$$\therefore \qquad I_{DC} = \frac{2V_m}{\pi R_B}$$

$$I' \text{ peak} = \frac{4V_m}{3\pi} \cdot \frac{1}{XL}.$$

$$\therefore \qquad \frac{2V_m}{\pi R_B} = \frac{4V_m}{3\pi} \cdot \frac{1}{X_L}$$

$$R_B = \frac{3.X_L}{2}; \qquad R_B \text{ is Bleeder Resistance.}$$

I_{DC} should be equal to or greater than I peak, since, X_L determines the peak value of the ripple component. If X_L is large I' peak can be $\leq I_{DC}$ and so the ripple is negligible or pure D.C. is obtained, where $X_L = 2\omega L$ corresponding to the second harmonic. Therefore, R should be at least equal to this value of R_B, the Bleeder Resistor.

2.9.3 SWINGING CHOKE

The value of the inductance in the LC circuits should be $> L_C$ the minimum value so that conduction angle of the diodes is 180^0 and ripple is reduced. But if the current I_{DC} is large, then the inductance for air cored inductors as $L = N\phi/I$ (flux linkages per ampere of current), as I increases, L decreases and may become below the critical inductance value L_C. Therefore, Iron cored inductors or chokes are chosen for filters such that the value of L varies within certain limits and when I is large, the core saturates, and the inductance value will not be $< L_C$. Such chokes, whose inductance varies with current within permissible limits, are called *Swinging Chokes*. (In general for inductors ϕ/I remains constant so that L is constant for any value of current I) This can be avoided by choosing very large value of L so that, even if current is large, Inductance is large enough $> L_C$ But this increases the cost of the choke. Therefore, swinging choke are used.

$$L_C \geq R_L / 3\omega$$
$$R_L = \text{ Load Resistance}$$
$$L_C = \text{ Critical Inductance}$$
$$\omega = 2\pi f$$

At no load $R_L = \infty$. Therefore, L should be ∞ which is not possible. Therefore, Bleeder Resistance of value $= 3X_L / 2$ is connected in parallel with R_L, so that even when R_L is ∞, the conduction angle is 180^0 for each diode.

The inductance of an iron - cored inductor depends up on the D.C. Current flowing through it, L is high at low currents and low at high currents. Thus its L varies or swings within certain limits. This is known as *Swinging Choke*.

Typical values are

$$L = 30 \text{ Henrys at } I = 20\text{mA} \text{ and}$$

$$L = 4 \text{ Henrys at } I   = 100\text{mA}.$$

Problem 2.5

Design a full wave rectifier with an LC filter (single section) to provide 9V DC at 100mA with a maximum ripple of 2%. Line frequency f = 60 HZ.

Solution

$$\text{Ripple factor } \gamma = \frac{\sqrt{3}}{2} \times \frac{1}{2\omega C} \times \frac{1}{2\omega L} = \frac{0.83}{LC} \mu$$

$$\therefore \qquad 0.02 = \frac{0.83}{LC} \text{ or} LC = \frac{0.83}{0.02} = 42.\mu$$

$$R_L = \frac{V_{DC}}{I_{DC}} = \frac{9V}{0.1} = 90\Omega.$$

$$LC \geq \frac{R_L}{3\omega} \geq \frac{R_L}{1130}.$$

But LC should be 25% larger. $\therefore$ for f = 60 Hz, the value of LC should be $\geq \dfrac{R_L}{900}$.

$$LC \geq \frac{R_L}{900} \cdot \geq \frac{90}{900} = 0.1 \text{ Henry.}$$

If L = 0.1H, then $C = \dfrac{42}{0.1} = 420\mu f.$ This is high value

If L = 1H, then $C = 42 \ \mu f.$

If the series resistance of L is assumed to be 50 Ω, the drop across L is

$$I_{DC} \times R = 0.1 \times 50 = 5V.$$

Transformer Rating :

$$V_{DC} = 9V + 5V = 14V$$

$$\therefore \qquad V_m = \frac{\pi}{2}(9+5) = 22V$$

$$\therefore \qquad \text{rms value is } \frac{22}{\sqrt{2}} = 15.5V$$

Therefore, a 15.5 – 0 –15.5 V, 100mA transformer is required. PIV of the diodes is $2V_m$ Because it is FWR Circuit.

$$\therefore \qquad PIV = 44V$$

So, diodes with 44V, 100mA ratings are required.

Problem 2.6

In a FWR with C filter circuit, V_rp-p = 0.8v and the maximum voltage is 8.8 volts (Fig. 2.24). $R_L = 100\Omega$ and C = 1050 µf. Power line frequency = 60 Hz. Determine the Ripple Factor and DC output voltage from the graph and compare with calculated values.

Solution

$$V_r' \text{ p-p} = 0.8V.$$

$$\therefore \quad V_{rms}' = \frac{0.8}{2\sqrt{3}} = 0.231V$$

$$V_{DC} = Vm - \frac{V_r'p-p}{2} = 8.8 - \frac{0.8}{2} = 8.4V.$$

$$\therefore \quad \gamma = \frac{V'rms}{V_{DC}} = \frac{0.2}{8.4} = 0.0238 \text{ or } 2.38\%$$

$$\text{Theoretical values, } \gamma = \frac{1}{4\sqrt{3}fR_L.C}$$

$$= \frac{1}{4\sqrt{3 \times 60 \times 100 \times 1050\mu f}}$$

$$= 0.057 \text{ or } 5.75\%$$

$$V_{DC} = \frac{4fR_L C}{1 + 4fR_L C} \times V_m = 8.46V$$

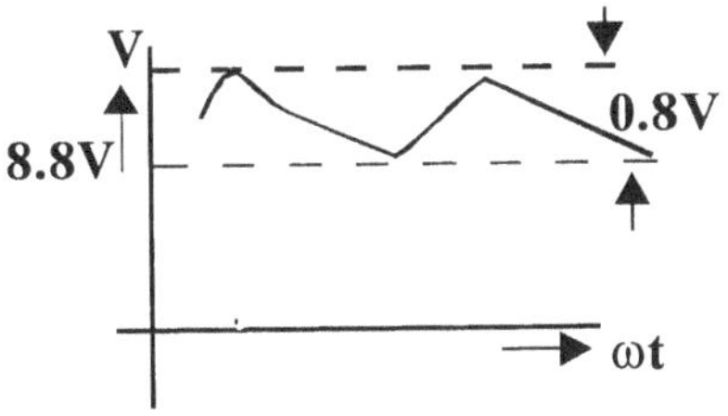

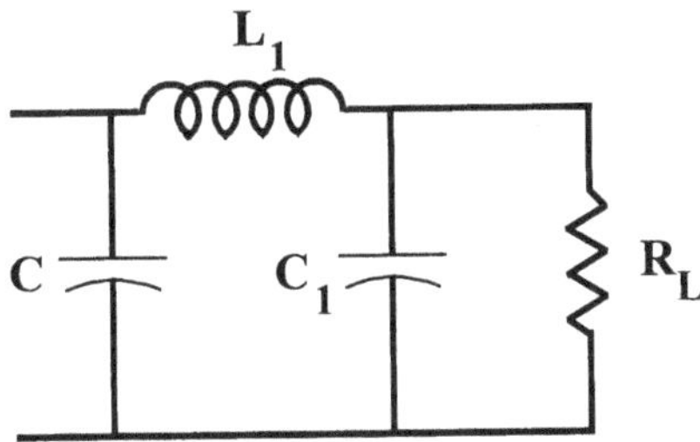

Fig 2.24 For Problem 2.6.

2.10 CLC OR π FILTER

In the LC filter or L section filter, there is some voltage drop across L. If this cannot be tolerated and more D.C. output voltage V_0 is desired, p filter or CLC filter is to be used. The ripple factor will be the same as that of L section filter, but the regulation will be poor. It can be regarded as a L section filter with L_1 and C_1, before which there is a capacitor C,i.e. the capacitor filter. The input to the L section filter is the output of the capacitor filter C. The output of capacitor filter will be a triangular wave superimposed over D.C.

Now the output V_0 is the voltage across the input capacitor C less by the drop across L_1. The ripple contained in the output of 'C' filter is reduced by the L section filter L_1C_1.

Fig 2.25 CLC or π Section Filter

2.10.1 π Section Filter with a Resistor Replacing the Inductor

Consider the circuit shown here. It is a π filter with L replaced by R. If the resistor R is chosen equal to the reactance of L (the impedance in Ω should be the same), the ripple remains unchanged.

V_{DC} for a single capacitor filter = $(V_m - I_{DC}/4fC)$ If you consider C_1 and R, the out put across C_1 is $(V_m - I_{DC}/4fC)$. There is some drop across R. Therefore, the net output is

$$V_m - \frac{I_{DC}}{4fC} - I_{DC} \times R$$

The Ripple Factor for π sector is

$$\gamma = \frac{\sqrt{2}X_c X_{c_1}}{R_L.X_{L_1}}. \qquad \qquad \dots\dots\dots (2.7)$$

So by this, saving in the cost, weight and space of the choke are made. But this is practical only for low current power supplies.

Suppose a FWR output current is 100 mA and a 20 H choke is being used in π section filter. If this curve is to be replaced by a resistor, $X_L = R$. Taking the ripple frequency as $2f = 100$ Hz,

$$X_L = \omega L = 2\pi(2f).L = 4\pi fL$$
$$= 4 \times 3.14 \times 50 \times 20 = 12560\,\Omega$$
$$\simeq 12\ K\Omega$$

Voltage drop across R = $12,000 \times 0.1 = 1200$V, which is very large.

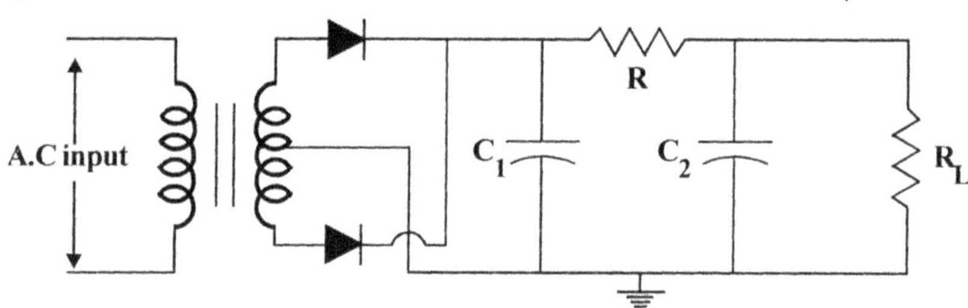

Fig 2.26 **FWR Circuit π Section / CRC Filter.**

2.10.2 Expression for Ripple Factor For π Filter

The output of the capacitor in the case of a capacitor filter is a Triangular wave. So assuming the output V_0 across C to be a Triangular wave, it can be represented by Fourier series as

$$V = V_{DC} - \frac{V'P - P}{\pi}(\text{Sin } 2\omega t - \frac{\sin 4\omega t}{2} + \frac{\sin 6\omega t}{3}\dots)$$

But the ripple voltage peak to peak V' p-p = $\frac{I_{DC}}{2fc}$·(for Δ wave)

Neglecting 4th and higher harmonic, the rms voltage of the Second Harmonic Ripple is

$$\frac{v'p - p}{\pi\sqrt{2}} = \frac{I_{DC}}{2\sqrt{2}\pi fc} = \frac{I_{DC} \times 2}{4\sqrt{2}\pi fc} = \frac{\sqrt{2}I_{DC}}{4\pi fc}$$
$$\frac{1}{4\pi fc} = x_c$$

It is the capacitive reactance corresponding to the second harmonic.

$$V'_{rms} = \sqrt{2}\,I_{DC}X_C. \qquad \text{(across 'C' filter only)}$$

The output of C filter is the input for the L section filter of L_1C_1. Therefore, as we have done in the case of L section filter,

$$\text{Current through the inductor (harmonic component)} = \frac{\sqrt{2}\,I_{DC}X_C}{X_{L_1}}$$

Output Voltage = Current (X_C) $(V_o = I.X_C)$

$$V'_{rms} = \frac{\sqrt{2}\,I_{DC}X_C}{X_{L_1}}.X_{C_1} \qquad \text{(after L section)}$$

Ripple factor, $$\gamma = \frac{V'_{rms}}{V_{DC}} = \frac{\left(\sqrt{2}.I_{DC}.X_C\right)}{V_{DC}}\frac{X_{C_1}}{X_{L_1}}$$

$$V_{DC} = I_{DC}.R_L$$

$$\therefore \qquad \gamma = \frac{\left(\sqrt{2}.X_C\right)}{R_L}.\left(\frac{X_{C_1}}{X_{L_1}}\right).$$

where all reactances are calculated at the 2nd harmonic frequency $\omega = 2\pi f$.

DC output voltage = (The DC voltage for a capacitor filter) – (The drop across L_1).

Problem 2.7

Design a power supply using a π-filter to give DC output of 25V at 100mA with a ripple factor not to exceed 0.01%. Design of the circuit means, we have to determine L,C, diodes and transformers.

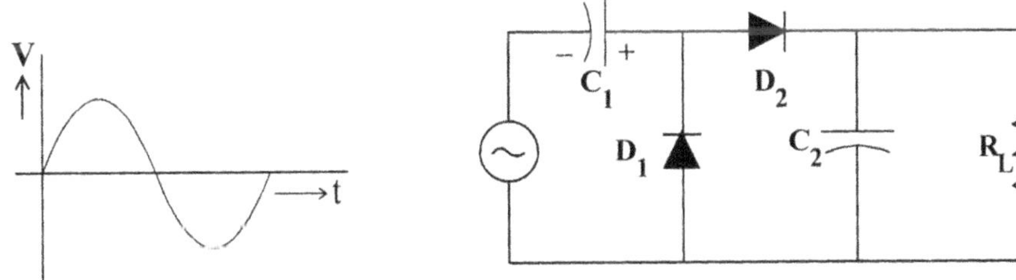

Fig 2.27 Peak to Peak detector.

Solution

Design of the circuit means, we have to determine L, C, Diodes and Transformer

$$R_L = \frac{V_{DC}}{I_{DC}} = \frac{25V}{I_{DC}} = \frac{25V}{100mA} = 250\Omega$$

Ripple factor $$\gamma = \sqrt{2}.\frac{X_c}{R_L}.\frac{x_{C1}}{X_{L1}}.$$

X_C can be chosen to be $= Xc_1$.

$$\therefore \qquad \gamma = \sqrt{2}.\frac{X_C^{\,2}}{R_L.X_{LI}.}$$

This gives a relation between C and L.

$$C^2 L = y$$

There is no unique solution to this.

Assume a reasonable value of L which is commercially available and   determine the corresponding value of capacitor. Suppose L is chosen as 20 H at 100mA with a D.C. Resistance of 375Ω (of Inductor).

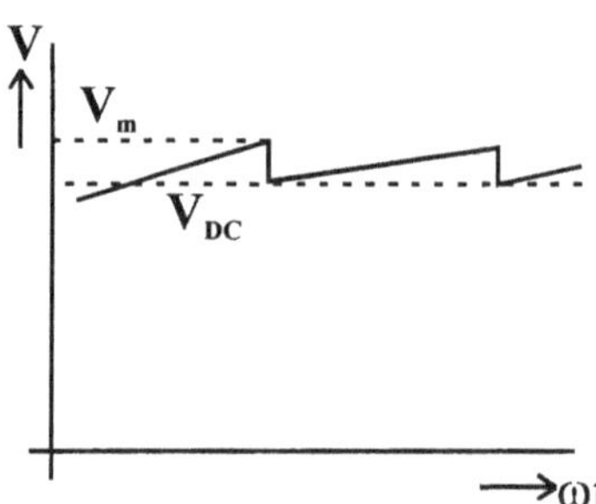

Fig 2.28   For Problem 2.7.

$$\therefore \qquad c^2 = \frac{y}{L}$$

$$\text{or} \qquad c = \sqrt{\frac{y}{L}}$$

$$V_{DC} = Vm - \frac{V_\gamma}{2}$$

$$V_\gamma = \frac{I_{DC}}{2fc}.$$

Now the transformer voltage ratings are to be chosen.

The voltage drop across the choke = choke resistance $\times I_{DC} = 375 \times 100 \times 10^{-3} = 37.5\text{V}$.

$$V_{DC} = 25\text{V}.$$

Therefore, voltage across the first capacitor C in the π-filter is

$$V_c = 25 + 37.5 = 62.5\text{V}.$$

The peak transformer voltage, to centre tap is

$$V_m = (V_C) + \frac{V_\gamma}{2} \text{ (for C filter)}$$

$$V_\gamma = \frac{I_{DC}}{2fc}.$$

$$\therefore \qquad V_m = 62.5 \text{ V} + \frac{0.1}{4 \times 50 \times c}$$

$$V_{rms} = \frac{V_m}{\sqrt{2}}. \cong 60\text{v}$$

Therefore, a transformer with 60 - 0 -60V is chosen . The ratings of the diode should be, current of 125mA. and voltage = PIV = $2V_m = 2 \times 84.6\text{V} = 169.2\text{V}$

Problem 2.8

A full wave rectifier with LC filter is to supply 250 v at 100m.a. D.C. Determine the ratings of the needed diodes and transformer, the value of the bleeder resistor and the ripple, if R_C of the choke = 400Ω. L = 10H and C = 20 μF.

Solution

$$R_L = \frac{V_{DC}}{I_{DC}}$$

$$R_L = \frac{250v}{0.1} = 2,500\Omega.$$

For the choke input resistor,

$$E_{DC} = \frac{2E_m}{\pi(1 + \dfrac{R_C}{R_L})}$$

and $\qquad I_{rms} \cong I_{DC}.$

$$E_m = \frac{\pi E_{DC}}{2}\left(1 + \frac{R_C}{R_L}\right) = \frac{\pi \times 250}{2}\left(1 + \frac{400}{2500}\right) = 455V$$

$$\therefore \qquad E_{rms} = \frac{455}{\sqrt{2}} = 322V$$

Therefore, the transformer should supply 322V rms on each side of the centre tap. This includes no allowance for transformer impedance, so that the transformer should be rated at about 340 volts 100 mA, D.C

$$\text{The Bleeder Resistance } R_B = \frac{3X_L}{2}$$

$$L = 10000\Omega.$$

$$I_B = \frac{2E_m}{3\pi\omega L} = \frac{2 \times 455}{3\pi \times 377 \times 10} = 0.0256A$$

Ripple factor $\qquad \gamma = \dfrac{0.47}{4\omega^2 LC - 1} = \dfrac{0.47}{4 \times 377^2 \times 20 \times 10^{-6}}$

The current ratings of each diode = 0.00413. Should be 50mA.

2.11 MULTIPLE LC FILTERS

Better filtering can be obtained by using two or more LC sections as shown in the Fig. 2.29.

Ripple factor $\qquad \gamma = \dfrac{\sqrt{2}}{3} \cdot \dfrac{XC_1}{XL_1} \cdot \dfrac{XC_2}{XL_2}.$

If we take $\qquad L = L_1 = L_2$

$\qquad\qquad\quad C = C_1 = C_2$ and f = 60 Hz.

$$\gamma = \frac{1.46}{L^2C^2}$$

Inductor value L in Henrys and Capacitor value C in μF(microfarads)

So ripple factor will be much smaller or better filtering is achieved.

$$\therefore \qquad I_1 = \frac{\sqrt{2}}{3} \times \frac{V_{DC}}{L_1} \text{ (RMS Value of the ripple component of current.)}$$

$$V_{A_2B_2} = I_1 \times C_1 \frac{\sqrt{2}}{3} \cdot \frac{V_{DC}}{X_{L_1}} \cdot XC_1$$

$$I_2 = \frac{V_{A_2B_2}}{X_{L_2}}$$

The value of XL_2 should be much larger than XC_2. Therefore, A.C. current should get dropped across C_2.

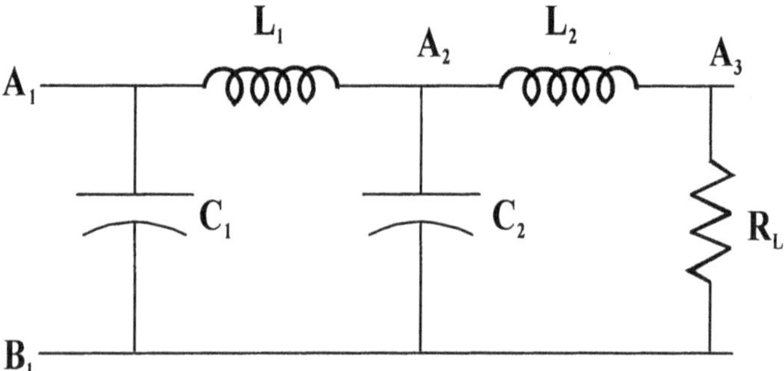

Fig 2.29 Multiple LC filter sections.

$$\therefore \qquad I_2 = V_{A_2B_2} / XL_2 \text{ and not } \frac{V_{A_2B_2}}{X_{L_2} + X_{L_2}}.$$

$$\therefore \qquad V_o \cong I_2 \, Xc_2 / R_L \text{ is in parallel with } C_2.$$

$$= \frac{\sqrt{2}}{3} \cdot V_{DC} \cdot \frac{X_{C_1}}{X_{L_1}} \cdot \frac{X_{C_2}}{X_{L_2}}$$

$$\text{Ripple factor} \quad \gamma = \frac{V_o}{V_{DC}} = \frac{\sqrt{2}}{3} \cdot \frac{X_{C_1}}{X_{L_1}} \cdot \frac{X_{C_2}}{X_{L_2}}$$

But the actual value of L_c should be 25% more than this because of the approximately made in the series used to represent V.

Problem 2.9

A full wave rectifier circuit with C-type capacitor filter is to supply a D.C. Current of 20 mA at 16V. If frequency is 50 Hz ripple allowed is 5% Calculate :

(a) *Required secondary voltage of the transformer.*

(b) *Ratio of I peak/I $_{max}$ through diodes.*

(c) *The value of C required.*

Solution

$$I_{DC} = 20mA \; V_{DC} = 16V.$$

$$\therefore \qquad R_L = \frac{V_{DC}}{I_{DC}} = \frac{16}{20mA} = 0.8K\Omega = 800\Omega.$$

$$\gamma = 5\% = \frac{1}{4\sqrt{3}fcR_L}$$

$$\therefore \quad 0.05 = \frac{1}{4\sqrt{3}\times 50 \times C \times 800}$$

$$\therefore \quad C = \frac{1}{4\times 1.732 \times 50 \times 0.05 \times 800} = \frac{1}{4\sqrt{3}\times 2.5 \times 800} = \frac{1}{4\sqrt{3}\times 2000}$$

$$C = \frac{\sqrt{3}}{8000 \times 3} = 72\mu F$$

V_{DC} with load

$$I_{DC} = \frac{V_m}{1+\dfrac{1}{4fcR_L}} \quad V_m = ?$$

$$V_{DC} = V_m \left(\frac{4fCR_L}{1+4fCR_L}\right)$$

$$V_{DC} = V_m \left(\frac{L_1 f_C R_L}{1+L_1 f_C R_L}\right)$$

$$V_{DC} = 16V \text{ at } I_{DC} = 20 \text{ mA.}$$

$$V_{DC} = V_m \left(\frac{0.05\sqrt{3}}{1+005\sqrt{3}}\right) 16$$

$$V_{DC} = V_m \left(\frac{11.52}{1+11.52}\right)$$

$$16 = V_m \left(\frac{11.52}{12.52}\right)$$

$$V_m = 16 \times \frac{12.52}{11.52} = 17.38V$$

Problem 2.10

A half wave rectifier allows current to flow from time $t_1 = \pi/6\omega$ to $t_2 = 5\pi/6\omega$, the A.C. Voltage being given by $V_s = 100 \sin \omega t$. Calculate the RMS and DC output voltage.

Solution

$$V_{DC} = \frac{1}{2\pi}\int_{\pi/6}^{5\pi/6} 100 \sin \omega t \cdot d\omega t$$

$$= \frac{1}{2\pi}\times 100 \times \left[-\cos \omega t\right]_{\pi/6}^{5\pi/6} = \frac{100}{2\pi}\left[-\cos\frac{5\pi}{6}+\cos\frac{\pi}{6}\right]$$

$$= \frac{100}{2\pi}\left[-\cos(150^0)+\cos(30^0)\right] = \frac{50}{\pi}\left[+\frac{\sqrt{3}}{2}+\frac{\sqrt{3}}{2}\right]$$

$$= \frac{50\times\sqrt{3}}{3.14} = \frac{50\times 1.732}{3.14} = 27.6V$$

$$V_{rms} = \sqrt{\frac{1}{2\pi}\int_{\pi/6}^{5\pi/6}(100\cdot\sin\omega t)^2\cdot d\omega t} = 48.5V.$$

Problem 2.11

A full wave rectifier operating at 50 Hz is to provide D.C. Current of 50mA at 30V. with a 80μF. C-type filter system used, calculate

  (a) V_m the peak secondary voltage of transformer.

  (b) Ratio of surge to mean currents of diodes.

  (c) The ripple factor of the output.

Solution

(a)
$$R_L = \frac{V_{dc}}{I_{dc}} = \frac{30\,V}{50mA} = 600\,\Omega.$$

$$V_{DC} = \left(\frac{V_m}{1 + \dfrac{I_{DC}}{4fcV_{DC}}}\right) = \frac{V_m}{1 + \dfrac{1}{4fcR_L}}$$

$$\frac{I_{DC}}{V_{DC}} = \frac{1}{R_L}.$$

$$V_m = V_{DC} + \frac{V_{DC}}{4fcR_L} = V_{DC} + \frac{I_{DC}}{4fc}$$

$$V_m = 30 + \frac{50 \times 10^{-3}}{4 \times 50 \times 80 \times 10^{-6}} = 33.125\,V.$$

$$V_{rms} = \frac{V_m}{\sqrt{2}} = \frac{33.125}{\sqrt{2}} = 23.4\,V$$

(b) Surge Current through the diode, I_{diode} surge is I_m.

$$\frac{V_m}{X_c} = V_m\omega C = 33.125 \times 2\,\pi \times 50 \times 80 \times 10^{-6} = 0.833\,A$$

$$I_{DC} = 50\ mA$$

   $= I_{diode\ mean}$ is the average or mean current through the diode.

$$\frac{I_{diode\ swing}}{I_{diode\ mean}} = \frac{0.833}{50 \times 10^{-3}} = 16.65.$$

(c)
$$R_L = \frac{V_{DC}}{I_{DC}} = 600\,\Omega.$$

$$\gamma = \frac{1}{4\sqrt{3}fcR_L} = 0.06.$$

Problem 2.12

For a FWR circuit, the A.C. Voltage input to transformer primary = 115V. Transformer secondary voltage is 50V. R_L = 25Ω. Determine

1. *Peak DC component, RMS and AC component of load voltage.*

2. *Peak DC component, RMS and AC component of load current.*

Solution

Since, the transformer develops 50V between secondary terminals, there must be 25V across each half of the secondary winding.

The peak value across each half of secondary is $25\sqrt{2}$ = 35.4V

Assuming ideal diodes, the rectified voltage across R_L also has a peak value of 35.4V(V_m).

$\therefore$ Average value V_{DC} for FWR circuit is

$$= \frac{2V_m}{\pi} = 0.636V_m$$

$$V_{DC} = 0.636 \times (35.4) = 22.5V.$$

A.C.Component load voltage is

$$V'_{rms} = \sqrt{V_m{}^2 - V_{DC}{}^2}$$

$$= \sqrt{(25)^2 - (22.5)^2} = 10.9V.$$

$$I_m = \frac{V_m}{R_L} = \frac{35.4}{25} = 1.41A.$$

$$I_{DC} = \frac{2I_m}{\pi} = 0.636I_m = 0.636(1.41) = 0.897A.$$

$$I_L(rms) = 0.707\, I_m$$

$$= 1.41(0.707) = 1A$$

$$I'_{rms} = \sqrt{(I_{rms})^2 - (I_{DC})^2}$$

$$= \sqrt{1^2 - (0.897)^2}$$

$$= 0.441\ A.$$

2.12 BREAK DOWN MECHANISM

There are three types of breakdown mechanisms in semiconductor devices.

    1. Avalanche Breakdown   **2.** Zener Breakdown   **3.** Thermal Breakdown

2.12.1 AVALANCHE BREAKDOWN

When there is no bias applied to the diode, there are certain number of thermally generated carriers. When bias is applied, electrons and holes acquire sufficient energy from the applied potential to produce new carriers by removing valence electrons from their bonds. These thermally generated carriers acquire additional energy from the applied bias. They strike the lattice and impart some energy to the valence electrons. So the valence electrons will break away from their parent atom and become free carriers. These newly generated additional carriers acquire more energy from the potential (since bias is applied). So they again strike the lattice and create more number of free electrons and holes. This process goes on as long as bias is increased and the number of free carriers gets multiplied. This is known as ***avalanche multiplication***, Since the number of carriers is large, the current flowing through the diode which is proportional to free carriers also increases and when this current is large, avalanche breakdown will occur.

2.12.2 ZENER BREAKDOWN

Now if the electric field is very strong to disrupt or break the covalent bonds, there will be sudden increase in the number of free carriers and hence large current and consequent breakdown. Even if thermally generated carriers do not have sufficient energy to break the covalent bonds, the electric field is very high, then covalent bonds are directly broken. This is ***Zener Breakdown***. A junction having narrow depletion layer and hence high field intensity will have zener breakdown effect. ($\cong 10^6$ V/m). If the doping concentration is high, the depletion region is narrow and will have high field intensity, to cause Zener breakdown.

2.12.3 THERMAL BREAKDOWN

If a diode is biased and the bias voltage is well within the breakdown voltage at room temperature, there will be certain amount of current which is less than the breakdown current. Now keeping the bias voltage as it is, if the temperature is increased, due to the thermal energy, more number of carriers will be produced and finally breakdown will occur. This is ***Thermal Breakdown***.

      In zener breakdown, the covalent bonds are ruptured. But the covalent bonds of all the atoms will not be ruptured. Only those atoms, which have weak covalent bonds such as an atom at the surface which is not surrounded on all sides by atoms will be broken. But if the field strength is not greater than the critical field, when the applied voltage is removed, normal covalent bond structure will be more or less restored. This is Avalanche Breakdown. But if the field strength is very high, so that the covalent bonds of all the atoms are broken, then normal structure will not be achieved, and there will be large number of free electrons. This is ***Zener Breakdown***.

      In Avalanche Breakdown, only the excess electron, loosely bound to the parent atom will become free electron because of the transfer of energy from the electrons possessing higher energy.

2.13 ZENER DIODE

This is a *p-n junction* device, in which zener breakdown mechanism dominates. ***Zener diode is always used in Reverse Bias***.

      Its constructional features are:

      1.   *Doping concetration is heavy on* p *and* n *regions of the diode, compared to normal* p-n junction *diode.*

      2.   *Due to heavy doping, depletion region width is narrow.*

3.   *Due to narrow depletion region width, electric field intensity* $E = \dfrac{V}{d}$

$= \dfrac{V_z}{W}$ *will be high, near the junction, of the order of $10^6 V/m$. So Zener Breakdown mechanism occurs.*

In normal *p-n junction* diode, avalanche breakdown occurs if the applied voltage is very high. The reverse characteristic of a p-n junction diode is shown in Fig. 2.29.

When the Zener diode is reverse biased, the current flowing is only the reverse saturation current I_0 which is constant like in a reverse biased diode. At $V = V_Z$, due to high electric field $\left(\dfrac{V_z}{W}\right)$, Zener breakdown occurs. Covalent bonds are broken and suddenly the number of free electrons increases. So I_Z increases sharply and V_Z remains constant, since, I_Z increases through Zener resistance R_Z decreases. So the product $V_Z = R_Z$. I_Z almost remains constant. If the input voltage is decreased, the Zener diode regains its original structure. (But if V_i is increased much beyond V_Z, electrical breakdown of the device will occur. The device looses its semiconducting properties and may become a short circuit or open circuit. *This is what is meant by device breakdown.*)

Applications

1.   *In Voltage Regulator Circuits*
2.   *In Clipping and Clamping Circuits*
3.   *In Wave Shaping Circuits.*

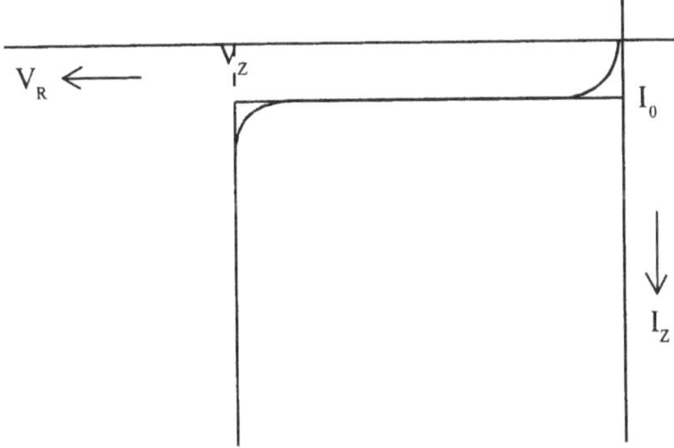

Fig 2.30   *Reverse characteristic of a Zener diode.*

2.14   INTRODUCTION TO REGULATORS

Voltage Regulator Circuits are electronic circuits which give constant DC output voltage, irrespective of variations in *Input Voltage V_i, current drawn by the load I_L* from output terminals, and *Temperature T*. Voltage Regulator circuits are available in discrete form using BJTs, Diodes etc and in IC (Integrated Circuit) form. The term voltage regulator is used when the output delivered is DC voltage. The input can be DC which is not constant and fluctuating. If the input is AC, it is converted to DC by Rectifier and Filter Circuits and given to I.C. Voltage Regulator circuit, to get constant DC output voltage. If the input is A.C 230 V from mains, and the output desired is constant DC, a stepdown transformer is used and then Rectifier and filter circuits are used, before the electronic regulator circuit. The block diagrams are shown in Fig. 2.31 and 2.32.

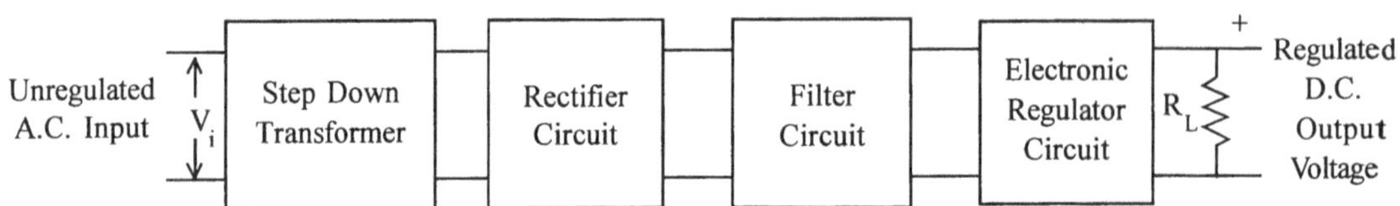

Fig. 2.31 Block diagram of voltage regulator with A.C input.

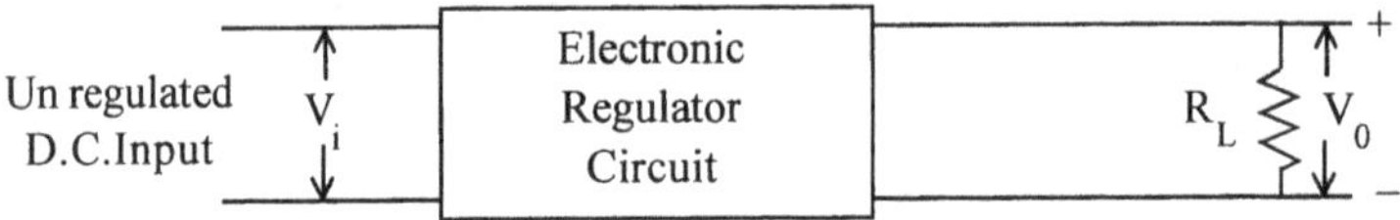

Fig. 2.32 Block diagram of voltage regulator with D.C input.

The term *Voltage Stabilizer* is used, if the output voltage is AC and not DC. The circuits used for voltage stabilizers are different. The voltage regulator circuits are available in IC form also. Some of the commonly used ICs are, μA 723, LM 309, LM 105, CA 3085 A.

7805, 7806, 7808, 7812, 7815 : Three terminal positive Voltage Regulators.

7905, 7906, 7908, 7912, 7915 : Three terminal negative Voltage Regulators.

The Voltage Regulator Circuits are used for electronic systems, electronic circuits, IC circuits, etc.

The specifications and Ideal Values of Voltage Regulators are :

Specifications		**Ideal Values**
1. Regulation (S_V)	:	0 %
2. Input Resistance (R_i)	:	∞ ohms
3. Output Resistance (R_0)	:	0 ohms
4. Temperature Coefficient (S_T)	:	0 mv/oc.
5. Output Voltage V_0	:	-
6. Output current range (I_L)	:	-
7. Ripple Rejection	:	0 %

Different types of Voltage Regulators are

1. Zener regulator
2. Shunt regulator
3. Series regulator
4. Negative voltage regulator
5. Voltage regulator with foldback current limiting
6. Switching regulators
7. High Current regulator

2.14.1 ZENER VOLTAGE REGULATOR CIRCUIT

A simple circuit without using any transistor is with a zener diode Voltage Regulator Circuit. In the reverse characteristic voltage remains constant irrespective of the current that is flowing through Zener diode. The voltage in the break down region remains constant. (Fig. 2.33)

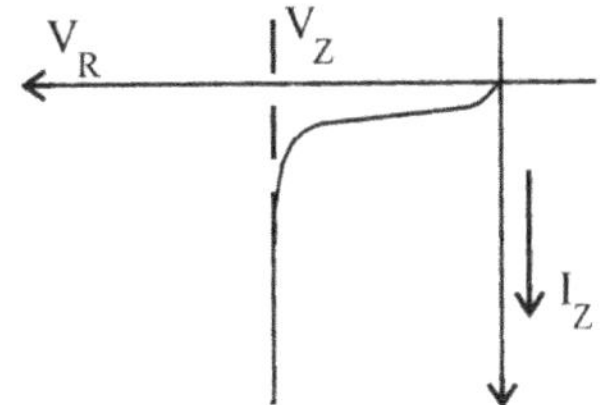

Fig. 2.33 Zener diode reverse characteristic.

Therefore in this region the zener diode can be used as a voltage regulator. If the output voltage is taken across the zener, even if the input voltage increases, the output voltage remains constant. The circuit as shown in Fig. 2.34.

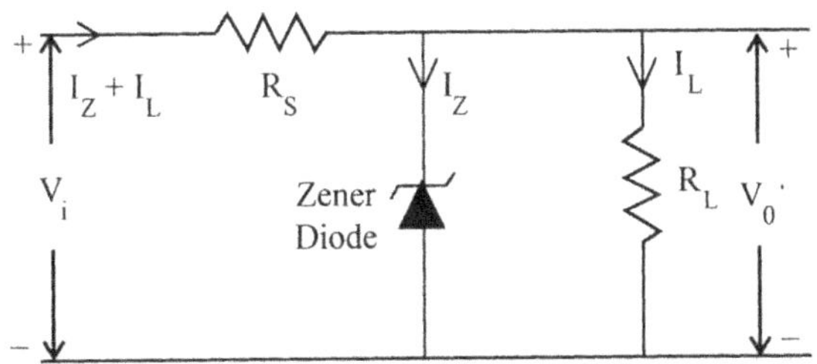

Fig. 2.34 Zener regulator circuit.

The input V_i is DC. Zener diode is reverse biased.

If the input voltage V_i increases, the current through R_s increases. This extra current flows, through the zener diode and not through R_L. Therefore zener diode resistance is much smaller than R_L when it is conducting. Therefore I_L remains constant and so V_0 remains constant.

The limitations of this circuits are

1. The output voltage remains constant only when the input voltage is sufficiently large so that the voltage across the zener is V_Z.

2. There is limit to the maximum current that we can pass through the zener. If V_i is increased enormously, I_Z increases and hence breakdown will occur.

3. Voltage regulation is maintained only between these limits, the minimum current and the maximum permissible current through the zener diode. Typical values are from 10m A to 1 ampere.

2.14.2 SHUNT REGULATOR

The shunt regulator uses a transistor to amplify the zener diode current and thus extending the Zener's current range by a factor equal to transistor h_{FE}. (Fig. 2.35)

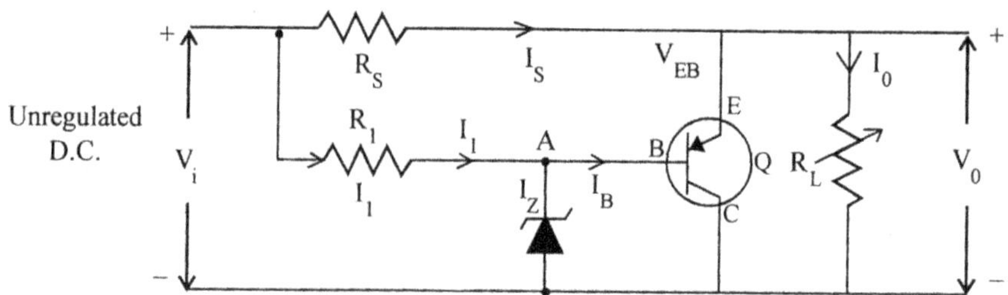

Fig. 2.35 Shunt Regulator Circuit

Zener current, passes through R_1

Nominal output voltage

$$= V_Z + V_{EB}$$

The current that gets branched as I_B is amplified by the transistor. Therefore the total current I_0 = $(\beta + 1) I_B$, flows through the load resistance R_L. Therefore for a small current through the zener, large current flows through R_L and voltage remains constant. In otherwords, for large current through R_L, V_0 remains constant. Voltage V_0 does not change with current.

Problem 2.13

For the shunt regulator shown, determine

1. The nominal voltage
· 2. Value of R_1,
3. Load current range
4. Maximum transistor power dissipation.
5. The value of R_S and its power dissipation.

V_i = Constant. Zener diode 6.3 V, 200 mW, requires 5 mA minimum current.

Transistor Specifications :

$$V_{EB} = 0.2V, \quad h_{FE} = 49, \quad I_{CBO} = 0.$$

1. The nominal output voltage is the sum of the transistor V_{EB} and zener voltage.

$$V_0 = V_{Eb} + V_Z = 0.2 + 6.3 = 6.5V$$

2. R_1 must supply 5 mA to the zener diode :

$$\therefore \qquad R_1 = \frac{8\,V - 6.3}{5 \times 10^{-3}} = \frac{1.7}{5 \times 10^{-3}} = 340\,\Omega$$

3. The maximum allowable zener current is

$$\frac{\text{Power rating}}{\text{Voltage rating}} = \frac{0.2}{6.3} = 31.8 \text{ mA}$$

The load current range is the difference between minimum and maximum current through the shunt path provided by the transistor. At junction A, we can write,

$$I_B = I_Z - I_1$$

I_1 is constant at 5 mA

$$\therefore \qquad I_B = I_Z - I_1$$

I_1 is constant at 5 mA

$$\therefore \qquad I_B = 5 \times 10^{-3} - 5 \times 10^{-3} = 0$$

$$I_B \text{ (min)} = I_{Z\,Max} - I_2$$
$$= 31.8 \times 10^{-3} - 5 \times 10^{-3} = 26.8 \text{ mA}$$

The transistor emitter current $I_E = I_B + I_C$

$$I_C = \beta I_B = h_{FE} I_B$$
$$\therefore \qquad I_E = (\beta + 1) I_B = (h_{FE} + 1) I_B$$

I_B ranges from a minimum of 0 to maximum of 26.8 mA

$\therefore$ Total load current range is $(h_{FE} + 1) I_B$

$$= 50 \, (26.8 \times 10^{-3}) = \textbf{1.34 A}$$

4. The maximum transistor power dissipation occurs when the current is maximum $I_E \simeq I_C$

$$P_D = V_o \, I_E = 6.5 \, (1.34) = \textbf{8.7 W}$$

5. R_S must pass 1.34 A to supply current to the transistor and R_L.

$$R_S = \frac{V_i - V_0}{1.34} = \frac{8 - 6.5}{1.34} = 1.12 \, \Omega$$

The power dissipated by R_S,

$$= I_S^2 \, R_S$$
$$= (1.34)^2 \cdot (1.12) = 3 \text{ W}$$

REGULATED POWER SUPPLY

An unregulated power supply consists of a transformer, a rectifier, and a filter. For such a circuit regulation will be very poor i.e. as the load varies (*load means load current*) [*No load means no load current or 0 current. Full load means full load current or short circuit*], we want the output voltage to remain constant. But this will not be so for unregulated power supply. The short comings of the circuits are :

1. Poor regulation

2. DC output voltage varies directly as the a.c. input voltage varies

3. In simple rectifiers and filter circuits, the d.c. output voltage varies with temperature also, if semiconductors devices are used.

An electronic feedback control circuit is used in conjuction with an unregulated power supply to overcome the above three short comings. Such a system is called a "*regulated power supply*".

Stabilization

The output voltage depends upon the following factors in a power supply.

1. Input voltage V_i

2. Load current I_L

3. Temperature

$\therefore$ Change in the output voltage ΔV_0 can be expressed as

$$\Delta V_0 = \frac{\partial V_0}{\partial V_i} \cdot \Delta V_i + \frac{\partial V_0}{\partial I_L} \cdot \Delta I_L + \frac{\partial V_0}{\partial T} \cdot \Delta T$$

$$\Delta V_0 = S_1 \, \Delta V_i + R_0 \, \Delta I_L + S_T \, \Delta T$$

Where the three coefficients are defined as,

(i)   Stability factors.

$$S_V = \frac{\Delta V_0}{\Delta V_i}\bigg|_{\Delta I_L = 0,\ \Delta T = 0}$$

This should be as small as possible. Ideally 0 since V_0 should not change even if V_i changes.

(ii)   *Output Resistance*

$$R_0 = \frac{\Delta V_0}{\Delta I_L}\bigg|_{\Delta V_i = 0,\ \Delta T = 0}$$

(iii)   *Temperature Coefficient*

$$S_T = \frac{\Delta V_0}{\Delta T}\bigg|_{\Delta V_i = 0,\ \Delta I_L = 0}$$

The smaller the values of the three coefficients, the better the circuit.

Series Voltage Regulator

The voltage regulation (i.e., change in the output voltage as load voltage varies (or input voltage varies) can be improved, if a large part of the increase in input voltage appears across the control transistor, so that output voltage tries to remain constant, i.e., increase in V_i results in increased V_{CE} so that output almost remains constant. But when the input increases, there may be some increase in the output but to a very smaller extent. This increase in output acts to bias the control transistor. This additional bias causes an increase in collector to emitter voltage which will compensate for the increased input.

If the change in output were amplified before being applied to the control transistor, better stabilization would result.

Series Voltage Regulator Circuit is as shown in Fig. 2.36.

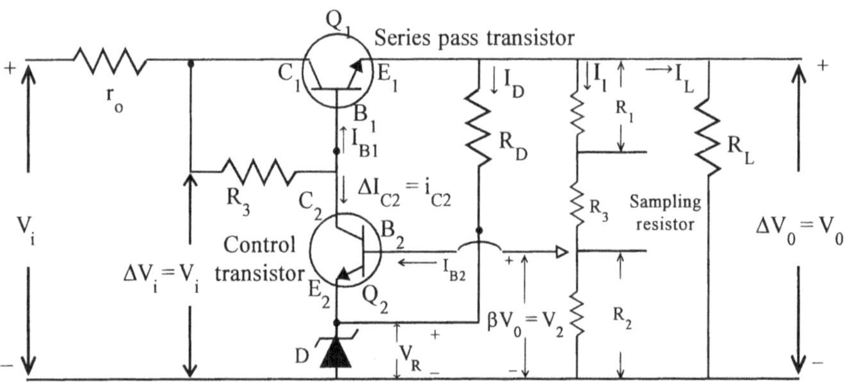

Fig. 3.36 Series Regulator Circuit

2.14.3   SERIES VOLTAGE REGULATOR CIRCUIT

Q_1 is the series pass element of the series regulator. Q_2 acts as the difference amplifier. D is the reference zener diode. A fraction of the output voltage bV_0 (b is a fraction, which is taken across R_2 and the potentiometer) is compared with the reference voltage V_R. The difference $(bV_0 - V_R)$ is amplified by the transistor Q_2. Because the emitter of Q_2 is not at ground potential, there is constant voltage V_R. Therefore the net voltage to the Base - Emitter of the transistor Q_2 is $(bV_0 - V_R)$. As V_0 increases, $(bV_0 - V_R)$ increases. When input voltage increases by ΔV_i, the base-emitter voltage of Q_2 increases. So Collector current of Q_2 increases and hence there will be large current change in R_3. Thus all the change in V_i will appear across R_3 itself. V_{BE} of the transistor Q_1 is small. Therefore the drop across $R_3 = V_{CB}$ of $Q_1 \simeq V_{CE}$ of Q_1 since V_{BE} is small. Hence the increase in the voltage appears essentially across Q_1 only. This type of circuit takes care of the increase in input voltages only. If the input decreases, output will also decrease. (If the output were to remain constant at a specified value, even when V_i decreases, buck and boost should be there. The tapping of a transformer should be changed by a relay when V_i changes). r_0 is the output resistance of the unregulated power supply which preceds the regulator circuit. r_0 is the output resistance of the rectifier, filter circuit or it can be taken as the resistance of the DC supply in the lab experiment.

The expression for S_V (Stability factor) $= \dfrac{\Delta V_0}{\Delta V_i}$

$$= \left[\dfrac{R_1 + R_2}{R_2}\right] \cdot \dfrac{\left(R_1 \text{ in parallel with } R_2\right) + h_{ie2} + \left(1 + h_{fe2}\right) R_2}{h_{fe2} R_3}$$

R_z = Zener diode resistance (typical value)

$$R_0 = \dfrac{r_0 + \dfrac{R_3 + h_{ie1}}{1 + h_{fe1}}}{1 + G_m \left(R_3 + r_0\right)}; \qquad G_m = \dfrac{\Delta I_{C2}}{\Delta V_0}$$

The preset pot or trim pot R_3 in the circuit is called as *sampling resistor* since it controls or samples the amount of feedback.

Preregulator

It provides constant current to the collector of the DC amplifier and the base of control element.

If R_3 is increased, the quantity $\dfrac{R_3 + h_{ie1}}{1 + h_{fe}}$ also increases, but then this term is very small, since it is being divided by h_{fe}.

2.14.4 NEGATIVE VOLTAGE REGULATOR

Sometimes it is required to have negative voltage viz., $-6V$, $-18V$, $-21V$ etc with positive terminal grounded. This type of circuit supplies regulated negative voltages. The input should also be negative DC voltage.

2.14.5 VOLTAGE REGULATOR WITH FOLDBACK CURRENT LIMITING

In high current voltage regulator circuits, constant load current limiting is employed. i.e. The load current will not increase beyond the set value. But this will not ensure good protection. So foldback current limiting is employed. When the output is shared, the current will be varying. The series pass transistors will not be able to dissipate this much power, with the result that it may be damaged. So when the output is shorted or when the load current exceeds the set value, the current through the series transistors decrease or folds back.

2.14.6 SWITCHING REGULATORS

In the voltage regulator circuit, suppose the output voltage should remain constants at $+6V$, the input can be upto $+12V$ or $+15V$ maximum. If the input voltage is much higher, the power dissipation across the transistor will be large and so it may be damaged. So to prevent this, the input is limited to around twice that of the input. But if, higher input fluctuations were to be tolerated, the voltage regulators IC is used as a switch between the input and the output. The input voltage is not connected permanently to the regulator circuit but ON/OFF will occur at a high frequency (50 KHz) so that output is constantly present.

A simple voltage regulator circuit using *CA3085A* is as shown in Fig. 2.37. It gives 6 V constant output upto 100 mA.

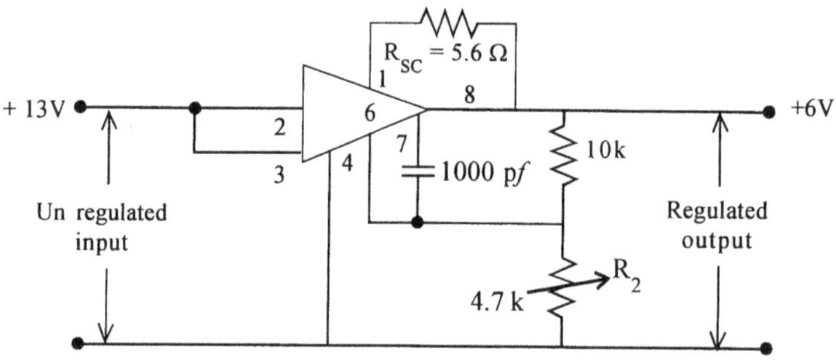

Fig. 3.37 CA 3085A IC voltage regulator.

$$V_0 = \frac{\beta A_V}{\left(1 + \beta A_V\right)}$$

$$\beta = \frac{R_2}{R_1 + R_2}$$

RCA (Radio Corporation of America) uses for the ICs, alphabets CA. CA3085A is voltage regulator. LM309 K – is another Voltage Regulator IC.

μA716C is head phone amplifier, delivers 50 mW to, 500-600Ω load

CA3007 is low power class AB amplifier, and delivers 30 mW of output power.

MC1554 is 20W class B power amplifier.

Preregulator

The value of the stability factor S_V of a voltage regulator should be very small. S_V can be improved if R_3 is increased (from the general expression) since $R_3 \simeq (V_i - V_0)/I$. We can increase R_3 by decreasing I, through R_3. The current I through R_3 can be decreased by using a Darlington pair for Q_1. To get even better values of S_V, R_3 is replaced by a constant current source circuit, so that R_3 tends to infinity ($R_3 \to \infty$). This constant current source circuit is often called *a transistor preregulator*. V_i is the maximum value of input that can be given.

Short Circuit Overload Protection

Overload means overload current (or short circuit). A power supply must be protected further from damage through overload. In a simple circuit, protection is provided by using a fuse, so that when current excess of the rated values flows, the fuse wire will blow off, thus protecting the components. This fuse wire is provided before r_0. Another method of protecting the circuit is by using diodes.

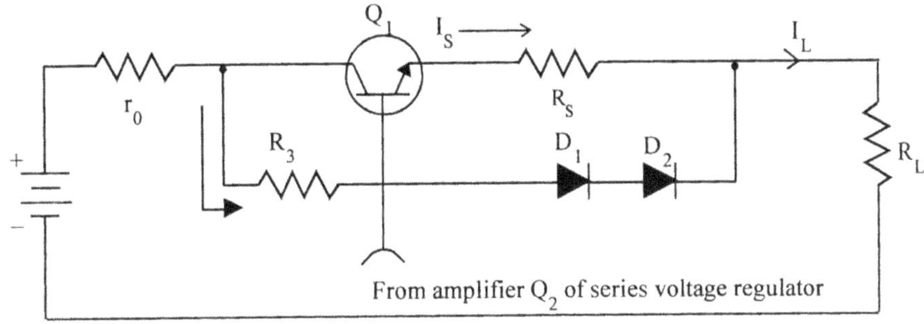

Fig. 3.38 Circuit for short circuit protection.

Zener diodes can also be employed, but such a circuit is relatively costly.

The diodes D_1 and D_2 will start conducting only when the voltage drop across R_S exceeds the current in voltage of both the diodes D_1 and D_2. In the case of a short circuit the current I_S will increase upto a limiting point determined by

$$I_S = \frac{V_{\gamma_1} + V_{\gamma_2} - V_{BE1}}{R_S}$$

When the output is short circuited, the collector current of Q_2 will be very high I_S . R_S will also be large.

∴   The two diodes D_1 and D_2 start conducting.

∴   The large collector current of Q_2 passes through the diodes D_1 and D_2 and not through the transistor Q_1.

∴   Transistor Q_1 will be safe, D_1 and D_2 will be generally si diodes, since cut in voltage is 0.6V. So I_S R_S drop can be large.

Problem 2.14

Design a series regulated power supply to provide a nominal output voltage of 25V and supply load current $I_L \leq 1A$. The unregulated power supply has the following specifications $V_i = 50 \pm 5V$, and $r_0 = 10\ \Omega$.

Given : $R_Z = 12\ \Omega$ at $I_Z = 10$ mA. At $I_{C2} = 10$ mA, $\beta = 220$, $h_{ie2} = 800\ \Omega$, $h_{fe2} = 200$, $I_1 = 10$ mA.

The reference diodes are chosen such that $V_R \simeq \dfrac{V_0}{2}$.

$$\frac{V_0}{2} = 12.5 \text{ V}$$

∴ Two Zener diodes with breakdown voltages of 7.5 V in series may be connected.

$$R_Z = 12\ \Omega \text{ at } I_Z = 20 \text{ mA}$$

Choose $I_{C2} \simeq I_{E2} = 10$ mA

At $I_{C_2} = 10$ mA, the h-parameters for the transistor are measured as,

$$\beta = 220, \quad h_{ie2} = 800\ \Omega, \quad h_{fe2} = 200$$

Choose $I_1 = 10$ mA, so that $I_Z = I_{D1} + I_{D2} = 20$ mA

$$R_D = \frac{V_0 - V_R}{I_D} = \frac{25 - 15}{10} = 1K\Omega$$

$$I_{B2} = \frac{I_{C_2}}{\beta} = \frac{10\,mA}{220} = 45\mu A$$

Choose I_1 as 10 mA for si transistors, $V_{BE} = 0.6$ V

$$V_2 = V_{BE2} + V_R = 15.6\text{ V}$$

$$\therefore \qquad R_1 = \frac{V_0 - V_2}{I_1} = \frac{25 - 15.6}{10 \times 10^{-3}} = 940\ \Omega$$

$$R_2 \simeq \frac{V_2}{I_1} = \frac{15.6}{10 \times 10^{-3}} = 1{,}560\ \Omega$$

For the transistor Q_1, choose I_2 as 1A and $h_{FE1} = 125$ (d.c. current gain β)

$$I_{B1} = \frac{I_L + I_1 + I_D}{h_{fe1}\,(\beta)} \qquad\qquad \therefore \quad I_{C_1} \simeq I_{E_1} = I_L + I_1 + I_D$$

$$\text{(DC Current gain)}$$

$$= \frac{1000 + 10 + 10}{125} \simeq 8\text{ mA}$$

The current through resistor R_3 is $I = I_{B1} + I_{C2} = 8 + 10 = 18$ mA

The value of R_3 corresponding to $V_i = 45$ V and $I_L = 1$A is (since these are given in the problem) $V_i = 50 + 5$, 45 V

$$R_3 = \frac{V_i - \left(V_{BE_1} + V_0\right)}{I} = \frac{50 - 25.6}{18 \times 10^{-3}} = 1{,}360\ \Omega.$$

Voltage regulator is a circuit which maintains constant output voltage, irrespective of the changes of the input voltage or the current.

Stabilizer - If the input is a.c, and output is also a.c, it is a stabilizer circuit.

2.15   TERMINOLOGY

Load Regulation : It is defined as the % change in regulated output voltage for a change in load current from minimum to the maximum value.

$$E_1 = \text{Output voltage when } I_L \text{ is minimum (rated value)}$$

$$E_2 = \text{Output voltage when } I_L \text{ is maximum (rated value)}$$

$$\%\text{ load regulation} = \frac{E_1 - E_2}{E_1} \times 100 \quad \% \; ; E_1 > E_2. \text{ This value should be small.}$$

Line Regulation : It is the % change in V_0 for a change in V_i.

$$= \frac{\Delta V_0}{\Delta V_i} \times 100\ \%$$

In the Ideal case $\Delta V_0 = 0$ when V_0 remains constant.

This value should be minimum.

Load regulation is with respect to change in I_L

Line regulation is with respect to change in V_i.

Ripple Rejection : It is the ratio of peak to peak output ripple voltage to the peak to peak input ripple voltage.

$$\frac{V_0'\,(p-p)}{V_i'\,(p-p)}$$

2.16 THE TUNNEL DIODE

In an ordinary p-n junction diode the doping concentration of impurity atoms is 1 in 10^8. With this doping, the depletion layer width, which constitutes barrier potential is $5\mu V$. If the concentration of the impurity atoms is increased to say 1 in 10^3(This corresponds to impurity density of $\approx 10^{19}/m^3$), the characteristics of the diode will completely change. Such a diode is called *Tunnel Diode*. This was found by *Esaki* in *1958*.

2.16.1 TUNNELING PHENOMENON

Barrier Potential V_B :

$$V_B = \frac{e.N_A}{2 \in} \cdot W^2$$

or
$$W = \sqrt{\frac{2\in .V_B}{e.N_A}} \, .$$

where $\qquad$ W = Width of Depletion Region in μ

$\qquad\qquad N_A$ = Impurity Concentration N_O/m^3.

So, the width of the junction barrier varies inversely as the square root of impurity concentration. Therefore as N_A increases W decreases.

Therefore, in tunnel diodes, by increasing N_A, W can be reduced from 5μ to 0.01μ. According to classical mechanics, a particle must possess the potential which is at least equal to, or greater than, the barrier potential, to move from one side to the other. When the barrier width is so thin as 0.01μ, according to Schrodinge equation, there is much probability that an electron will penetrate when a forward bias is applied to the diode, so that potential barrier decreases below E_0. The *n-side* levels must shift upward with respect to those on the *p-side*. So there are occupied states in the conduction band of the n material, which are at the same energy level as allowed empty states in the valence band of the *p-side*. Hence electrons will tunnel from the *n-side* to the *p-side* giving rise to forward current. As the forward bias is increased further, the number of electrons on *n-side* which occupy the same energy level as that *vacant energy* state existing on *p-side*, also increases. So more number of electrons tunnel through the barrier to empty states on the left side giving rise to peak current I_p. If still more forward bias is applied, the energy level of the electrons on the *n-side* increases, but the empty states existing on the *p-side*, reduces. So the tunneling current decreases. In addition to the *Quantum Mechanical Tunneling Current*, there is regular *p-n junction* injection current also. The magnitude of this current is considerable only beyond a certain value of forward bias voltage.

Therefore, the current again starts beyond V_V. The graph is shown in Fig. 2.39.

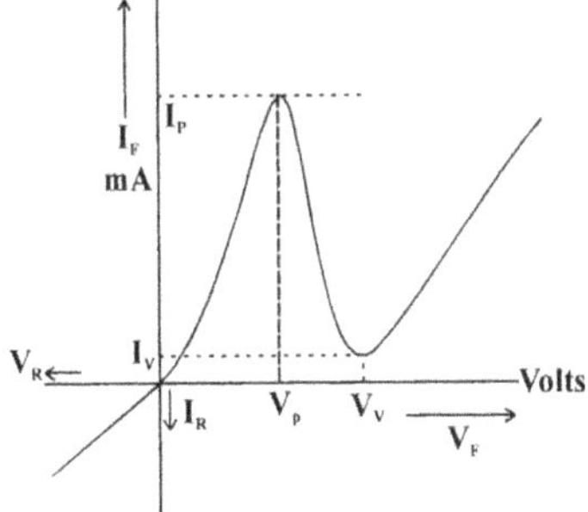

Fig 2.39 V - I Characteristics of a Tunnel diode.

$$I_p = \text{Peak current}$$

$$I_V = \text{Valley current}$$

$$V_F = \text{Peak forward voltage}$$

$$V_p \approx 50 \text{ mV}$$

2.16.2   CHARACTERISTICS OF A TUNNEL DIODE

For small forward voltage (ie., $V_p \approx 50$mV for Ge), forward resistance is small $\approx 5\Omega$ and so current is large. At the **Peak Current** I_p, corresponding to voltage V_p, $\dfrac{dI}{dV}$ is zero. If V is beyond V_p, the current decreases. So the diode exhibits **Negative Resistance Characteristics** between I_p and I_v called the **Valley Current**. The voltage at which the forward current again equals I_V is called as *peak* forward voltage V_V. Beyond this voltage, the current increases rapidly.

The symbol for tunnel diode is shown in Fig. 2.40(a). Typical values of a tunnel diode are :

$$V_p = 50 \text{ mV}, \qquad V_V = 350 \text{ mV},$$

$$V_F = 0.5\text{V}, \qquad \frac{I_P}{I_V} = 8$$

$$I_p = 10 \text{ mA}, \qquad \text{Negative resistance } R_n = -30\Omega.$$

Series Ohmic resistance $R_s = 1\Omega$. The series inductance L_s depends upon the lead length and the geometry of the diode package. $L_s \approx 5$ nH. Junction capacitance $C = 20$ pF. Its circuit equivalent is shown in Fig. 2.40(b).

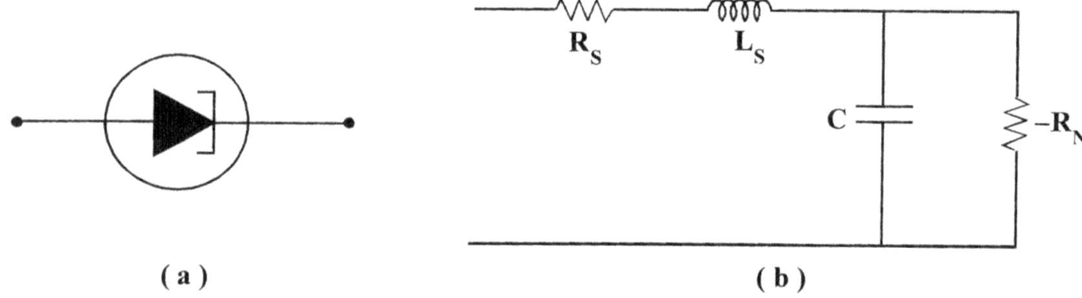

(a)   (b)

Fig 2.40 (a) **Symbol of Tunnel Diode**   (b) **Equivalent Circuit**

Advantages :

1. Low Cost
2. Low Noise
3. Simplicity
4. High Speed
5. Low Input Power
6. Environmental Immunity.

Disadvantages :

1. Low output voltage swings, even for small voltage, while the current goes to large values. So the swing is limited.

2. Circuit design difficulty, since it is a two terminal device and there is no isolation between input and output.

Applications :

1. As a high frequency oscillator (GHz)

2. As a fast switching device. Switching time is in nano-seconds. Since tunneling is a quantum mechanical phenomenon, there is no time lag between the application of voltage and consequent current variation. So it can be used for high frequencies.

If the barrier width is $\approx 3A^\circ$, the probability that electrons will tunnel through the barrier is large. The barrier width will be $\approx 3A^\circ$, when impurity doping concentration is $\geq 10^{19}/cm^3$. If width $\approx 0.01\mu$ electrons will tunnel without the application of field. But if width $\approx 3A^\circ$, very small applied voltage is sufficient for the electron to tunnel.

The two conditions to be satisfied for tunneling phenomenon to take place are :

Necessary Condition :

1. The effective depletion region width near the junction must be small, of the order of $3A^\circ$ by heavy doping.

Sufficient Condition :

2. There must be equivalent empty energy states on the *p-side* corresponding to energy levels of electrons on the *n-side*, for these electrons to tunnel from *n-side* to *p-side*.

When the tunnel diode is reverse biased, there will be some empty states on *n-side* corresponding to filled states on the *p-side*. So electrons will tunnel through the barrier from the p side to n side [since when the diode is reverse biased the energy levels on n side will decrease]. As bias voltage is increased the number of electrons tunneling will increase, so forward current increases. When the diode is forward biased, electrons from *n-side* will increase, and tunnel through the barrier to *p-side*. This is called quantum mechanical current. Apart from this, normal p-n diode current will also be there. Beyond valley voltage V_V, it is normal diode forward current.

2.17 VARACTOR DIODE

Barrier of transition capacitance C_T varies with the value of reverse bias voltage. The larger the reverse voltage, the larger the W.

$$C_T = \frac{\in \times A}{W}$$

So C_T of a *p-n junction* diode varies with the applied reverse bias voltage.

Diodes made especially for that particular property of variable capacitance with bias are called *Varactors*, *Varicaps* or *Voltacaps*. These are used in LC Oscillator Circuits.

The Symbol is shown in Fig. 2.41

Varactor diodes are used in high frequency circuits.

In the case of abrupt junction,

$$C = \frac{\in A}{W}$$

Fig 2.41 Varactor Diode

But　　　　$W \propto (V)^{\frac{1}{2}}$

$\therefore$　　　　$C \propto (V)^{-\frac{1}{2}}$

In the case of linearly graded junction,

$$W \propto (V)^{1/3}$$
$$\therefore \qquad C \propto (V)^{-1/3}$$

The variation of impurity atom concentration with ω for different types of junctions is shown in Fig. 2.42.

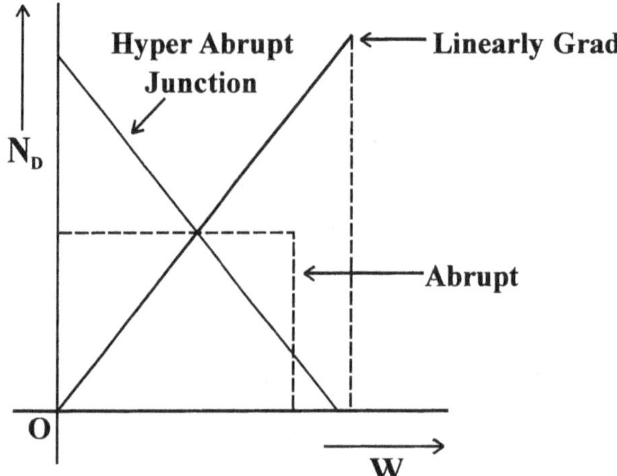

Fig 2.42 Impurity concentration variation in the junction region.

But 'C' can be made proportional to V^2 by changing the doping profile. Such a junction is called *Hyper Abrupt Junction*. This junction can be obtained by *Epitaxial Process*. Therefore, C changes rapidly with voltage. Varactors are used in Reverse Bias condition, since C_T proportional to V and $C_D = \tau \times g$ No direct relation with V . So Varactors are used in Reverse Bias condition.

$$\text{Capacitance,} \qquad C = \frac{\in A}{W}$$

As the reverse bias voltage increases, width of the depletion region increases. So W increases $C \propto \dfrac{1}{W}$. Hence C decreases. C is maximum under no reverse bias condition, since W is minimum C is minimum under maximum reverse bias, since W is maximum.

$$\text{Figure of merit} = \frac{C_{Max}}{C_{Min}}$$

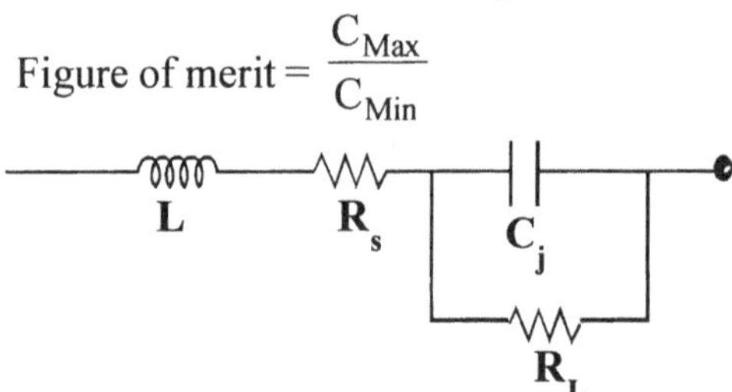

Fig 2.43 Equivalent circuit of Varactor diode

L   :   Lead inductance.
R_s :   Series resistance of the diode due to the semiconductor material
C_j :   Junction Capacitance
R_L :   Leakage resistance of the diode since the capacitor can not be ideal one with
        out any leakage.

Problem 2.14

For a Zener Diode, the breakdown voltage $V_z = \in E_z^2/\, 2eNA$. Prove that, the breakdown voltage for a Ge diode is $51/\sigma_p$ if $E_z = 2 \times 10^7$ v/m where σ_p = conductivity of the p material is $(\Omega\text{-cm})^{-1}$. Assume $N_A \ll N_D$. If p material is intrinsic, calculate V_z

Solution

$$V_z = \frac{\in \times E_z^2}{2eN_A} \qquad\qquad \dots\dots\dots (1)$$

$$\sigma_p = N_A \times e \times m_p \qquad\qquad \text{and} \qquad\qquad p \cong N_A$$

or
$$e \times N_A = \sigma_p / \mu_p$$

Substituting for $e \times N_A$ in equation (1)

$$\therefore \qquad V_z = \frac{\in . E\mu_p}{2\sigma_p}$$

$$\in \; = \frac{16}{36\pi \times 10^9} \text{ F/m}$$

$$V_z = \frac{16}{36\pi \times 10^9} \times 1800 \times \frac{4 \times 10^{14}}{2} \times 10^{-6} \times \frac{1}{\sigma_p}$$

$$= \frac{51}{\sigma_p} \qquad \text{where } \sigma_p \text{ is in } (\Omega\text{-cm})^{-1}.$$

For intrinsic Germanium, $\sigma_e = \dfrac{1}{45}$.

$$\therefore \qquad V_z = 51 \times 45 = 2300V$$

Problem 2.15

The transition capacitance of an abrupt junction diode is 20 pF. at 5V. Compute the value of decrease in capacitance for a 1.0 volt increase in the bias.

Solution

$$C_T \; \alpha \; \frac{1}{\sqrt{V}} \; ; \; C_T = 20 \text{ pF; when } V = 5V$$

$$\therefore \qquad 20 = \frac{k}{\sqrt{5}} \; ; \; k \text{ is a constant}$$

$$\therefore \qquad k = 20\sqrt{5}$$

when $V = 6V$, $C_T = ?$

$$\therefore \qquad C_T = \frac{20\sqrt{5}}{\sqrt{6}} = 18.25 \text{ pF}$$

Therefore, decrease in the value of capacitance is $20 - 18.25 = 1.75$ pF.

Problem 2.16

(a) The Zener Diode regulates at 50V over a range of diode current from 5 mA to 40 mA. Supply voltage $V = 200V$. Calculate the value of R to allow voltage regulation from a load current $I_L = 0$ upto I_{max}; the maximum possible value of I_L. What is I_{max}? (Fig. 2.44)

(b) If R is set as in part (a) and $I_L = 25mA$, what are the limits between which V may vary without loss of regulation in the circuit ?

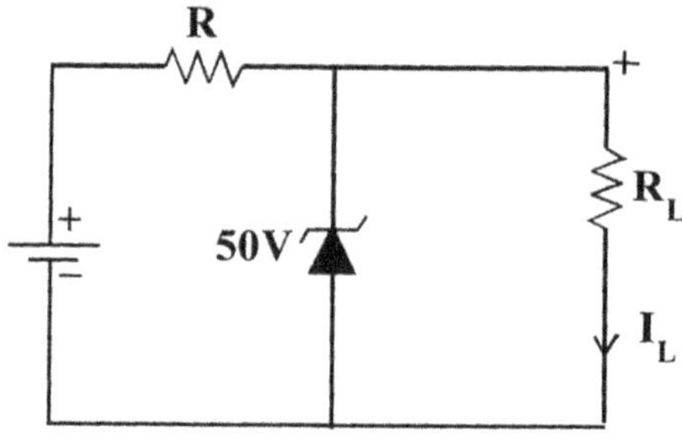

Fig 2.44 For Problem 2.37

Solution

(a) If $I_L = 0$, $\quad R = \dfrac{200 - 50}{40 \times 10^{-3}} = 3750\ \Omega$

I_{max} occurs when $I_0 = 5mA$

$\therefore \qquad I_{max} = (40mA - 5mA) = 35\ mA$

(b) For V_{min}

$\qquad I_L + I_D = 25 + 5 = 30\ mA$

$\therefore \qquad V_{min} = 50 + (30 \times 10^{-3})(3750)$

$\qquad\qquad = 162.5V$

For V_{max}

$\qquad I_L + I_D = 25 + 40 = 65\ mA$

$\therefore \qquad V_{max} = 50 + (65 \times 10^{-3})(3750) = 293.8V$

Problem 2.17

A resistor R is placed parallel to a Ge tunnel diode. The tunnel diode has

$$\left|\frac{di_d}{dv}\right|_{max} = \frac{1}{10}.$$

Find the value of R so that the combination does not exhibit negative resistance region in its volt ampere characteristic.

Solution

The combination is called as tunnel resistor. If the V-I characteristics were to exhibit no negative resistance region, the slop of the curve

$$\left|\frac{di}{dv}\right| \geq 0, \qquad \text{for all V.}$$

$$I = i_D + i_R$$

$$I = i_D + \frac{V}{R}$$

$$\frac{di}{dv} = \frac{di_D}{dv} + \frac{1}{R} \geq 0$$

Fig 2.45 For Problem 2.38.

$$\therefore \qquad \frac{1}{R} \geq \left| \frac{di_D}{dv} \right|$$

But it is given that $\left| \dfrac{di_D}{dv} \right|_{max} = \dfrac{1}{10}$ mhos

Therefore, R should be as least 10 Ω, so that there is no negative resistance region in the characteristic.

Problem 2.18

The Zener Diode can be used to prevent overloading of sensitive meter movements without affecting meter linearity. The circuit shown in Fig. 2.46 represents a D.C volt meter which reads 20V full scale. The meter resistance is 560 Ω and $R_1 + R_2 = 99.5 k\Omega$. If the diode is a 16V Zener, find R_1 and R_2 so that when $V_i > 20V$, the Zener Diode conducts and the overload current is shunted away from the meter.

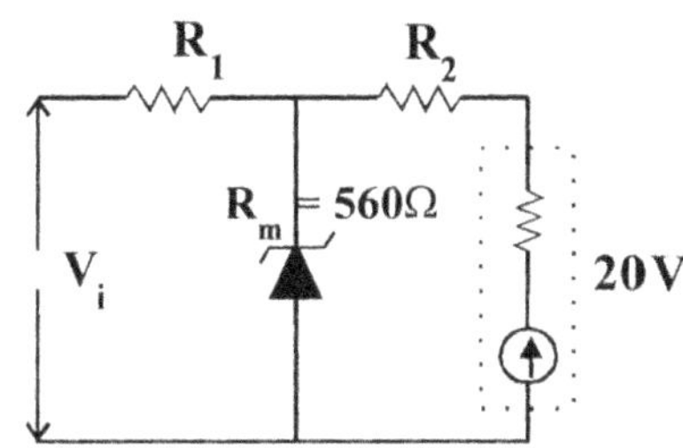

Fig 2.46 For Problem 2.39.

Solution

When $V_i = 20V$, the Zener should not conduct.

$$\therefore \qquad 20V = (R_1 + R_2 + R_m) \times I = (99.5 + 0.56) \times 10^3 \times I$$

$$\therefore \qquad I \approx \frac{20V}{100 K\Omega} = 200 \ \mu A \text{ on full scale.}$$

When $V_i > 20V$, the voltage across R_1 and R_m must be equal to the Zener voltage $V_z = 16V$.

or $\qquad (R_1 + R_m) \times I = 16V$

$$\therefore \qquad R_1 + R_m = \frac{16V}{200 \times 10^{-6}} = 80 K\Omega$$

$$\therefore \qquad R_1 = 80 K\Omega - 500 \Omega \simeq 79.5 K\Omega$$

$$\therefore \qquad R_2 = 99.5 K\Omega - 79.5 K\Omega = 20 K\Omega$$

Problem 2.19

What is the ratio of the current for a forward bias of 0.05V to the current for the same magnitude of reverse bias for a Germanium Diode?

Solution

For Forward Bias, V $\qquad = + 0.05V = +50mV.$

For Reverse Bias, V $\qquad = - 0.05V = -50mV.$

$\qquad V_T \qquad = + 0.026V = +26mV$

$\qquad \eta \qquad = 1$ for Germanium Diode.

$$\frac{e^{\frac{V}{V_T}} - 1}{e^{\frac{V}{V_T}} - 1} = \frac{e^{50/26} - 1}{e^{-50/26} - 1} = \frac{e^{1.92} - 1}{e^{-1.92} - 1} = \frac{6.82 - 1}{0.147 - 1} = -6.83.$$

Negative sign is because, the direction of current is opposite when the diode is reverse biased.

Problem 2.20

It is predicted for Germanium the reverse saturation current should increase by $0.11°C^{-1}$. It is found experimentally in a particular diode that at a reverse voltage of 10V, the reverse current is 5 mA and the temperature dependence is only $0.07\ °C^{-1}$. What is the resistance shunting the diode ?

Solution

$$I = I_0 + I_R$$
$$\frac{dI}{dT} = \frac{dI_0}{dT}$$

$\therefore$ I_R the reverse saturation current, will not change with temperature.

$$I = I_0 + \frac{10}{R}$$

For Germanium, $\quad \dfrac{1}{I_0} \times \dfrac{dI_0}{dT} = 0.11.$

$$\frac{1}{I} \times \frac{dI}{dT} = 0.07.$$

Taking the ratio of

$$\frac{\dfrac{1}{I_0} \times \dfrac{dI_0}{dT}}{\dfrac{1}{I} \times \dfrac{dI}{dT}} = \frac{I}{I_0} = \frac{0.11}{0.07}$$

or
$$I_0 = 0.636\ I.$$
$$I_R = I - 0.636$$

$$I = 0.364\ I\ = \frac{0.364}{0.636}I_0 = 0.572\ I_0$$

$$I_0 = 5mA$$
$\therefore \qquad I_R = 0.572(5mA) = 2.68\ mA.$

$\therefore \qquad R = {V}\!/\!{I_R} = \dfrac{10}{2.86mA} = 3.5\ K\Omega$

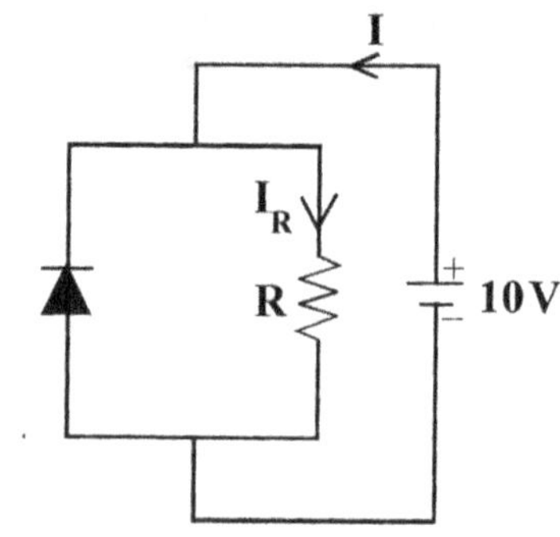

Fig 2.47 For Problem 2.41.

Problem 2.21

Over what range of input voltage will the zener regulator circuit maintains 30V across $2K\Omega$ resistor, assuming $R_S = 200\Omega$ and max. zener current is 25 mA

Solution

$$I_L = \frac{30V}{2K\Omega} = 15mA$$

Max. Zener Current $= 25mA$

$\therefore \qquad$ Total Current $= 25 + 15 = 40mA$

$\therefore \qquad V_{i\ max} = 30\ V + R_S \times I = 30V + 200(15 + 25) = 38V$

SUMMARY

♦ Rectifier circuit converts AC to Unidirectional Flow.

♦ Filter Circuit converts Unidirectional flow to DC. It minimises Ripple.

♦ The different types of filter circuits are

    1. Capacitor Filter

    2. Inductor Filter

    3. L-Section (LC) Filter.

    4. π-Section Filter

    5. CRC and CLC Filters

♦ Ripple Factor $= \dfrac{I'_{rms}}{I_{DC}}$. For Half Wave Rectifier, $\gamma = 1.21$, for Full Wave Rectifier $\gamma = 0.482$

♦ Expression for Ripple Factor for C Filter,

$$\gamma = \frac{1}{4\sqrt{3}\,f\,CR_L}$$

♦ Expression for Ripple Factor for L Filter,

$$\gamma = \frac{R_L}{4\sqrt{3}\,\omega L}$$

♦ Expression for Ripple Factor for LC Filter,

$$\gamma = \frac{\sqrt{2}}{3} \times \frac{1}{2\omega C} \times \frac{1}{2\omega L}$$

♦ Expression for Ripple Factor for π Filter,

$$\gamma = \frac{\sqrt{2}X_C}{R_L}\left(\frac{X_{C_1}}{X_{L_1}}\right)$$

♦ Critical Inductance $L_C \geq \dfrac{R_L}{3\omega}$

♦ Bleeder Resistance $R_B = \dfrac{3X_L}{2}$

OBJECTIVE TYPE QUESTIONS

1. Rectifier Converts .... to ....

2. Filter Circuit Converts .... to ....

3. Ripple Factor in the case of Full Wave Rectifier Circuit is ....

4. The characteristics of a swinging choke is ....

5. In the case of LC Filter Circuits, Bleeder Resistance ensures

    (1) ....

    (2) ....

6. In the case of Half wave Rectifier (HWR), expressions for V_{OC} in terms of V_m, the maximum value of A.C input voltage is (neglecting R_f).

7. In the case of HWR circuit, the expreesion for rms value of ripple voltage of V_m is, V_V (rms) = ...

8. For HWR circuit, riple factor γ as a % is, ...

9. Expression for V_{OC} in the case of Full Wave Rectifier (FWR) circuit, negleting R_f the forward resistance of the diobe, in terms of V_m is ....

10. RMS value of ripple voltage V_r in terms of V_m in the case of FWR circuiit is,

11. Ripple factor γ as a % in the case of FWR circuits ....

12. In the case of capacitor filter, exprssion for rms value of ripple voltage V_P (rms) in terms of I_{OC}, f and c is ....

13. Expression for ripple factor in the case of capacitor filter in terms of R_L and C is ....

14. Value of Transformer utilization factor (TUF) for HWR circuit is (if R_f is small) ....

15. Value of TUF for FWR circuit, negleting R_f is using centre tapped transformer is ....

ESSAY TYPE QUESTIONS

1. Obtain the expression for ripple factor in the case of Full Wave Rectifier Circuit with Capacitor Filter.

2. Explain the terms Swinging Choke and Bleeder Resistor.

3. Compare C, L, L-Section, π-Section (CLC and CRC) Filters in all respects.

MULTIPLE CHOICE QUESTIONS

1. **Special types of diodes in which transition time and storage time are made small are called ...**

   (a) Snap diodes   (b) Rectifier diodes (c) Storage diodes   (d) Memory diodes

2. **The Circuit which converts undirectional flow to D.C. is called**

   (a) Rectifier circuit   (b) Converter circuit

   (c) filter circuit   (d) Eliminator

3. **For ideal Rectifier and filter circuits, % regulations must be ...**

   (a) 1%   (b) 0.1 %   (c) 5%   (d) 0%

4. **The value of current that flows through R_L in a 'π' section filter circuit at no load is**

   (a) ∞ (b) 0.1 mA   (c) 0   (d) few mA

CONCEPTUAL QUESTIONS (INTERVIEW QUESTIONS)

1. When a circuit is on 'No Load', what is the current value ?

Ans: It is zero. 'No Load' means NO LOAD CURRNT. I_L is zero. The impedance value Z_L or R_L is ∞. So no current flows or I_L is zero at no load.

2. What is the difference between a Rectifier and Filter Circuit?

Ans: Rectifier circuit converts A.C to unidirectional flow. Filter converts unidirectional flow to D.C.

Transistor Characteristics

3

In this Chapter,

- The principle of working of Bipolar Junctions Transistors, (which are also simply referred as Transistors) and their characteristics are explained.

- The Operation of Transistor in the three configurations, namely Common Emitter, Common Base and Common Collector Configurations is explained.

- The variation of current with voltage, in the three configurations is given.

- The structure of Junction Field Effect Transistor (JFET) and its V-I characteristics are explained.

- The structure of Metal Oxide Semiconductor Field Effect Transistor (MOSFET) and its V-I characteristics are explained.

- JFET amplifier circuits in Common Source (CS) Common Drain (CD) and Common Gate (CG) configurations are given.

I.   INTRODUCTION

3.1   BIPOLAR JUNCTION TRANSISTORS (BJT'S)

The device in which conduction takes place due to two types of carriers, electrons and holes is called a Bipolar Device. As *p-n junctions* exist in the construction of the device, it is a junction device. When there is transfer of resistance from input side which is Forward Biased (low resistance) to output side which is Reverse Biased (high resistance), it is a **Trans Resistor or Transistor Device**. There are two types of transistors NPN and PNP. In NPN Transistor, a *p-type* Silicon (Si) or Germanium (Ge) is sandwiched between two layers of *n-type* silicon. The

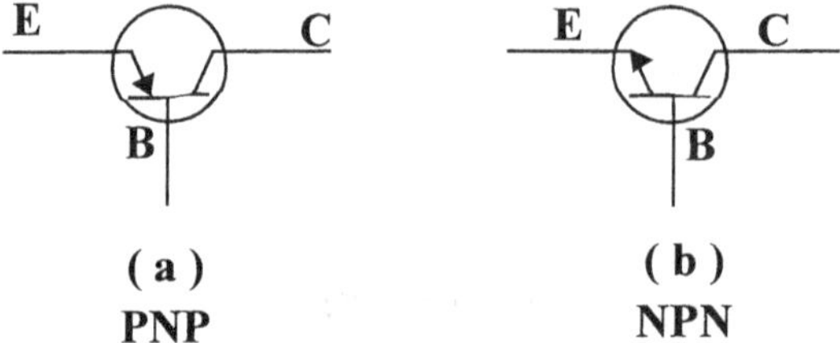

Fig. 3.1 Transistor symbols.

symbol for PNP transistor is as shown in Fig. 3.1(a) and for NPN transistor, is as shown in Fig 3.1 (b).

The three sections of a transistor are Emitter, Base and Collector. If the ***arrow mark is towards the base***, it is PNP transistor. If it is away then it is NPN transistor. The arrow mark on the emitter specifies the direction of current when the emitter base junction is forward biased. When the PNP transistor is forward biased, holes are injected into the base. So the holes move from emitter to base. The conventional current flows in the same direction as holes. So arrow mark is towards the base for PNP transistor. Similarly for NPN transistor, it is away. DC Emitter Current is represented as I_E, Base Current as I_B and Collector Current as I_C. These currents are assumed to be positive when the currents flow into the transistor. V_{EB} refers to Emitter - Base Voltage.   Emitter (E) Voltage being measured with reference to base B (Fig. 3.2). Similarly V_{CB} and V_{CE}.

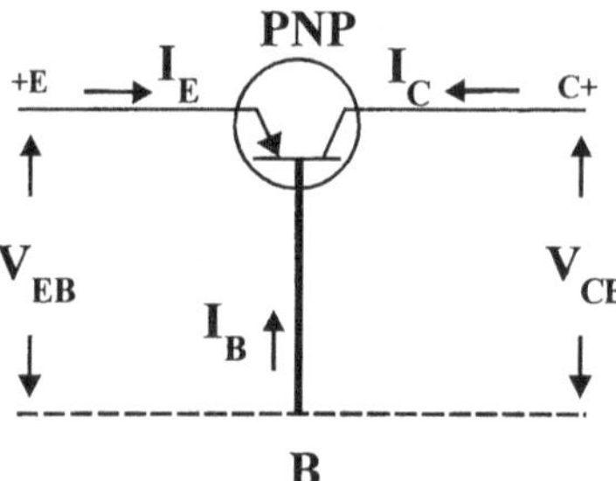

Fig. 3.2 Current components in a pnp transistor.

3.1.1   POTENTIAL DISTRIBUTIONS THROUGH A TRANSISTOR

Fig. 3.3 shows a circuit for a PNP transistor. Emitter Base Junction is Forward Biased Collector - Base Junction is reverse biased. Fig. 3.4 (a) shows potential distribution along the transistor (x).

When the transistor is open circuited, there is no voltage applied. The potential barriers adjust themselves to a height V_o. V_o will be few tenths of a volt. It is the barrier potential. So when the transistor is open circuited holes will not be injected into the collector. (Potential barrier exists all along the base width, since the holes from emitter have to reach collector and not just remain in base).

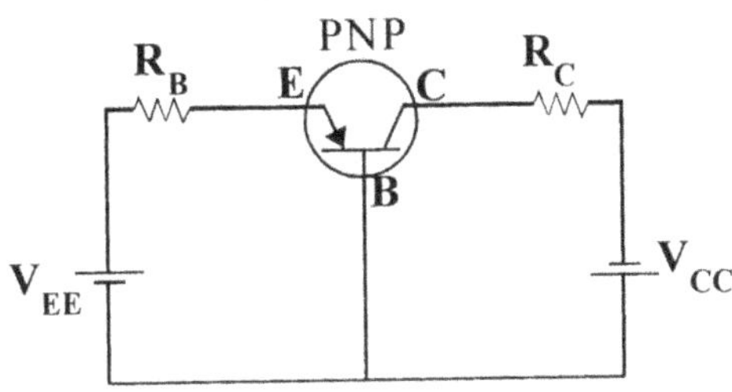

Fig 3.3 PNP transistor circuit connections.

But when transistor bias voltages are applied, *emitter base junction is forward biased* and the *collector base junction is reverse biased*. The base potential almost remains constant. But if the reverse bias voltage of collector - base junction is increased, *effective base width decreases*. Since EB junction is forward biased, in a PNP transistor, the emitter base barrier is lowered. So holes will be injected into the base. The injected holes diffuse into the collector across the *n-type* material (base). When EB junction is forward biased emitter - base potential is increased by $|V_{EB}|$. Similarly collector base potential is reduced by $|V_{CB}|$. Collector base junction is reverse biased. So collector extends into the base or depletion region width increases. Emitter - Base Junction is to be Forward Biased because, the carriers must be injected and Collector - Base Junction must be reverse biased, because the carriers must be attracted into the Collector Region. Then only current flow results through the transistor.

3.1.2 TRANSISTOR CURRENT COMPONENTS

Consider a PNP transistor. When Emitter - Base junction is forward biased, holes are injected into the base. So this current is I_{pE} (holes crossing form Emitter to base). Also electrons can be injected from base to emitter (which also contribute for emitter current I_E). So the resulting current is I_{nE}. Therefore the total emitter current $I_E = I_{pE} + I_{nE}$.

The ratio of $\dfrac{I_{pE}}{I_{nE}}$ is proportional to the ratio of conductivity of the 'p' material to that of 'n' material. In commercial transistor the *Doping of the Emitter is made much large than the Doping of the Base*. So 'p' region conductivity will be much larger than 'n' region conductivity. So the emitter current mainly consists of holes only. (Such a condition is desired since electron hole recombination can take place at the emitter junction). *Collector is lightly doped. Basewidth is narrow* (Fig. 3.4(a)).

All the holes injected from emitter to base will not reach collector, since some of them recombine with electrons in the base and disappear.

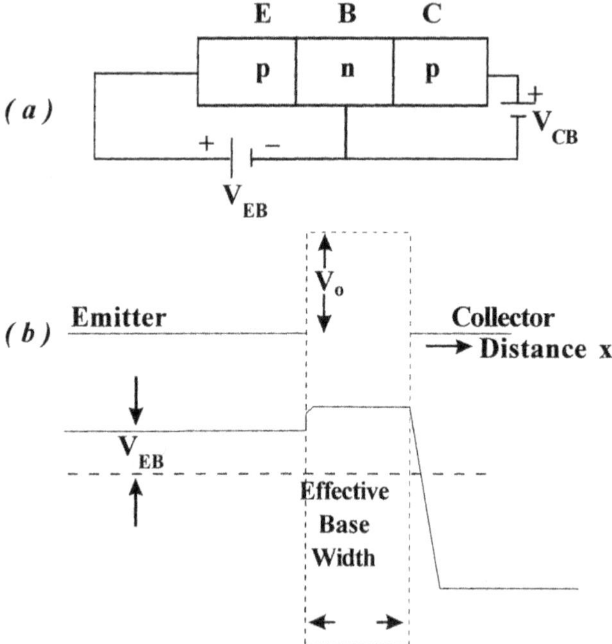

Fig 3.4 Potential barrier levels in a transistor.

If I_{PE} is the current due to holes at the Emitter Junction, $I_{pE} > I_{pC}$ since some holes have recombined with electrons in the base. The recombination current is equal to ($I_{PE} - I_{PC}$). If the emitter is open circuited, so that $I_E = 0$, then I_{pC} would be zero. Under these conditions, junction acts as a reverse biased diode. So the Collector Current is approximately equal to reverse saturation Current I_{CO}. If $I_E \neq 0$, then $I_C = I_{CO} + I_{pC}$.

For a PNP transistor, I_{CO} consists of holes moving from left to right (base to collector) and electrons crossing in the opposite direction. In the base, which is *n-type*, holes are the minority carriers. In the collector, which is 'p' type, electrons are the minority carriers. I_{co} is due to minority carriers. Currents entering the transistor are positive. For NPN transistor, I_{CO} consists of electrons, the minority carriers in the base moving to collector and holes from collector moving to base. So the direction of I_{CO} for NPN transistor is the same as the conventional current entering into the transistor. *Hence I_{CO} is positive for NPN transistor.*

Emitter Efficiency (γ) :

$$\gamma = \frac{\text{Current due to Injected at the Emitter - Base Junction}}{\text{Total Emitter Current}}$$

For a PNP transistor,

$$\gamma = \frac{I_{pE}}{I_{pE} + I_{nE}} = \frac{I_{pE}}{I_E}$$

Value of γ is always less than 1.

I_{pE} = Current due to injected holes from emitter to base

I_{nE} = Current due to injected electrons from base to emitter.

Transportation Factor (β^*) :

$$\beta^* = \frac{\text{Current due to injected carriers reaching B - C junction}}{\text{Injected carrier current at emitter - base junction}}$$

For a PNP transistor,

$$\beta^* = \frac{I_{pc}}{I_{pE}} = \frac{I_p \text{ in the Collector}}{I_p \text{ in the Emitter}} = \frac{\text{Hole Current in the Collector}}{\text{Hole Current in the Emitter}}$$

Large Signal Current Gain (α)

$$\alpha = \frac{\text{Collector Current}}{\text{Emitter Current}}$$

$$\alpha = \frac{I_{pC}}{I_E}$$

Multiplying and dividing by I_{pE}.

$$\alpha = \frac{I_{pC}}{I_{pE}} \times \frac{I_{pE}}{I_E}$$

$$\frac{I_{pc}}{I_{pE}} = \beta^* ; \qquad\qquad \frac{I_{pE}}{I_E} = \gamma$$

$$\therefore \qquad \boxed{\alpha = \beta^* \times \gamma}$$

3.1.3 TRANSISTOR AS AN AMPLIFIER

A small change ΔV_i of Emitter - Base Voltage causes relatively large Emitter - Current Change ΔI_E. α' is defined as the ratio of the change in collector current to the change in emitter current. Because of change in emitter current ΔI_E and consequent change in collector current, there will be change in output voltage ΔV_o.

$$\Delta V_0 = \alpha' \times R_L \times \Delta I_E \qquad\qquad \text{Since } \alpha' = \frac{\Delta I_C}{\Delta I_E}$$

$$\Delta V_i = r_e' \times \Delta I_E$$

R_L is load resistance. Therefore, the voltage amplification is,

$$A_V = \frac{\Delta V_0}{\Delta V_i} = \frac{\alpha' R_L . \Delta I_E}{r_e' \Delta I_E}$$

$$A_V = \frac{\alpha'.R_L}{r_e'} ; \alpha' = \left.\frac{\Delta I_C}{\Delta I_E}\right|_{V_{CB}=K}$$

$$R_L > r_e'$$

$$A_V > 1$$

α' = Small Signal Forward - Circuit Current Gain.

r_e' = Emitter Junction Resistance.

The term, r_b' is used since this resistance does not include contact or lead resistances. r_b' value is small, because emitter - base Junction is forward biased.

So a small change in emitter base voltage produces large change in collector base voltage. Hence, the transistor acts as an amplifier. (This is for common base configuration. Voltages are measured with respect to base).

The input resistance of the circuit is low (typical value 40Ω) and output resistance is high (3,000Ω). So current from low resistance input circuit is transferred to high resistance output circuit (R_L). *So it is a transfer resistor device and abbreviated as transistor.* The magnitude of current remains the same ($\alpha' \approx 1$).

3.2 TRANSISTOR CONSTRUCTION

Two commonly employed methods for the fabrication of diodes, transistors and other semiconductor devices are :

    1. Grown type **2.** Alloy type

3.2.1 GROWN TYPE

It is made by drawing a single crystal from a melt of Silicon or Germanium whose impurity concentration is changed during the crystal drawing operation.

3.2.2 ALLOY TYPE

A thin wafer of *n-type* Germanium is taken. Two small dots of indium are attached to the wafer on both sides. The temperature is raised to a high value where indium melts, dissolves Germanium beneath it and forms a saturator solution. On cooling, indium crystallizes and changes Germanium to *p-type*. So a PNP transistor formed. The doping concentration depends upon the amount of indium placed and diffusion length on the temperature raised.

3.2.3 MANUFACTURE OF GROWN JUNCTION TRANSISTOR

Grown Junction Transistors are manufactured through growing a single crystal which is slowly pulled from the melt in the crystal growing furnace Fig. 3.5. The purified polycrystalline semiconductor is kept in the chamber and heated in an atmosphere of N_2 and H_2 to prevent oxidation. A seed crystal attached to the vertical shaft which is slowly pulled at the rate of (1mm / 4hrs) or even less. The seed crystal initially makes contact with the molten semiconductors. The crystal starts growing as the seed crystal is pulled. To get *n-type* impurities, to the molten polycrystalline solution, impurities are

1. **Furnace**

2. **Molten Germanium**

3. **Seed Crystal**

4. **Gas Inlet**

5. **Gas Outlet**

6. **Vertical Drive**

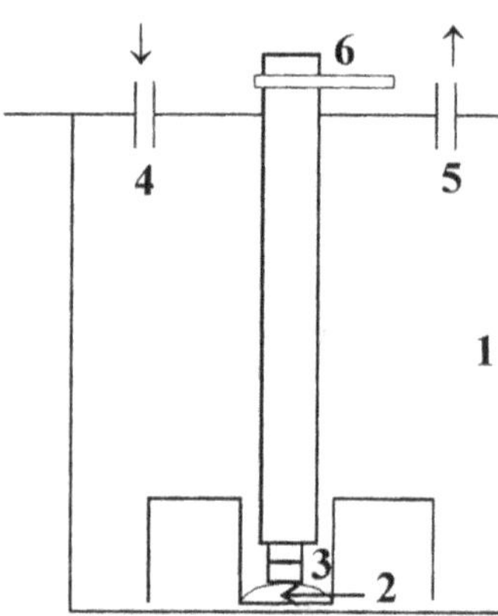

Fig 3.5 Crystal growth.

added. Now the crystal that grows is ***n-type*** semiconductor. Then ***p-type*** impurities are added. So the crystal now is ***p-type***. Like this NPN transistor can be fabricated.

3.2.4 DIFFUSION TYPE OF TRANSISTOR CONSTRUCTION

To fabricate NPN transistor, ***n-type*** 'Si' (or 'Ge') wafer is taken. This forms the collector. Base-collector junction areas is determined by a mask and diffusion length. ***p-type*** impurity (Boron) is diffused into the wafer by diffusion process. The wafer to be diffused and the Boron wafer which acts as ***p-type*** impurity, are placed side by side. Nitrogen gas will be flowing at a particular rate, before the Boron is converted to Boron Oxide and this gets deposited over Siliconwafer. Baron is driven inside by drive in diffusion process. So the base-collector junction will be formed. Now again ***n-type*** impurity (Phosphorus) is diffused into the Base Area in a similar way. The different steps in process are :

1. *Oxidation*
2. *Photolithography*
3. *Diffusion*
4. *Metallization*
5. *Encapsulation*

3.2.5 EPITAXIAL TYPE

In this process, a very thick, high purity (high resistance), single crystal layer of Silicon or Germanium is grown as a heavily doped substrate of the same material. (n on n^+) or (p on p^+). This forms the collector. Over this base and emitter are diffused. *Epitaxi* is a greek word. *Epi* mean 'ON' *Taxi* means 'ARRANGEMENT'. This technique implies the growth of a crystal on a surface, with properties identical to that of the surface. By this technique abrupt step type p-n junction can be formed. The structure is shown in Fig 3.6 (a).

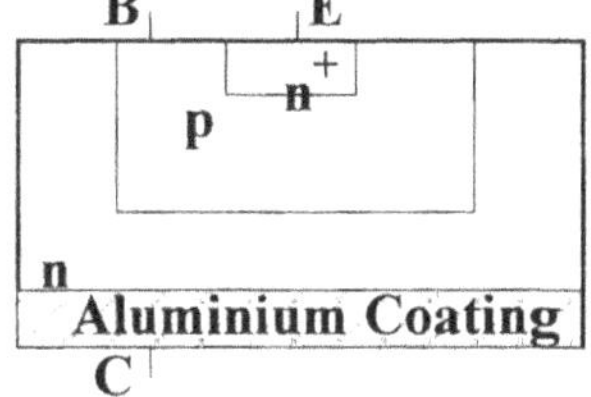

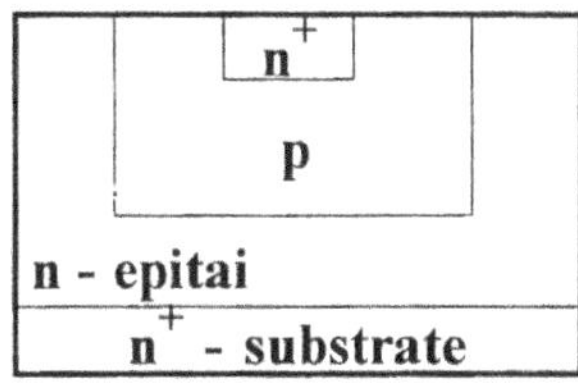

Fig 3.6 (a) Epitaxial type transistor.

3.3   THE EBERS - MOLL EQUATION

EBERS - MOLL MODEL OF A TRANSISTOR

A transistor can be represented by two diodes connected back to back. Such a circuit for a PNP transistor shown is called ***Eber - Moll Model***. The transistor is represented by the diodes connected back to back. I_{EO} and I_{CO} represent the reverse saturation currents. $\alpha_N I_E$ is the current source. $\alpha_I I_C$ represents the current source in inverted mode (Fig. 3.7).

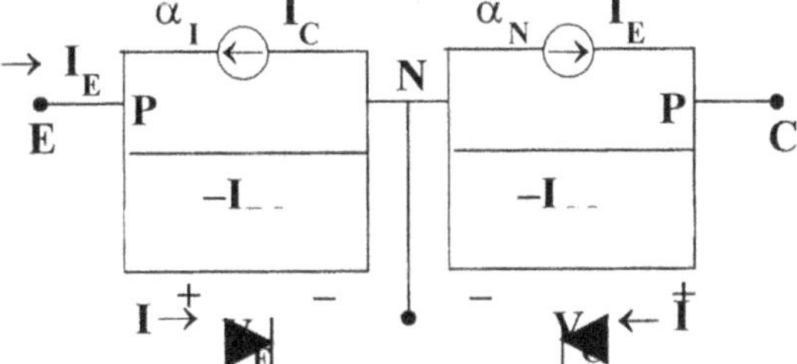

Fig 3.7 Eber-Moll Model of PNP transistor.

A transistor cannot be formed by simply connecting two diodes since, the base region should be narrow. Otherwise the majority carriers emitted from the emitter will recombine in the base and disappear. So transistor action will not be seen.

$$I_E = I_{pE} + I_{nE} = I_{pn}(o) + I_{np}(o)$$

$$I_{pn}(o) = \frac{Ae\,D_p\,p_{no}}{w}\left[\left(e^{V_C/V_T}-1\right)-\left(e^{V_E/V_T}-1\right)\right]$$

$$I_{np}(o) = \frac{Ae\,D_n\,n_{EO}}{L_E}\left(e^{V_E/V_T}-1\right)$$

$$\therefore \qquad I_E = a_{11}\left(e^{V_E/V_T}-1\right) + a_{12}\left(e^{V_C/V_T}-1\right) \qquad \dots\dots\dots\dots(1)$$

$$a_{11} = A_e\left(\frac{D_p\cdot p_{no}}{w} + \frac{D_n\,n_{EO}}{L_E}\right)$$

$$a_{12} = \frac{-Ae\,D_p\,p_{no}}{w}$$

Similarly

$$I_C = a_{21}\left(e^{V_E/V_T}-1\right) + a_{22}\left(e^{V_C/V_T}-1\right) \qquad \dots\dots\dots\dots(2)$$

$$a_{21} = \frac{-Ae\,D_p\,p_{no}}{w} \;\;;\;\; a_{22} = Ae\left(\frac{D_p\,p_{no}}{w} + \frac{D_n\,n_E}{L_C}\right)$$

$$\therefore \qquad a_{11} = a_{21} \;; \text{ Equations (1) and (2) are called } \textit{\textbf{Eber-Moll Equations}}.$$

Nomanclature used for Transistors (BJTs)

BC107	B :	Silicon Device
(NPN Transistor)	C :	Audio Amplifier application
	107 :	Type Number (No Significance)
AF 114	A :	Germanium Device
	F :	High Frequency Applications
	114 :	Type No.
2N 2222	2 :	Two Junction semiconductor device
(NPN Transistor)		
AC 128	:	Germanium Audio Frequency range BJT
BC 147	:	Silicon Audio Frequency range Transistor
2N 3866	:	NPN RF Power Transistor, 5W, 30V 0.4A.
2 N6253	:	NPN High power, 115W, 45V, 15A.
2 N 1481	:	NPN 5W, 40V, 1.5A

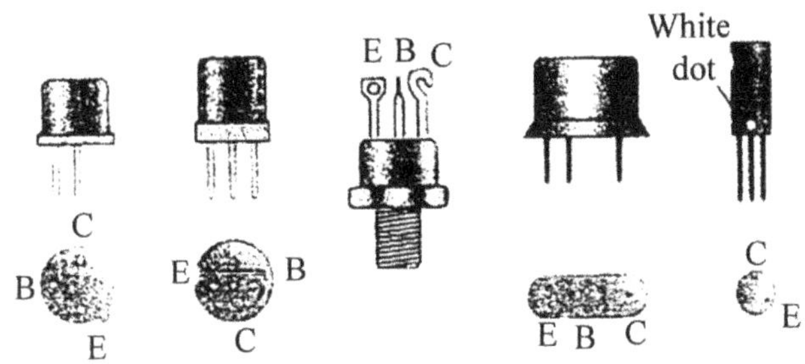

Fig. 3.7(a) Transistor terminal identification

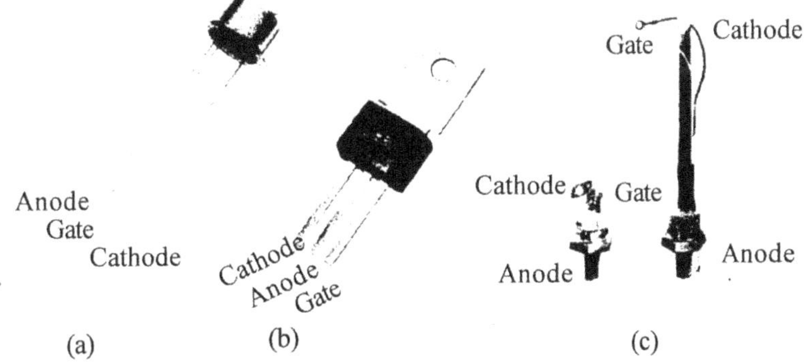

Fig. 3.7(b) SCR case construction and terminal identification [(a) courtesy general electric company; (b) and (c) courtesy international rectifier corporation]

Specifications of a Bipolar Junction Transistor (BJT)

Sl.No.	Parameter	Symbol	Typical value	Units
1.	Collector-Emitter Voltage	V_{CE}	40	V
2.	Collector-Base Voltage	V_{CB}	60	V
3.	Emitter-Base Voltage	V_{EB}	6	V
4.	Collector-Current	I_C	500	mA
5.	Total Power Dissipation	P_D	0.4	W
6.	Operating Temperature	T_{op}	25-80	°C
7.	Collector- Emitter Break down voltage	BV_{CEO}	40	V
8.	Emitter-Base Break down voltage	BV_{EBO}	50	V

9.	Current - Gain Bandwidth product	f_T	250	MHz
10.	Output Capacitance	C_0	8	p_f
11.	Input Capacitance	C_{in}	20	p_f
12.	Small Signal Current gain	h_{fe}	100	-
13.	Output admittance	h_{oe}	10	m
14.	Voltage feedback ratio	h_{re}	8×10^{-4}	-
15.	Input impedance	h_{ie}	3	$k\Omega$
16.	Delay time	t_d	10	nSec.
17.	Rise time	t_r	20	nSec.
18.	Fall time	t_f	50	nSec.
19.	Storage time	t_s	200	nSec.
20.	Noise Figure	NF	4	dB

3.4 TYPES OF TRANSISTOR CONFIGURATIONS

Transistors (BJTs) are operated in three configuration namely,

    1.   Common Base Configuration (C.B)

    2.   Common Emitter Configuration (C.E)

    3.   Common Collector Configuration (C.C)

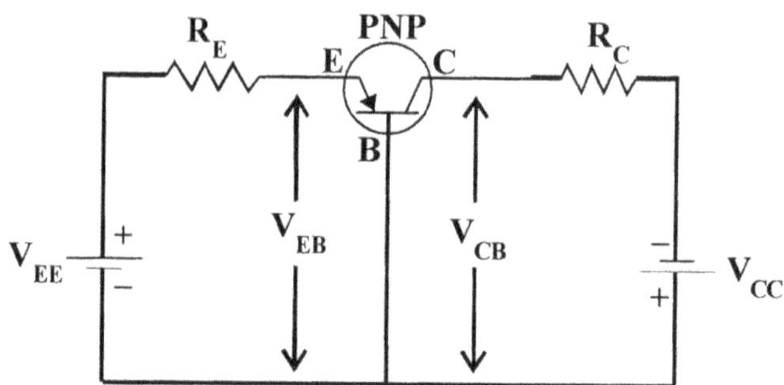

Fig. 3.8 Common base amplifier circuit.

3.4.1 COMMON BASE CONFIGURATION (C.B)

The base is at ground potential. So this is known as ***Common Base Configuration*** or ***Grounded Base Configuration.*** Emitter and Collector Voltages are measured with respect to the base. The convention is currents, entering the transistor are taken as positive and those leaving the transistor as negative. For a PNP transistor, holes are the   majority carriers. Emitter Base Junction is

Forward Biased. Since holes are entering the base or I_E is positive. Holes through the base reach collector. I_C is flowing out of transistor. Hence I_C is negative; similarly I_B is negative. The characteristics of the transistor can be described as,

$$V_{EB} = f_1(V_{CB}, I_E)$$

Dependent Variables are Input Voltage and Output Current.

$$I_C = f_2(V_{CB}, I_E)$$

Independent Variables are Input Current and Output Voltage. For a Forward Biased Junction V_{EB} is positive, for reverse bias collector junction V_{CB} is negative.

Transistor Characteristics in Common Base Configuration

The three regions in the output characteristics of a transistor are

1.   *Active region*

2.   *Saturation region*

3.   *Cut off region*

The input and output characteristics for a transistor in C.B configuration are as shown in Fig. 3.9 (a) and (b) When $I_E = 0$, the emitter is open circuited. So the transistor is not conducting and the I_{CBO} is not considered. Generally the transistor is not used in this mode and it is regarded as cut off; Even though I_{CBO} is present it is very small for operation. During the region OA, for a small variation of V_{BC} there is very large, change in current I_C with I_E. So current is independent of voltage. There is no control over the current. Hence the transistor can not be operated in this region. So it is named as Saturation region. The literal meaning of saturation should not be taken. The region to the right is called active region. For a given V_{CB}, with I_E increased, there is no appreciable change in I_C. It is in saturation. When transistor is being used as a switch, it is operated between cut off and saturation regions.

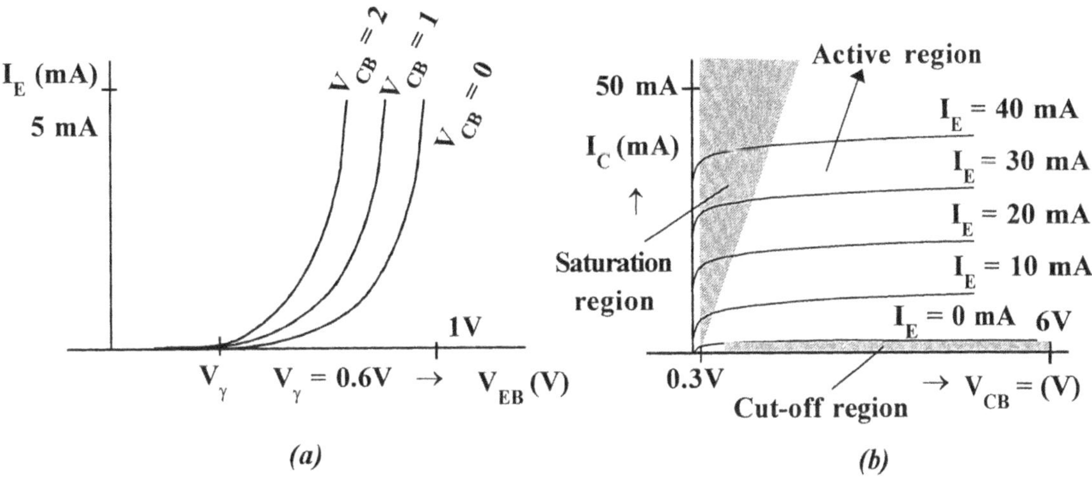

Fig 3.9 (a) Input characteristics in C.B configuration

(b) Output charactersitics in C.B configuration.

EARLY EFFECT

The collector base junction is reverse biased. If this reverse bias voltage is increased, the space charge width at the *n-p region* of Base - Collector Junction of PNP Transistor increases. So effective base width decreases (Fig. 3.10)

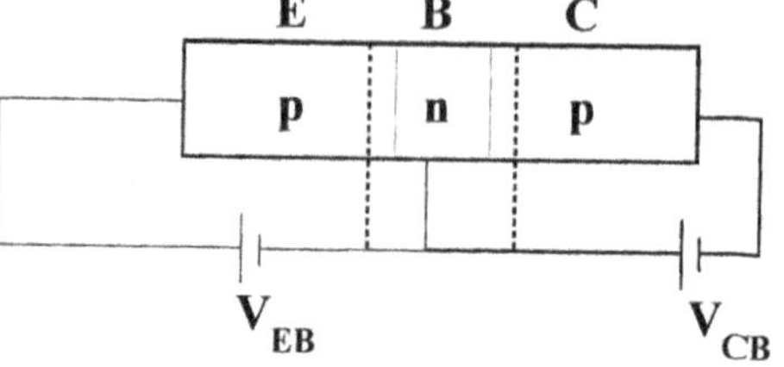

$$V_B = \frac{e.N_A}{2\in}.W^2$$

Fig 3.10 Early effect.

W = Space Charge Region Width.

The decrease in base width with increase in collector reverse bias voltage is known as Early Effect. When the base width decreases, the probability of recombination of holes and electrons in the base region is less. So the transportation factors β^* increases.

3.4.2 COMMON EMITTER CONFIGURATION (C.E)

In this circuit Fig 3.11, emitter is common to both base and collector. So this is known as CE configuration or grounded emitter configuration. The input voltage V_{BE}, and output current I_C are taken as the dependent variables. These depend upon the output voltage V_{CE} and input current I_B.

$$\therefore \quad V_{BE} = f_1 (V_{CE}, I_B)$$
$$I_C = f_2(V_{CE}, I_B)$$

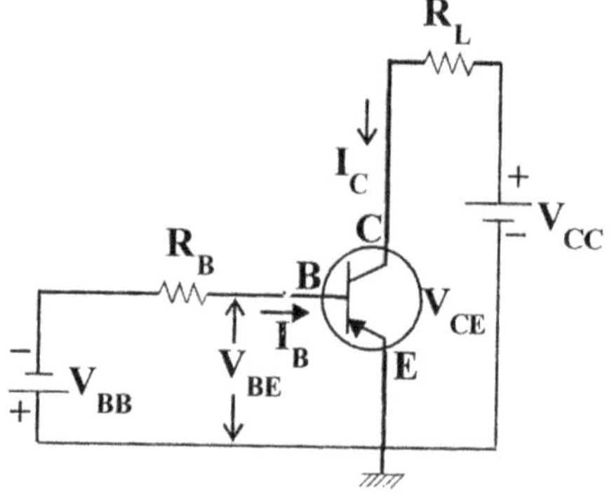

Fig 3.11 Common Emitter pnp transistor amplifier circuit.

INPUT CHARACTERISTICS

If the collector is shorted to the emitter, the transistor is similar to a Forward Biased Diode. So I_B increases as V_{BE} increases. The input characteristics are as shown in Fig. 3.12. I_B increases with V_{BE} exponentially, beyond cut-in voltage V_γ. The variation is similar to that of a Forward Biased Diode. If $V_{BE} = 0, I_B = 0$, since emitter and collector junctions are shorted. If V_{CE} increases, base width decreases (by Early effect), and results in decreased recombination. As V_{CE} is increased, from -1 to -3 I_b decreases. V_γ is the cut in voltage.

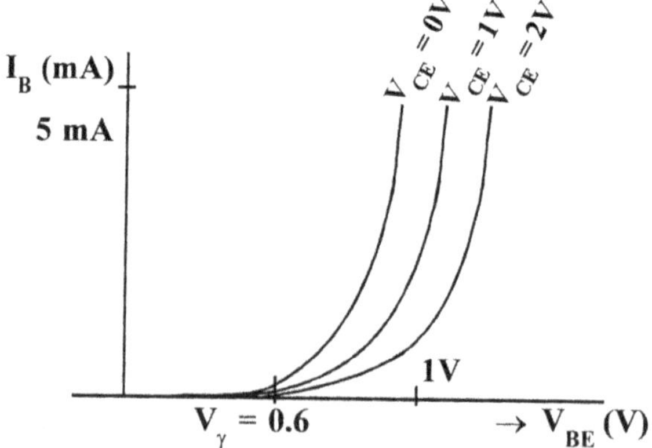

Fig 3.12 Input characteristics in C.E configuration.

$$I_E + I_B + I_C = 0$$
$$I_E = -(I_B + I_C)$$

As V_{CE} increases, I_C increases, therefore I_B decreases. Therefore, base recombination will be less. The input characteristics are similar to a forward biased p-n junction diode.

$$I_B = I_0\left(e^{\frac{V_{BE}}{nV_T}} - 1\right) \qquad ....(3.3)$$

The input characteristics I_B vs V_{BE} follow the equation,

$$I_B = I_0\left(e^{\frac{V_{BE}}{nV_T}} - 1\right) \qquad ....(3.4)$$

OUTPUT CHARACTERISTICS

The transistor is similar to a Collector Junction Reverse Biased Diode. The output characteristics is divided into three regions namely,

    1.   *Active Region.*

    2.   *Saturation Region*

    3.   *Cut-Off Region.*

These are shown in Fig 3.13. If I_B increases, there is more injection into the collector. So I_C increases. Hence characteristics are as shown.

The output characteristics are similar to a reverse biased *p-n junction.* (Collector Base junction is reverse biased). They follow the equation.

$$I_C = I_0\left(e^{\frac{V_{CE}}{nV_T}} - 1\right) \qquad .....(3.5)$$

The three different regions are

    1. Active Region   2. Saturation Region   3. Cut-off Region.

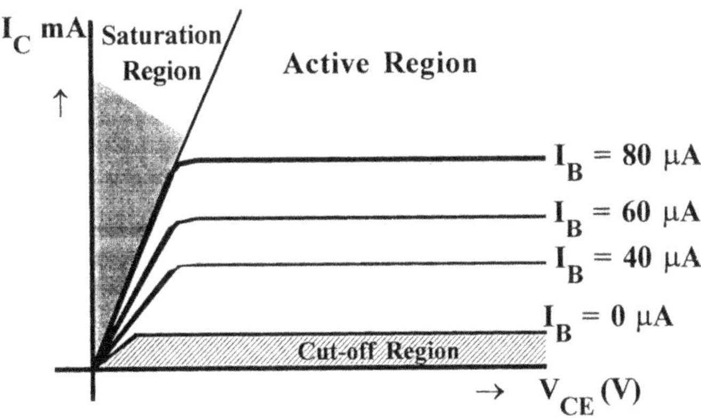

Fig 3.13 Output characteristics in C.E Configuration.

$$I_B + I_C + I_E = 0$$

$\therefore \quad I_E = -(I_B + I_C)$

$$-I_C = +I_{CO} - \alpha . I_E$$

But $\quad I_E = -(I_B + I_C)$

$\therefore \quad I_C = -I_{CO} + \alpha . I_B + \alpha . I_C$

$$I_C (1 - \alpha) = -I_{CO} + \alpha . I_B$$

If the transistor were to be at cut off, I_B must be equal to zero.

TEST FOR SATURATION

1. $$|I_B| \geq \left| \frac{I_C}{\beta} \right|$$

2. If V_{CB} is positive for PNP transistor and negative for NPN transistor, the transistor is in saturation.

COMMON EMITTER CUT OFF REGION

If a transistor was to be at cut off, $I_C = 0$. To achieve this, if the emitter-base junction is open circuited, there cannot be injection of carriers from E to C, through B. Hence $I_C = 0$. (Any reverse saturation current flowing because of thermal agitation is very small. So the transistor can not be operated). But in the case of C.E configuration, even if I_B is made 0, the transistor is not cut off. I_C will have a considerable value even when $I_B = 0$. The circuit is shown in Fig. 3.14.

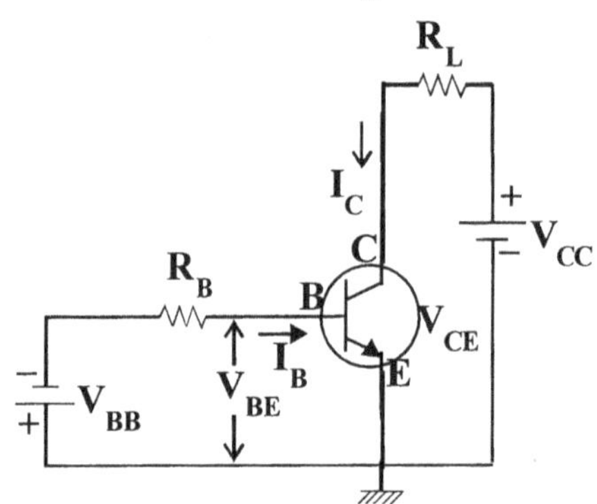

Fig 3.14 Common Emitter Cut-off region.

Because, $I_C = -\alpha I_E + I_{CO}$.

α = Current gain ;

I_{CO} = Reverse saturation current in the collector.

But $I_E = -(I_B + I_C)$.

If $I_B = 0, I_E = -I_C$

$\therefore \quad I_C = -\alpha(-I_C) + I_{CO}$

or $I_C (1 - \alpha) = I_{CO}$

or $I_C = \dfrac{I_{CO}}{1 - \alpha} = I_{CBO}$

In order to achieve cut-off in the Common Emitter Configuration, Emitter Base Junction should be reverse biased. α is constant for a given transistor. So the actual collector current with collector junction reverse biased and base open circuited is designated by I_{COB}. If $\alpha \approx 0.9$, then $I_{CBO} \approx 10\, I_{CO}$. So the transistor is not cut off. Therefore, in the C.E configuration to cut of the transistor, it is not sufficient, if $I_B = 0$. The Emitter Base Junction must be reverse biased, so that I_E is equl to 0. Still I_C will not be zero, but will be I_{CO}. Since I_{CO} is small, the transistor can be regarded as at cut off.

3.4.2.1 Reverse Collector Saturation Current I_{CBO}

I_{CBO} is due to leakage current flowing not through the junction, but around it. The collector current in a transistor, when the emitter current is zero, is designated by I_{CBO}. $|I_{CBO}|$ can be $> I_{CO}$. I_{CO} is collector reverse saturation current (when collector-base junction is reverse biased).

3.4.2.2 Saturation Resistance

For a transistor operating in the saturation region, saturation resistance for Common Emitter configuration is of importance. It is denoted by R_{CES} or R_{CE}'s (saturation) or R_{CS}

$$R_{CS} = \frac{V_{CE}(\text{sat})}{I_C}$$

The operating point has to be specified for R_{CS}.

3.4.2.3 *Saturation Voltage :*

Manufacturers specify this for a given transistor. This is done in number of ways. R_{CS} value may be given for different values of I_B or they may supply the output and input characteristic for the transistor itself. $V_{CE}(\text{Sat})$ depends upon the operating point and also the semiconductor material, and on the type of transistor construction. Alloy junction and epitaxial transistors give the lowest values of V_{CE}. Grown junction transistor yield the highest V_{CE}.

3.4.2.4 DC Current Gain β_{dc}

β_{dc} is also designated as h_{FE} (DC forward current transfer ratio).

$$\beta = \frac{I_C}{I_B}$$

$$I_C = \frac{V_{CC}}{R_L}$$

So if β is known, $\quad I_B = \left(\frac{I_C}{\beta}\right)$

Therefore, $\dfrac{I_C}{\beta}$ gives the value of base current to operate the transistor in saturation region.

β varies with I_C for a given transistor.

3.4.2.5 Test for Saturation

To know whether a transistor is in saturation or not,

1. $\quad |I_B| \geq \dfrac{I_C}{h_{fe}(\text{or}\beta)}$

2. V_{CB} should be positive for PNP transistor and negative for NPN transistor. In the C.E configuration,

$$I_C = -(I_B + I_E)$$

$$I_C = -\alpha I_E + I_{CO}$$

$$\therefore \quad I_E = \frac{I_{CO} - I_C}{\alpha}$$

$$\therefore \quad I_C = \frac{I_{CBO}}{1-\alpha} + \frac{\alpha.I_B}{1-\alpha}$$

But, $$\beta = \frac{\alpha}{1-\alpha}$$

$$\therefore \quad (1-\alpha) = \frac{\alpha}{\beta}$$

If we replace I_{CO} by I_{CBO},

$$I_C = I_{CBO} \times (\beta + 1) + \frac{\alpha.I_B.\beta}{\alpha}$$

$$\boxed{I_C = (1 + \beta) I_{CBO} + \beta I_B}$$

Transistor is cut off when $I_C = 0$, $I_C = I_{CBO}$ and $I_B = -I_{CBO}$. So at cut-off $\beta = 0$.

3.4.2.6 THE RELATIONSHIP BETWEEN α AND β

$$I_E = \text{Emitter Current}$$

$$I_B = \text{Base Current}$$

$$I_C = \text{Collector Current}$$

Since in a PNP transistor, I_E flows into the transistor. I_B and I_C flow out of the transistor. The convention is, currents leaving the transistor are taken as negative.

$$\therefore \quad I_E = -(I_B + I_C)$$

or $$I_E + I_B + I_C = 0$$

or $$I_C = -I_E - I_B$$

But $$\alpha = \frac{-I_C}{I_E}$$

or for PNP transistor taking the convention,

$$I_E = \frac{-I_C}{\alpha}$$

$$\therefore \quad I_C = +\frac{I_C}{\alpha} - I_B$$

$$I_C\left(1 - \frac{1}{\alpha}\right) = -I_B$$

$$I_C\frac{(\alpha - 1)}{\alpha} = -I_B$$

or $\qquad I_C \dfrac{(1-\alpha)}{\alpha} = + I_B$

$$\dfrac{I_C}{I_B} = \dfrac{\alpha}{1-\alpha}$$

$$\dfrac{I_C}{I_B} = \beta = \text{Large Signal Current Gain} = h_{FE}$$

$\therefore \qquad \boxed{\beta = \dfrac{\alpha}{1-\alpha}.}$

3.4.2.7 The Expression for Collector Current, I_C :

The current flowing in the circuit, when E - B junction is left open i.e., $I_B = 0$ and collector is reverse biased. The magnitude of I_{CEO} is large. Hence the transistor is not considered to be cut off.

The general expression for I_C in the active region is given by,

$$I_C = -\alpha I_E + I_{CBO} \qquad\qquad .....(1)$$
$$I_E = - (I_C + I_B) \qquad\qquad .....(2)$$

Substitute eq. (2) in eq. (1)

$$I_C = + \alpha (I_C + I_B) + I_{CBO}$$
$$I_C [1 - \alpha] = \alpha I_B + I_{CBO}$$

$$I_C = \dfrac{\alpha}{1-\alpha} I_B + \dfrac{I_{CBO}}{1-\alpha}$$

$$I_C = \beta I_B + (\beta + 1) I_{CBO}$$
$$I_C = \beta I_B + I_{CEO}$$

where $\qquad I_{CEO} = \dfrac{I_{CBO}}{1-\alpha}$

I_{CEO} is the reverse collector to emitter current when base is open.

3.4.2.8 RELATIONSHIP BETWEEN I_{CBO} AND I_{CEO}

$$I_C = - \alpha I_E + I_{CBO}$$

This is the general expression for I_C in term of a I_E and I_{CBO}.

when $\qquad I_B = 0$, in Common-Emitter (C.E.) Configuration,

$$I_C = - I_E = I_{CEO}$$
$\therefore \qquad I_C + I_B + I_E = 0$

$\qquad\qquad I_B = 0 \qquad$ (Since E-B Junction is left open)

$\therefore \qquad I_C = - I_E$

$\therefore \qquad I_{CEO} = +\alpha I_{CEO} + I_{CBO}$

or $\qquad I_{CEO} (1 - \alpha) = I_{CBO}$

$\therefore \qquad I_{CEO} = \dfrac{I_{CBO}}{(1-\alpha)}; \ \dfrac{1}{(1-\alpha)} = (\beta + 1)$

$\therefore \qquad \boxed{I_{CEO} = (\beta + 1) I_{CBO}}$

In Common Emitter Configuration, for NPN transistor, holes in the collector and electrons in the base are the minority carriers. When one hole from the collector is injected into the base, to neutralize this, the emitter has to inject many more electrons. These excess electrons which do not combine with the hole, travel into the collector resulting in large current. Hence I_{CEO} is large compared to I_{CBO}. When E B junction is reverse biased, the emitter junction is reduced and hence I_{CEO} is reduced.

Therefore, the expression for β gives the ratio of change in collector current $(I_C - I_{CBO})$ to the increase in base current from cut off value I_{CBO} to the I_B. So β is *large signal current gain of a Transistor in Common Emitter Configuration*.

3.4.3   THE COMMON COLLECTOR CONFIGURATION (C.C)

Here the load resistor R_L is connected in the emitter circuit and not in the collector circuit. Input is given between base and ground. The drop across R_L itself acts as the bias for emitter base junction. The operation of the circuit similar to that of Common Emitter Configuration. When the base current is I_{CO}, emitter current will be zero. So no current flows through the load. Base current I_B should be increased so that emitter current is some finite value and the transistor comes out of cut-off region.

Input characteristics I_B vs V_{BC} Output characteristics I_C vs V_{EC}

The circuit diagram is shown in Fig. 3.15 (a) and the characteristics are shown in Fig. 3.15 (b) and (c).

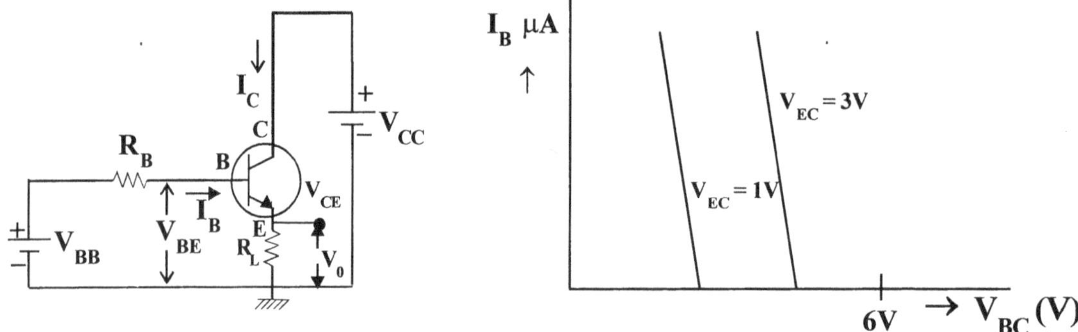

(a) Cirucuit for C.C configuration   *(b) Input characteristics in C.C configuration*

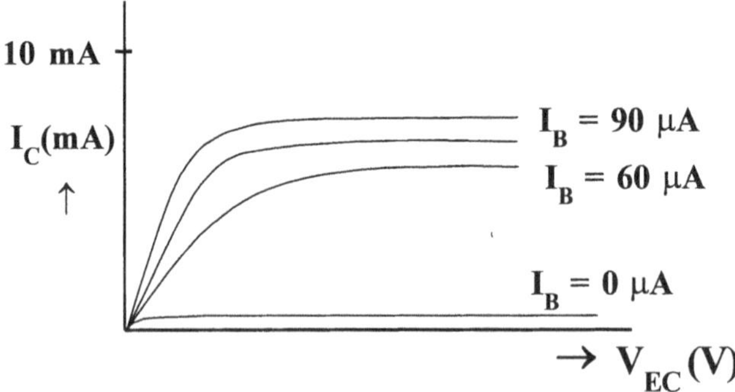

(c) Output characteristics in C.C configuration

Fig. 3.15

The current gain in CC, 'γ' :

It is defined as emitter current to the base current

$$\gamma = -\frac{I_E}{I_B} \qquad \qquad .....(1)$$

Relationship between 'α' and 'γ' :

$$\alpha = -\frac{I_C}{I_B} \qquad \qquad .....(2)$$

$$I_E + I_B + I_C = 0, \quad \therefore I_B = (I_C + I_E) \qquad \qquad .....(3)$$

Substitute eq. (3) in eq. (1)

$$\gamma = \frac{I_E}{+(I_C + I_E)} = \frac{1}{+\dfrac{I_C}{I_E} + \dfrac{I_E}{I_E}}$$

$$\boxed{\gamma = \frac{1}{1-\alpha}} \qquad \left[\because \alpha = \frac{-I_C}{I_E} \right]$$

Expression for emitter current, I_E :

$$I_C = -\alpha I_E + I_{CBO} \qquad \qquad .....(1)$$
$$I_E = -(I_B + I_C) \qquad \qquad .....(2)$$

Substitute eq. (3.16) in eq. (3.17)

$$I_E = -(I_B - \alpha I_E + I_{CBO})$$
$$I_E [1 - \alpha] = -I_B - I_{CBO}$$

$$\boxed{I_E = \frac{-I_B}{1-\alpha} - \frac{I_{CBO}}{1-\alpha}}$$

$$I_E = -\gamma I_B + I_{ECO}$$

$$\boxed{I_{ECO} = -\frac{I_{CBO}}{1-\alpha}}$$

Where I_{ECO} is reverse emitter to collector current when base is open.

PARAMETERS

$$\gamma = \text{Emitter Efficiency}$$
$$= I_{pE}/I_E \text{ for PNP transistor.}$$
$$\beta^* = \text{Transportation factor}$$

$$= \frac{I_{pC}}{I_{pE}} = \frac{\text{No. of holes reaching collector}}{\text{No. of holes emitter}}$$

$$\alpha = \text{Small signal current gain}$$

$$= \frac{I_C}{I_E} \; ; \alpha = \beta^*\gamma \qquad \alpha < 1$$

$$\alpha' = \text{Small signal current gain}$$

$$= \partial I_C / \partial I_E = \frac{\Delta I_C}{\Delta I_E} \qquad \alpha' < 1$$

$$h_{FE} = \beta = \text{Large signal current gain}$$

$$= \frac{I_C}{I_B} \qquad \beta \gg 1$$

$$h_{FE} = \beta' = \text{Small signal current gain}$$

$$= \frac{\partial I_C}{\partial I_B} = \frac{\Delta I_C}{\Delta I_B} \qquad \beta' \gg 1$$

$$I_{CBO} = \text{Reverse saturation current when E-B junction is open circuited.}$$

$$I_{CO} = \text{Normal Reverse Saturation Current}$$

$$\boxed{\alpha = \frac{\beta}{1+\beta}}$$

$$\boxed{\beta = \frac{\alpha}{1-\alpha}}$$

3.5   CONVENTION FOR TRANSISTORS AND DIODES

Two letters followed by a number

First Letter :

A $\rightarrow$ Ge devices. B : Silicon devices C : Ga As devices

D $\rightarrow$ Indium Antimonide

Second Letter :

A $\rightarrow$ Detection diode ; B $\rightarrow$ Variable capacitance device

C $\rightarrow$ Transistor for A.F application : D $\rightarrow$ Power Transistor :

Y $\rightarrow$ Rectifying diode F : Transistor for R.F application

Example :   AF 139 : It is Ge transistor for R.F applications No. 139 is design Number

BY 127 : Silicon rectifying diode.

Old System : Letter 'O' indicates that it is semiconductor device.

Second letter A for diode, C for transistor. AZ for zener diode

OA 81 : It is semiconductor diode

OC 81 : It is transistor

OA Z 200 : zener diode

Problem 3.1

Given an NPN Transistor for which $\alpha_N = 0.98$, $I_{CO} = 2mA$. A common emitter connection is used and $V_{CC} = 12V$, $R_L = 4.0$ KΩ. What is the minimum base current required in order that the transistor enter its saturation region? (Fig. 3.16)

Solution

$$I_C \text{(saturation)} = \frac{12v}{R_L}$$

$$= \frac{12}{4K} = 3mA.$$

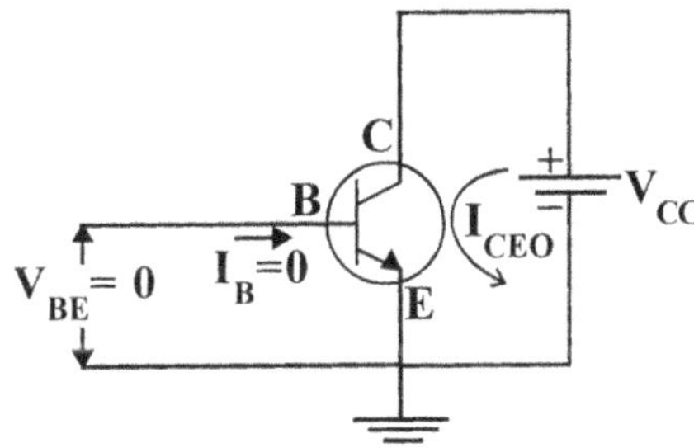

Then

$$I_{B(min)} = \frac{I_C(sat.)}{\beta}$$

$$\beta = \frac{\alpha}{1-\alpha} = \frac{0.98}{1-0.98} = 49$$

$$\therefore \quad I_{B(min)} = \frac{3}{49} = 61.3 \text{ mA}$$

Fig. 3.16   For Problem 3.1.

I_{CBO} :   This is the reverse saturation current in the C.B junction due to minority carriers when $I_E = 0$. (in Covalent Bond Configuration when E - B, n is left open), $V_{BE} = 0$

I_{CEO} :   It is the collector current when $I_B = 0$ or when E - B junction is oscillator circuited in C.E configuration. (Fig. 3.17)

Fig. 3.17   For Problem 3.1.

Problem 3.2

Show that the ratio of the hole to electron currents I_{pE}/I_{nE} crossing the emitter junction of a pnp transistor is proportional to the ratio of the conductivity of the p - type material to that of the n - type material.

Solution

Assume $e^{-\frac{V_C}{V_T}}$ is much smaller than 1 and $e^{V_E/V_T} \gg 1$.

$$I_{np}(o) = \frac{Ae\,D_n\,n_{po}}{Ln} \left(e^{V_E/V_T} - 1\right) = I_{nE}$$

$$I_{np}(o) = \text{Current due to injection of electrons from the base,}$$

where $I_{nE} = \text{The electron current component of emitter current.}$

Since, the collector is reverse biased, V_C is negative $= \left(e^{-V_C/V_T} - 1\right)$.

If $\qquad \dfrac{V_C}{V_T} \gg 1,\ e^{-V_C/V_T}$ is negligible.

$$\therefore \qquad I_{pn}(o) = \frac{-Ae\, Dp\, p_{no}}{w}\left[\left(e^{-V_C/V_T} - 1\right) - \left(e^{\frac{V_E}{V_T}} - 1\right)\right]$$

$$I_{pn}(o) = \frac{+Ae.Dp.p_{no}}{w}\, e^{V_E/V_T} = I_{pE}$$

$$\therefore \qquad \frac{I_{pE}}{I_{nE}} = \frac{Ae\, Dp.p_{no}\, e^{V_E/V_T}}{w} \cdot \frac{Ln}{Ae\, Dn\, n_{po}e^{V_E/V_T}}$$

$$\frac{I_{pE}}{I_{nE}} = \frac{L_n.D_p.p_{no}}{w.D_n.n_{po}} \qquad\qquad\qquad \text{........(3)}$$

$$\sigma_p = \mu_p \times e \times p_{po}$$

But $\qquad p_p = \dfrac{n_i^2}{n_p}$

$$\sigma_p = \mu_p \times e \times \frac{n_i^2}{n_{po}} \quad \text{and}$$

$$\sigma_n = \mu_n \times e\, n_{no} = \mu_n = \mu_n \times n \times \frac{n_i^2}{p_{no}}$$

$$\therefore \qquad n_{po} = \mu_p \times e \times \frac{n_i^2}{\sigma_p} \qquad\qquad\qquad \text{...........(1)}$$

$$P_{no} = \mu_n \times e \times \frac{n_i^2}{\sigma_n} \qquad\qquad\qquad \text{.......... (2)}$$

Substituting these two values in the (eq .3)we get.

$$\frac{I_{pE}}{I_{nE}} = \frac{L_n.D_p}{wDn}\, \frac{\mu_n.en_i^2\sigma_p}{\sigma_n.\mu_p.e.n_i^2} = \frac{\mu_n.\sigma_p}{\sigma_n.\mu_p} \cdot \frac{L_n}{w} \cdot \frac{D_p}{D_n}$$

But $\qquad \dfrac{D_p}{\mu_p} = \dfrac{D_n}{\mu_n} = V_T$

$$\therefore \qquad \frac{I_{pE}}{I_{uE}} = \frac{L_n}{w} \cdot \frac{\sigma_p}{\sigma_n}$$

Problem 3.3

Assuming that $\beta' = h_{fe}$ and $\beta \approx h_{FE}$. Show that

$$h_{fe} = \frac{h_{FE}}{1 - \left(I_{CBO} + I_B\right)\dfrac{\partial h_{FE}}{\partial I_C}}$$

β' = Small Signal C.E current gain = h_{fe}

β = Large Signal C.E current gain = h_{FE}

Solution :

The relation between β' and β is,

$$\beta' = \beta + (I_{CBO} + I_B) \cdot \frac{\partial \beta}{\partial I_B}$$

$$\frac{\partial \beta}{\partial I_B} = \frac{\partial \beta}{\partial I_C} \cdot \frac{\partial I_C}{\partial I_B} \quad \text{(Multiplying and dividing by } \partial I_C)$$

But
$$\beta' = \frac{\partial I_C}{\partial I_B}$$

$$\therefore \quad \frac{\partial \beta}{\partial I_B} = \beta' \cdot \frac{\partial \beta}{\partial I_C}$$

But
$$\beta' = \beta + (I_{CBO}) \beta' \cdot \frac{\partial \beta}{\partial I_C}$$

$$\boxed{\beta' \left[1 - \left(I_{CBO} + I_B\right) \cdot \frac{\partial \beta}{\partial I_C}\right] = \beta}$$

$$\therefore \quad \beta' = h_{fe} = \frac{h_{FE}}{1 - \left(I_{CBO} + I_B\right) \cdot \dfrac{\partial h_{FE}}{\partial I_C}} \qquad \because \beta = h_{FE}$$

Problem 3.4

If $I_B \gg I_{CBO}$. Show that

$$\frac{h_{fe} - h_{FE}}{h_{fe}} = \frac{I_C}{h_{FE}} \cdot \frac{\partial h_{FE}}{\partial I_C}$$

Solution :

$$h_{FE} = \beta = \text{Large signal current gain}$$
$$h_{fe} = \beta' = \text{Small signal current gain}$$

$$\beta' = \beta + (I_{CBO} + I_B) \cdot \frac{\partial \beta}{\partial I_B}$$

$$\because \quad I_{CBO} \text{ is small, compared to } I_B,$$

$$\beta' = \beta + I_B \cdot \frac{\partial \beta}{\partial I_B} \; ;$$

$$h_{fe} = h_{FE} + I_B \cdot \frac{\partial h_{FE}}{\partial I_B}$$

$$h_{fe}\left(1 - \frac{I_B}{h_{fe}} \cdot \frac{\partial h_{FE}}{\partial I_B}\right) = h_{FE} = h_{fe}\left(1 - \frac{I_B}{\partial I_C} \cdot \partial I_B \times \frac{\partial h_{FE}}{\partial I_C}\right) \quad \because h_{fe} = \frac{\partial I_C}{\partial I_B}$$

$$\therefore \qquad \frac{h_{fe} - h_{FE}}{h_{fe}} = I_B \cdot \frac{\partial h_{FE}}{\partial I_C} = \frac{I_C}{h_{FE}} \cdot \frac{\partial h_{FE}}{\partial I_C} \qquad (\text{ proved })$$

$$\frac{I_C}{I_B} = \beta = h_{FE}$$

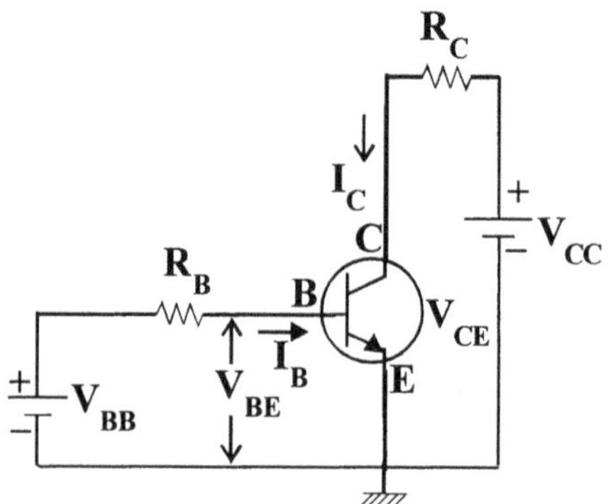

Fig 3.18   For Problem 4.4.

In the Fig. 3.18 NPN transistor in Common Emitter Configuration is shown. In order to cut off, it is not sufficient if I_E is made zero, but the E – B junction should be reverse biased. *The reverse saturation current that is flowing even when $I_B = 0$ is called I_{CBO}.* For a transistor in C.E configuration to cut off, the condition is

$$V_{BE} = - V_{BB} + R_B \cdot I_{CBO} \leq - 0.1$$

This is valid for Si or Ge devices. Note V_{BE} should be negative.

Problem 3.5

The reverse saturation current of the Germanium transistor shown in the Fig 3.18. is 2 µA at room temperature of (25^0C) and increases by a factor of 2 for each temperature increase of 10^0C. The bias $V_{BB} = 5$V. Find the maximum allowable value of R_B if the transistor is to remain cut-off at a temperature of 75^0 C.

Solution :

$$I_{CBO} \text{ at } 25^0\text{C} = 2\mu\text{A}$$

It gets doubled for every 10^0 rise in temperature.

Therefore, I_{CBO} is at 75^0C.

Increase in temperature is $75 - 25 = 50^0$ C.

$$\therefore \qquad I_{CBO} \text{ at } 75^0C = 2 \times 10^{-6} \times 2^{\frac{50}{10}} = 2 \times 2^5 \times 10^{-6} = 64\mu A$$

If V_{BE} is $\leq 0.1V$, the transistor will be at cut-off, since $V_{BE} < V_\gamma$. The equation of transistor to be at cut-off is

$$+V_{BE} = -V_{BB} + R_b \cdot I_{CBO} \leq 0.1V. \quad V_{BB} = 5V \quad I_{CBO} = 64\mu A \, ;$$
$$R_B = ? \quad V_{BE} = -0.1V.$$
$$\therefore \qquad -0.1 = -5 + 64 \times 10^{-6} \times R_B$$
$$\therefore \qquad R_B = \frac{4.9}{64 \times 10^{-6}} = 76.5 k\Omega$$

Problem 3.6

If $V_{BB} = -1.0V$, and $R_B = 50K\Omega$, how high may the temperature increase before the transistor comes out of cut-off?

Solution

As the temperature increases, conduction takes place in the transistor. But because of high value of R_B, transistor remains cut off till the temperature increases above a particular value.

$$V_{BB} = -1.0V \, ; \, R_B = 50k$$
$$-1 + 50 \times 10^3 \times I_{CBO} = -0.1$$
$$\therefore \qquad I_{CBO} = 18\mu A$$

I_{CBO} doubles for every 10^0 rise in temperature.

$$\therefore \qquad 18 = 2 \times 2^{\frac{\Delta T}{10}} \, ; \, 2^{\frac{\Delta T}{10}} = 9 \qquad \text{or} \qquad \frac{\Delta T}{10} = 3.2$$
$$\therefore \qquad \Delta T = 32^0C$$
$$\text{Hence} \qquad T = 25 + 32 = 57^0C$$

Problem 3.7

For the transistor AF 114 connected as shown in Fig. 3.19, determine the values of I_B, I_C and V_{BC} given that $V_{CE} = -0.07V$ and $V_{BE} = -0.21V$.

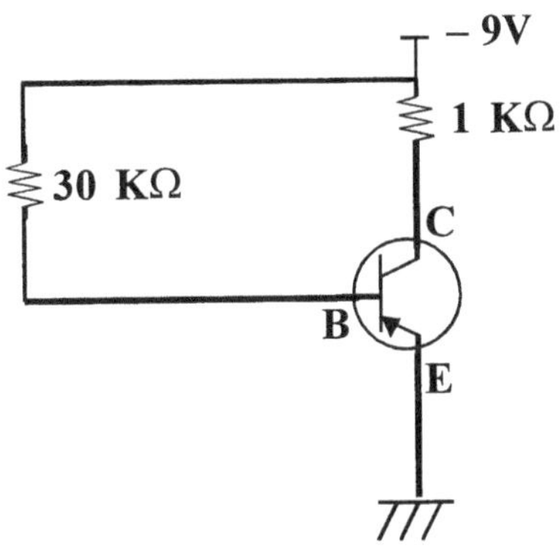

Fig 3.19 For Problem 4.7

Solution

PNP transistor collector is reverse biased. E-B junction is open circuited. So the current flowing is the reverse saturation current.

$$I_C = \frac{-9V - V_{CE}}{1K} = \frac{-9 + 0.07}{1 \times 10^3} = -8.93 \text{ mA}$$

$$I_B = \frac{-9 - V_{BE}}{30K}$$

$$= \frac{-9 + 0.21}{30K} = -0.293 \text{ mA}.$$

$$V_{BC} = V_{BE} - V_{CE} = -0.21 + 0.07 = -0.14V$$

Problem 3.8

If $\alpha = 0.98$, and $V_{BE} = 0.6$ V, find R_1 in the circuit shown in Fig. 3.20 an Emitter Current $I_E = -2$ mA; Neglect reverse saturation current.

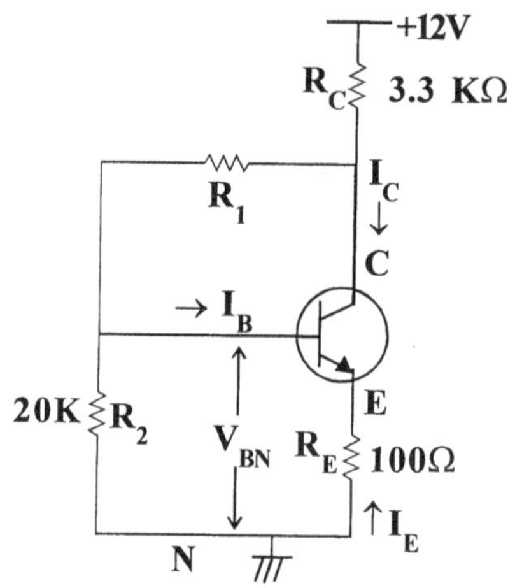

Fig 3.20 For Problem 3.8.

Solution

I_E is negative because it is NPN transistor. So actually I_E is flowing out of transistor. The convention is currents, entering into the transistor are positive. So I_E is negative here. Though there is no separate bias for the emitter, the voltage drop between emitter and ground itself acts as the forward bias voltage.

$$I_E = -2mA \; ; \; I_C = -\alpha . I_E = 0.98 \times 2 = 1.96 \text{ mA}$$

I_C is positive since it is entering into transistor.

$$I_B = (1 - \alpha) I_E = (1 - 0.98)(-2) = 40 \text{ μA}$$

The voltage between base and ground $V_{BN} = +V_{BE} + I_E R_E$ and (considering only magnitude)

$$V_{BE} = +0.6 : I_E = 2 \text{ mA} : R_E = 100\Omega$$

$$\therefore \quad V_{BN} = 0.6 + (2 \times 0.1) = 0.8V$$

$$V_{BN} \text{ is} > V_{BE}.$$

$$\therefore \qquad I_{R2} = \frac{V_{BN}}{R_2} = \frac{0.8}{20k} = 40\mu A$$

$$I_{R1} = I_B + I_{R_2} = 40 + 40 = 80\mu A$$

$$I_{Rc} = I_{R_1} + I_C \; ; \; I_{R_1} = 80\mu A \; ;$$

$$I_C = 1.96mA$$

$$I_{RC} = 80 \; \mu A + 1.96 \; mA = 2.04\mu A$$

$$V_{R_1} = V_{CC} - V_{RC} - V_{BN} = 12 - (2.09 \times 3.3) - 0.8 = 4.47V$$

$$\therefore \qquad R_1 = \frac{V_{R_1}}{I_{R_1}} = \frac{4.47}{0.08} = 56K$$

Problem 3.9

Show that for an NPN silicon transistor of the alloy junction type, in which the resistivity ρ_B of the base is much larger than that of the collector, the voltage V is given by

$$V = \frac{6.34 \times 10^3 . W^2}{\rho_B} \, ,$$

where ρ_B is in Ω – cm and base width W in *mills*.

Solution

$$\in = \frac{12}{36\pi \times 10^9} \; F/m = \frac{12}{36\pi \times 10^{11}} \; F/cm$$

$$\mu_p = 500 \; cm^2/V\text{ - sec.}$$

The expression for $W^2 = 2\in . \dfrac{V_B}{eN_A}$,

where $\qquad \in = 12/36\pi \times 10^{11}$ F/cm.

In NPN transistor, the base is 'P' type.

$$\sigma_p = \sigma_B \; (base)$$

$$\therefore \qquad \sigma_B = e . N_A . \mu_p$$

or $\qquad e . N_A = \dfrac{\sigma_B}{\mu_p} = \dfrac{1}{\rho_B . \mu_p}$

Thus $\qquad e . N_A = \dfrac{1}{\rho_B . \mu_p}$

$$\therefore \qquad V_B = \frac{e.N_A.W^2}{2\in} \qquad V_B \text{ is called the Barrer Potential or Contact Potential.}$$

Substitute the value of $\in .N_A$ in the expression for V_B.

$$V_B = \frac{W^2}{2\in .\mu_p \rho_B}$$

Substitute the value of $\in$ of Silicon as $\dfrac{12}{36\pi \times 10^{11}}$ F/cm.

Since   W is in mills,

$$1 \text{ mil} = \dfrac{1}{1000"} \quad (\text{one thousand of an inch})$$

To convert this to centimeters, multiply by 2.45

$\therefore$ Barrier Potential,

$$V_B = \dfrac{W^2(\text{mills}) \times (2.54)^2 \times \left(\dfrac{1}{1000}\right)^2}{\dfrac{2\times 12}{36\pi \times 10^{11}} \times \dfrac{500}{V-\text{sec}}\rho_B}$$

$$\mu_p = 500 \text{ cm}^2/V \text{ - sec} = \text{Mobility of Holes}$$

$$V_B = 6.1 \times 10^3 \times \dfrac{W^2}{\rho_B}$$

Problem 3.10

A transistor having $\alpha = 0.96$ is placed in Common Base Configuration with a load resistance of 5 KΩ. If the emitter to base junction resistance is 80Ω, find the values of amplifier current, voltage and power gain.

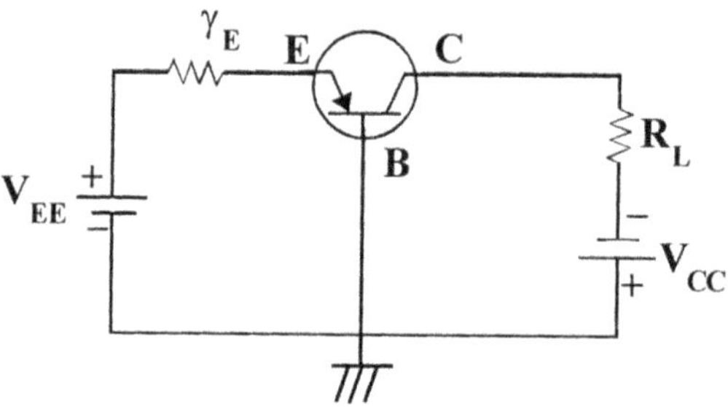

Fig 3.21   For Problem 3.10

Solution

$$A_I = \text{current gain} = \dfrac{I_C}{I_E} = \alpha = 0.96$$

$$A_V = \dfrac{\alpha . R_L}{\gamma_{BE}}$$

$$= \dfrac{0.9 \times 5 \times 10^3}{80} = 60$$

$\therefore$ Power Gain $A_P = A_V \times A_I = 60 \times 0.96 = 57.6$

Problem 3.11

If a transistor, with $\alpha = 0.96$ and emitter to base resistance 80Ω is placed in Common Emitter Configuration, find A_I, A_V and A_P.

Solution

$$A_I = \frac{I_C}{I_B} = \beta$$

$$\therefore \quad \beta = \frac{\alpha}{1-\alpha}$$

$$= \frac{0.96}{1-0.96} = 24$$

$$A_V = \frac{-\beta R_L}{\gamma_{BE}}$$

$$= \frac{24 \times 7.5 \times 10^4}{80} = -22{,}500 \quad A_P = A_I \times A_V = 24 \times 22{,}500 = 5{,}40{,}000.$$

Problem 3.12

A transistor is operated at a forward current of $2\mu A$ and with the collector open circuited. Calculate the junction voltages V_C and V_E, the collector to emitter voltage V_{CE} assuming $I_{CO} = 2\mu A$, $I_{EO} = 1.6\mu A$ and $\alpha_N = 0.98$.

Solution

$$I_C = 0 \quad \text{Since Collector Junction is open circuited.}$$

$$I = I_0 \left(e^{V_N / V_T} - 1 \right)$$

$$I_E = I_{EO} \left(e^{\frac{V_E}{nV_T}} - 1 \right)$$

$$\frac{I_E}{I_{EO}} = e^{\frac{V_E}{nV_T}} - 1 \quad \text{or} \quad e^{\frac{V_E}{nV_T}} = 1 + \frac{I_E}{I_{EO}}$$

$$\therefore \quad \frac{V_E}{nV_T} = \ln\left(1 + \frac{I_E}{I_{EO}}\right) \qquad \therefore \quad V_E = \eta \cdot V_T \ln\left(1 + \frac{I_E}{I_{EO}}\right)$$

$$\therefore \quad V_E = 0.1853V$$

$$V_C = V_T \ln \cdot \left(1 + \frac{I_C}{I_{CO}}\right)$$

$$= V_T \ln\left(1 - \frac{\alpha . I_E}{I_{CO}}\right)$$

$$= 0.179V$$

$$V_{CE} = V_C - V_E = -0.006 \text{ V}$$

Problem 3.13

A load draws current varying from 10 to 100 mA at a nominal voltage of 100V. A regulator consists of $R_S = 200\Omega$ and zener diode represented by 100 V, and $R_Z = 20\Omega$. For $I_L = 50$ mA. Determine variation in V_i corresponding to a 1 % variation in V_L.

Solution

The zener diode is represented by a voltage source with it. V_L can vary with in 1 % so what is the allowable range in which V_i can be permitted to vary ? In order that the zener diode is operated near the breakdown region, the current through the zener diode is operated near the breakdown region, the current through the zener should be at least 0.1 times the load current.

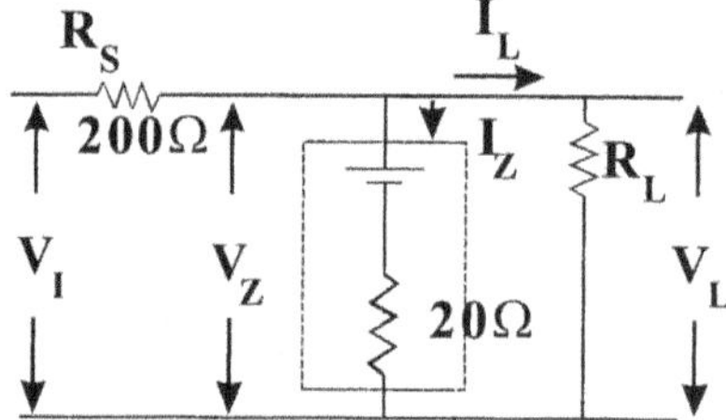

Fig 3.22   For Problem 3.13

$$I_{Zmin} = 0.1\,(I_{Lmax}) \qquad\qquad I_{Lmax} = 100 \text{ mA.}$$

$\therefore$
$$I_Z = 0.1 \times 100 = 10 \text{ mA.}$$
$$V_L = V_Z + I_Z R_Z$$
$$= 100 + 0.01 \times 20 = 100.2V$$
$$V_i = V_L + (I_L + I_Z)\, R_S$$
$$= 100.2 + (0.05 + 0.01)\, 200$$
$$= 112.2 \text{ V.} \qquad\qquad V_L = 100.2V$$

1% increase is $V_L \approx 1V.$

$$V'_L = 100.2 + 1 = 101.2V$$

Therefore, the increase in

$$I_Z, I'_Z = \frac{V_C' - V_Z}{R_Z}$$

$$= \frac{101.2 - 100}{20}$$

$$= \frac{1.2}{20}\,\frac{V}{\Omega} = 0.06A$$

Total Current $= 0.05 + 0.06 = 0.1$ A

$\therefore$
Total Voltage $V_i = V_Z + R_S \times (I_Z + I_L)$
$$= 101.2 + 200 \times (0.05 + 0.06)$$
$$V_i = 123.2V \qquad\qquad \text{Change in } V_i = 123.2V - 112.2V = 11V$$

A change of 11V in V_i on the input side produces a change of 1V on the output side due to zener diode action.

3.6   INTRODUCTION

In this chapter, the definitions of h - parameters, expressions of A_v, A_I etc., in terms of h - parameters are given.

These are also known as hybrid parameters (h for hybrid), so called because, the units of these parameters are different, Ohms (Ω), mhos ($\mho$), constants etc. Because the units are different for different parameters, these are called *Hybrid Parameters.* In defining these parameters short circuit and open circuit conditions are used.

In order to analyze different amplifier circuits and compare their merits and demerits, equivalent circuits of the amplifier must be drawn. These circuits are drawn in terms of the h - parameters. The transistor equivalent circuits assume that the operating point is chosen correctly and does not involve biasing resistors.

An amplifier circuit is one which will increase the level or value of the input signal. If the output voltage (V_0) is greater than input voltage (V_i), it is called voltage amplification. Similarly if $P_0 > P_i$, it is power amplification. The voltage amplification factor is represented as A_v and

$$A_v = V_0 / V_i$$

The other parameters of an amplifier circuit are:

Input resistance R_i (or input impedance Z_i)

Output resistance R_0 (or output impedance Z_0 or out admittance Y_0)

The resistance associated with the amplifier circuit, as seen from the input terminals of the amplifier, into the circuit is called input resistance (R_i). For a given circuit, R_i must be high. (Ideal value of $R_i = \infty$). Because, when the input signal source V_S is connected to the amplifier circuit, the amplifier circuit will draw current (I_i) from V_S. ($I_i = V_S /R_i$). If R_i is small, I_i will be large, which V_S the signal source may not be able to supply. So V_S may fall from the set value (due to loading effect). So R_i must be large for a given amplifier circuit.

Similarly, the output resistance R_0 of the amplifier must be small Because, the amplified signal V_0 must be fully delivered to the load R_L connected to the amplifier. The signal voltage drop in the amplifier must be as small as possible. So R_0 must be small. Ideally R_0 must be zero. Output resistance is the resistance of the amplifier circuit, as seen through the output terminals into the amplifier circuit.

3.7   BLACK BOX THEORY

Any four terminal network (two terminals for input and two terminals for output), no matter how complex it is, can be represented by an equivalent circuit, if there is a connection between input and output, as shown in the Fig 3.23

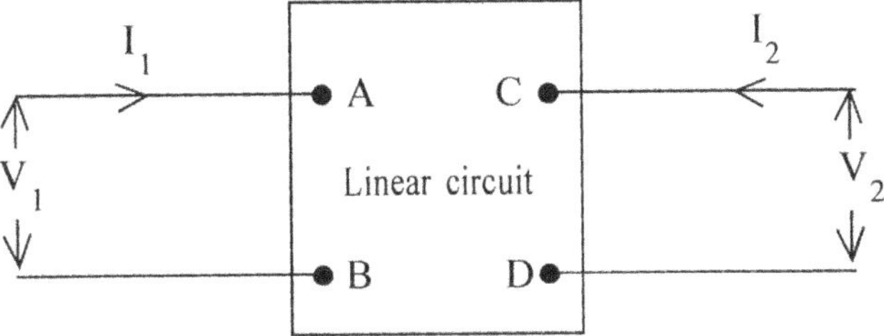

Fig 3.23 Four terminal network.

If a voltage V_2 is applied at the output terminals and current I_2 is allowed to flow into the circuit, the Thevenins equivalent circuit looking into the input terminals is given by a voltage source Vo_1 in series with an impedance Zo_1. Similarly, if a voltage V_1 is at the input, the equivalent circuit can be represented as a voltage source Vo_2 in series with an impedance Zo_2 so the Thevenins equivalent circuit reduces as shown in Fig 3.24.

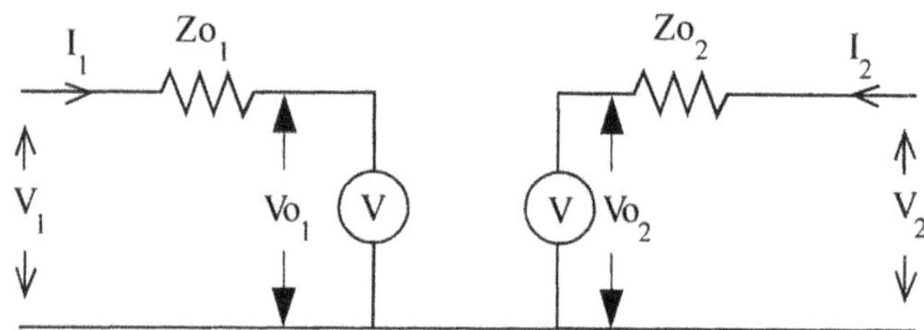

Fig 3.24 Equivalent circuit of a four terminal network.

But it is convenient to represent the output equivalent circuits by following Nortons Theorem as a current generator $\dfrac{Vo_2}{Zo_2}$ in parallel with an impedance of Zo_2. Therefore, the final equivalent circuit reduces to as shown in Fig 3.25.

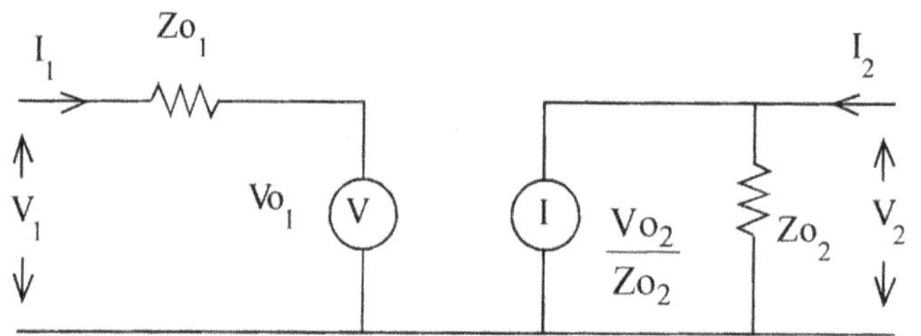

Fig 3.25 Equivalent circuit with the current source.

What is the relationship between the Variables V_1, I_1, V_2, I_2, and the constants of the circuits Zo_1, Zo_2, Vo_1, Vo_2, etc? The general symbols that are used in these equivalent circuits are defined as :

h_{11} : If voltage V_1 is applied to the device, the resulting input current is I_1 then the ratio $\dfrac{V_1}{I_1}$ represents the resistance. This parameter represents input resistance. Units are Ohms (Ω).

h_{12} : Vo_1 is related to V_2 as $Vo_1 = h_{12} V_2$. It is called as Reverse voltage gain.

h_{21} : The current in the output circuit depends upon the forward current gain and magnitude of the input current I_1. Hence I_1 and I_2 are related as $I_2 = h_{21} I_1$.

h_{22} : The output of a device has some output resistance or admittance which is represented as h_{22}. This is output admittance parameter. Units are in mhos (℧)

Therefore, for transistors we use, h_{11} for Zo_1 or h_{22} for $1/Zo_2$ and so on.

The equivalent circuits for a transistor can be represented as shown in Fig 3.26.

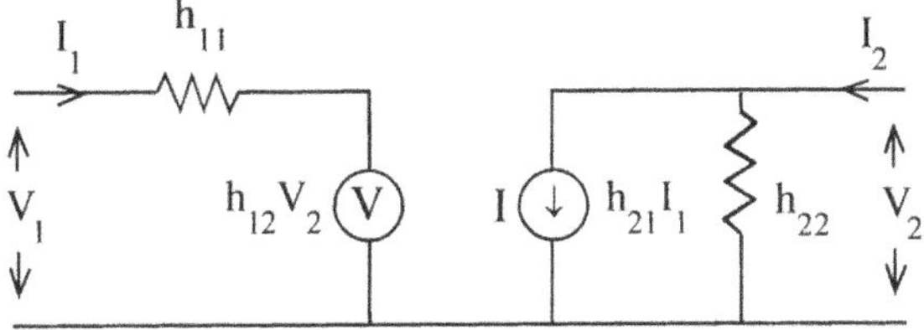

Fig 3.26 Equivalent circuit of a transistor.

This is the hybrid equivalent circuit for a transistor. From the equivalent circuit,

$$V_1 = h_{11}I_1 + h_{12}V_2$$
$$I_2 = h_{21}I_1 + h_{22}V_2$$

These h-parameters are constant for a given circuit but they depend upon the type of Configuration i.e. parameters vary depending upon , whether the circuit is in Common Emitter, or Common Base or Common Collector Configuration.

So these are represented as h_{ie}, h_{oe}, h_{fe} and h_{re} in Common Emitter Configuration. The second subscript 'e' indicating Common Emitter. In Common Base Configuration, these are represented as h_{ib}, h_{ob}, h_{fb} and h_{rb}. In Common Collector Configuration, the h - parameters are represented as h_{ic}, h_{oc}, h_{fc} and h_{rc}. The second subscript 'b' and 'c' indicating common base and common collector.

The general equations are

$$V_1 = h_{11}I_1 + h_{12}V_2 \qquad \dots\dots\dots (3.1)$$
$$I_2 = h_{21}I_1 + h_{22}V_2 \qquad \dots\dots\dots (3.2)$$

From the first equation 3.1,

$$h_{11} = \frac{V_1}{I_1} \text{ when } V_2 = 0 \qquad \dots\dots\dots (3.3)$$

It has the units of Resistance and $V_2 = 0$ indicates that output is shorted. Therefore, h_{11} is *the input resistance when output is shorted.* It is represented as h_{ie}, in Common Emitter Configuration.

$$h_{21} = \frac{I_2}{I_1}\Bigg|_{V_2 = 0} \qquad \dots\dots\dots (3.4)$$

This is *Forward Current Transfer Ratio or Forward Current Gain, or Forward Short Circuit Current Gain* $V_2 = 0$ means, output is shorted. Second subscript 'e' in h_{ie} represents C.E. Configuration.

$$h_{12} = \frac{V_1}{V_2}\Bigg|_{I_1 = 0} \qquad \dots\dots\dots (3.5)$$

It is the ratio of input voltage to the output voltage. Hence it is called as *Reverse Voltage Gain*, with *Input Open Circuited*, since $I_1 = 0$.

$$h_{22} = \frac{I_2}{V_2}\Bigg|_{I_1 = 0} \qquad \dots\dots\dots (3.6)$$

This parameter is called the *Output Conductance when the Input is open circuited.*

3.7.1 Output Conductance

In the audio frequency range, the h - parameters are real numbers (Integers). But beyond this frequency range, they become complex.

The reason is that in the audio frequency range, the effect of Capacitance and Inductance are negligible. Capacitive reactance, $X_C = \dfrac{1}{2\pi fC}$. If f is small, $X_C \simeq \infty$ and can be neglected (Open Circuit). Inductive Reactance $X_L = 2\pi fL$. If f is small $X_L \simeq 0$ and can be neglected (Short Circuit). Beyond, audio frequency range, they can't be neglected and reactances must also be considered. Transistor can be represented as a two port active device . It is specified by two voltages input and output and two currents input and output . The Representation can be shown as in the Fig 3.27.

Fig 3.27 Block diagram.

Current flows only when there is closed path. Input current enters the two port device and the output current flows through the return path back into the device. Hence the direction of I_2 is as shown. Now of these 4 quantities (i_1, i_2, v_1, v_2) two can be chosen as independent parameters and the remaining two can be expressed as variables.

A transformer is a two port device. But V_1 and V_2 cannot be chosen as independent parameters because $\dfrac{V_1}{V_2} = \dfrac{I_2}{I_1}$. Hence we cannot choose independent parameters on our choice.

If v_2 and i_1 are chosen as Independent Variables, then v_1 and i_2 can be expressed in terms of v_2 and i_1.

$$v_1 = h_{11} i_1 + h_{12} v_2 \qquad \qquad \cdots\cdots\cdots (3.7)$$
$$i_2 = h_{21} i_1 + h_{22} v_2 \qquad \qquad \cdots\cdots\cdots (3.8)$$

h_{11} is in Ω, h_{12} is constant, h_{21} is constant and h_{22} is in mhos. All these are not the same dimensionally. Hence these are known as **Hybrid Parameters**. Suffix 1 indicates input side and Suffix 2 denotes output side. h-parameters are defined as,

$$h_{11} = \left. \dfrac{v_1}{i_1} \right|_{v_2 = 0} \qquad \text{Input Resistance with Output Shorted. } (\Omega)$$

$$h_{12} = \left. \dfrac{v_1}{v_2} \right|_{i_1 = 0} \qquad \text{Reverse Voltage Gain with Input open circuited (Dimension Less)}$$

$$h_{21} = \left. \dfrac{i_2}{i_1} \right|_{v_2 = 0} \qquad \text{Forward Current Gain with Output Shorted (Dimension Less).}$$

$$h_{22} = \left. \dfrac{i_2}{v_2} \right|_{i_1 = 0} \qquad \text{Output Conductance with Input Open Circuited (\mho)}$$

But, according to IEEE Standards, the notation used is 'i' for 11 (input) 'o' for 22 (output) 'f' for 21 (Forward Transfer) and 'r' = 12 (Reverse Transfer). The subscripts b, c or e are used to denote Common Base, Common Collector or Common Emitter Configurations. Thus for Common Emitter Configuration,

the equation are

$$v_1 = h_{ie} \, i_1 + h_{re} \cdot v_2 \qquad \dots\dots\dots (3.9)$$

$$i_2 = h_{fe} \, i_1 + h_{oe} \, v_2. \qquad \dots\dots\dots (3.10)$$

If the device has no reactive components, the hybrid parameters are real. But if the device consists of reactive elements also, h - parameters will be functions of frequency. The four h - parameters can be used to construct a mathematical model of an Amplifier Circuit. The Circuit representing the two equations relating v_1, v_2, i_1 and i_2 is shown in the Fig 3.28. $h_{12} \, v_2$ is a Voltage Source. $h_{21} \, i_2$ is a Current Source.

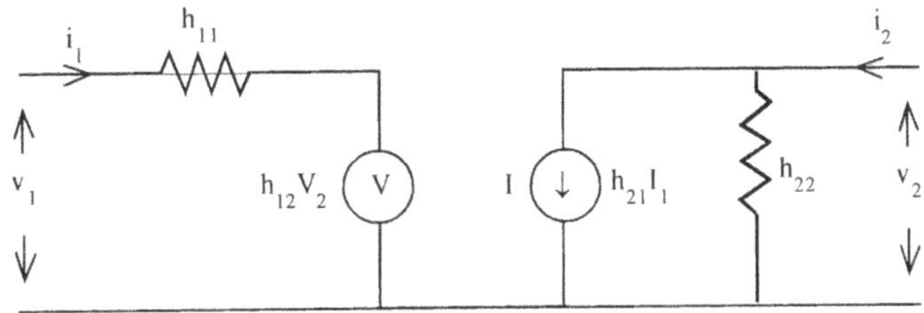

Fig 3.28   h-parameter equivalent circuit

3.8   TRANSISTOR HYBRID MODEL

In drawing the equivalent circuit of a transistor, we assume that the Transistor Parameters are constant. With the variation in operating points, transistor parameters will also vary. But we assume that the Quiescent or operating point variation is small.

The advantages of considering h-parameters for transistor are

1.   *They are easy to measure.*

2.   *Can be obtained from Transistor characteristics.*

3.   *They are real numbers at audio frequencies.*

4.   *Convenient to use in circuit design and analysis.*

Manufactures specify a set of ***h-parameters*** for transistors.

3.9   TRANSISTOR IN COMMON EMITTER CONFIGURATION

The variables are i_b, i_c, v_b and v_c. We can select i_b and v_c as independent parameters.

$$v_b = f_1 \, (\, i_b \text{ and } v_c) \qquad \dots\dots (3.11)$$

These are DC quantities Incremental variation (Elementary Changes) in DC is considered as a.c. quantity. So small letters v and i are used..

Therefore, V_b depends upon i_b, and also the Voltage across Collector and Emitter is V_{CE}

$$i_c = f_2 \, (\, i_b, \, v_c) \qquad \dots\dots (3.12)$$

Making a Taylor Series Expansion of (1) and (2) and neglecting Higher Order Terms,

$$\Delta v_b = \left.\frac{\partial f_1}{\partial i_B}\right|_{V_c} \Delta i_b + \left.\frac{\partial f_1}{\partial v_C}\right|_{I_b} \Delta v_C \qquad .... (3.13)$$

$$\Delta i_C = \left.\frac{\partial f_2}{\partial i_B}\right|_{V_c} \Delta i_b + \left.\frac{\partial f_2}{\partial v_C}\right|_{i_b} \Delta v_C \qquad .... (3.14)$$

Δv_c, Δi_C are A.C. quantities and (Incremental Variations in D.C. quantities). The quantities Δv_b, Δi_b, Δv_C and Δi_C represent small signal (incremental) base and collector voltages and currents. They are represented as v_b, i_b, v_C, and i_C (A.C. quantities, so small letters are used)

$$\therefore \qquad v_b = h_{ie} i_b + h_{re} v_C \qquad .... (3.15)$$
$$i_C = h_{fe} i_b + h_{oe} v_C \qquad .... (3.16)$$

where
$$h_{ie} = \frac{\partial f_1}{\partial i_b} = \left.\frac{\partial v_b}{\partial i_B}\right|_{V_C = K} \qquad v_C \text{ is kept constant} \qquad .... (3.17)$$

$$h_{re} = \left.\frac{\partial v_B}{\partial v_C}\right|_{I_B = K} \qquad\qquad .... (3.18)$$

DC values of $i_b = K$ or $v_C = K$

$$h_{fe} = \left.\frac{\partial i_C}{\partial i_B}\right|_{V_C = K} \qquad\qquad .... (3.19)$$

AC $i_b = 0$ or $v_C = 0$

$$h_{oe} = \left.\frac{\partial i_C}{\partial v_C}\right|_{i_b = K} \qquad (i_b \text{ is kept constant }) \qquad .... (3.20)$$

Parameters 3.17 to 3.20 represent **h-parameters** for C.E configuration from equations (5) and (6), we can draw the equivalent circuit.

3.10 DETERMINATION OF h-PARAMETERS FROM THE CHARACTERISTICS OF A TRANSISTOR

The output characteristics of a Transistors in Common Emitter Configuration are as shown in Fig. 3.29.

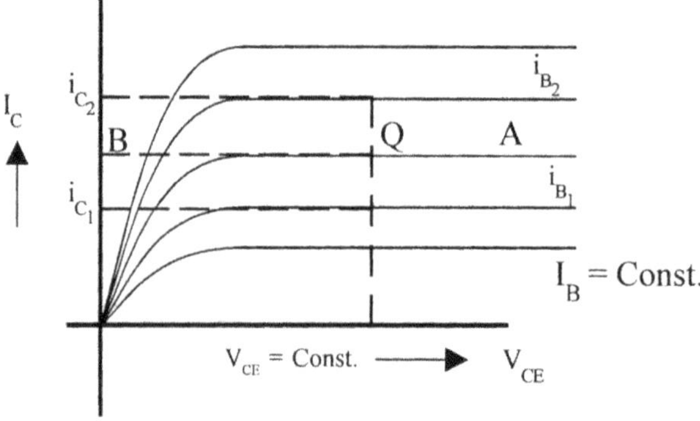

Fig 3.29 Transistor output characteristics (To determine h_{fe}, h_{oe} graphically)

$$h_{fe} = \left.\frac{\partial i_C}{\partial i_B}\right|_{V_C=K} = \left.\frac{\Delta i_c}{\Delta i_b}\right|_{V_c=0}$$

h_{fe}, h_{oe} can be determined from output characteristics. h_{ie}, h_{re} can be determined from input characteristics.

Suppose Q is the Quiescent Point around the operating point h_{fe} is to be determined. For h_{fe}, we have to take the ratio of incremental change in i_b keeping V_e constant. From the output characteristics two values i_{c1}, and i_{c2} are taken corresponding to i_{b_1} and i_{b_2}. Then,

$$\mathbf{h_{fe}} = \left(i_{c_2} - i_{c_1}\right)/\left(i_{b_2} - i_{b_1}\right)$$

h_{fe} is also known as **Small-Signal Beeta** (β) of the transistors. Since we are determining the incremental values of Current Ratio of i_c and i_b. h_{fe} is represented as β.

β is represented as h_{fe} and is called large signal β. $\beta = i_c / i_b$, since here no incremental values are being considered.

$$h_{oe} = \left.\frac{\partial i_C}{\partial v_C}\right|_{i_b = K}$$

Output characteristics of a transistor are plotted between i_c and v_c. Therefore, h_{oe} is the slope of the output characteristics, corresponding to a give n value of I_b at the operating point. For Common Emitter Configuration, in exactly the same way we have drawn for amplifier system, the h-parameter equivalent circuit can be drawn. The Transistor in Common Emitter Mode is shown in Fig. 3.30. The h-parameter equivalent is shown in Fig. 3.31.

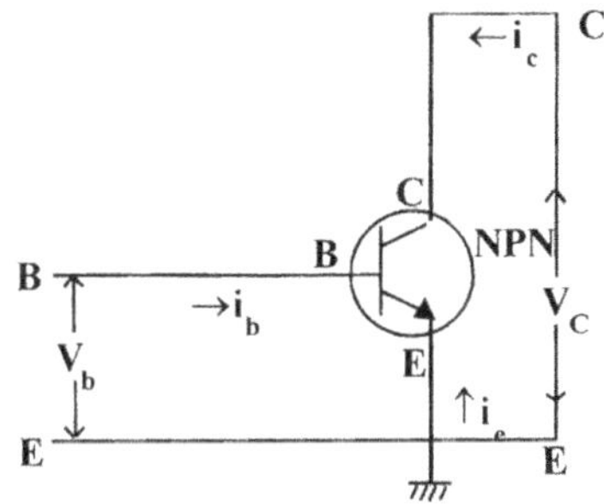

Fig 3.30 Common emitter configuration.

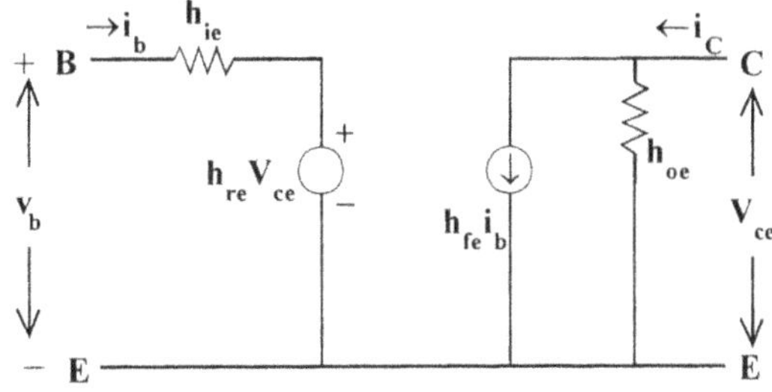

Fig 3.31 Equivalent circuit in C.E. configuration.

Equations are $\quad v_b = h_{ie} i_b + h_{re} v_C$ $\qquad\qquad$ .... (3.21)

$\qquad\qquad\qquad\quad i_C = h_{fe} i_b + h_{oe} v_C$ $\qquad\qquad$ .... (3.22)

3.11   COMMON COLLECTOR CONFIGURATION (CC)

The NPN Transistor in Common Collector Configuration is shown in Fig 3.32. The h - parameter equivalent circuit is shown in Fig. 3.33.

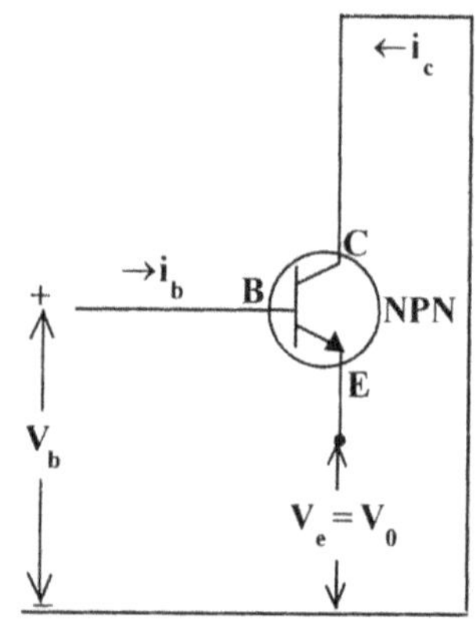

Fig 3.32   *Common collector configuration.*

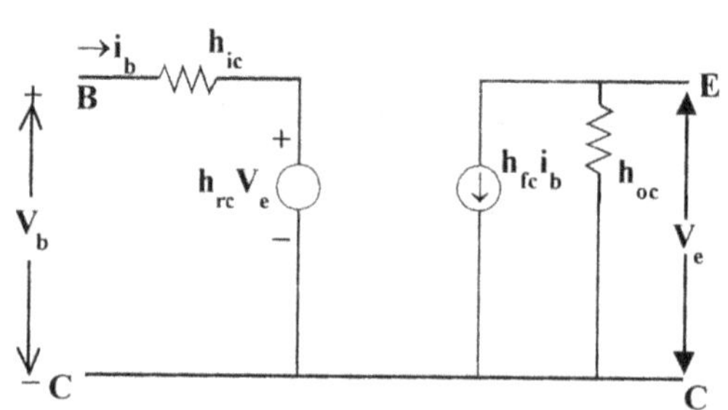

Fig 3.33   *Equivalent circuit in common collector.*

Equations are, $\quad v_b = h_{ic} i_b + h_{rc} v_e$ $\qquad\qquad$ .... (3.23)

$\qquad\qquad\qquad\quad i_e = h_{fc} i_b + h_{oc} v_e$ $\qquad\qquad$ .... (3.24)

Common Base Configuration (CB)

Transistor in Common Base Configuration is shown in Fig. 3.34. The h-parameter equivalent circuit is shown in Fig. 3.35.

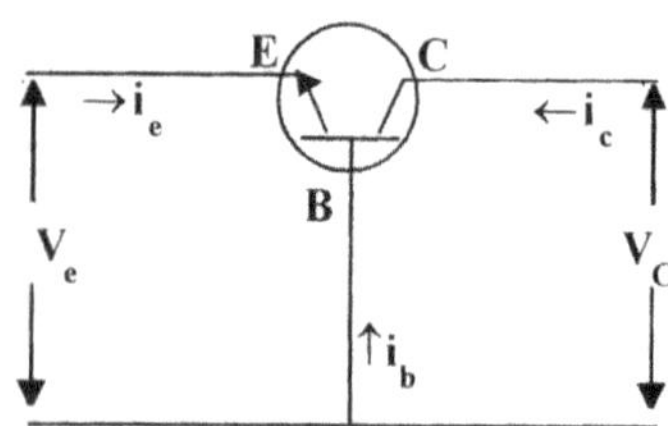

Fig 3.34   C.B. Configuration.

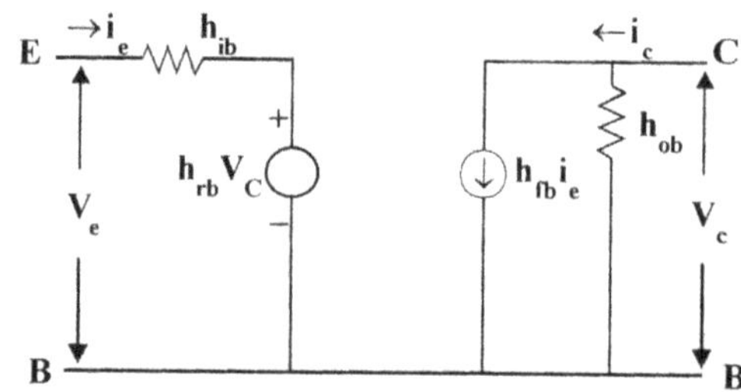

Fig 3.35   Equivalent Circuit in C.B.

The equations are, $v_e = h_{ib} i_e + h_{rb} v_c$

$\qquad\qquad\qquad\quad i_c = h_{fb} i_e + h_{ob} v_c$

The above circuits are valid for NPN or PNP Transistors. For PNP Transistors, the direction of current will be different from NPN Transistor.

Therefore, at the operating point, Q, a tangent AB is drawn The slope of AB gives h_{oe}.

$$h_{ie} \qquad h_{ie} = \frac{\partial v_b}{\partial I_b}\bigg|_{v_C=K}$$

$$= \frac{\Delta v_b}{\Delta i_b}$$

This can be obtained from the input characteristics. The input characteristics are drawn as shown in the Fig 3.36. Then a tangent drawn at the operating point and its slope directly gives h_{oe}.

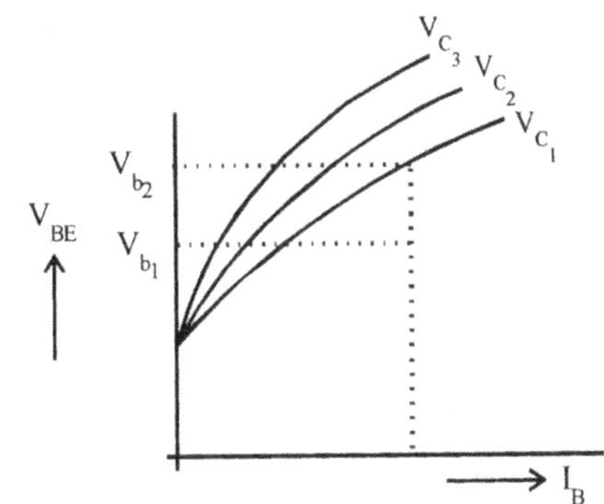

$$h_{re} \qquad h_{re} = \frac{\partial v_b}{\partial v_C}$$

Fig 3.36 To determine h_{ie}, h_{re} graphically.

$$= \frac{\Delta v_b}{\Delta v_C}\bigg|_{ic=K}$$

From the input characteristics corresponding to two values V_{C_1} and V_{C_2} V_{B_1} and V_{B_2} are

noted then the ratio $\dfrac{V_{b_2} - V_{b_1}}{V_{c_2} - V_{c_1}}$ gives h_{re}.

Here we have seen the methods of determining 'h' parameters for Common Emitter Configuration. The same is true for Common Base and Common Collector Configuration.

3.12 HYBRID PARAMETER VARIATIONS

From the above analysis, it is clear that, when 'Q' the operating point is given, from input and output characteristics of the given transistors, we can determine the h-parameters. Conversely if the operating point is changing, the 'h' parameters will also change. I_C changes with temperature. Hence 'h' parameters of a given transistor also change with temperature because the output and input characteristics change with temperature . Hence when the manufactures specify typical h-parameters for a given transistor, they also specify the operating point and temperature. *Specifying the operating point is giving the values of I_c, β and V_{CE} (DC Values).*

h_{fe} the small signal current amplification factor is very sensitive to I_C. Its variations is as shown in Fig 3.37. The variation of h_{fe} with I_E and Temperature T are shown in Fig. 3.38.

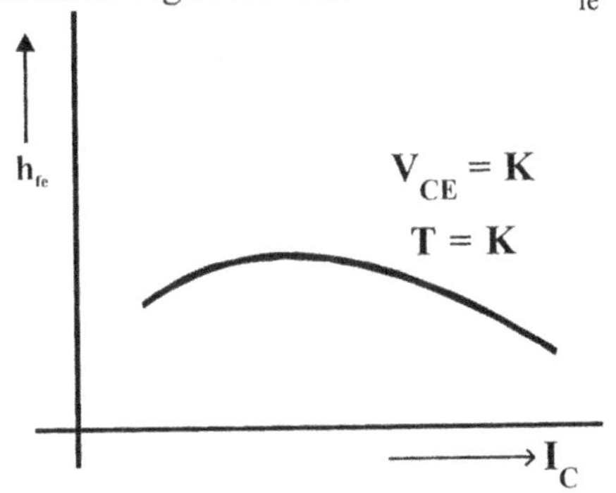

Fig 3.37 Variation of h_{fe} with I_C.

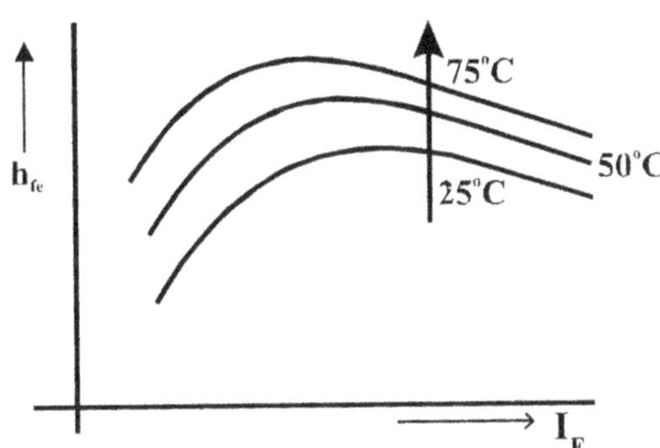

Fig 3.38 Variation of h_{fe} with I_E and T.

We are determining these h parameters from the static D.C characteristics of the transistors But the 'h' parameters are defined with respect to A.C voltages and currents. So AC voltage V_{be} may be represented as small changes in D.C. levels of V_{BE}. Hence the above graphical analysis is correct . We are taking incremental values of DC to represent AC quantities. Typical values of h-parameters at room temperature, in the three transistor configurations, namely Common Emitter, Common Collector and Common Base are given in the Table 3.1.

Table 3.1 *Comparison of h-parameters in the three configurations (at I_E = 1.3 mA)*

Parameter	CE	CC	CB
h_i	$1100\,\Omega$	$1100\,\Omega$	$21.6\,\Omega$
h_r	2.5×10^{-4}	~1	2.9×10^{-4}
h_f	50	-51	-0.98
h_0	$24\mu A/V$	$25\mu A/V$	$0.49\mu A/V$

3.13 CONVERSION OF PARAMETERS FROM C.B. TO C.E.

The circuit shown in Fig. 3.39 is the h-parameter equivalent circuit in Common Base Configuration. This circuit is to be redrawn in the Common Emitter Configuration. In Common Emitter Configuration, Emitter is the common point In C.B, base is common point Hence the above circuit is to be redrawn so as to get 'E' emitter, as the common point. Hence invert the voltage source $h_{rb} V_{cb}$. From the -ve of $h_{rb} V_{cb}$, we go to $h_{fb} I_e$ point:

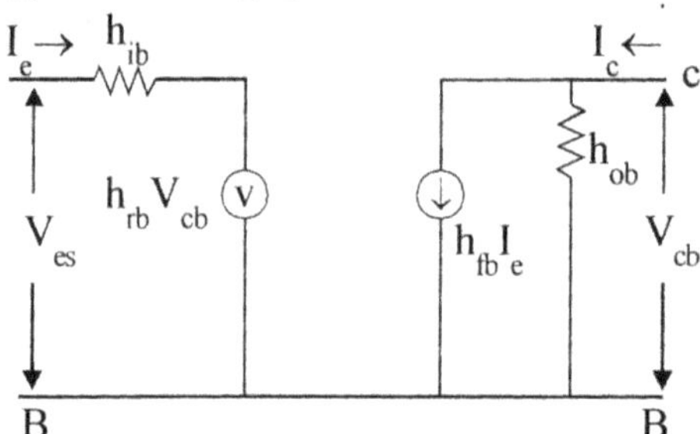

Fig 3.39 h - parameters circuit in C.B. configuration

$h_{fb} I_e$ is as shown. h_{ob} is in parallel with $h_{fb} I_e$, V_{cb} is across h_{ob}. I_c is entering into $h_{fb} I_e$ and h_{ob}. There is no connection between E and C terminals. Hence the circuit has to be drawn as shown in the Fig. 3.39. The circuit in Fig. 3.40 is exactly the same as Fig. 3.39, But the figure in 3.40 is in Common Emitter Configuration.

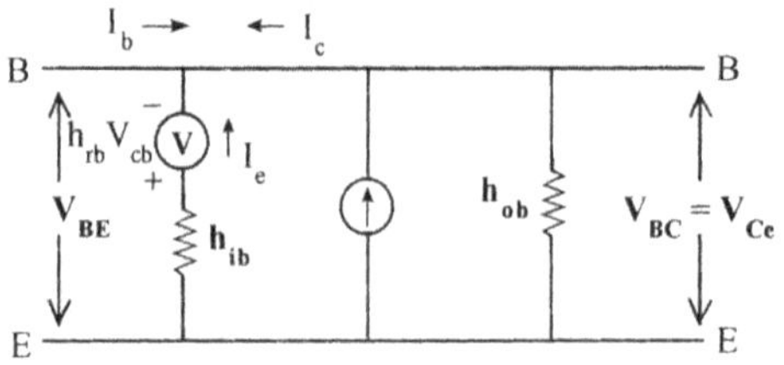

Fig 3.40 (a) Equivalent Circuits

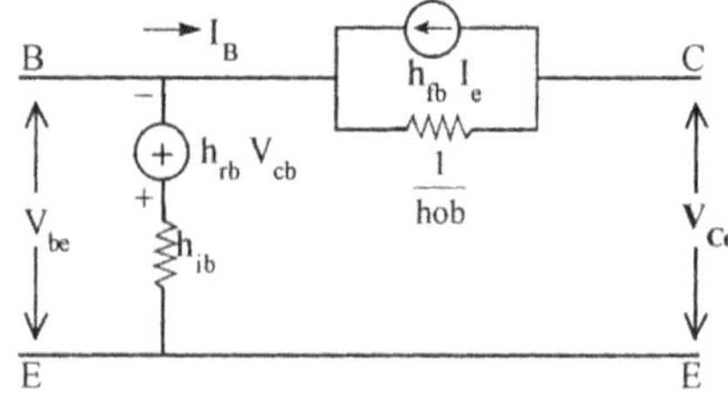

Fig. 3.40 (b)

$$h_{re} = \left.\frac{V_{be}}{V_{ce}}\right|_{I_{b=0}} = \text{Reverse Voltage Gain}$$

But $$V_{be} = V_{bc} + V_{ce} \text{ (of redrawn equivalent circuit)}$$

$$\therefore \quad h_{re} = \left.\frac{V_{bc} + V_{ce}}{V_{ce}}\right|_{I_{B=0}}$$

$$h_{re} = \left.\left(1 + \frac{V_{bc}}{V_{ce}}\right)\right|_{I_{B=0}}$$

If $I_b = 0$, $I_c = -I_e$. Therefore, $I_B + I_C + I_E = 0$ for any Transistor. But the current Through h_{ob} is I, (I is in opposite direction. I_C gets branched through h_{ob}).

Hence $$h_{fb} I_e - I = I_c$$

$$\therefore \quad h_{fb} I_e - I = -I_e$$

or $$I = h_{fb} I_e + I_e = I_e (1 + h_{fb})$$

Since h_{ob} represents conductance, $I = h_{ob} \cdot V_{bc}$

$$I = (1 + h_{fb}) I_e \qquad\qquad \dots\dots\dots\ (\,3.25\,)$$

$$\therefore \quad h_{ob} V_{bc} = (1 + h_{fb}) I_e$$

or $$I_e = \frac{h_{ob} V_{bc}}{1 + h_{fb}}$$

Applying KVL to the output mesh,

$$h_{ib} \times I_c + h_{rb} V_{cb} + V_{bc} + V_{ce} = 0$$

$$= \frac{(h_{ib} . h_{ob}) V_{bc}}{1 + h_{fb}} = -h_{rb} V_{bc} + V_{bc} + V_{ce} = 0$$

$$\frac{h_{ib} h_{ob}}{1 + h_{fb}} V_{bc} - h_{rb} V_{bc} + V_{bc} + V_{ce} = 0$$

$$= V_{bc} \frac{h_{ib} h_{ob}}{1 + h_{fb}} - h_{rb} + 1 = -V_{ce}$$

or $$\frac{V_{bc}}{V_{ce}} = -\frac{-(1 + h_{fb})}{h_{ib} . h_{ob} + (1 - h_{rb})(1 + h_{fb})}$$

Hence $$h_{re} = 1 + \frac{V_{bc}}{V_{ce}} = \frac{h_{ib} h_{ob} - (1 + h_{fb}) h_{rb}}{h_{ib} h_{ob} + (1 - h_{rb})(1 + h_{fb})}$$

But $$h_{rb} \ll 1, \text{ and } h_{ob} h_{ib} \ll (1 + h_{fb}) = \frac{h_{ib} h_{ob} - (1 + h_{fb}) h_{rb}}{(1 + h_{fb})} .$$

Hence, $$h_{re} \cong \frac{h_{ib} h_{ob}}{1 + h_{fb}} - h_{rb}. \qquad\qquad \dots\dots\dots\ (\,3.26\,)$$

$$h_{ie} = \frac{V_{be}}{I_b}\bigg|_{V_{ce=0}}$$

If we connect terminals C and B together, we get the circuit as (the same redrawn circuit in Common Base Configuration with C and B shorted) is shown in Fig. 3.18. Applying KVL around the I mesh on the left hand side,

$$V_{be} + h_{ib} I_e - h_{rb} V_{cb} = 0.$$
$$h_{ib} I_e = + h_{rb} V_{be} - V_{be}$$
$$\therefore \qquad V_{bc} = V_{be}$$

Combining these two equations, $I_e = - \dfrac{(1 - h_{rb})}{h_{ib}} V_{be}$

Applying KCL to node B,

$$I_b + I_e + h_{fb} I_e - h_{ob} V_{be} = 0$$

or $\qquad I_b = (1 + h_{fb}) \dfrac{1 - h_{rb}}{h_{ib}} V_{be} + h_{ob} V_{be}$

Hence $\qquad h_{ie} = \dfrac{V_{be}}{I_b} = \dfrac{h_{ib}}{h_{ib}h_{ob} + (1 - h_{rb})(1 + h_{fe})}$

$\qquad\qquad h_{rb} \ll 1, \text{ and } h_{ob} h_{ib} \ll (1 + h_{fb})$

then $\qquad h_{ie} \cong \dfrac{h_{ib}}{1 + h_{fb}}.$ .... (3.27)

3.14 MEASUREMENT OF h-PARAMETERS

In the circuit shown in Fig. 3.41 to determine the h-parameter, PNP transistors are taken into account. The operating point is chosen by suitably adjusting R_2, V_{EE} and V_{CC}, there by adjusting the values of I_C and V_{CB}. The h-parameters depend upon the frequency of operation. So the

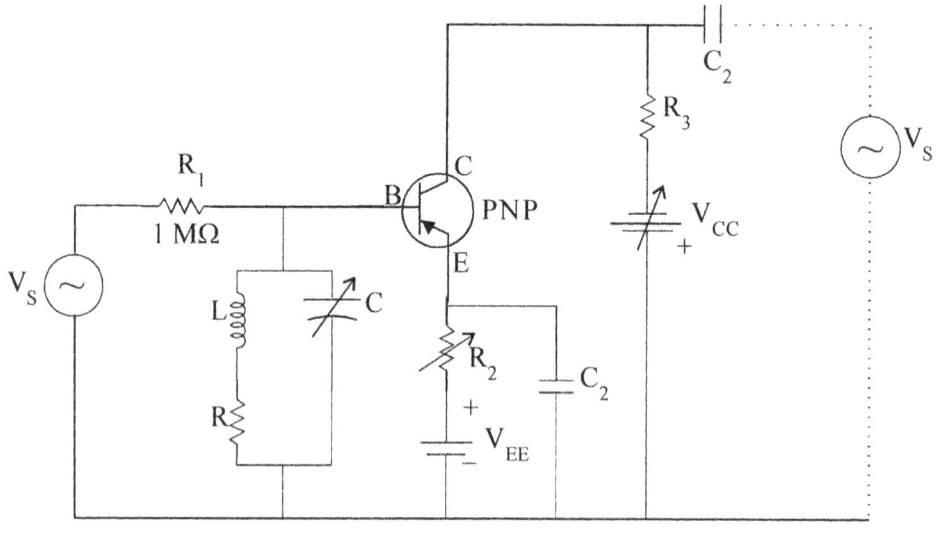

Fig 3.41 Circuit for determining h - parameters.

signal generator is adjusted to be 1KHz, at which frequency normally all the transistor h parameters are specified by the manufactures. 1 KHz is in the audio frequency range. *So the h-parameters are assumed to be real at audio frequency*. The tank Circuit impedance will be very high at this frequency, of the order of 500kΩ. This is very high compared to the input resistance of the transistor. The tank circuit is tuned to the same frequency of the signal generator (1KHz). This will prevent any stray pick up. The stray signal will pass through the LC tank circuit, since it offers least impedance. So spurious signal will not affect the net voltages and currents of the transistor, while determining the h-parameter.

The name, tank circuit, is used for LC Parallel circuit, because energy is stored in these elements, as water is stored in a tank.

$$h_{ie} = \left.\frac{V_b}{I_b}\right|_{V_c} \quad ; \quad I_b = \frac{V_S}{R_1}$$

Since the input resistance of the transistor is small compared to the Tank Circuit Resistance, all the current I_b flows into the transistors alone. R_S is negligible compared to R_T (1MΩ).

$$\therefore \qquad I_b = \frac{V_S}{R_1}$$

Now h_{ie} and h_{fe} are determined when $V_c = 0$ or the collector is shorted to ground. But if the collector is directly connected to the ground, I_c will be different and V_{CE} will be different. The operating point will change h parameters. Depending on the operating point h - parameters differ. Hence AC short circuit is to be done . C_2 will block D.C. Hence the D.C values of V_{CE}, I_c will not change at all. But the A.C. Voltage of V_c is $= 0$ or the collector is at ground potential for A.C. voltages . So this is called AC short circuit.

$$\text{Now} \qquad h_{ie} = \left.\frac{V_b}{I_b}\right|_{V_c = 0} \qquad \therefore \ I_b = \frac{V_S}{R_1}$$

$$h_{fe} = \left.\frac{I_c}{I_b}\right|_{V_c = 0} \qquad I_b = \frac{V_S}{R_1}$$

$$\therefore \qquad h_{fe} = \frac{I_C R_1}{V_S}$$

where V_o is the voltage across R_3.

h_{re} and h_{oe} are defined when $I_b = 0$ or input is open circuited. Therefore, the tank circuit impedance is very large. The input side can be regarded as open circuit. The signal generator is connected on the output side.(as shown by the dotted lines).

$$h_{re} = \left.\frac{V_b}{V_c}\right|_{I_b = 0} = \frac{V_b}{V_c}$$

$$h_{oe} = \left.\frac{I_c}{V_c}\right|_{I_b = 0} = \frac{I_c}{V_c} = \frac{V_O}{R_L V_C} \qquad \therefore \ I_C = \frac{V_O}{R_L}$$

3.15 GENERAL AMPLIFIER CHARACTERISTICS

An amplifier generally produces an enlarged version of an input signal. The amplifier may be thought of as a black box, that has two input terminals; and two output terminals for the connection of the load and a means of supplying power to the amplifier. One of the input and output terminals will be common for the input and output (Fig. 3.42).

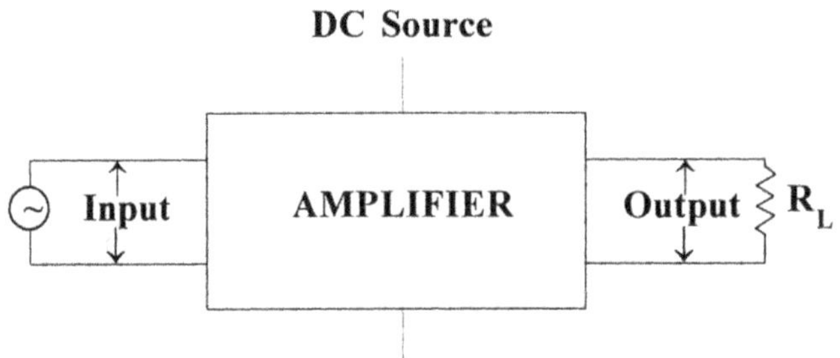

Fig 3.42 Block schematic of amplifier.

The signal to be amplified may be AC or DC. Input signal is a low level voltage such as the one obtained from a tape head, or a transducer like thermocouple, pressure gauge etc. The output load can be a loud speaker in an audio amplifier, a motor in a servo amplifier or a relay in control applications. In any case, the output of the amplifier is an enlarged version of the input to amplify means to increase the size. So in the case of electrical signals, if it is voltage or power amplifications, there is some additional energy being gained by the output signal. Energy can be neither created nor destroyed. So this additional energy is being gained from the D.C. bias supply. So without V_{CC} or DC Bias Voltage, (in the case of transistor amplifier circuit), the circuit will not work.

The notations that are used in the case of amplifier circuit are (Fig. 3.43)

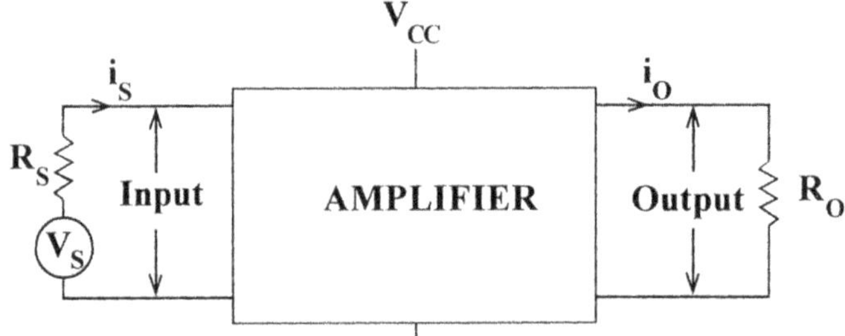

Fig 3.43 Block schematic.

V_s = Open circuit signal voltage

i_s = Source or signal current.

R_s = Internal resistance of the source in Ω.

R_i = Amplifier input resistance.

V_i = Amplifier input voltage

V_o = Amplifier output voltage

i_o = Amplifier output current

R_o = Load Resistance.

$$\text{Current Gain} \qquad A_i = i_o \mid i_s = \frac{i_o \text{ peak to peak}}{i_s \text{ peak to peak}}$$

$$\text{Voltage Gain} \qquad A_V = V_o \mid V_i = \frac{V_o \text{ peak to peak}}{V_i \text{ peak to peak}}$$

$$\text{Power Gain} \qquad A_p = \frac{P_0}{P_i} = \frac{V_o i_o}{V_i i_s} = A_v . A_i.$$

An amplifier may or may not exhibit both voltage and current gains, but in general, it exhibits power gain.

3.15.1 AMPLIFIER INPUT RESISTANCE R_i

The amplifier circuit presents a load R_i to the source which can be 50 Ω to few thousand Ω for transistor circuits.

The input voltage to the amplifier will be reduced from V_s by an amount that depends on both R_s and R_i (Fig. 3.44)

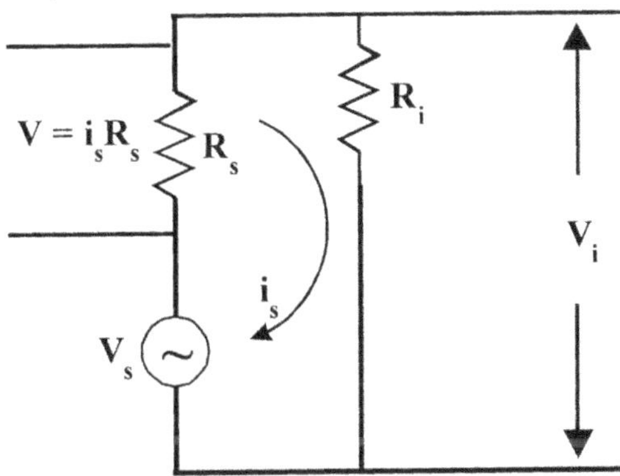

Fig 3.44 Input side equivalent circuit for an amplifier.

$$V_i = V_s \times \frac{R_i}{R_i + R_s} = V_s \times \frac{1}{1 + \dfrac{R_s}{R_i}}$$

If $R_i \gg R_s$ the input voltage to the amplifier is same as that of V_s. or if R_s is small then also $V_i = V_s$ usually $R_s = 50\ \Omega$. From the above equation, we can determine the input resistance of a given amplifier circuit by noting the open circuit voltage of the voltage source V_s and then voltage V_i after connections to the amplifier. If R_s of the voltage source is known, R_i can be calculated.

3.15.2 AMPLIFIER OUTPUT RESISTANCE R_o

R_L is used to denote the load resistance. So R_o is the output resistance of the amplifier circuit. (Fig. 3.45).

The amplifier can be represented as a voltage source in series with internal resistance of R_o in parallel with R_L is the external load resistance K is the amplification factor.

$$\therefore \qquad V_0 = KV_i \, \frac{R_L}{R_L + R_O} = KV_i \times \frac{1}{1 + \dfrac{R_o}{R_L}}$$

Output voltage V_0 can be made large by increasing the value of R_L But the output current will be small and so the current gain will fall. Hence power gain will also be less. In order to get maximum power gain, R_L should be matching with R_o. It should be complex conjugate of R_o according to Maximum Power Transfer Theorem. For this purpose, matching transformers are used $\left(\dfrac{R_1}{R_2} = \dfrac{N_1}{N_2} \right)$ in the output stage of an amplifier. Similarly matching transformers are also used to couple one stage of an amplifer to the second stage, if the input resistance of the second stage amplifier is very small.

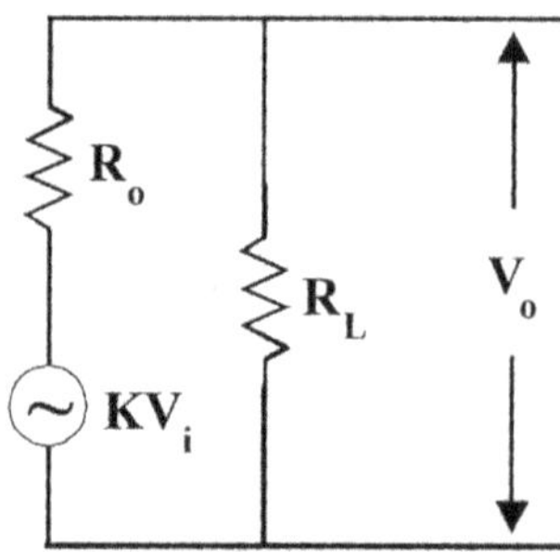

Fig 3.45   Output side equivalent circuit.

Conversion Efficiency :

An amplifier circuit draws energy from the D.C. source and amplifies the A.C. signal. The D.C. supply provides the major portion of the extra energy.

$$\therefore \qquad \text{Conversion } \eta = \frac{\text{A.C. signal power delivered to the load}}{\text{D.C. input power to the active device}} \times 100 = \frac{P_o}{P_{DC}}.$$

This is also known as ***collector circuit efficiency*** in the case of transistor and plate circuit efficiency in the case of tube amplifier circuits.

Problem 3.14

A single stage amplifier employing one active device, is powered by a 9V battery which has a current drain of 20 mA. If the load voltage is 3V at 12 m A, determine the conversion η.

$$P_o = V_o I_o = 3 \times 12 \times 10^{-3} = 36 \text{mW}$$

$$P_{DC} = V_{DC} I_{DC} = 9 \times 20 \times 10^{-3} = 180 \text{mW}$$

$$\eta = \frac{36}{180} \times 100 = 20\% .$$

3.16 ANALYSIS OF TRANSISTOR AMPLIFIER CIRCUIT USING h-PARAMETERS

To form transistor amplifier configuration, we connect a load impedance Z_L and a signal source as shown in Fig. 3.46.

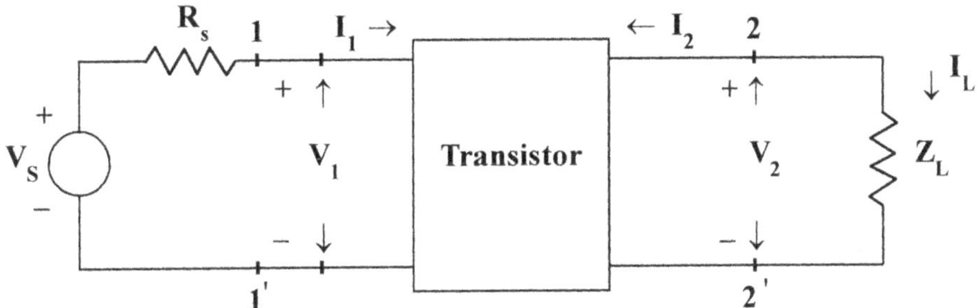

Fig 3.46 Amplifier circuit.

V_s is the signal source, R_s is the resistance of the signal source, and Z_L is the load impedance. Transistor can be connected in C.E, C.B. and C.C. Configuration. To analyse these circuits i.e. to determine the current gain A_I, Voltage gain A_v, input impedance, output impedance etc, we can use the h parameters. So the equivalent circuit for the above transistor amplifier circuit in general form without indicating C.E, C.B, or C.C. Configuration, can be denoted as in the Fig. 3.47.

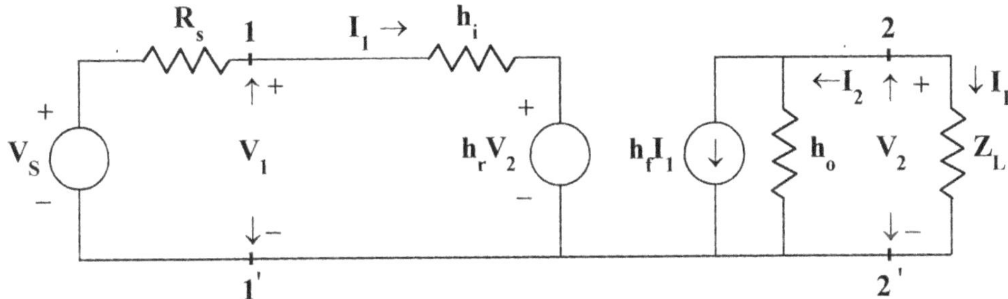

Fig 3.47 h-parameter equipvalent circuit (general repersentation).

Current Amplification,

$$A_I = \frac{I_L}{I_1};$$

But

$$I_L = -I_2$$

$$\therefore \quad A_I = -\frac{I_2}{I_1};$$

Negative sign is because I_2 is always represented as flowing into the current source. For a transistor PNP or NPN type, if it is flowing out of the transistor, I_2 is represented as $-I_2$. Now voltage across Z_2 is taken with 2 as +ve and 2^1 as -ve The directions of I_L are as shown and $I_L = -I_2$.

From the above circuit we have,

$$I_2 = h_f I_1 + h_o V_2$$

But

$$V_2 = I_L Z_L = -I_2 \cdot Z_L \quad (\text{Since } I_L = -I_2)$$

$$\therefore \qquad I_2 = h_f I_1 - h_o \cdot Z_L \cdot I_2$$

$$\text{or} \qquad I_2 + h_o Z_L I_2 = h_f I_1$$

$$I_2 (1 + h_o Z_L) = h_f I_1$$

$$\therefore \qquad \boxed{A_I = -\frac{I_2}{I_1} = \frac{-h_f}{1 + h_o Z_L}} \qquad \dots\dots (3.28)$$

Negative sign indicates that I_1 and I_2 are out of phase by 180^0.

3.16.1 INPUT IMPEDANCE : Z_i

Input impedance of any circuit is the impedance we measure looking back into the amplifier circuit. Now amplifier terminals are 1 and 1^1. R_s is the resistance of the signal source. So to determine the output z of the amplification alone, it need not be considered.

$$Z_i = \frac{V_1}{I_1}$$

But $\qquad V_1 = h_i I_1 + h_r V_2.$

Hence $\qquad Z_i = \frac{V_1}{I_1} = h_i + h_r \cdot \frac{V_2}{I_1}. \qquad \dots\dots (3.29)$

But $\qquad V_2 = -I_2 \cdot Z_L, \qquad \therefore \frac{V_2}{I_1} = \left(\frac{-I_2}{I_1}\right) Z_L = A_I Z_L \qquad \dots\dots (3.30)$

Substituting the values of V_2 in Equation (3.29),

$$\boxed{Z_i = h_i + h_r \cdot A_I \cdot Z_L}$$

But $\qquad A_I = \frac{-h_f}{1 + h_o z_L}$

$$\therefore \qquad Z_i = h_i - \frac{h_f \cdot h_r z_L}{1 + h_o z_L}$$

$$= h_i - \frac{h_f h_r}{\dfrac{1}{z_L} + h_o}$$

$$\boxed{Z_i = h_i - \frac{h_f \cdot h_r}{Y_L + h_o}} \qquad \dots\dots (3.31)$$

Y_L is load admittance. Therefore, Input impedance is a function of load impedance.

If $\qquad Z_L = 0, Y_L = \infty,$

$$\therefore \qquad Z_i = h_i.$$

3.16.2 VOLTAGE GAIN

$$A_v = \frac{V_2}{V_1}$$

But $\qquad V_2 = + I_L \cdot Z_L = -I_2 \cdot Z_L \qquad\qquad \text{Since, } I_2 = -I_L$

But
$$I_2 = -A_I \cdot I_1$$

$\therefore$
$$V_2 = +A_I \cdot I_1 Z_L.$$

$$\frac{v_2}{v_1} = \frac{A_I I_1 Z_L}{V_1}$$

But
$$\frac{V_1}{I_1} = Z_i$$

$$z_i = h_i - \frac{h_f \cdot h_r}{\dfrac{1}{R_L} + h_o}$$

$\therefore$
$$A_V = \frac{A_I Z_L}{Z_i}$$

$$\boxed{A_V = -\frac{-h_f \cdot z_L}{h_i + z_L (h_i h_o - h_f h_r)}}$$
.... (3.32)

3.16.3 OUTPUT ADMITTANCE, Y_o

The output impedance can be determined by using two assumptions $Z_L = \infty$, and $V_s = 0$

Y_o is defined as $\dfrac{I_2}{V_2}$ with $Z_L = \infty$

But
$$I_2 = h_f I_1 + h_o v_2.$$

Dividing by V2,

$$\frac{I_2}{V_2} = Y_O = \frac{h_f \cdot I_1}{V_2} + h_o.$$
.... (3.33)

From the equivalent circuit with $V_s = 0$,
$$R_s \cdot I_1 + h_i I_1 + h_r V_2 = 0.$$

Dividing by V_2 Through out, we get

$$\frac{R_S \cdot I_1}{V_2} + \frac{h_i I_1}{V_2} + h_r = 0$$

or
$$\frac{I_1}{V_2} = \frac{-h_r}{h_i + R_S}$$
.... (3.34)

Substitute this value in the equation for Y_o or equation 3.23.

$$Y_o = h_f \left(\frac{-h_r}{h_i + R_s} \right) + h_o$$

$$\boxed{Y_o = h_o - \frac{h_f \cdot h_r}{h_i + R_s}}$$

Therefore, $Z_0 = 1/Y_0$

Therefore, output admittance is a function of Source Resistance R_s where as Z_i is a function of Y_L. If $R_s = 0$, $Y_0 \cong h_0$. Since ($h_f h_r / h_i$) is very small.

3.16.4 Voltage Gain A_{vs} Considering Source Resistance, R_s

$$\therefore \qquad A_{vs} = \frac{V_2}{V_S} = \frac{V_2}{V_1} \times \frac{V_1}{V_S} = A_v . \frac{V_1}{V_S}.$$

This is the equivalent circuit of the input side of the amplifier circuit.

$$\therefore \qquad V_1 = \frac{V_S . Z_i}{z_i + R_S}$$

$$\therefore \qquad A_{vs} = \frac{A_v . Z_i}{Z_i + R_S}$$

$$A_v = \frac{v_2}{v_1}.$$

$$\therefore \qquad A_v Z_i = \frac{V_2}{V_1} . Z_i$$

But

$$V_2 = I_L . Z_L \; ; \; V_1 = I_1 Z_i$$

$$\therefore \qquad A_v . z_i = \frac{I_L . Z_L . Z_i}{I_1 . Z_L} = A_I . Z_L$$

$$\therefore \qquad \boxed{A_{vs} = \frac{A_I Z_L}{R_S + Z_i}}. \qquad \text{If } R_s = 0, A_{vs} = A_v . \qquad \dots\dots (3.35)$$

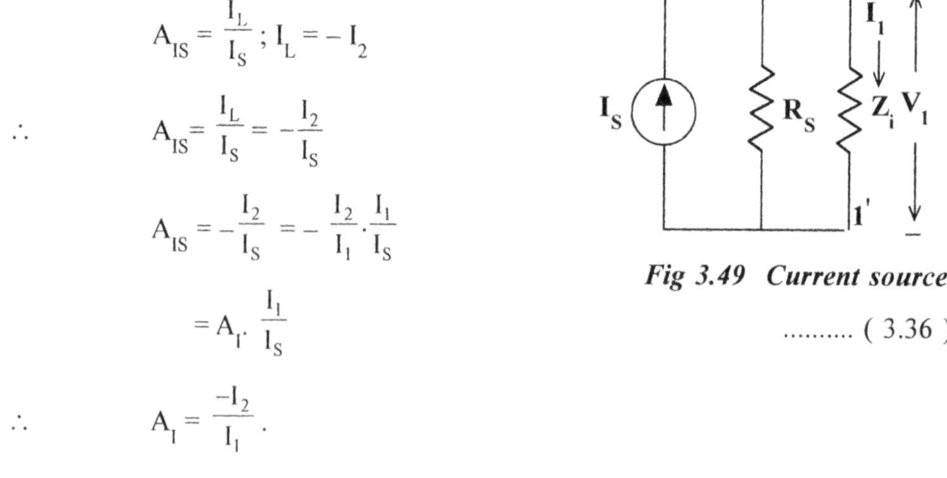

Fig 3.48 Voltage source.

Hence A_v is the voltage gain of an ideal voltage source with zero terminal resistance.

3.16.5 Current Gain A_{IS} Considering Source Resistance

The input source can also be represented as a current source I_s in parallel with resistance R_s or voltage source V_s in series with resistance R_s. Now let us consider the input source as a current source in parallel with R_s. The equivalent circuit is

$$A_{IS} = \frac{I_L}{I_S} \; ; \; I_L = -I_2$$

$$\therefore \qquad A_{IS} = \frac{I_L}{I_S} = -\frac{I_2}{I_S}$$

$$A_{IS} = -\frac{I_2}{I_S} = -\frac{I_2}{I_1} . \frac{I_1}{I_S}$$

$$= A_I . \frac{I_1}{I_S}$$

$$\therefore \qquad A_I = \frac{-I_2}{I_1}.$$

Fig 3.49 Current source.

$$\dots\dots (3.36)$$

From the Fig 3.27, $I_1 = \dfrac{I_S.R_S}{R_S + Z_i}$

$$\therefore \quad A_{IS} = \dfrac{A_1 \times R_S}{(R_S + Z_i)} = \dfrac{A_1.R_S}{(R_S + Z_i)}$$

If $R_S = \infty$, $A_1 = A_{IS}$. Therefore, A_1 is the current gain for an ideal source

Now

$$A_{VS} = \dfrac{A_1.Z_L}{Z_i + R_s} \qquad \dots\dots\dots (\ 3.37\ )$$

$$A_{VS} = \dfrac{A_1.R_S}{Z_i + R_S} \qquad \dots\dots\dots (\ 3.38\ )$$

Dividing 1 and 2, $\dfrac{A_{vs}}{A_{IS}} = \dfrac{Z_L}{R_S}$.

or

$$\boxed{A_{vs} = A_{IS}.\ \dfrac{Z_L}{R_S}} \qquad \dots\dots\dots (\ 3.39\ )$$

This equation is independent of the transistor parameters This is valid if the equivalent current and voltage sources have the same resistance.

3.16.6   POWER GAIN A_p

$$A_P = \dfrac{P_2}{P_1} = \dfrac{V_2 I_L}{V_1 I_1}$$

$$= A_v.\ A_1$$

$$\because \quad A_v = \dfrac{A_1.Z_L}{Z_i},$$

$$\therefore \quad \boxed{A_P = \dfrac{A_1^2 Z_L}{Z_i}} \qquad \dots\dots (\ 3.40\ )$$

3.17 COMPARISON OF THE CE, CB, CC CONFIGURATIONS

C.E :   Of the three, it is the most versatile. Its voltage and current gain are > 1. Input and output resistance vary least with R_s and R_L. R_i and R_o values lie between maximum and minimum for all the three configurations. Phase shift of 180^0 between V_i and V_o. Power Gain is Maximum.

C.B : $A_1 < 1\ A_V > 1$. R_i is the lowest and R_o is the highest . It has few applications. Sometimes it is used to match low impedance source V_i and V_o. No phase shift.

C.C : $A_V < 1$. A_1 is very high. It has very ***high input Z and low output Z. So it is used as a buffer*** between high Z source and low impedance load. It is also called as ***emitter follower***.

To analyse circuit consisting of a number of Transistors, each Transistor should be replaced by its equivalent circuit in h-parameters. The emitter base and collector points are indicated and other circuit elements are connected without altering the circuit Configuration. This way the circuit analysis becomes easy.

Example 3.15

Find the Common Emitter Hybrid parameters in terms of the Common Collector Hybrid parameters for a given transistor.

Solution

We have to find h_{ie}, h_{re}, h_{fe} and h_{oe} in terms of h_{ic}, h_{rc}, h_{fc} and h_{oc}.

The transistor circuit in Common Collector Configuration is shown in Fig. 3.50. The h-parameter equivalent circuit is shown in Fig. 3.51.

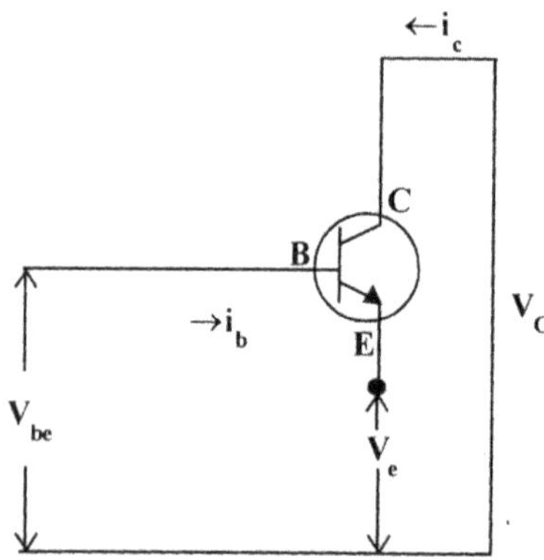

Fig 3.50 For Example 5.2.

Fig 3.51 Equivalent circuit.

C.C :

$$v_b = h_{ic} \cdot i_b + h_{rC} v_e$$

$$i_e = h_{fc} i_b + h_{oC} v_e$$

$$i_e = h_{fc} i_b - i_l$$

$$= (h_{fe} i_b - V_{ce} h_{oc})$$

$$v_{be} = i_b \cdot h_{ic} - h_{rc} v_{ce} + v_{ce} = i_b \cdot h_{ic} \qquad \text{Since } v_{ce} = 0$$

But
$$h_{ie} = \left. \frac{V_{be}}{I_b} \right|_{V_{ce}=0}$$

$\therefore$
$$\boxed{h_{ie} = h_{ic}} \qquad\qquad \text{........ (3.41)}$$

$$h_{fe} = \left. \frac{i_c}{i_b} \right| V_{ce} = 0$$

$$i_c = -i_b - i_e = -i_b - h_{fc} \cdot i_b + h_{oc} v_{ce}$$

But $\qquad v_{ce} = 0$

$$= -i_b - (h_{fc}\, i_b - h_{oc}\, v_{ce})$$

$$= -i_b - h_{fc}\, i_b + h_{oc}\, v_{ce} \qquad\qquad v_{ce} = 0$$

$$h_{fe} = \frac{-i_b - h_{fc} i_b}{i_b} = -(1 + h_{fc})$$

$$\boxed{h_{fe} = -(1 + h_{fc})} \qquad\qquad .... (\,3.42\,)$$

$$h_{re} = \frac{V_{be}}{V_{ce}}\bigg|_{i_b = 0}$$

$$V_{be} = i_b\, h_{ie} - h_{rc}\, V_{ce} + V_{ce}.$$

$$i_b = 0$$

$\therefore \qquad$
$$h_{re} = -\frac{h_{rc} v_{ce} + V_{ce}}{V_{ce}} = 1 - h_{rc}$$

$$\boxed{h_{re} = 1 - h_{rc}} \qquad\qquad .... (\,3.43\,)$$

$$h_{oe} = \frac{i_c}{V_{ce}}\bigg|_{i_b = 0}$$

$$i_c = -i_e - i_b$$

But $i_b = 0$.

$$i_c| = i_e$$

$$i_e = V_{ce}\, hoc$$

$\therefore \qquad$
$$\frac{i_e}{V_{ce}} = \frac{i_e}{V_{ce}} = h_{oc}$$

$\therefore \qquad$
$$i_c = -i_e = -h_{fc}\, i_b + V_{ce}\, h_{oc}$$

$\therefore \qquad$
$$\boxed{h_{oe} = h_{oc}} \qquad\qquad .... (\,3.44\,)$$

Since, $\qquad i_b = 0$

$\therefore \qquad i_c = V_{ce} \times h_{oc}$

$\therefore \qquad h_{oe} = h_{oc} \qquad\qquad$ is admittance in mhos ($\mho$).

3.18 SMALL SIGNAL ANALYSIS OF JUNCTION TRANSISTOR

Small Signal Analysis means, we assume that the input AC signal peak to peak amplitude is very small around the operating point Q as shown in Fig. 3.30. The swing of the signal always lies in the active region, and so the output is not distorted. In the ***Large Signal Analysis***, the swing of the input signal is over a wide range around the operating point. The magnitude of the input signal is very large. Because of this the operating region will extend into the cut-off region and also saturation region.

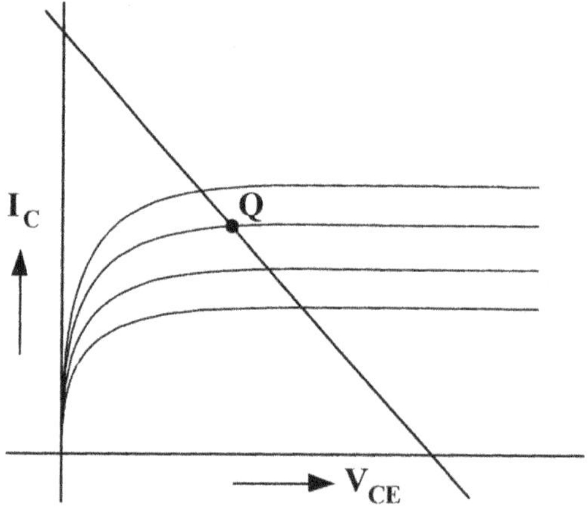

Fig 3.52   Operating point Q.

3.18.1   COMMON EMITTER AMPLIFIER

Common Emitter Circuit is as shown in the Fig. 3.31. The DC supply, biasing resistors and coupling capacitors are not shown since we are performing an ***AC Analysis***.

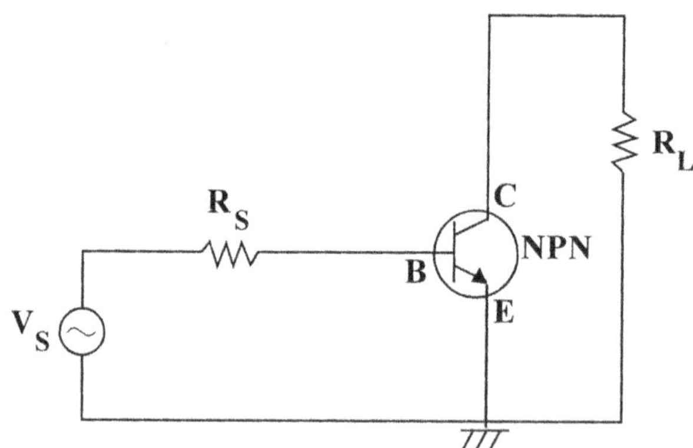

Fig 3.53   C.E. Amplifier circuit.

V_S is the input signal source and R_S is its resistance. The ***h-parameter*** equivalent for the above circuit is as shown in Fig. 3.32.

$$h_{ie} = \left. \frac{V_{BE}}{I_B} \right|_{V_{CE=0}} \qquad\qquad h_{re} = \left. \frac{V_{BE}}{V_{CE}} \right|_{I_{B=0}}$$

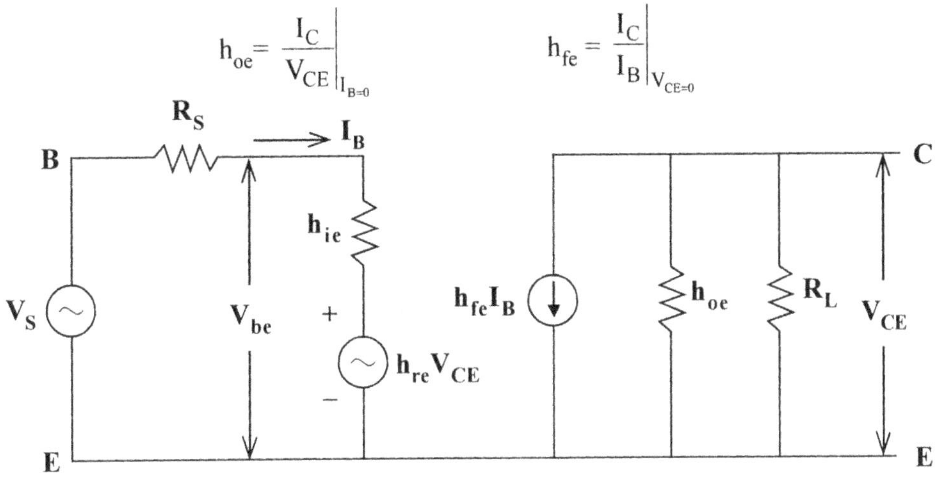

Fig 3.54   h-parameter equivalent circuit.

The typical values of the **h-parameter** for a transistor in Common Emitter Configuration are,

$$h_{ie} = 4 \text{ K}\Omega,$$

Since, $\qquad h_{ie} = \dfrac{V_{BE}}{I_B} \; ;$

V_{BE} is a fraction of volt 0.2V, I_B in µA, 100 µA and so on.

$$\therefore \qquad h_{ie} = \dfrac{0.2V}{50 \times 10^{-6}} = 4K\Omega$$

$$h_{fe} = \dfrac{I_C}{I_B} = 100 \; .$$

I_C is in mA and I_B in µA.

$$\therefore \qquad h_{fe} \gg 1 \simeq \beta$$

$h_{re} = 0.2 \times 10^{-3}$. Because, it is the Reverse Voltage Gain.

$$h_{re} = \dfrac{V_{BE}}{V_{CE}} \quad \text{and} \quad V_{CE} > V_{BE}; \qquad h_{re} = \dfrac{\text{Input}}{\text{Output}}$$

Output is $\gg$ input, because amplification takes place. Therefore $h_{re} \ll 1$.

$$h_{oe} = 8 \; \mu \, \mho \quad \text{and} \quad h_{oe} = \dfrac{I_C}{V_{CE}} \; .$$

INPUT RESISTANCE OF THE AMPLIFIER CIRCUIT (R_i)

The general expression for R_i in the case of Common Emitter Transistor Circuit is

$$R_i = h_{ie} - \dfrac{h_{fe}h_{re}}{h_{oe} + \dfrac{1}{R_L}}$$

For Common Emitter Configuration,

$$R_i = h_{ie} - \frac{h_{fe}h_{re}}{h_{oe} + \dfrac{1}{R_L}}$$

R_i depends on R_L. If R_L is very small, $\dfrac{1}{R_L}$ is large, therefore the denominator in the second term is small or it can be neglected.

$$\therefore \qquad R_i \cong h_{ie} .$$

If R_L increases, the second term cannot be neglected.

$$R_i = h_{ie} - (\text{ finite value })$$

Therefore, R_i decreases as R_L increases. If R_L is very large, $\dfrac{1}{R_L}$ will be negligible compared to h_{oe}. Therefore, R_i remains constant . The graph showing R_i versus R_L is indicated in Fig. 3.33. R_i is not affected by R_L if $R_L < 1$ KΩ and $R_L > 1$ MΩ as shown in Fig. 3.33.

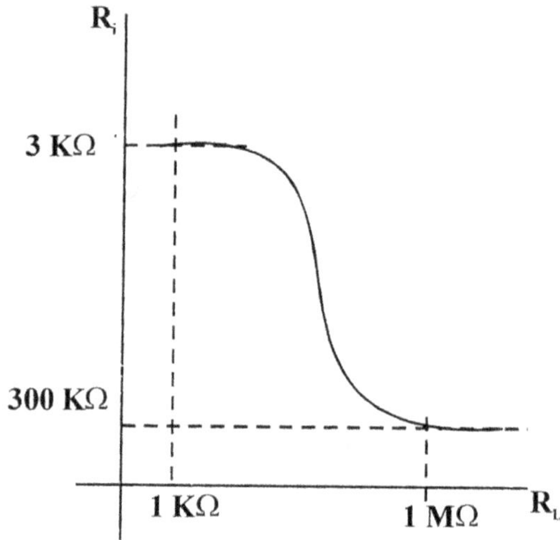

Fig 3.55 Variation of R_i with R_L.

R_i varies with frequency f because **h-parameters** will vary with frequency. h_{fe}, h_{re} will change with frequency f of the input signal.

OUTPUT RESISTANCE OF AN AMPLIFIER CIRCUIT (R_o)

For Common Emitter Configuration,

$$R_o = \frac{1}{h_{oe} - \left(\dfrac{h_{re}h_{fe}}{h_{ie} + R_S} \right)}$$

R_S is the resistance of the source. It is of the order of few hundred Ω.

R_o depends on R_S. If R_S is very small compared to h_{ie},

$$R_o = \cfrac{1}{h_{oe} - \cfrac{h_{re}h_{fe}}{h_{ie}}} \quad (\text{ independent of } R_s)$$

Then, R_o will be large of the order of few hundred KΩ. If R_s is very large, then

$$R_o \simeq \frac{1}{h_{oe}} \simeq 150 \text{ K}\Omega.$$

The graph is as shown in Fig. 3.34.

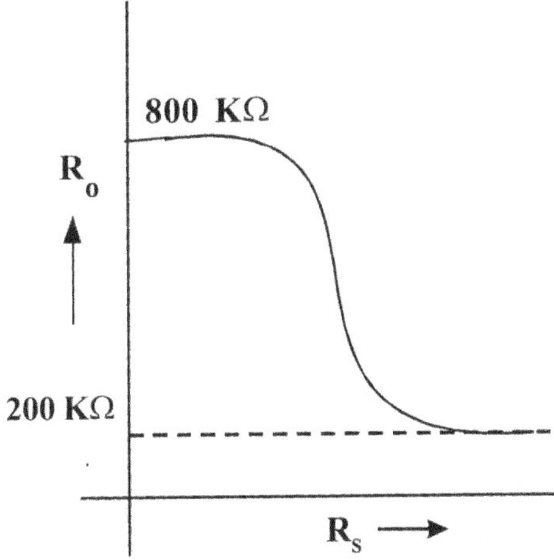

Fig 3.56 Variation of R_o with R_S.

CURRENT GAIN (A_i)

$$A_i = \frac{h_{fe}}{1 + h_{oe}R_L}$$

If R_L is very small, $A_i \simeq h_{fe} \simeq 100$. So, Current Gain is large for Common Emitter Configuration. As R_L increases, A_i drops and when $R_L = \infty$, $A_i = 0$. Because, when $R_L = \infty$, $I_o = 0$. Therefore, $A_i = 0$. Variation of A_i with R_L is shown in Fig. 3.35.

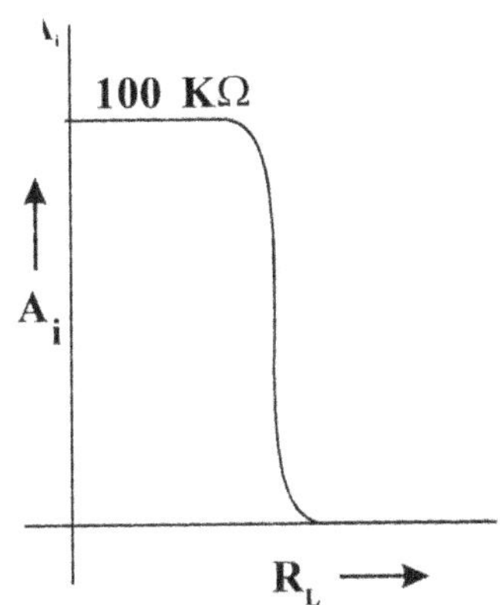

Fig 3.57 Variation of A_i with R_L.

VOLTAGE GAIN (A_V)

$$A_V = \frac{-h_{fe}R_L}{h_{ie} + R_L(h_{ie}h_{oe} - h_{fe}h_{re})}$$

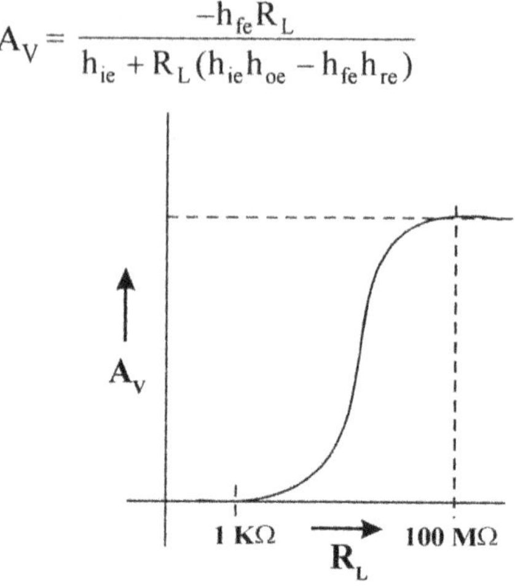

Fig 3.58 Variation of A_v with R_L.

If R_L is low, most of the output current flows through R_L. As R_L increases, output voltage increases and hence A_V increases. But if $R_L \gg \dfrac{1}{h_{oe}}$, then the current from the current generator in the *h-parameters* equivalent circuit flows through h_{oe} and not R_L. Then the,

$$\text{Output Voltage} = h_{oe} \cdot I_b \cdot \frac{1}{h_{oe}}$$

(R_L is in parallel with h_{oe}. So voltage across h_{oe}= voltage across R_L). Therefore, V_o remains constant as output voltage remains constant (Fig.3.36).

POWER GAIN

As R_L increases, A_I decreases. As R_L increases, A_V also increases.

Therefore, Power Gain which is the product of the two, A_V and A_I varies as shown in Fig. 3.37.

$$A_P = A_V \, A_I$$

Fig 3.59 Variation of A_p with R_L.

Power Gain is maximum when R_L is in the range 100 KΩ – 1MΩ. ie. when R_L is equal to the output resistance of the transistor. Maximum power will be delivered, under such conditions.

Therefore, it can be summarised as, Common Emitter Transistor Amplifier Circuit will have,

1. *Low to Moderate Input Resistance (300Ω – 5KΩ).*

2. *Moderately High Output Resistance (100KΩ – 800KΩ).*

3. *Large Current Amplification.*

4. *Large Voltage Amplification.*

5. *Large Power Gain.*

6. *180^0 phase-shift between input and output voltages.*

As the input current i_B, increases, i_C increases. Therefore Drop across R_C increases. $V_0 = V_{CC}$ – (drop and across R_C). Therefore, there is a phase shift of 180^0.

The amplifier circuit is shown in Fig. 3.38.

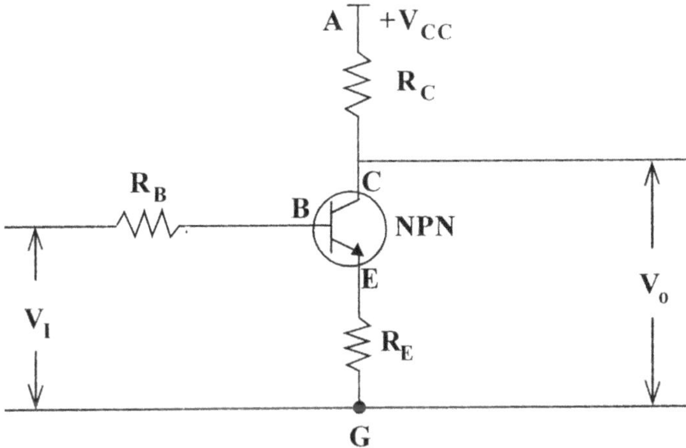

Fig 3.60   C.E. amplifier circuit.

3.18.2   COMMON BASE AMPLIFIER

The circuit diagram considering *only A.C.* is shown in Fig. 3.39.

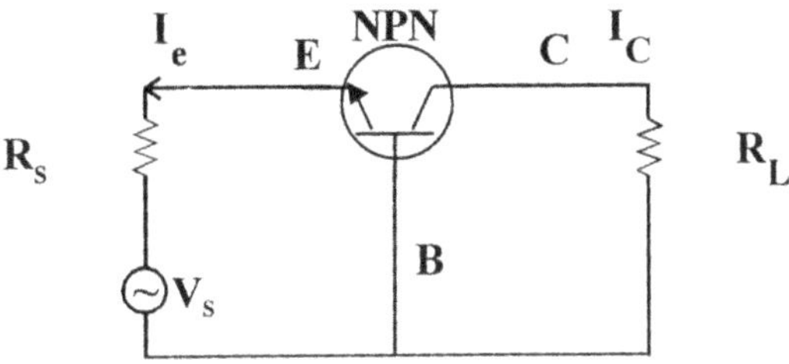

Fig 3.61   C.B. Amplifier circit.

$$h_{ib} = \left. \frac{V_{eb}}{I_e} \right|_{V_{cb}=0}$$

V_{eb} is small fraction of a volt. I_e is in mA. So, h_{ib} is small.

$$h_{fb} = \frac{I_c}{I_e}\Big|_{V_{cb}=0} = -0.99 \text{ (Typical Value)}$$

$$I_c < I_e \qquad \therefore \qquad h_{fb} < 1$$

$$h_{ob} = \frac{I_c}{V_{cb}}\Big|_{I_e=0} = 6.7 \times 10^{-8} \text{ mhos (Typical Value)}$$

I_c will be very small because $I_e = 0$. This current flows in between base and collector loop.

$$h_{rb} = \frac{V_{eb}}{V_{cb}}\Big|_{I_e=0} = 37 \times 10^{-6} \text{ (Typical Value)}$$

h_{rb} is small, because V_{eb} will be very small and V_{cb} is large.

INPUT RESISTANCE (R_I)

$$R_i = h_{ib} - \frac{h_{fb}.h_{rb}}{h_{ob} + \dfrac{1}{R_L}} \quad ; \quad h_{fb} \text{ is } -ve$$

when R_L is small < 100 KΩ, the second term can be neglected.

$$\therefore \qquad R_i = h_{ib} \simeq 30\Omega.$$

when R_L is very large, $\dfrac{1}{R_L}$ can be neglected.

$$R_i = h_{ib} - \frac{h_{fb}.h_{rb}}{h_{ob}}$$

So $R_i \simeq 500\Omega$ (Typical value) $\quad [\because h_{fb}$ is negative]

$$\therefore \qquad R_i = h_{ib} + \frac{h_{fb}h_{rb}}{h_{ob}}$$

The variation of R_i with R_L is shown in Fig. 3.40. R_i varies from 20Ω to 500Ω.

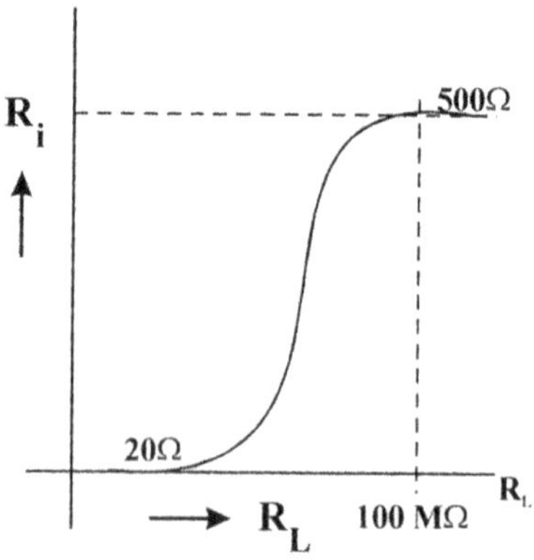

Fig 3.62 Variation of R_i with R_L.

OUTPUT RESISTANCE (R_0)

$$R_o = \cfrac{1}{h_{ob} - \cfrac{h_{rb}h_{fb}}{h_{ib} + R_s}}$$

If R_s is small, $R_o = 1 \Big/ \left(h_{ob} - \dfrac{h_{rb}h_{fb}}{h_{ib}} \right)$. But h_{fb} is negative.

$$\therefore \qquad R_o = \cfrac{1}{h_{ob} + \cfrac{h_{rb}h_{fb}}{h_{ib}}}$$

This will be sufficiently large, of the order of 300 KΩ. Therefore, value of h_{ob} is small. As R_s increases, $R_o = \dfrac{1}{h_{ob}}$ also increases. [This will be much larger because, in the previous case, in the denominator, some quantity is subtracted from h_{ob}.]

$$\therefore \qquad R_o = 12 M\Omega$$

The variation of R_o with R_s is shown in Fig. 3.41.

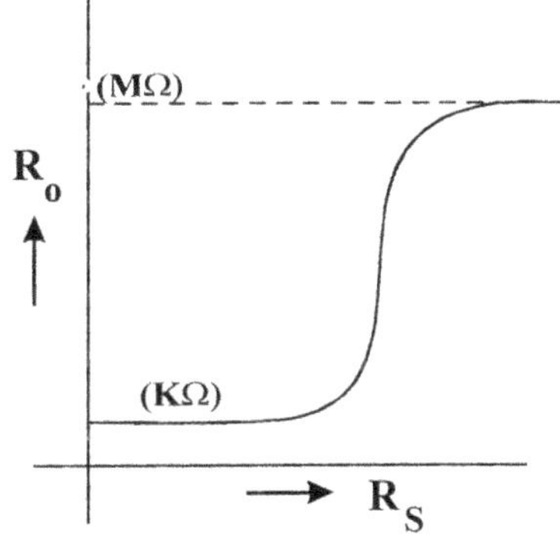

Fig 3.63 Variation of R_o with R_S.

CURRENT GAIN (A_i)

$$A_i = \cfrac{-h_{fb}}{1 + h_{ob}R_L}$$

A_i is < 1. Because h_{fe} < 1. As R_L increases, A_i decreases. A_i is negative due to h_{fb}. The variation of A_i with R_L is shown in Fig. 3.42.

VOLTAGE GAIN (A_V)

$$A_v = \cfrac{-h_{fb}R_L}{h_{ib} + R_L(h_{ib}h_{ob} - h_{fe}h_{rb})}$$

As R_L increases, A_V also increases. If R_L tends to zero, A_V also tends to zero. ($A_V \rightarrow 0$, as $R_L \rightarrow 0$). The variation of Voltage Gain A_V with R_L is shown in Fig. 3.43.

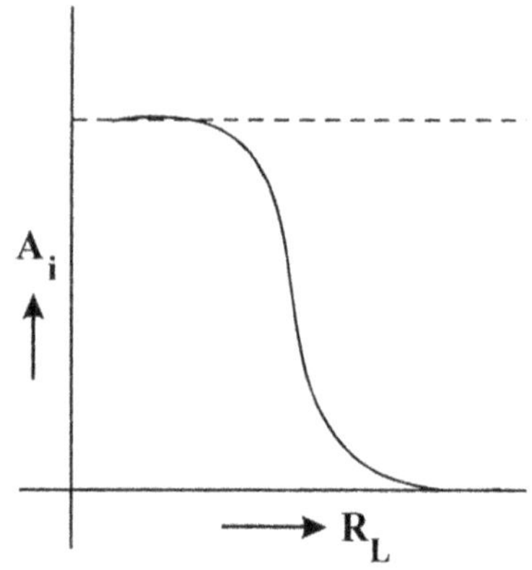

Fig 3.64 Variation of A_i with R_L

Fig 3.65 Variation of A_v with R_L

POWER GAIN (A_P)

Power Gain $A_P = A_V \cdot A_I$

A_V increases as R_L increases. But A_I decreases as R_L increases. Therefore, Power Gain, which is product of both, varies with R_L as shown in Fig. 3.44.

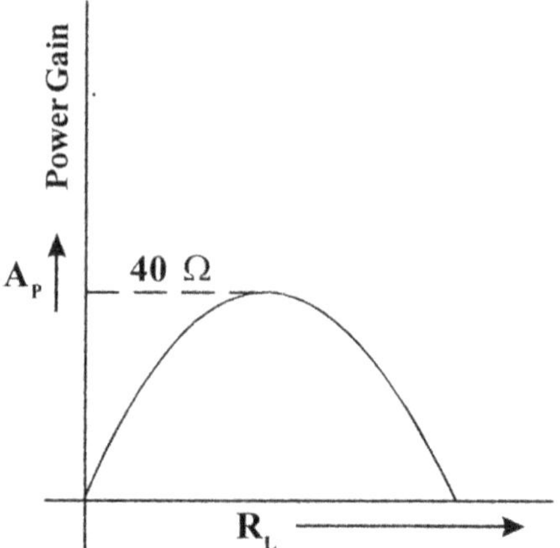

Fig 3.66 Variation of A_P with R_L.

The characteristics of Common Base Amplifier with typical values are as given below.

1. *Low Input Resistance (few 100 Ω).*
2. *High Output Resistance (MΩ).*
3. *Current Amplification $A_i < 1$.*
4. *High Voltage Amplification and No Phase Inversion*
5. *Moderate Power Gain (30).* $\because$ $A_i < 1$.

3.18.3   COMMON COLLECTOR AMPLIFIER

The simplified circuit diagram for A.C. of a transistor (BJT) in Common Collector Configuration is as shown in Fig. 3.45 (a).

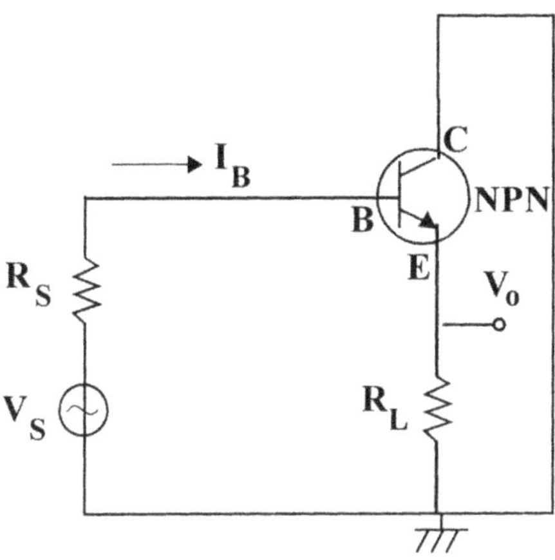

Fig 3.67 (a) Common collector amplifier circuit.

The ***h-parameters*** equivalent circuit of transistor in Common Collector Configuration is shown in Fig. 3.45 (b).

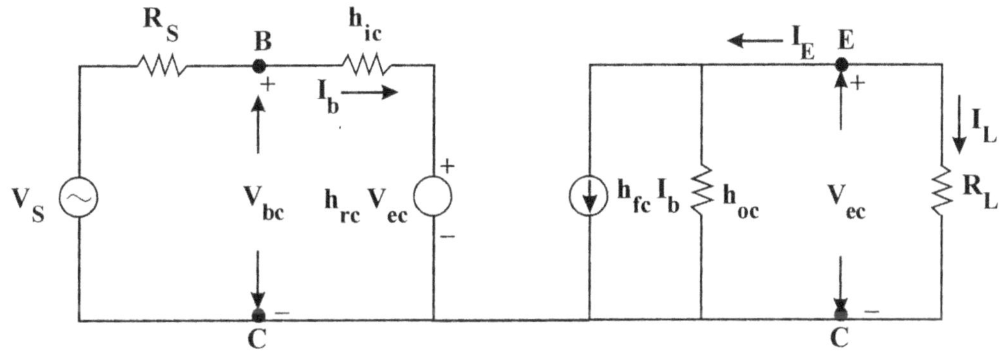

Fig 3.67 (b) h-parameter equivalent circuit.

$$h_{ic} = \left.\frac{V_{BC}}{I_B}\right|_{V_{CE}=0} = 2{,}780 \ \Omega \ (\text{Typical Value})$$

$$h_{oc} = \left.\frac{I_E}{V_{EC}}\right|_{I_B=0} = 7.7 \times 10^{-6} \ \text{mhos} \ (\text{Typical Value})$$

$$h_{fc} = -\left.\frac{I_E}{I_B}\right|_{V_{CE}=0} = 100 \qquad \because \ I_E \gg I_B. \ (\text{Typical Value})$$

h_{fc} is negative because, I_E and I_B are in opposite direction.

$$h_{rc} = \left.\frac{V_{BC}}{V_{EC}}\right|_{I_B=0} ; \ V_{BC} = V_{EC} \ (\text{Typical Value})$$

Because, $I_B = 0$, E-B junction is not forward biased.

$\therefore$ $V_{EB} = 0.$

For other circuit viz Common Base. and Common Emitter, h_r is much less than 1.

For Common Collector Configuration, $h_{rc} \simeq 1$.

The graphs (variation with R_C) are similar to Common Base Configuration.

Characteristics

1. *High Input Resistance $\simeq$ 3 KΩ (R_i)*

2. *Low Output Resistance 30 Ω. (R_o)*

3. *Good Current Amplification. $A_i >> 1$*

4. *$A_v \leq 1$*

5. *Lowest Power Gain of all the configurations.*

Since, A_v is < 1, the output voltage (Emitter Voltage) follows the input signal variation. Hence it is also known as ***Emitter Follower***. The graphs of variation with R_L and R_S are similar to Common Base amplifier.

INPUT RESISTANCE (R_i)

R_i input resistance looking into the base is h_{ie} only

The expression for R_i of the transistor alone $= h_{ie} - \left(\dfrac{h_{fe} h_{re}}{h_{oe} + \dfrac{1}{R_L}} \right)$.

R_L is very small and h_{re} is negligible. Therefore, the second term can be neglected. So R_i of the transistor alone is h_{ie}. Now R_i of the entire amplifier circuit, considering the bias resistors is,

$$R_i = h_{ie} \parallel R_1 \parallel R_2$$

$$\therefore \qquad \frac{R_1 R_2}{R_1 + R_2} = \frac{100 \times 100}{100 + 100} = 50 \text{ K}\Omega$$

$$\therefore \qquad R_i = \frac{2 \times 50}{2 + 50} = 1.925 \text{ K}\Omega$$

OUTPUT RESISTANCE (R_o)

$$R_o = \frac{1}{h_{oe} - \left[\dfrac{h_{re} h_{fe}}{h_{ie} + R_s} \right]}$$

Because, h_{re} is negligible, R_o of the transistor alone in terms of ***h-parameters*** of the transistor $= \dfrac{1}{h_{oe}}$. Now R_o of the entire amplifier circuit is,

$$\frac{1}{h_{oe}} \parallel R_4 = (2.1 \times 10^{-3}) \parallel (100 \text{ K}\Omega)$$

$$= 2 \text{K}\Omega.$$

CURRENT GAIN (A_I)

To determine A_i the direct formula for A_i in transistor in Common Emitter Configuration is,

$\dfrac{h_{fe}}{1+h_{oe}R_L}$. But this cannot be used because the input current I_i gets divided into I_1 and I_b. There is some current flowing through the parallel configuration of R_1 and R_2. So the above formula cannot be used.

Problem 3.1

$$V_{be} = I_b \cdot h_{ie}$$
$$h_{ie} = 10^{-4} \times (2000) = 0.2V. \text{ (This is AC Voltage not DC)}$$

Voltage across R_1 R_2 parallel configuration is also V_{be}.

$$\therefore \qquad \text{Current } I_i = \frac{V_{be}}{50 \times 10^3} = \frac{0.2}{50k\Omega} = 4 \ \mu A.$$

$$\therefore \qquad \text{Total Input Current } I_i = I_1 + I_b = 100 + 4 = 104 \mu A.$$

I_0 is the current through the $1K\Omega$ load.

$$\frac{1}{h_{oe}} = 100 \ K\Omega \text{ is very large compared with } R_4 \text{ and } R_L. \text{ Therefore, all}$$

the current on the output side, $h_{ie} I_b$ gets divided between R_4 and R_L only.

$$\therefore \qquad \text{Current through } R_L = I_0 = h_{fe} \ I_b \cdot \frac{R_4}{R_4 + R_L}$$

$$I_0 = 100 \times 10^{-4} \ \frac{2.1 \times 10^3}{(2.1 \times 10^3 + 10^3)} = 6.78 \text{ mA.}$$

$$\therefore \qquad \text{Current amplification} = A_i = \frac{I_0}{I_i} = \frac{6.78 \times 10^{-3}}{104 \times 10^{-6}} = 65.$$

$$A_v = \frac{V_0}{V_i} ; \qquad V_i = V_{be}$$

$$V_0 = -I_0 \cdot R_L$$
$$= (- 6.78 \times 10^{-3}) \times (10^{-3})$$
$$= - 6.78V$$

Because, the direction of I_0 is taken as entering into the circuit. But actually I_0 flows down, because V_0 is measured with respect to ground.

$$\therefore \qquad A_v = \frac{-6.78}{0.2} = -33.9$$

Negative sign indicates that there is phase shift of 180^0 between input and output voltages, i.e. as base voltage goes more positive, (it is NPN transistor), the collector voltage goes more negative.

Problem 3.3

For the circuit shown in Fig.3.46 estimate A_i, A_v, R_i and R_o using reasonable approximations. The ***h-parameters*** for the transistor are given as

$$h_{fe} = 100 \quad h_{ie} = 2000\ \Omega\ h_{re} \text{ is negligible and } h_{oe} = 10^{-5}\text{mhos.}$$

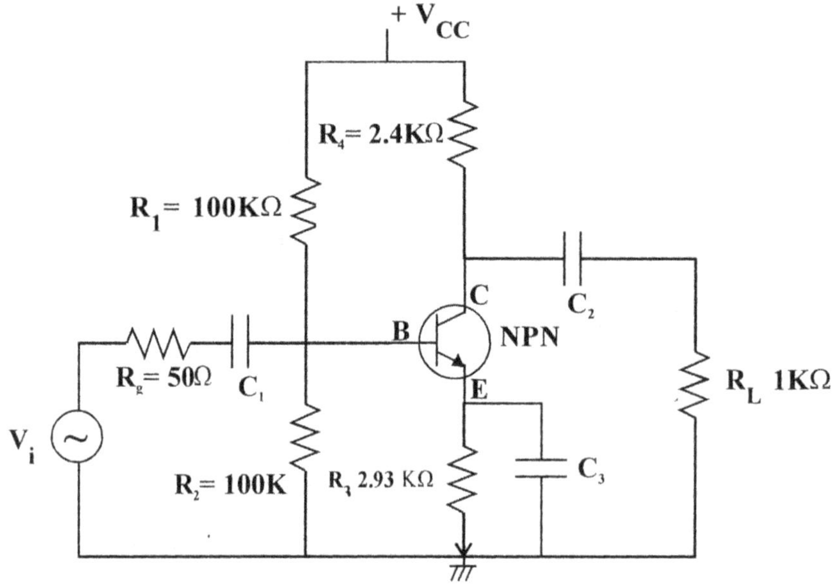

Fig 3.68 For Problem 6.3.

Solution

At the test frequency capacitive reactances can be neglected. V_{CC} point is at ground because the AC potential at $V_{CC} = 0$. So it is at ground. R_1 is connected between base and ground for AC. Therefore, $R_1 \parallel R_2$. R_4 is connected between collector and ground. So R_4 is in parallel with $1/h_{oe}$ in the output.

The A.C. equivalent circuit in terms of ***h-parameters*** of the transistor is shown in Fig.3.47.

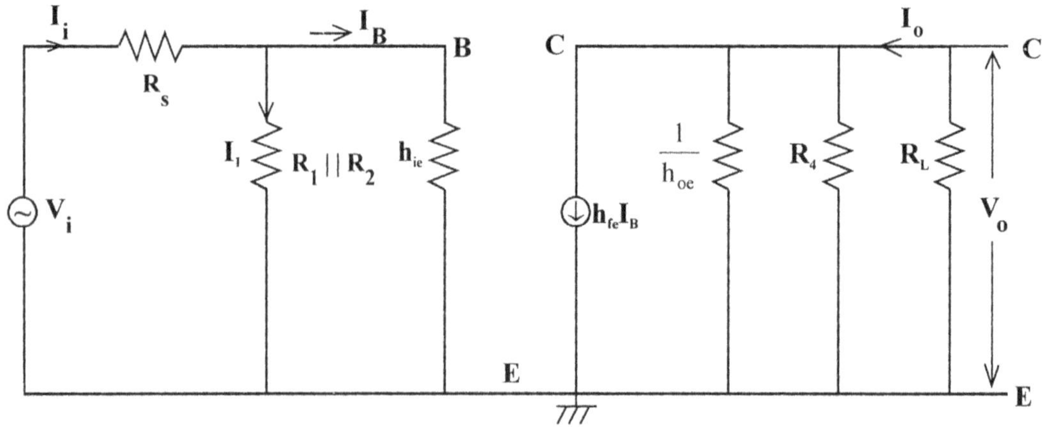

Fig 3.69 Equivalent circuit.

The voltage source $h_{re} V_c$ is not shown since, h_{re} is negligible. At the test frequency of the input signal ,the capacitors C_1 and C_2 can be regarded as short circuits. So they are not shown in the AC equivalent circuit. The emitter is at ground potential. Because X_{C_3} is also negligible, all the AC passes through C_3.   Therefore, emitter is at ground potential and this circuit is in Common Emitter Configuration.

Problem 3.4

For the circuit shown, in Fig. (3.48), estimate A_v and R_i . $\dfrac{1}{h_{oe}}$ is large compared with the load seen by the transistor. All capacitors have negligible reactance at the test frequency.

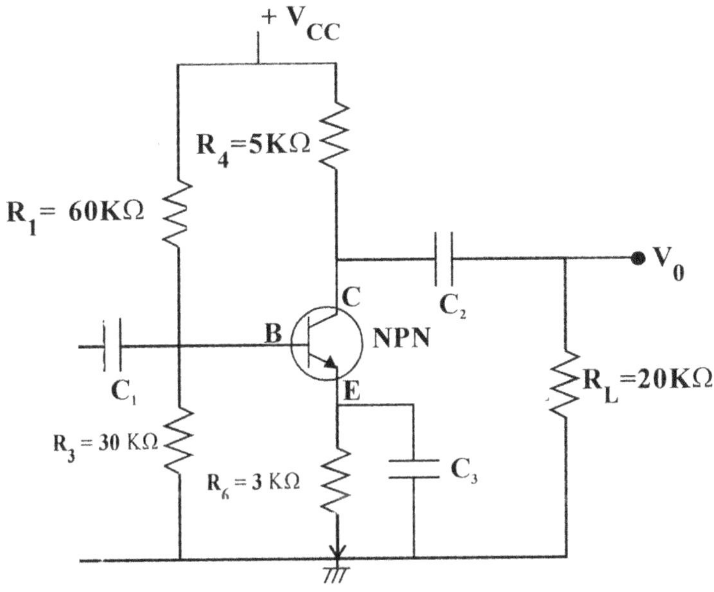

Fig 3.70 For Problem 6.4.

h_{ie} = 1KΩ, h_{fe} = 99 and h_{re} is negligible.

Solution :

The same circuit can be redrawn as shown in Fig. 3.49.

In the second circuit also, R_4 is between collector and positive of V_{CC}. R_2 is between $+V_{CC}$ and base. Hence both the circuits are identical. Circuit in Fig. 3.48 is same as circuit in Fig. 3.49. In the AC equivalent circuit, the direct current source should be shorted to ground. Therefore, R_4 is between collector and ground and R_2 is between base and ground. Therefore, R_4 is in parallel with R_7 and R_2 is in parallel with R_3 (Fig. 3.49).

$$R_2 \| R_3 = \frac{60 \times 30}{60 + 30} = \frac{1800}{90} = 20K\Omega.$$

$$R_4 \| R_7 = R_L = \frac{5 \times 20}{20 + 5} = 4K\Omega.$$

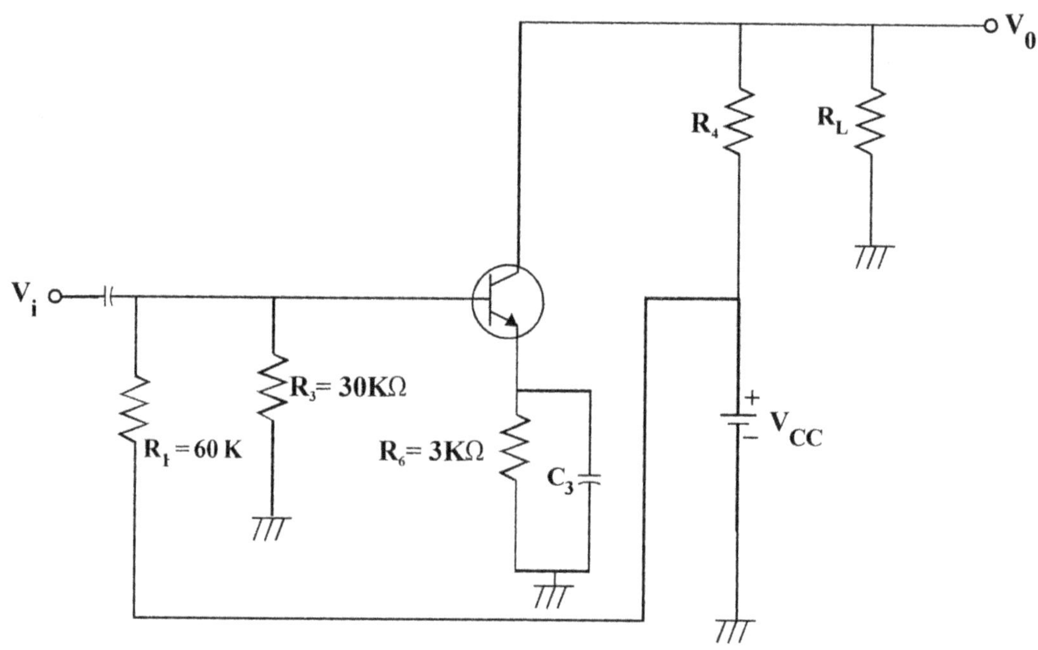

Fig 3.71 Redrawn circuit.

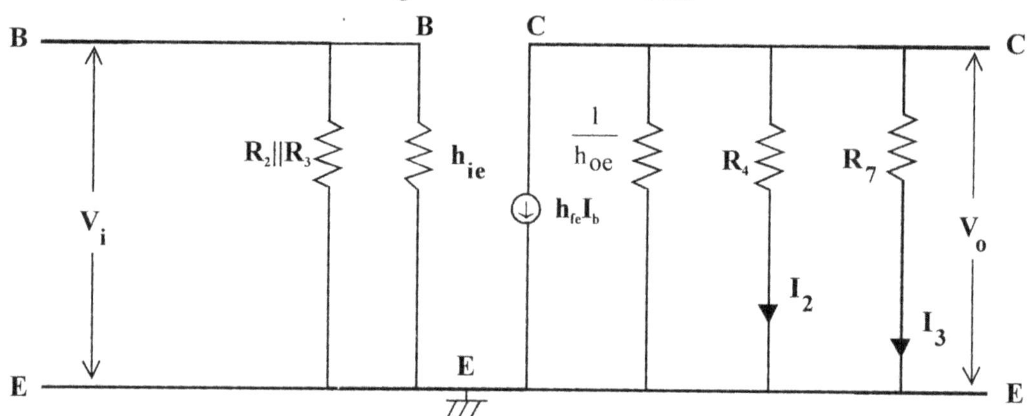

Fig 3.72 Equivalent circuit.

Therefore, the circuit reduces to, (as shown in Fig. 3.51)

$$I_b = \frac{V_i}{h_{ie}} \quad (\because h_{re} \text{ is negligible})$$

$$I_c = h_{fe}\, I_b = \frac{h_{fe}.\, V_i}{h_{ie}}$$

$$\therefore \quad I_b = \frac{V_i}{h_{re}} \; ; V_o = I_3.$$

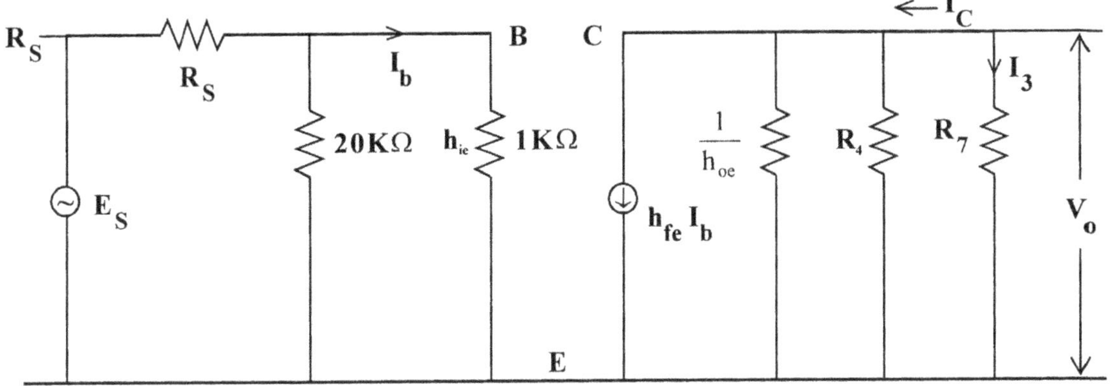

Fig 3.73 Simplified circuit.

$$\therefore \qquad V_o = -I_c \cdot R_L = -\frac{h_{fe}\,V_i \cdot R_L}{h_{ie}}$$

$$A_v = \frac{V_o}{V_i} = -\frac{h_{fe}R_L}{h_{ie}} = \frac{-99(4\times10^3)}{10^3}$$

$$A_v = -400$$

R_i is the parallel combination of $20\,K\Omega$ and h_{ie}

$$\frac{20\times1k\Omega}{20+1} = 950\,\Omega$$

While calculating R_o, R_2 is in parallel with R_3 and R_S will come in parallel with these two. But since R_s value is very small (R_2 parallel R_3), effective value will be R_S only.

Problem 3.5

Given a single stage transistor amplifier with *h - parameter* as $h_{ic} = 1.1\ K\Omega$, $h_{rc} = 1$, $h_{fc} = -51$, $h_{oc} = 25\ \mu A/v$. Calculate A_I, A_V, A_{Vs}, R_i, and R_o for the Common Collector Configuration, with $R_s = R_L = 10K$.

Solution

$$A_I = \frac{-h_{fc}}{1+h_{oc}R_L} = \frac{51}{1+25\times10^{-6}\times10^4} = 40.8$$

$$R_i = h_{ic} + h_{rc}\,A_I\,R_L = 1.1\times10^3 + 1\times40.8\times10^4 = 409.1\ K\Omega$$

$$A_v = \frac{A_I \cdot R_L}{R_i} = \frac{40.8\times10^4}{409.1} = 0.998$$

$$A_{vs} = \frac{A_V \cdot R_i}{R_i + R_s} = \frac{0.998\times409.1}{419.1} = 0.974$$

$$R_o = \frac{1}{h_{oC} - \dfrac{h_{fc}.h_{rc}}{h_{ic}+R_s}} = \frac{1}{25\times10^{-6} + \dfrac{51\times1}{(1.1+10)10^3}} = \frac{1}{4.625\times10^{-3}}$$

$$R_o = 217\,\Omega$$

Problem 3.6

For any transistor amplifier prove that

$$R_i = \frac{h_i}{1 - h_r A_v}$$

Solution

$$R_i = h_i - \frac{h_f . h_r}{h_o + \dfrac{1}{R_L}}$$

But

$$A_I = \frac{-h_f}{1 + h_o . R_L}$$

$$\therefore \quad R_i = h_i + h_r A_I R_L \qquad\qquad .....(1)$$

$$R_L = \frac{A_v . R_i}{A_I}$$

$$\therefore \quad A_v = \frac{A_I . R_L}{R_i}$$

Substituting this value of R_L in equation (1)

$$R_i = h_i + \frac{h_r . A_I . A_v . R_i}{A_I} = h_i + h_r . A_v . R_i , \therefore R_i = \frac{h_i}{1 - h_r A_v}$$

Problem 3.7

For a Common Emitter Configuration, what is the maximum value of R_L for which R_1 differs by no more than 10% of its value at $R_2 = 0$?

$$h_{ie} = 1100\Omega \; ; \; h_{fe} = 50$$
$$h_{re} = 2.50 \times 10^{-4}; \; h_{oe} = 25\mu\mho.$$

Solution

Expression for R_i is,

$$R_i = h_{ie} - \frac{h_{fe} . h_{re}}{h_{oe} + \dfrac{1}{R_L}} .$$

If $R_L = 0$, $R_i = h_{ie}$. The value of R_L for which $R_i = 0.9\, h_{ie}$ is found from the expression,

$$0.9\, h_{ie} = h_{ie} - \frac{h_{fe} . h_{re}}{h_{oe} + \dfrac{1}{R_L}}$$

or

$$\frac{h_{fe} . h_{re}}{h_{oe} + \dfrac{1}{R_L}} = h_{ie} - 0.9\, h_{ie} = 0.1\, h_{ie}$$

$$\frac{h_{fe} . h_{re}}{0.1\, h_{ie}} = h_{oe} + \frac{1}{R_L}$$

$$\frac{1}{R_L} = \frac{h_{fe}.h_{re}}{0.1h_{ie}} - h_{oe} = \frac{h_{fe}h_{re} - 0.1h_{oe}h_{ie}}{0.1h_{ie}}$$

or

$$R_L = \frac{0.1h_{ie}}{h_{fe}h_{re} - 0.1h_{oe}h_{ie}} = \frac{0.1 \times 1100}{50 \times 2.5 \times 10^{-4} - 0.1 \times 1100 \times 25 \times 10^{-6}}$$

$$R_L = 11.3 \text{K}\Omega$$

SUMMARY

- Hybrid Parameters are h - parameters, so called because the units of the four parameters are different. These are used to represent the equivalent circuit of a transistor.

- The h - parameters and their typical values, in Common Emitter Configuration are,

$$h_{ie} = 500~\Omega; \qquad h_{re} = 1.5 \times 10^{-4}; \qquad h_{oe} = 6~\mu\Omega; \qquad h_{fe} = 250;$$

- The second subscript denotes the configuration. similarly b and c represent Common Base and Common Collector Configurations.

- The Transistor Amplifier Circuit Analysis can be done, using h-parameters, replacing the transistor by h-parameter equivalent.

- Expressions for Voltage Gain A_V, Current Gain A_I, Input Resistance R_i, Output Resistance R_o etc., can be derived in terms of h-parameters.

- $r_{bb'}$ is called the base spread resistance. It is the resistance between the fiction base terminal B′ and the outside base terminal B.

- C.E. Amplifier circuit characteristics are :

  (a) Low to moderate R_i (300 Ω – 5 kΩ)

  (b) Moderately high R_0 (100 kΩ – 800 kΩ)

  (c) Large current amplification

  (d) Large voltage amplification

  (e) Large power gain

  (f) 180° phase shift between V_i and V_0.

- For C.B. amplifier :

  Low R_i. High R_0, $A_i < 1$, $A_v > 1$, No phase inversion, Moderate A_p

- For C.C. amplifier,

  High R_i, Low R_0, Large A_I, $A_V < 1$, Low A_p

- In Darlington pair circuit, the two BJTs are in C.C. configuration. The characteristics are :

  High R_i, Large A_I, Very low R_0, $A_v < 1$

♦ In CASCODE amplifier configuration, one BJT in C.E. configuration is in series with another BJT amplifier in C.B. configuration. Its characteristics are, Very large A_v, Large A_I and thigh R_0.

♦ In the case of R-C coupled amplifier, the output is coupled to the load through R and C, hence the name. The lower cut-off frequency f_1 is also called lower 3-db point or lower half power point. The upper cut-off frequency f_2 is also called upper 3-db point or upper half power frequency. In the low frequency range and high frequency range, the gain decreases. In the mid frequency range, the gain remains constant, $(f_2 - f_1)$ is called Band width.

♦ The relation between Emitter Efficiency γ, Transportation Factor β^* and Current Gain α is,
$$\alpha = \beta^* \times \gamma$$

♦ Transistor is an acronym for the words Transfer Resistor. As the input side in forward biased and output side is reverse biased, there is transfer of resistance from a lower value on input side to a higher value on the output side.

♦ Transitor can be used as an amplifiers, when operated in the Active Region. It is also used as a Switch, when operated in the cut-off and saturation regions.

♦ The three configurations of Transistor are Common Emitter, Common Base and Common Collector.

♦ The proper name for this device being referred as transistor is Bipolar Junction Transistor (BJT).

♦ The three regions of the output characteristics of a transistor are
    1.   Active Region
    2.   Saturation Region
    3.   Cut-off Region

OBJECTIVE TYPE QUESTIONS

1. The units of h-parameters are ....

2. h-parameters are named as hybrid parameters because ....

3. The general equations governing h-parameters are
$$V_1 = \text{............................}$$
$$I_2 = \text{............................}$$

4. The parameter h_{re} is defined as $h_{re} = $ ....

5. h-parameters are valid in the .... frequency range.

6. Typical values of h-parameters in Common Emitter Configuration are ....

7. The units of the parameter h_{rc} are ....

8. Conversion Efficiency of an amplifier circuit is ....

9. Expression for current gain A_I in terms of h_{fe} and h_{re} are $A_I =$ ....

10. In Common Collector Configuration, the values of $h_{rc} \simeq$ ....

11. In the case of transistor in Common Emitter Configuration, as R_L increases, R_i ....

12. Current Gain A_I of BJT in Common Emitter Configuration is high when R_L is ....

13. Power Gain of Common Emitter Transistor amplifier is ....

14. Current Gain A_I in Common Base Configuration is ....

15. Among the three transistor amplifier configurations, large output resistance is in .... configuration.

16. Highest current gain, under identical conditions is obtained in .... transistor amplifier configuraiton.

17. C.C Configuration is also known as .... circuit.

18. Interms of h_{fe}, current gain in Darlington Pair circuit is approximately ....

19. The disadvantage of Darlington pair circuit is ....

20. Compared to Common Emitter Configuration R_i of Darlington piar circuit is ....

21. In CASCODE amplifier, the transistors are in .... configuration.

22. The salient featuers of CASCODE Amplifier are ....

23. For identical construction, ... types of bipolar junction transistors (BJTs) have faster switching times.

24. The arrow mark in the symbols of BJTs indicates ....

25. For NPN Bipolar Junction Transistor Emitter Efficiency $\gamma =$ ....

26. For PNP Transistor (BJT), the transportation factor $\beta^* =$ ....

27. Relation between α, β^* and γ for a transistor (BJT). $\alpha =$ ....

28. The expression for α in terms of β for a BJT is ....

29. The expression for β in terms of α for a BJT is ....

30. Expression for I_C in terms of β, I_B and I_{CBO} is ....

31. Expression for I_{CEO} in terms of I_{CBO} and α is ....

32. β' of a BJT is defined as ....

33. β' of a transistor (BJT) is ....

34. Relation between β' and β is ....

35. β' is synonymous to .... and β is same as ....

36. β is .... (for DC Currents).

37. β' is .... (for AC Currents).

38. Base width modulation is ....

ESSAY TYPE QUESTIONS

1. Write the general equations in terms of h-parameters for a BJT in Common Base Amplifiers configuration and define the h-parameters.

2. Convert the h-parameters in Common Base Configuration to Common Emitter Configuration, deriving the necessary equations.

3. Compare the transistor (BJT) amplifiers circuits in the three configurations with the help of h-parameters values.

4. Draw the h-parameter equivalent circuits for Transistor amplifiers in the three configurations.

5. With the help of necessary equations, discuss the variations of $A_v, A_I, R_i R_o, A_p$ with R_S and R_L in Common Emitter Configuration.

6. Discuss the Transistor Amplifier characteristics in Common Base Configuration and their variation with R_S and R_L with the help of equations.

7. Compare the characteristics of Transistor Amplifiers in the three configurations.

8. Draw the circuit for Darlington piar and derive the expressions for A_I, A_v, R_i and R_o.

9. Draw the circuit for CASCODE Amplifier. Explain its working, obtaining overall values of the circuit for h_i, h_f, h_o and h_r.

10. With the help of a neat graph qualitatively explain the Potential distribution through a transistor (BJT).

11. Explain about the different current components in a transistor.

12. Derive the relation between β and β' .

13. Derive the relation between α, β^* and γ

14. Differentiate between the terms h_{FE} and h_{fe}. Derive the relationship between them.

15. Qualitatively explain the input and output.

16. Explain the V-I characteristics in Common Emitter Configuration.

17. Describe the V-I characteristics of a transistor in Common Collector Configuration and Explain.

18. Draw the Eber-Moll Model of a transistor for NPN transistor and explain the same.

19. Compare the input and output characteristics of BJT in the three configurations, critically.

MULTIPLE CHOICE QUESTIONS

1. The h-parameters have

(a)   Same units for all parameters   (b)   different units for all parameters
(c)   do not have units   (d)   dimension less

2. The expression defining the h-parameter h_{22} is,

(a) $h_{22} = \dfrac{I_1}{I_2}\Big|\, I_1 = 0$

(b) $h_{22} = \dfrac{I_2}{I_1}\Big|\, I_1 = 0$

(c) $h_{22} = 12.11$

(d) $h_{22} = \dfrac{I_2}{V_2}\Big|\, I_1 = 0$

3. h - Parameters are valid over a ...... frequency range

(a) R.F.

(b) For DC only

(c) Audio frequency range

(d) upto 1 M Hz

4. By definition the expression for h_{oe} is,

(a) $\dfrac{\partial I_C}{\partial I_B}\Big|_{I_E=K}$

(b) $\dfrac{\partial I_C}{\partial V_C}\Big|_{I_B=k}$

(c) $\dfrac{\partial V_C}{\partial I_C}\Big|_{I_B=k}$

(d) $\dfrac{\partial I_B}{\partial V_B}\Big|_{I_C=k}$

5. Typical value of h_{fb} is

(a) -0.98

(b) -101

(c) $+2.3$

(d) $26.6\,\Omega$

6. The expression for h_{ie} interms of h_{fb} and h_{ib} is, $h_{ie} \simeq$

(a) $h_{ie} \simeq \dfrac{h_{fb}}{1-h_{fb}}$

(b) $\dfrac{h_{fb}}{1+h_{fb}}$

(c) $\dfrac{h_{ib}}{1+h_{fb}}$

(d) $\dfrac{h_{ib}}{1-h_{fb}}$

7. The ratio of AC signal power delivered to the load to the DC input power to the active device as a percentage is called

(a) conversion   (b) Rectification η (c) power η (d) utilisation factor

8. The general expression for A_I interms of h_f h_o and Z_L is ... $A_I =$

(a) $-\dfrac{h_o}{1+h_f Z_L}$

(b) $-\dfrac{h_f}{1+h_o Z_L}$

(c) $\dfrac{h_o}{1-h_f Z_L}$

(d) $\dfrac{h_f}{1-h_o Z_L}$

9. Expression for Avs in terms of A_s, R_s, Z_1, Z_i, Z_o is .....

(a) $\dfrac{A_I}{R_s + Z_i}$

(b) $\dfrac{A_I \cdot Z_L}{R_s - Z_i}$

(c) $\dfrac{Z_L}{R_s + Z_i}$

(d) $\dfrac{A_I Z_L}{R_s + Z_i}$

10. In the case of BJT, the resistance between fictious base terminal B' and outside base terminal B is, called

(a) Base resistance

(b) Base drive resistance

(c) Base spread resistance

(d) Base fictious resistance

11. In large signal analysis of amplifiers ....

(a) The swing of the input signal is over a wide range around the operating point.
(b) Operating point swimgs over large range
(c) stability factor is large
(d) power dissipation is large

12. **The units of the parameter hoe ...**

   (a)   (b) Ω (c)   constant   (d) V-A

13. **Typical value of hre is**

   (a) 10^{-4} V/A   (b) 10^{-4} (c) 10^{-6} A/V   (d) 10^4

14. **CASCODE transistor amplifier configuration consists of**

   (a)   CE-CE amplifier stages   (b)   CE-CC stages
   (c)   CE-CB stages   (d)   CB-CC stages

15. **For CASCODE amplifier, on the input side, the amplifier stage is in ....**

   (a)   C.B configuration   (b)   C.C. configuration
   (c)   emitor follower   (d)   C.E. configuration

16. **CASCODE amplifier characteristics are**

   (a)   Low voltage gain, large current are
   (b)   Low voltage gain, low current gain
   (c)   large voltage gain, large current gain, high output resistance
   (d)   Large current gain, Large voltage gain Low putput resistance

17. **Characteristics of Darlington circuit are typical values**

   (a) $Ri = M\Omega$, $A_I = 10$ k, $R_0 = M\Omega$, Av = 100
   (b) $Ri = M\Omega$, $A_I = 10$ k, $R_0 = 10\Omega$, Av < 1
   (c) $Ri = 10\Omega$, $A_I = 1$, $R_0 = 1.0\Omega$, Av > 1
   (d) $Ri = M\Omega$, $A_I = 1$, Ro = 10 Ω, Av < 1

18. **The disadvantage of Darlington pair circuit is ...**

   (a)   low current gain   (b)   low output resistance
   (c)   leakage current is more   (d)   high input resistance

19. **Compared to C.C. configuration, Darlington pair circuit has**

   (a)   low current gain   (b)   low voltage gain
   (c)   large current gain   (d)   large p output resistance

20. **In the case of Darlington circuit, A_I is approximately**

   (a) $2 h_{fe}$ (b) $4 h_{fe}$ (c) h_{fe} (d) $(h_{fe})^2$

21. **As reverse bias voltage is increased, for a diode, the base width at the junction**

   (a)   decreases   (b)   increases   (c)   remains same   (d)   none of these

22. **Expression for 'α' of a BJT in terms of I_{pc}, I_E, I_c etc is**

   (a) $\dfrac{I_{pc}}{I_E}$ (b) $\dfrac{I_{pE}}{I_E}$ (c) $\dfrac{I_{nc}}{I_E}$ (d) $\dfrac{I_{nE}}{I_C}$

23. **Expression for β^* is,**

(a) $\dfrac{I\,pc}{I\,p_B}$ (b) $\dfrac{I\,pc}{In_B}$ (c) $\dfrac{I\,pc}{Ip_E}$ (d) $\dfrac{I\,p_E}{I\,pc}$

24. **The relation between I_{CEO}, I_{CBO} and α is, $I_{CEO} =$**

(a) $\dfrac{I_{CBO}}{\alpha}$ (b) $\dfrac{I_{CBO}}{(1+\alpha)}$ (c) $\dfrac{I_{CBO}}{(1-\alpha)^2}$ (d) $\dfrac{I_{CBO}}{(1-\alpha)}$

25. **For a transistor in C.E configuration to be at cut-off, the condition is,**

(a) $V_{BE} = -V_{BB} + R_B.\,I_{CBO} \leq -0.4$ (b) $V_{BE} = +V_{BB} - R_B.\,I_{CBO} \leq 0.4$

(c) $V_{BE} = -V_{BB} + R_B.\,I_{CBO} \leq -0.1$ (d) $V_{BE} = -V_{BB} - R_B.\,I_{CBO} \leq -0.1$

26. **The condition for test for saturation in a BJT is**

(a) $\left|I_B\right| \leq \left|\dfrac{I_C}{\alpha}\right|$ (b) $\left|I_B\right| \leq \left|\dfrac{I_C}{\beta}\right|$ (c) $\left|I_B\right| \geq \left|\dfrac{I_C}{\alpha}\right|$ (d) $\left|I_B\right| \leq \left|\dfrac{I_C}{\alpha}\right|$

27. **Expression for saturation resistance Rcs of a transistor (BJT) is ...**

(a) $R\,c\,s = \dfrac{V_{CE}}{I_C(sat)}$ (b) $R\,c\,s = \dfrac{V_{CE}}{I_C(sat)}$

(c) $R\,c\,s = \dfrac{V_{CE}\,(sat)}{I_C}$ (d) $R\,c\,s = \dfrac{V_{CE}}{I_C}$

28. **The symbol shown is ⋊⋉ is that of a .....**

(a) DIAC (b) GTO (c) SCR (d) TRIAC

29. **DIAC is a ..... layer device**

(a) 4 layer (b) 3 layer (c) 6 layer (d) Two layers

30. **The forward break over voltage is symbolically represented as, (For SCR)**

(a) V_{BO} (b) V_{BOO} (c) V_{BR} (d) V_{FBO}

CONCEPTUAL QUESTIONS (Interview Questions)

1.   Why low frequency equivalent circuit is different from High frequency equivalent circuit?

Ans :   At low frequencies, capacitive reactance X_c is open circuit. Because, $X_C = \dfrac{1}{2\pi fC}$, when f is low, X_C is high or like open circuit. So junction capacitances of a transistor can be neglected. At high frequencies, when f is large, X_c is finite and can not be neglected. So the junction capacitances are to be considered. As such high frequency equivalent circuit contains capacitances, where as these are neglected (open circuited) in low frequency equivalent circuits.

Specifications of Amplifiers

Parameters		Typical Values
1. Type of Amplifier	:	Audio / Typical Values / Power / RF / Video
2. Frequency Range	:	15 Hzs - 100 KHzs
3. Output Power	:	200 mW
4. Voltage gain	:	20 db
5. Current gain	:	100
6. Power gain	:	9
7. Input impedance	:	10 kΩ 5 pf
8. Output impedance	:	500Ω 1 pf
9. Band width	:	100 KHzs

Transistor Biasing and Stabilization

In this Chapter,

- The need for biasing and its significance in amplifier circuits is explained. The Quiescent point, or operating point or 'Q' point is explained.

- Different types of biasing circuits are given and the expressions for stability factors are derived.

- Variation of 'Q' point with temperature and temperature compensating circuits are given.

I. THE OPERTING POINT

4.1 TRANSISTOR BIASING

The transistor output characteristics in Common Emitter Configuration are as shown in Fig. 4.1. The point in the characteristics of the transistors indicated by V_{CE}, I_C is called **Operating Point** or **Quiescent Point** Q. This is shown in Fig. 4.1. The operating point must be in the active region for

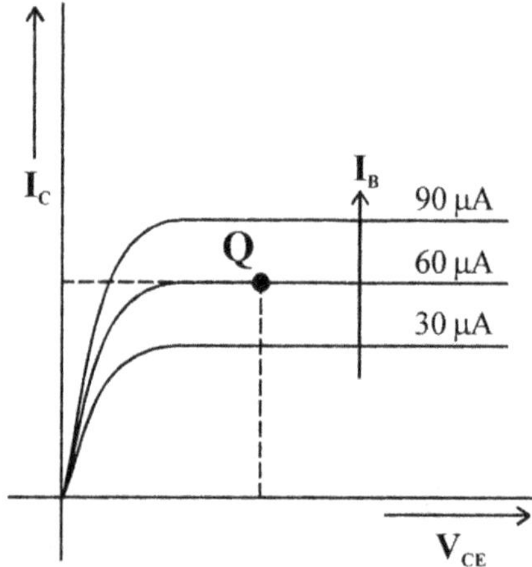

Fig 4.1 Output characteristics of BJT.

transistor to act as an amplifier.

Transistor when being used as an amplifier is operated in the active region. As I_B (in μA) increases, I_C (in mA) increases. So for a small change in I_B, there is corresponding large change in I_C and thus transistor acts as an amplifier. The operating point or the quiescent point has to be fixed when any transistor is being used as an amplifier.

$$A_V = \frac{\alpha' R_L}{r_C} \qquad\qquad R_L \ll r_c.$$

R_L = Load Resistance

r_c = Collector resistance

α' = Small signal current gain (less than 1)

A_V = Voltage gain

Therefore, transistor acts as an amplifier.

$$\alpha' = \frac{\Delta I_C}{\Delta I_E}$$

$$\Delta V_C = \alpha' \times R_L \times \Delta I_E$$

$$\Delta V_i = r_C \Delta I_E$$

$$\therefore \quad A_V = \frac{\Delta V_C}{\Delta V_i} = \frac{\alpha' R_L \times \Delta I_E}{r_C \times \Delta I_E} = \frac{\alpha' R_L}{r_C}$$

$$R_L \gg r_c. \text{ So } A_V > 1$$

The quiescent point changes with temperature T, because it depends on the parameters β, I_{CO} or V_{BE} etc. β and I_{CO} depend upon temperature. Hence operating point changes with temperature. So by suitably biasing the transistor circuit, the quiescent point may be made stable or not to change with temperature. Compensation techniques are also used in order to make the operating point stable with temperature.

It is very essential to choose the operating point properly so that the transistor is operated in the active region and that the operating point does not change with temperature. Since if the input is sinusoidal wave, the output will also be a true replica of the input or it will also be a perfect sine wave. Otherwise, the output will be distorted, clipped etc., if the quiescent point is not chosen correctly or it is changing with temperature. Hence it is essential to choose the operating point correctly. Operating point is fixed with respect to the output characteristic of a transistor only.

So now the question is how to choose the operating point? Before this let us consider as to how the quiescent point is fixed. Consider the circuit as shown in Fig. 4.2 and 4.3. It is Common Emitter Transistor circuit with a load resistance R_L. PNP transistor is being considered.

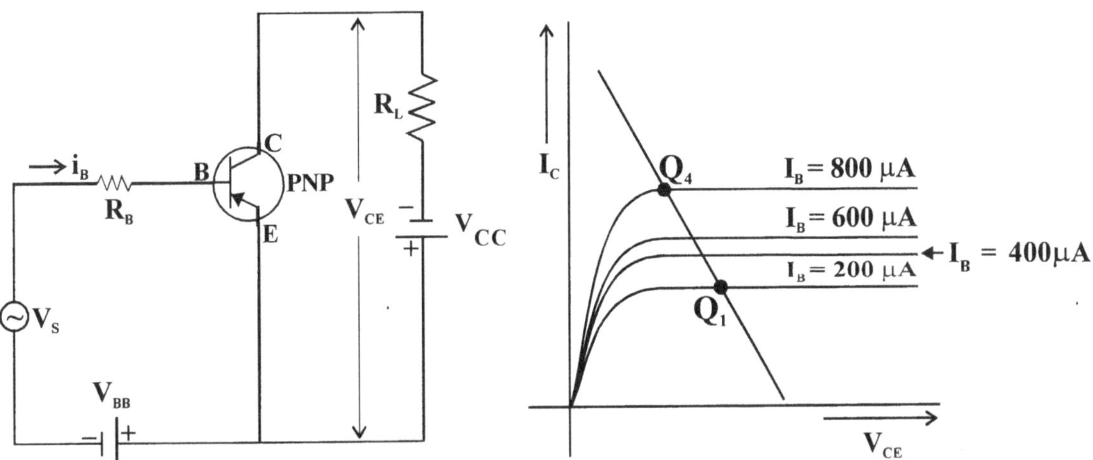

Fig 4.2 Amplifier circuit.   *Fig 4.3 Variation of operating point.*

From the above circuit,
$$V_{CE} = V_{CC} - i_C \times R_L \qquad\qquad \ldots\ldots (4.1)$$
V_{CC} is DC Bias Supply. i_C is instantaneous value of AC Collector Current.
$$V_{CE} = \text{Instantaneous Voltage}$$
From the Equation (4.1), the load line can be drawn. To draw the load line,

Let $\quad i_C = 0$ then $V_{CE} = V_{CC}$. This is one point of the load line.

Let $\quad V_{CE} = 0$ then $i_C = \dfrac{V_{CC}}{R_L}$. This another point of the load line.

The value of r_C is very small compared to R_L. Hence it can be neglected.

$$i_C = \left(-\dfrac{V_{CC}}{R_L + r_C} \right)$$

Neglecting, r_C, $\quad i_C = \left(-\dfrac{V_{CC}}{R_L} \right)$

So by setting $i_C = 0$, $V_{CE} = 0$, the load line is drawn. It is a straight line independent of the transistor parameters. The slope of the line depends on R_L the load resistance only. V_{CC} is a fixed quantity. For a given value of I_B, I_C is also fixed when the transistor is being operated in the active region. Therefore for $I_B = 200\mu A$, the quiescent point is Q_1. For $I_B = 800\ \mu A$, the operating point is Q_4 and so on. Now, when the input signal applied between base and emitter is a symmetrical signal, varying both on positive and negative sides equally, within the range of the transistor characteristics, then the operating point is chosen to lie in the centre of the output characteristics (i.e., corresponding to $I_B = 300\ \mu A$), so that the output signal is a true replica of the input without distortion (V_{CC} and V_{BB} are bias voltages. Input signal is the a.c signal between B and E). When the input signal is symmetrical, the *quiescent point is chosen at the centre of the load line*.

The load line as explained above, if drawn, is called as the *dynamic (AC) load line*. If the output impedance in the circuit is reactive, then the load line will be elliptical. Since the voltage and current are shifted in phase, and the resulting equation will be that of an ellipse. For a resistance the load line is a straight line.

While drawing a load line, if only the collector resistance R_C is considered and load resistance

R_L is infinity (∞), then the load line is drawn with the points $\left(\dfrac{V_{CC}}{R_C} \text{ and } V_{CE} \right)$. It is called as *Static*

Load Line or *DC Load Line*

But if the load line is drawn considering a finite load resistance R_L, and the resistance considered is R_L in parallel with R_C, it is called *AC Load Line or Dynamic Load Line*, with AC Input.

There are three types of Biasing Circuits

1. Fixed Bias circuit or Base Bias circuit

2. Collector to Base Bias circuit

3. Self Bias or Emitter Bias circuit (Universal bias or Voltage divider bias circuit).

4.2   FIXED BIAS CIRCUIT OR (BASE BIAS CIRCUIT)

Consider the NPN Transistor Circuit as shown in Fig. 4.4. R_C is collector resistance. C_C is coupling capacitor. It blocks DC since capacitor is open circuit for D.C. So the output signal will be pure a.c signal. R_C. limits the current I_C . R_B provides the bias voltage for the base. Since this is NPN transistor, V_{CC} should be positive, because the collector should be reverse biased. The drop across R_b, ($V_{CC} - I_B \times R_B$) is negative and it provides positive bias for the base. Since R_L is connected, dynamic load line has to be considered. Suppose the output characteristics of the transistor are as shown in Fig. 4.5.

Suppose Q_1 is the centre point, and it is chosen as the operating point. But if the input signal is greater than 40 μA on the negative side, with Q_1, the transistor will be cut off since for that swing $I_B = 0$, and the transistor is cut off, so another operating point is to be chosen. Choosing the operating point means, a correct value of I_B, I_C, V_{CE}. (if the transistor is in Common Emitter Configuration) so that we get undistorted output.

Since from the above circuit,

$$I_B = \frac{V_{CC} - V_{BE}}{R_B}$$

$$I_C = \frac{V_{CC} - V_{CE}}{R_C}$$

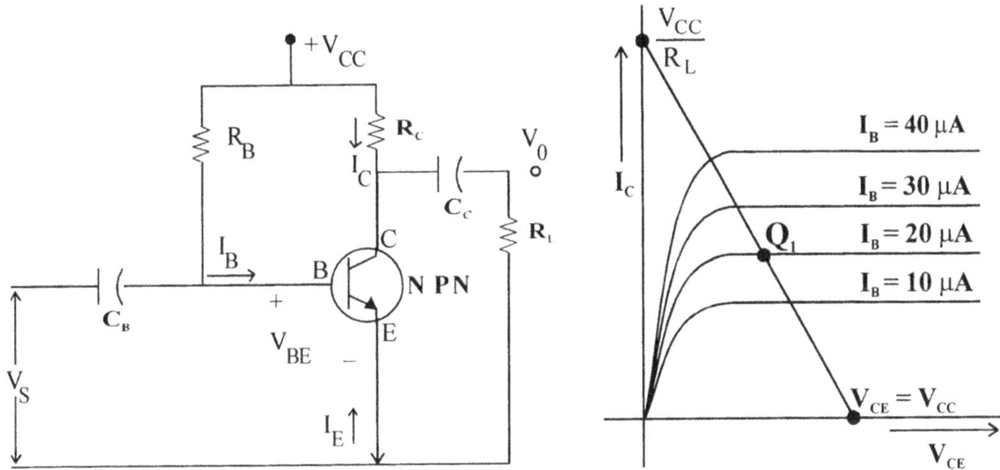

Fig 4.4 Fixed bias circuit   *Fig 4.5 Operating point variation*

V_{BE} is the base emitter voltage for a forward biased EB Junction. This will be 0.2V for Germanium and 0.6 for Silicon. In order to get a large value of I_B or the change in the operating point, V_{CC} or R_B has to be changed. Since once V_{CC}, R_B etc., are fixed, I_B is fixed. So the above circuit is called as *Fixed Bias Circuit*.

I_B is fixed when R_B and V_{CC} are fixed. Because in the expression, for I_B, β is not appearing. So once R_B and V_{CC} are fixed, the biasing point is also fixed. Hence the name Fixed Bias Circuit.

4.3 BIAS STABILITY

The circuit shown in Fig. 4.4 is called as a *Fixed Bias Circuit*, because I_B remains constant for given values of V_{CC} and V_{BB} and R_B. ($I_B \simeq V_{BB}/R_B$). So the operating point must also remain fixed. Suppose the transistor in the circuit is AC128, and we replace this transistor by another AC 128 transistor. The characteristics of these two transistors will not be exactly the same. There will be slight difference. It is because of the limitations in the fabrication technology.

The doping concentration, diffusion length etc., in fabrication cannot be controlled precisely. If we say impurity concentration is 1 in $10^6/cm^3$, it need not be the same. The diffusion length will also be not the same since the temperature inside the furnace during drive in diffusion may vary. The furnace may not have constant temperature profile along the length of the furnace. Because of these reasons, transistors of the same type may not have exactly the same characteristics. Hence in the above circuit, if one AC 128 transistor is replaced by another transistor, the operating point will change, since the characteristics of the transistor will be slightly different.

Hence for the same circuit when one transistor is replaced by the other, operating point changes. In some cases, because of the change in the operating point, the transistor may be cut off. In the above circuit, the operating point is not fixed, since I_B is fixed, and β changes. On the other hand I_B should be changed to account for the change in β so that the operating point is fixed or I_C and V_{CE} are fixed.

4.4 THERMAL INSTABILITY

Change in temperature also adversely effects the operating point. I_{CO} doubles for every 10^0C rise in temperature.

$$I_C = (\beta + 1) I_{CO} + \beta \times I_B \qquad\qquad \ldots\ldots (4.2)$$

$$I_C = \frac{I_{CO}}{1-\alpha} + \frac{\alpha.I_B}{1-\alpha} \qquad \dots\dots\dots (4.3)$$

The collector current causes the collector junction temperature to rise. Hence I_{CO} increases. So I_C also increases. This in turn, increases the temperature and so I_{CO} still increases, and then I_C. So the transistor output characteristics will shift upwards (because I_C increases). Then the operating point changes. So in certain cases, even if the operating point is fixed in the middle of active region, because of the change in temperature, the operating point will be shifted to the saturation region.

4.5 STABILITY FACTOR 'S' FOR FIXED BIAS CIRCUIT

In the previous circuit, if I_B is fixed, operating point will shift with changes in the values of I_C with temperature.

$$I_C = \frac{I_{CO}}{1-\alpha} + \frac{\alpha.I_B}{1-\alpha}$$

I_{CO} gets doubled for every 10°C rise in temperature.

I_C also increases with temperature. So if I_B is fixed, the operating point will shift. In order to keep operating point fixed, I_C and V_{CE} should be kept constant. There are two methods to keep I_C constant.

1. *Stabilization Technique :*

   Here resistive biasing circuits are used to allow I_B to vary to keep I_C relatively constant, with variation in I_{CO}, β and V_{BE} i.e., I_B decreases if I_{CO} increase, to keep I_C constant.

2. *Compensation Technique :*

   In this method, temperature sensitive devices such as diodes, transistor, thermistors etc., which provide compensating voltage and currents to maintain the operating point constant are used.

   Stability Factor S *is defined as the rate of change of I_C with respect to I_{CO} the reverse saturation current, keeping β and V_{BE} constant.*

$$S = \left.\frac{\partial I_C}{\partial I_{CO}}\right|_{\beta,\ V_{BE}\ =\ K}$$

$$= \frac{\Delta I_C}{\Delta I_{CO}}$$

The ***value of S should be small***. If it is large, it indicates that the circuit is thermally unstable.

$\dfrac{\partial I_C}{\partial \beta}$, $\dfrac{\partial I_C}{\partial V_{BE}}$ are also some times called as Stability Factors.

Expression for I_C,

$$I_C = (1 + \beta)\,I_{CO} + \beta I_B \qquad \dots\dots\dots (4.4)$$

$$S = \left.\frac{\partial I_C}{\partial I_{CO}}\right|_{\beta\ =\ K,\ V_{BE}\ =\ K}$$

Therefore, differentiating Equation (4.4) with respect to I_C, with β as constant,

$$1 = (1+\beta) \times \frac{\partial I_{CO}}{\partial I_C} + \beta \times \frac{\partial I_B}{\partial I_C}$$

$$1 = \frac{(1+\beta)}{S} + \beta \cdot \frac{\partial I_B}{\partial I_C}$$

I_{CO} cannot be measured directly. I_C and I_B can be measured directly. So I_C is not differentiated directly with I_{CO}.

$$\left(1 - \beta \cdot \frac{\partial I_B}{\partial I_C}\right) = \frac{1+\beta}{S}$$

$S = 0$ implies that I_C is independent of I_{CO}. This is ideal case. $S = 5$ is a reasonable value for practical circuit. General Expression for S is,

$$S = \frac{1+\beta}{1 - \beta\left(\dfrac{\partial I_B}{\partial I_C}\right)} \qquad\qquad ....(4.5)$$

$$\therefore \qquad I_B = \frac{V_{CC} - V_{BE}}{R_B}$$

For a fixed bias circuit, I_B is independent of I_C.

$$\therefore \qquad \frac{\partial I_B}{\partial I_C} = 0.$$

$$\therefore \qquad \boxed{S = 1 + \beta} \qquad\qquad ....(4.6)$$

If $\beta = 40$, $S = 41$. It is a very large value or thermal instability is more. So some circuits are available which make I_C more independent of I_{CO} and hence S will be low.

4.6   COLLECTOR TO BASE BIAS CIRCUIT

The circuit shown in Fig. 4.4 is fixed bias circuit, since I_B is fixed, if V_{CC}, R_C and V_{BB} are fixed. But an improvement over this circuit is the circuit shown is Fig. 4.6.

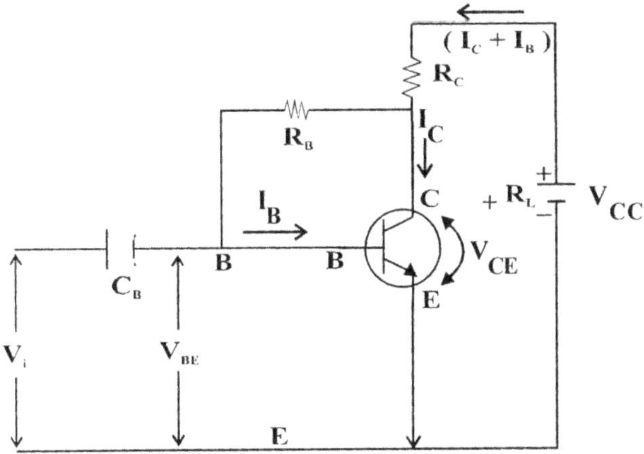

Fig 4.6 Collector to base bias circuit.

In this circuit, R_B is connected to the collector point C, instead of V_{CC} as in Fig. 4.4. V_i is input, the capacitor will block DC input and allow only AC signal to the base. Since it is NPN transistor, collector junction is reverse biased. R_C is a resistor in the collector circuit, which controls the reverse bias voltage of the collector junction with respect to emitter.

$$\because \quad V_{CE} = V_{CC} - I\,R_C$$
$$I = I_C + I_B$$

This circuit is an improvement over the Fig. 4.4 fixed bias circuit because,

$$V_{CE} = V_{CC} - (I_C + I_B)\,R_C$$
$$I = I_C + I_B$$

With temperature I_C increases. Since V_{CC} is same, I is same and so I_B decreases.

As I_C increases with temperature because of the increases in I_{CO}, $(I_C + I_B)\,R_C$ product increase. So V_{CE} reduces. That means the reverse bias voltage of the collector junction is reduced or number of carriers collected by the collector reduces. Hence I_B also reduces. Therefore, because of the reduction in the bias current I_B, the increase in I_C will not be as much as it would have been considering only increase in I_{CO} with temperature.

We can calculate the stability factor for this circuit. Applying KVL,

$$V_{BE} - V_{CC} + (I_B + I_C)\,R_C + I_B \cdot R_B = 0$$
$$\text{or} \quad V_{CC} - V_{BE} = I_B\,(R_C + R_B) + I_C \cdot R_C$$

$$I_B = \frac{V_{CC} - V_{BE} - I_C R_C}{R_C + R_B}$$

V_{BE} is the cut in voltage and it is 0.6V for Silicon, and 0.2V for Germanium. It is independent of I_C. So differentiate this expression with respect to I_C, we get

$$\frac{\partial I_B}{\partial I_C} = \frac{0 - R_C - 0}{R_C + R_B} = \frac{-R_C}{R_C + R_B}$$

The expression for stability factor S

$$S = \frac{1 + \beta}{1 - \beta\left(\dfrac{\partial I_B}{\partial I_C}\right)}$$

$$\text{But} \quad \frac{\partial I_B}{\partial I_C} = \frac{-R_C}{R_C + R_b}$$

$$\therefore \quad \boxed{S = \frac{1 + \beta}{1 + \beta \times \dfrac{R_C}{R_C + R_B}}}$$

Therefore, the value of S is less than $(1 + \beta)$.

$(1 + \beta)$ is the maximum value for the fixed bias circuit. Therefore collector bias circuit, is an improvement over the fixed bias circuit. Now let us consider the effect of β on the stability of the circuit, discussed above.

$$I_C = (1+\beta)\, I_{CO} + \beta \times I_B \qquad \qquad \dots\dots\dots\ (\,4.7\,)$$

$$I_B = \frac{V_{CC} - I_C R_C - V_{BE}}{R_C + R_B} \qquad \qquad \dots\dots\dots\ (\,4.8\,)$$

From Equation (4.7), $I_B = \dfrac{I_C - (1+\beta)I_{CO}}{\beta}$ $\qquad\qquad \dots\dots\dots\ (\,4.9\,)$

Equating (4.8) and (4.9) and transferring I_C to one side

$$\frac{I_C - (1+\beta)I_{CO}}{\beta} = \frac{V_{CC} - I_C R_C - V_{BE}}{R_C + R_B}$$

$$I_C (R_C + R_B) - (1+\beta)(R_C + R_B)\, I_{CO} = \beta[V_{CC} - V_{BE}] - \beta I_C R_C$$

$$I_C (R_C + R_B) + \beta\, I_C R_C = \beta\,([V_{CC} - V_{BE}] + (R_C + R_B)\, I_{CO}\,(1+\beta))$$

$$\beta \gg 1 = 1 + \beta \simeq \beta$$

$$\therefore \quad I_C[R_C + \beta R_C + R_B] = \beta[V_{CC} - V_{BE}] +)(R_C + R_B)\, I_{CO}\,\beta$$

$$I_C\,[(\beta + 1)\, R_C + R_B] \simeq \beta\,[V_{CC} - V_{BE} + (R_C + R_B)I_O]$$

Again $(\beta + 1) \simeq \beta$.

$$\therefore \quad I_C \simeq \frac{\beta[V_{CC} - V_{BE} + (R_C + R_B)I_{CO}]}{(\beta R_C + R_B)} \qquad \dots\dots\dots\ (\,4.10\,)$$

This expression, in this form is derived to get the condition to make I_C independent of β. Therefore, with the changes in the value of β, with temperature, I_C changes so the operating point also changes.

The above circuit, is an improvement over the fixed bias circuit, I_B is being decreased, with increase in I_C (Because I_{CO} increases with temperature) to keep operating point constant. Since I_C is dependent on the β from the above expression, and since β changes with temperature, I_C also changes and thereby operating point changes. Therefore from the above expression, it is essential to make I_C insensitive to β, so that the modified circuit works satisfactorily to keep the operating point fixed. The assumption to be made in the above expression (4.10) is,

$$\beta \times R_C \gg R_B.$$

Therefore, the above expression becomes,

$$I_C \simeq \frac{\beta\left[V_{CC} - V_{BE} + (R_C + R_B)I_{CO}\right]}{\beta R_C}$$

$$\therefore \quad \beta\, R_C \gg R_B$$

$$\therefore \quad I_C \simeq \frac{V_{CC} - V_{BE} + (R_C + R_B)I_{CO}}{R_C}$$

So if we choose R_B very small compared to βR_C, the circuit will work satisfactorily. But in all cases, this may not work out. If R_B value is ***not so small***, then the above assumption is not valid and hence the circuit won't work satisfactorily.

For a given circuit, to determine the operating point, draw the load line on the output characteristics of the given transistor. Load line is drawn as given below :

$$V_{CC} = I_C R_L + V_{CE}$$

If $I_C = 0$, then, $\qquad V_{CC} = V_{CE}$

If $V_{CE} = 0$, then $I_C = \dfrac{V_{CC}}{R_L}$.

With these points, a straight. line is drawn on the output characteristics of the transistor. Corresponding to a given value of I_B, where this line cuts the output characteristic, it is called as the operating point.

In the collector to base bias circuit we said that the stability is good since, I_B is decreased with increase in I_C to keep operating point fixed. In that circuit, there is feedback from the output to the input through the resistor R_B. If the signal voltage causes an increase in I_B, I_C increases. But V_{CE} decreases. So because of this, the component of base current coming through R_B decreases. Hence the increase in I_B is less than what it would have been without feedback. In feedback circuits the amplification factor will be less than what it would have been without feedback. But the advantage is stability is improved. Such an amplifier circuit with feedback is called as *Feedback Amplifier Circuit*. This type of circuits are discussed in subsequent chapters.

4.7   SELF BIAS OR EMITTER BIAS CIRCUIT

In the fixed bias circuit, since I_C increases with temperature, and I_B is fixed, operating point changes. In collector to base bias circuit, though I_C changes with temperature, I_B also changes to keep operating point fixed. But since β changes with temp. and I_C depends on β, the assumption has to be made that $\beta R_C \gg R_B$ to make I_C independent of β. But if in one circuit, R_C is small, the above assumption will not hold good. [In transformer coupled circuits R_C will be small].

So collector to base circuit is as bad as fixed bias circuit. A circuit which can be used even if there is zero DC resistance in series with the collector terminal, is the *Self Bias on Emitter Bias Circuit*. The circuit is as shown. This circuit is also known as *Voltage Divider Bias or Universal Bias Circuit.*

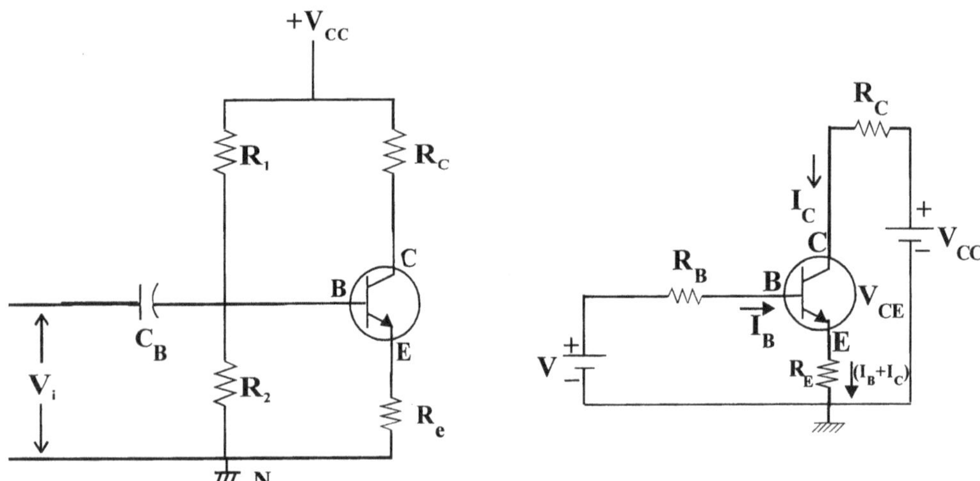

Fig 4.7 *Self Bias circuit.*   Fig 4.8   *Thevenin's equivalent.*

The current in the emitter lead causes a voltage drop across R_e in such a direction, that it forward biases the emitter-base junction. It is NPN transistor emitter is *n-type*. Negative polarity should be at the emitter. If I_C increases due to increase in I_{CO} with temperature the current in R_C also increases.

$$I = I_C + I_B; \quad V_{BE} = V_{BN} - V_{EN}$$

As I_B increases, I_C also increases.

So current I or emitter current increases. Therefore IR_e drop increases. Since the polarity of this voltage is to reverse bias the $E - B$ junction, I_B decrease. Therefore I_C will increase less than it would have been, had there been no self biasing resistor.

[The voltage drop across R_C provides the self bias for the emitter]. Hence stability is good.

With reference to Fig. 4.8, it is NPN transistor. So the conventional current is flowing out of the transistor, from the emitter. That is why the symbol is with arrow mark pointing outwards. So emitter point is at positive with respect to N. Hence the drop $(I_B + I_C) R_C$ has the polarity such as to reverse bias or to oppose the forward bias voltage V_{BE} junction. R_B can be regarded as parallel combination of R_1 and R_2.

4.8 STABILITY FACTOR 'S' FOR SELF BIAS CIRCUIT

S is calculated assuming that no AC signal is impressed and only DC voltages are present. In the Fig 4.8, the voltage V is the drop across R_2. Combination acts as a potential divider. So the drop across R_2 which is equal to V, is given by the expression,

$$V = \frac{V_{CC}.R_2}{R_1 + R_2} .$$

R_B is the effective resistance looking back from the base terminal

$$\therefore \qquad R_B = \frac{R_1 R_2}{R_1 + R_2}$$

Applying Kirchhoff's Voltage Law around the base circuit,

$$V = I_B R_B + R_e (I_B + I_C) + V_{BE}$$

Assuming V_{BE} to be independent of I_C, differentiating I_B with respect to I_C, to get 'S'.

$$I_B, R_B + I_B R_e = + V - V_{BE} - I_C \times R_e$$

$$I_B (R_B + R_e) = V - V_{BE} - I_C \times R_e$$

$$\therefore \qquad I_B = \frac{(V - V_{BE} - I_C R_e)}{(R_B + R_e)}$$

$$\frac{\partial I_B}{\partial I_C} = - \frac{R_e}{R_e + R_B}$$

The expression for stability factor $S = \dfrac{1 + \beta}{1 - \beta \left(\dfrac{\partial I_B}{\partial I_C} \right)}$

$$\therefore \qquad S = \frac{1 + \beta}{1 + \beta \left(\dfrac{R_e}{R_e + R_B} \right)}$$

If $\dfrac{R_B}{R_e}$ is very small, $S \simeq 1$.

The fixed bias circuit, collector to base bias circuit and Emitter bias circuits provide stabilisation of I_C against variations in I_{CO}. But I_C also varies with V_{BE} and β. V_{BE} *decreases at the rate of* *2.5 mV/°C for both Germanium and Silicon transistors*. Because, with the increase in T, the potential barrier is reduced and more number of carriers can move from one side (p) to the other side (n). β also increases side with temperature. Hence I_C changes.

4.9 STABILITY FACTOR S$'$

The variation of I_C with V_{BE} is given by stability factors S'.

$$S' = \left. \frac{\partial I_C}{\partial V_{BE}} \right|_{\beta = K}$$

where both I_{CO} and β are considered constant.

$$V = I_B \times R_B + V_{BE} + (I_B + I_C) R_e \qquad \ldots\ldots\ldots (4.11)$$
$$I_C = (1 + \beta) I_{CO} + \beta \cdot I_B \qquad \ldots\ldots\ldots (4.12)$$

Therefore, From equation (4.12),

$$I_B = \frac{I_C - (1+\beta)I_{CO}}{\beta}$$

From equation (4.11),

$$V_{BE} = V - I_B R_B \times (I_B + I_C) R_e \qquad \ldots\ldots\ldots (4.13)$$

But

$$I_B = \frac{I_C - (1+\beta)I_{CO}}{\beta}$$

Substituting this value in equation (4.13),

$$V_{BE} = V - \left\{ \frac{I_C - (1+\beta)I_{CO}}{\beta} \right\} R_B - \left\{ \frac{I_C - (1+\beta)I_{CO}}{\beta} + I_C \right\} R_e$$

$$V_{BE} = V - \frac{I_C \cdot R_B}{\beta} + \frac{R_b \cdot (1+\beta)}{\beta} \times I_{CO} +$$

$$\left(\frac{1+\beta}{\beta} \right) R_e \times I_{CO} - \frac{I_C}{\beta} \times R_e - I_C \times R_e$$

$$V_{BE} = V - I_C \left\{ \frac{R_B}{\beta} + \frac{R_e}{\beta} + R_e \right\} + I_{CO} \left\{ \frac{R_B \cdot (1+\beta)}{\beta} + \left(\frac{1+\beta}{\beta} \right) R_e \right\}$$

$$V_{BE} = V + (R_B + R_e) \left\{ \frac{1+\beta}{\beta} \right\} I_{CO} - \left\{ \frac{R_B + R_e(1+\beta)}{\beta} \right\} I_C \quad \ldots\ldots\ldots (4.14)$$

$$S' = \frac{\partial I_C}{\partial V_{BE}}$$

$$\frac{\partial V_{BE}}{\partial I_C} = -\left\{ \frac{R_B + R_e(1+\beta)}{\beta} \right\}$$

or
$$S' = \frac{\partial I_C}{\partial V_{BE}} = \frac{-\beta}{R_B + R_e(1+\beta)}$$

But
$$S = \frac{(1+\beta)(R_e + R_B)}{R_B + R_e(\beta+1)}$$

$$\therefore \quad \boxed{S' = \frac{-S.\beta}{(R_B + R_e)(\beta+1)}} \qquad \qquad .... (4.15)$$

$$\beta + 1 \simeq \beta$$

$\dfrac{R_B}{R_e}$ if it small can be neglected

$$\therefore \quad \boxed{S' = -\frac{S}{R_e}} \qquad\qquad .....(4.16)$$

$S = 1$, implies that the stability of the circuit is very good. Or to a better accuracy,

$$1 + \beta + \frac{R_B}{R_e} = 1 + \frac{R_B}{R_e} = 1 + \beta$$

when $\dfrac{R_B}{R_e}$ is small,

$$\therefore \quad S = \frac{(1+\beta)\left(1 + \dfrac{R_B}{R_e}\right)}{(1+\beta)} = 1 + \frac{R_B}{R_e}$$

If $\dfrac{R_B}{R_e}$ is very large, then $S = 1 + \beta$. Now for a fixed $\dfrac{R_B}{R_e}$, S increases with increases β.

If β is small, S is almost independent of β. If a graph is plotted between S and $\dfrac{R_B}{R_e}$, then we get, the graph as shown in Fig. 4.9.

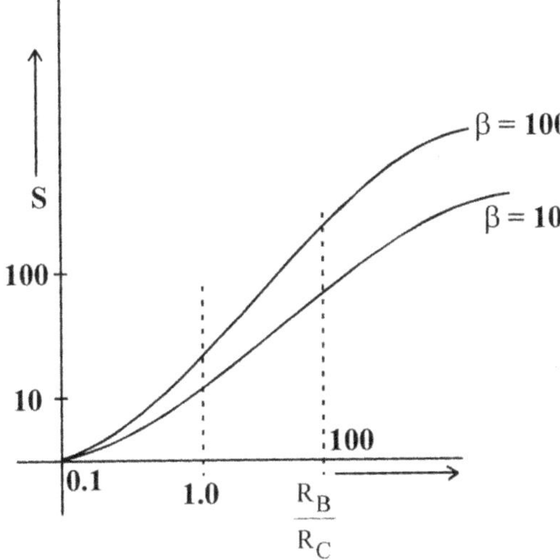

Fig 4.9 Variation of stability factor

The smaller the value of R_B the better the stabilisation $R_B = R_1$ in parallel with R_2. So R_1 and R_2 should be small.

The smaller, the value of R_B, the better the stabilisation. Because of feedback through R_C, the AC gain is reduced. To prevent this, R_e is shunted with a capacitor whose reactance at the operating frequencies is very small so that AC signal passes through the capacitor only.

S' has the units of $\dfrac{1}{\Omega}$ or mhos. S' is negative indicate that as V_{BE} decreases, I_C increases.

$$S' \simeq \frac{-S}{R_E}$$

Therefore, if S is made unity, than S' will also be less. We get stability of I_C with respect to both V_{BE} and I_{CO}. For self bias circuit, variation of S'' with V_{BE} is also less. S'' is approximately equal to 1.

4.10 STABILITY FACTOR S''

The variation of I_C with respect to β is given by the stability factors S''.

$$S'' = \frac{\partial I_C}{\partial \beta}$$

where both I_{CO} and V_{BE} are assumed to be constant.

From equation (4.14),

$$V_{BE} = V + (R_B + R_e)\frac{\beta+1}{\beta} \times I_{CO} - \frac{R_B + R_e(1+\beta)}{\beta} \times I_C$$

$$V_{BE} - V - (R_B + R_e)\frac{\beta+1}{\beta} \times I_{CO} = -\frac{R_B + R_e(1+\beta)}{\beta} \times I_C \times (1 + \beta)$$

$$I_C = \frac{-\beta.V_{BE}}{(R_B + R_e)(1+\beta)} + \frac{V.\beta}{(R_B + R_e)(1+\beta)} \quad \frac{I_{CO}(R_B + R_e)(1+\beta)}{R_B + R_e(1+\beta)}$$

$$I_C = \frac{V.\beta - \beta V_{BE}}{(R_B + R_e)(1+\beta)} + I_{CO}$$

$$\frac{\partial I_C}{\partial \beta} = \frac{(R_B + R_e)(1+\beta)(V - V_{BE}) - \beta(V - V_{BE})(R_e + R_B)}{(R_B + R_e)^2(1+\beta)^2}$$

$$= \frac{(R_B + R_e)(V - V_{BE})}{(R_B + R_e) + (1+\beta)^2}$$

$$\boxed{S'' = \frac{\partial I_C}{\partial \beta} = \frac{I_C.S}{\beta(1+\beta)}}$$

4.11 PRACTICAL CONSIDERATIONS

Silicon transistors are superior to Germanium transistor as far as thermal stability is concerned. The variation in the value of I_C for Silicon transistor is 30% for a typical circuit when the temperature varies from -65°C to $+175^\circ$C. But for Germanium transistor, the variation will be 30% when the temperature is varied from $-65°C$ to $+75°C$. Therefore, *the thermal stability is poor for Germanium transistor*. Silicon transistors are never used above $175°C$ and Germanium transistors above $+75°C$. Such high temperature will be encountered in the electronic control of temperature for furnaces, space applications etc.

4.12 BIAS COMPENSATION

The collector to base bias circuit and self bias circuit are used for stability of I_C with variation in I_{CO}, V_{BE} and β. But we said that there is feed back from the output to the input. Hence the amplification will be reduced, even though the stability is improved.

Problem 4.1

An NPN transistor with $\beta = 50$, is used in Common Emitter Circuit with $V_{CE} = 10V$ and $R_C = 2K\Omega$. The bias is obtained by connecting a $100\ k\Omega$ resistance from collector to base. Assume $V_{BE} = 0$. Find

      (a)   *The Quiescent Point*

      (b)   *The Stability Factor*

Solution

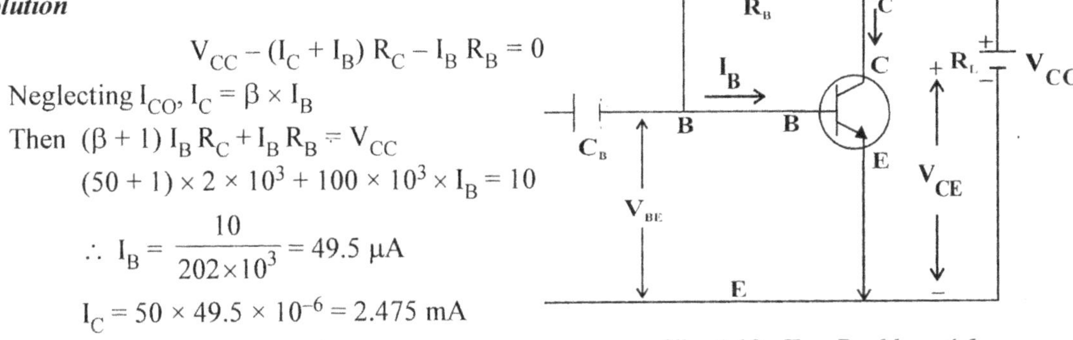

$$V_{CC} - (I_C + I_B)\,R_C - I_B\,R_B = 0$$

Neglecting I_{CO}, $I_C = \beta \times I_B$

Then $(\beta + 1)\,I_B\,R_C + I_B\,R_B = V_{CC}$

$$(50 + 1) \times 2 \times 10^3 + 100 \times 10^3 \times I_B = 10$$

$$\therefore\ I_B = \frac{10}{202 \times 10^3} = 49.5\ \mu A$$

$$I_C = 50 \times 49.5 \times 10^{-6} = 2.475\ mA$$

$$V_{CE} = I_B \times R_B = 4.95$$

Fig 4.10   For Problem 4.1.

$$S = \frac{\beta + 1}{1 + \dfrac{\beta.R_C}{R_C + R_B}} = \frac{51}{1 + \dfrac{50 \times 2}{102}} = 25.6$$

Problem 4.2

A transistor with $\beta = 100$ is to be used in Common Emitter Configuration with collector to base bias. $R_C = 1K\Omega$ $V_{CC} = 10V$. Assume $V_{BE} = 0$. Choose R_B so that quiescent collector to emitter voltage is $4V$. Find Stability Factor.

Solution

$$V_{CC} - (\beta + 1)\,I_B \times R_C - V_{CE} = 0$$

$$S = \frac{101}{1 + \dfrac{100 \times 1}{68.3}} = 41$$

$$10 - 101 \times 10^3\,I_B - 4 = 0$$

$$I_B = \frac{6}{101 \times 10^3} = 59.4\mu A$$

$$V_{CE} = I_B R_B = 4V$$

$$\therefore \quad R_B = \frac{4}{59.4 \times 10^{-6}} = 67.3K$$

Problem 4.3

One NPN transistor is used in the self biasing arrangement. The circuit components values are V_{CC} = 4.5V, R_C = 1.5K, R_C = 0.27 KΩ, R_2 = 2.7 KΩ and R_1 = 27K. If β = 44. Find the Stability Factor. Also determine the Quiescent point $Q(V_{CE}, I_C)$.

Solution

The Quiescent Point

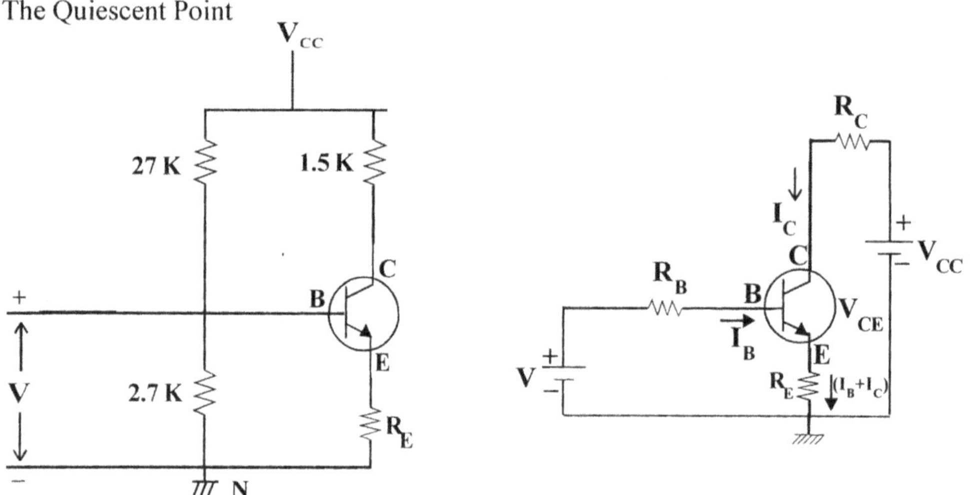

Fig 4.11 For Problem 4.3.   *Fig 4.12 For Problem 4.3.*

$$V = \frac{R_2 \cdot V_{CC}}{R_1 + R_2} = \frac{2.2 \times 4.5}{27 + 2.7} = 0.409V$$

$$R_B = \frac{R_1 R_2}{R_1 + R_2} = \frac{27 \times 2.7}{27 + 2.7} = 2.46k\Omega$$

$$S = \frac{(1 + \beta)1 + \dfrac{R_B}{R_e}}{1 + \beta + \dfrac{R_B}{R_e}} = 8.4$$

Applying KVL over the loop.

$$+V = I_B \times R_B + V_{BE} + (1 + \beta) R_C I_B$$

$$+0.409 = I_B \times (2.46 + 45 \times 0.27) + 0.2$$

$$\text{or} \qquad I_B = \frac{0.409 + 0.2}{14.6} = 1.1 \text{mA}$$

$$I_C = \beta \times I_B = 44 \times 1.1 = 48.4 \text{ mA}$$

$$V_{CE} = V_{CC} - I_C \times R_C - (I_B + I_C)\, R_C = 2.4V$$

If the reduction in gain is not tolerable then compensation techniques are to be used. Thus both stabilization and compensation technique are used depending upon the requirements.

BIAS COMPENSATION

4.12.1 DIODE COMPENSATION FOR V_{BE}

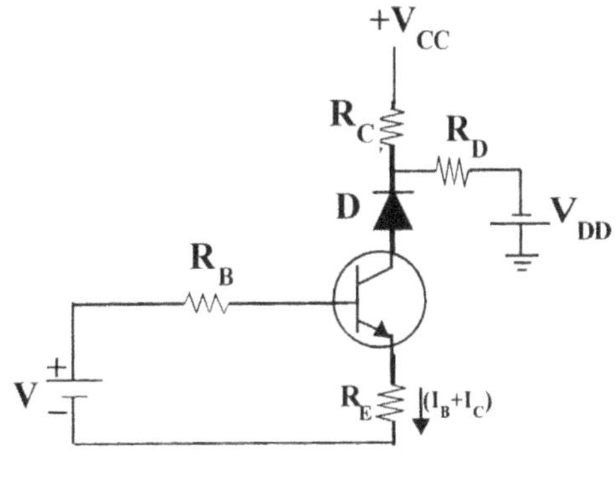

The circuit shown in Fig. 4.13 employs self biasing technique for stabilization. In addition a diode is connected in the emitter circuit and it is forward biased by the source V_{DD}. Resistance R_d limits the current through the diode. The diode should be of the same material as the transistor. The negative voltage V_0 across the diode will have the same temperature coefficient as V_{BE}. Since the diode is connected as shown, if V_{BE} decreases with increase in temperature (V_B is cut in voltage), V_0 increases with temperature. Since with temperature, the mobility of carriers increases, so current through the diode increases and V_0 increases. Therefore, as V_{BE} decreases, V_0 increases, so that I_C will be insensitive to variations in V_{BE}.

Fig 4.13   Diode compensation for V_{BE}

The decrease in V_{BE} is nullified by corresponding increase in V_D. So the net voltage more or less remains constant. Thus the diode circuit compensates for the changes in the values of V_{BE}.

4.12.2 DIODE COMPENSATION FOR I_{CO}

I_{CO} the reverse saturation current of the collector junction increases with temperature. So I_C also increases with temperature. Therefore, diode compensation should also be given against variation in I_{CO}. A circuit to achieve this is as shown in the Figures 4.14.

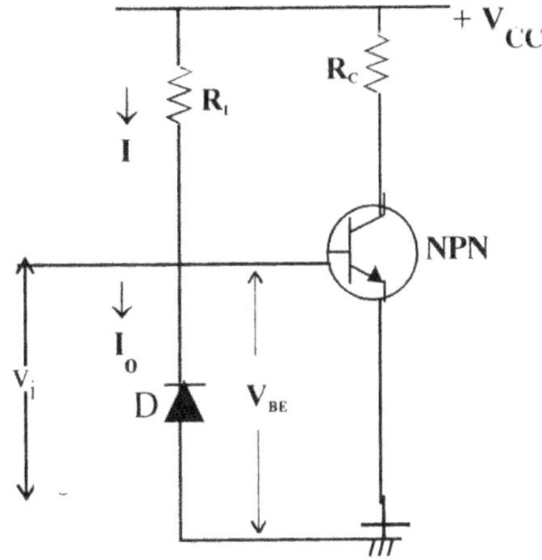

Fig 4.14   Diode compensation for I_{C0}

The diode D is reverse biased since the voltage V_{BE} which also appears across the diode is so as to reverse bias the diode. If the diode and the transistor are of the same material, the reverse saturation current I_0 of diode will increase at the same rate as the transistor collector saturation current I_{CO}.

$$I = \frac{V_{CC} - V_{BE}}{R_1}$$

Therefore, V_{CC} is measured with respect to ground.

V_{BE} is small compare to V_{CC}.

$$V_{CC} \simeq 15V,$$
$$V_{BE} = 0.2 \text{ or } 0.6V.$$

$$\therefore \qquad I \simeq \frac{V_{CC}}{R_1}$$

The diode D is reverse biased by an amount 0.2V for Germanium. Current through the diode is the reverse saturation current I_0.

$$\therefore \qquad I_B = I - I_0$$

But we know that $I_C = (1 + \beta) I_{CO} + \beta \times I_B$

Substituting the value of I_B,

$$I_C = \beta \times I - \beta \times I_0 + (1 + \beta) I_{CO}$$

But $\qquad \beta \gg 1,$

$$\therefore \qquad \beta + 1 \simeq \beta$$

$$\therefore \qquad I_C = \beta \times I - \beta \times I_0 + \beta \times I_{CO}$$

$-\beta \times I_0$ and $+\beta \times I_{CO}$ are of opposite signs. If the diode and the transistor are of the same material, the rate of increase of I_0 and I_{CO} will be the same. Therefore, I_C is essentially is constant. I_0 and I_{CO} need not be the same. Only the changes in I_{CO} and I_0 will be of the same order. So net value of I_C will remain constant.

4.13 BIASING CIRCUITS FOR LINEAR INTEGRATED CIRCUITS

With the advancements in semiconductor technology, the active components made with this technology are no more costlier than the passive components like resistors, capacitors etc. Moreover, incorporating the passive components particularly capacitors in ICs is difficult. With today's technology, in biasing and stability circuits, transistors and diodes can be used. Hence certain circuits incorporating transistors in biasing compensation are developed.

One such circuit is shown in Fig. 4.15. Why should transistor be used for compensation when it is being used as a diode only? It is because of convenience in fabrication technology. In IC fabrication for a diode, a transistor is being used as a diode. It is one step process and fabrication is easy.

Transistor Q_1 is used as a diode between base and emitter because, collector and base are shorted. Q_1 is connected as a diode across the emitter base junction of transistor Q_2. So effectively the circuit is same as circuit in Fig. 4.14. Instead of a diode, a transistor is used. Now the collector current through transistor Q_1 is,

$$I_{C1} = \frac{V_{CC} - V_{BE}}{R_1} - I_{B1} - I_{B2}$$

$$V_{BE} \ll V_{CC}, \text{ and } (I_{B1} + I_{B2}) \ll I_{C1}$$

$$\therefore \quad I_{C1} = \frac{V_{CC}}{R_1} = \text{Constant.}$$

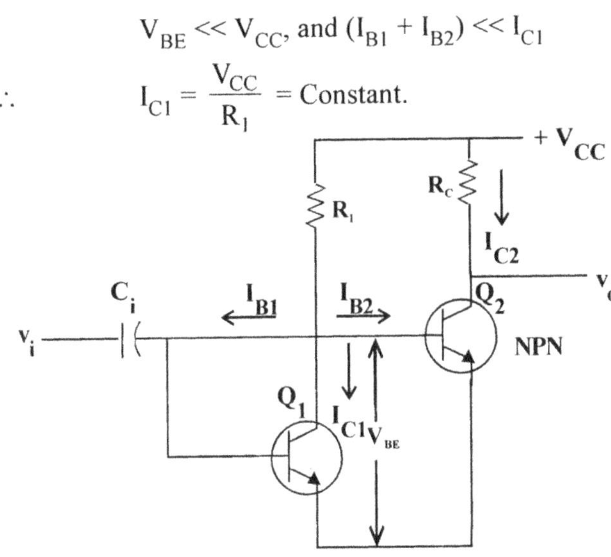

Fig. 4.15 Biasing circuit for ICs.

If transistors Q_1 and Q_2 are identical, their V_{BE}, (cut in voltage) will be the same. Hence $I_{C1} = I_{C2} = $ Constant if $R_1 = R_C$. These two transistors are being driven by the base voltage V_{BE}. Since, V_{BE} is the same, I_C will not increase. V_{CC} and R_C are also fixed. Hence the stability will be good. Even if the two transistors are not exactly identical, practical experience has shown that, the stability factor will be around 5. Hence in ICs, such circuits are employed for biasing stability.

4.14 THERMISTOR AND SENSISTOR COMPENSATION

By using temperature sensitive resistive elements rather than diodes or transistors, compensation can be achieved. ***Thermistor has negative temperature coefficient i.e., its R decrease exponentially with temperature.***

The circuit is as shown in Fig. 4.16. R_T is the thermistor. As temperature increases, R_T decreases. So th current through R_T increases. As this current passes through R_e, there is voltage drop across R_e. Since, it is PNP transistor. Collector is to be reverse biased and hence – V_{CC} is given. So the potential at the emitter (E) point is negative. Since the emitter is grounded, this acts as the reverse bias for the emitter. As the drop across R_e increases, the emitter reverse bias increases and hence I_C decreases. Thus the effect of temperature on I_C i.e., due to increase in temperature, there will be increase in β, V_{BE} and I_{CO}. The corresponding increase in I_C is nullified by the increase in the reverse bias voltage at the emitter. Thus compensation is achieved.

$V_{BE} = $ Drop Across R_E. As current through R_E increases, V_{BE} or the forward bias voltage of E - B junction reduces. Hence I_C reduces. The materials used for thermistors are:

Mixtures of such oxides as NiO (Nickle Oxide), Mn_2O_3 (Manganese Oxide), (CO_2) O_3 (Cobalt Oxide).

Instead of a thermistor, a temperature sensitive resistor with positive temperature coefficient like a metal or sensistor can be used. ***Sensistor is a heavily doped semiconductor material.*** Its temperature coefficient will be $+ 0.7\% /^\circ C$. When a semiconductor is heavily doped, as temperature increase, mobility of carriers decrease since there are more number of thermally generated free carriers (n is large).

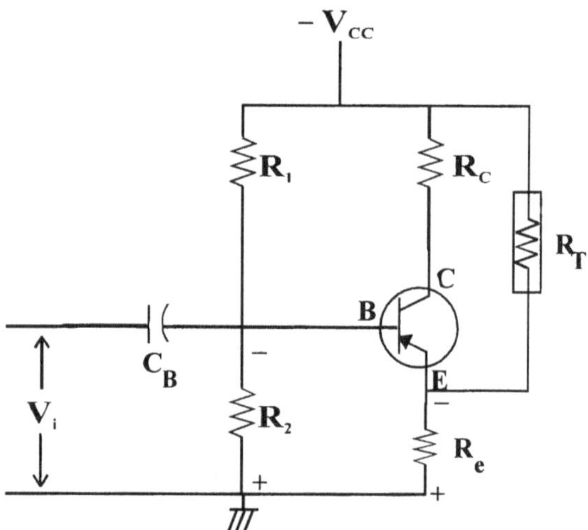

Fig 4.16 Thermistor compensation

So temperature compensation can be obtained by placing a sensistor in parallel with R_1 or R_e. Sensistor should be placed in parallel with R_e because, to start with, R_e will have the same value of sensistor. So the current will get branched equally. As T increases, I_C increases. Also the resistance of sensistor increases. Therefore, more current will pass through R_e now, since current takes the path of least resistance. Hence the reverse bias of emitter increases and so I_C decrease. (Since V_{BE} = Drop across R_2 – Drop across R_C). As Drop across R_e increases, V_{BE} decreases hence I_C decreases. So the effect of temperature on I_C is nullified.

4.15 THERMAL RUNAWAY

The max power which a transistor can dissipate (power that can be drawn from the transistor) will be from mW to a few hundred Watts. The maximum power is limited by the temperature also that the collector to base junction can withstand. For Silicon transistor the max temperature range is 150°C to 205°C and for Germanium it is 60°C to 100°C. For Germanium it is low since the conductivity is more for Germanium compared to Silicon. The junction temperature will increase either because of temperature increase or because of self heating.

As the junction tempetature rises, collector current increases. So this results in increased power dissipation. So temperature increases and I_C still further increases. This is known as *Thermal Runaway*. This process can become cumulative to damage the transistor. Now

$$\Delta T = T_J - T_A = \theta \times P_D$$

where T_J and T_A are the temperature of the junction and ambient. P_D is the power dissipated at the junction in Watts. θ is the so called *Thermal Resistance* of the transistor varying from 0.2° C/W for a high power device to 1000°C/W for a low power device. The condition for thermal stability is that the rate at which heat is released at the collector junction must not exceed the rate at which heat can be dissipated. So

$$\frac{\partial P_C}{\partial T_J} < \frac{\partial P_D}{\partial T_J}$$

$$\text{or} \qquad \frac{\partial P_C}{\partial T_J} < \frac{1}{\theta}$$

$$\therefore \qquad \theta = \frac{\partial T_J}{\partial P_D}$$

Heat Sink limit the temperature rise. Heat sink is a good conducting plate which absorbs heat from the transistor and will have large area for dissipation.

Problem 4.4

Calculate the value of thermal resistance for a transistor having P_C (max) = 125 mW at free air temperature of 25°C and T_J (max) = 150°C. Also find the junction temperature if the collector dessipation is 75 mW.

(i) $\quad T_J - T_A = \theta \times P_D \qquad$ or $\qquad 150 - 25 = \theta \times 125 \qquad$ or $\qquad \theta = 1°C/mW$

(ii) $\quad T_J - 25 = 1 \times 75 \; ; \; T_J = 100°C$

Solution

It has been found that, if the operating point is so chosen that

$$V_{CE} = \frac{V_{CC}}{2}$$

then the effect of thermal runaway is not cumulative. But if

$$V_{CE} > \frac{V_{CC}}{2}$$

then temperature increases with a increase in I_C. Again with increase in I_C, T increases and hence the effect is cumulative. As I_C increases, P_C increases. But it follows a certain law. If I_C is so chosen that

$$V_{CE} = \frac{V_{CC}}{2}$$

rate of increase of P_C with I_C is not large and Thermal Runaway problem will not be there.

Problem 4.5

A silicon type 2N780 transistor with $\beta = 50$, $V_{CC} = 10V$ and $R_C = 250\Omega$ is connected in collector to base bias circuit. It is desired that the quiescent point be apporoximately at the middle of the load line, so I_B and I_C are chossen as 0.4 mA and 21mA respectively. Find R_B and Calculate 'S', the stability factor.

Solution

$$V_{CE} = V_{CC} - (I_C + I_B) R_C$$

I_B is very small compared to I_C. So it can be neglected.

$$\therefore \qquad V_{CE} = V_{CC} - I_C \cdot R_C$$
$$= 10V - (21 + 0.4) \, 250$$

$$V_{CE} \simeq 4.6 \text{ V}$$

Since it is Silicon transistor,

$$V_{BE} = V_{\gamma} = 0.6V$$

$$\therefore \quad R_B = \frac{V_{CE} - V_{BE}}{I_B}$$

$$= \frac{4.6 - 0.6}{0.4} = 10 \text{ k}\Omega$$

$$S = \frac{\beta + 1}{1 + \beta \cdot \dfrac{R_c}{R_c + R_b}}$$

$$S = \frac{(50 + 1)}{1 + 50 \times \dfrac{250}{(250 + 10 \times 10^3)}} = 23$$

$$\therefore \quad S = 23$$

Problem 4.6

A transistor with $\beta = 100$ is to be used in Common Emitter Configuration with collector to base bias. The collector circuit resistance is $R_C = 1\text{k}\Omega$ and $V_{CC} = 10V$. Assume $V_{BE} = 0$.

(a) *Choose R_B so that the quiescent collector to emitter voltage is 4V.*

(b) *Find the stability factors.*

Solution

$$\beta = 100; \quad R_C = 1\text{k}\Omega \quad V_{CC} = 10V \quad V_{BE} = 0$$
$$R_b = ? \quad S = ?$$
$$V_{CE} \text{ should be 4V}$$

Applyining KVL around the loop,

$$V_{CC} - (I_C + I_B) R_C - V_{CE} = 0.$$

V_{CC} is voltage these negative to positive. V_{CE} is voltage drop positive to negative. But I_C value is not given.

$$\frac{I_C}{I_B} = \beta$$

$$\therefore \quad I_C = \beta \times I_B$$
$$\therefore \quad V_{CC} - (\beta + 1)I_B R_C - V_{CE} = 0.$$
$$V_{CC} = 10V \quad V_{CE} = 4V \quad R_C = 1\text{k}\Omega \quad \beta = 100 \quad I_B = ?$$
$$10 - (101) I_B \times 1 \times 10^3 - 4 = 0$$

$$\text{or} \quad I_B = \frac{6}{101 \times 10^3} = 59.4 \text{ } \mu A$$

$$\text{But} \quad V_{CE} = I_B \cdot R_B$$
$$\therefore \quad 4 \text{ V} = 59.4 \text{ } \mu A \times R_B$$

$$\therefore \qquad R_B = \frac{4}{59.4 \times 10^{-6}} = 67.3 k\Omega$$

Therefore, Stability Factor, $\qquad S = \dfrac{\beta + 1}{1 + \beta . \dfrac{R_C}{R_C + R_B}}$

$$S = \frac{101}{1 + 100 \times \dfrac{1 \times 10^3}{10^3(1 + 67.3)}} = 41$$

$$\therefore \qquad S = 41$$

Problem 4.7

Two identical Silicon transistor with $\beta = 48$, $V_{BE} = 0.6V$ at $T = 25^0 C$, $V_{CC} = 20.6V$, $R_1 = 10K$ and $R_C = 5K$ are connected as shown. Find the currents I_{B1}, I_{B2}, I_{C1} and I_{C2} at $T = 25^0C$

   (b) Find I_{C2} at $T = 175^0 C$ when $\beta = 98$ and V_{BE} 0.22V

Solution

Since the two transistors are identical, we can assume that $I_{B1} = I_{B2}$.

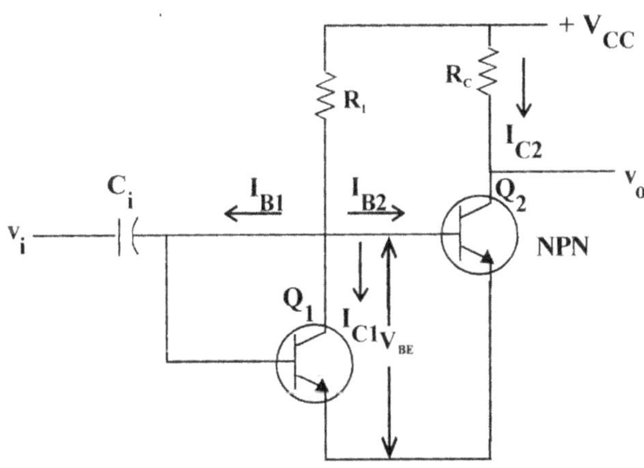

Fig 4.17 For Problem 4.7.

$$\therefore \qquad I = \frac{V_{CC} - V_{BE}}{R_1}$$

$$I = \frac{20.6 - 0.6}{10k} = 2mA$$

$$\because \qquad V_{BE} = 0.6V \text{ at } 25^0 C$$

Since $I_{B1} = I_{B2} = I_B$, then $I = 2I_B + I_{C1} = 2I_B + \beta I_B$ or $2mA = (2 + 48)I_B$

$$\therefore \qquad I_B = \frac{2}{50} \text{ mA} = 40\text{mA}$$

$$\therefore \qquad I_{C1} = I_{C2} = \beta \times I_B = 48 \times 40 \times 10^{-3} = 1.92\text{mA}$$

β is same for the two transistors. I_B is same. Hence I_C should be the same.

(b) At 175°C, V_{BE} decrease with temperature. Hence V_{BE} at 175^0 C = 0.22V and $\beta = 98$

$$\therefore \qquad I = \frac{V_{CC} - V_{BE}}{10k} = 2.038\text{mA} = (2 + \beta)I_B$$

or

$$I_B = \frac{2.038}{2 + 98} \text{ mA} = 20.38\mu\text{A}.$$

$$I_{C1} = I_{C2} = \beta \times I_B = 98 \times 20.38 \times 10^{-6} = 2\text{mA}$$

Problem 4.8

Determine the quiescent currents and the collector to emitter voltage for a Ge transistor with $\beta = 50$ in the self biasing arrangements. The circuit component values are $V_{CC} = 20V$, $R_C = 2K$, $R_e = 0.1k$, $R_1 = 100k$ and $R_2 = 5k\Omega$. Find the stability factor S.

Solution

$$V = \frac{R_2 \cdot V_{CC}}{R_1 + R_2} = \frac{5 \times 20}{100 + 5} = 0.952 \text{ V}$$

$$R_B = \frac{R_1 R_2}{R_1 + R_2} = \frac{5 \times 100}{100 + 5} = 4.76 \text{ K}\Omega$$

For Germanium transistor, $V_{BE} = 0.2V$

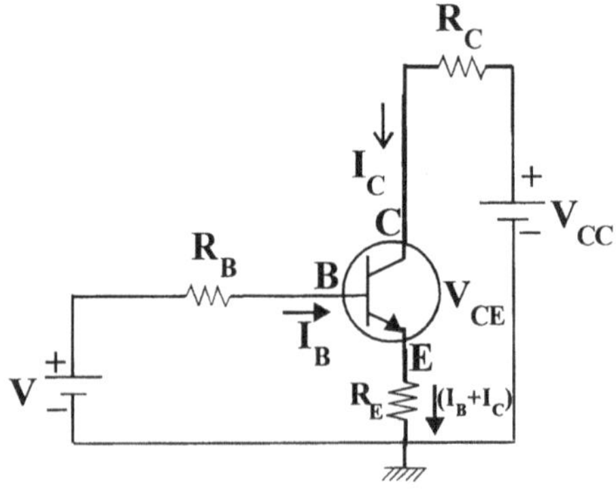

Fig 4.18 For Problem 4.8.

$$\therefore \qquad V = I_B \times R_B + V_{BE} + (I_B + I_C)R_e$$

$$V = I_B \times R_B + V_{BE} + (\beta + 1) I_B R_e$$

$$= I_B [R_B + R_e(1+\beta)\,] + V_{BE}$$

$$\therefore \qquad 0.952 = I_B[4.76 \times 10^3 + 51 \times 0.1] + 0.2$$

$$\therefore \qquad I_B = \frac{0.952 - 0.2}{9.86 \times 10^3} = 76.2 \text{ mA}$$

$$I_C = \beta \times I_B = 3.81 \text{ mA}$$

$$I_E = (I_B + I_C) = 3.81 \text{ mA}$$

$$V_{CE} = V_{CC} - I_C R_C - I_E R_e$$

$$= 20 - 3.81 \times 2 - 3.89 \times 0.1 = 12V$$

$$S = (1+\beta) \times \left(\frac{1 + \dfrac{R_B}{R_e}}{1 + \beta + \dfrac{R_B}{R_e}} \right) = 51 \times \frac{1 + \dfrac{4.76}{0.1}}{51 + \dfrac{4.76}{0.1}} = 25$$

Problem 4.9

Design a circuit with Ge transistor in the self biasing arrangement with $V_{CC} = 16V$ and $R_C = 1.5K$. The quiescent point is chosen to be $V_{CE} = 8V$ and $I_C = 4mA$. Stability factor $S = 12$ is desired $\beta = 50$.

Solution

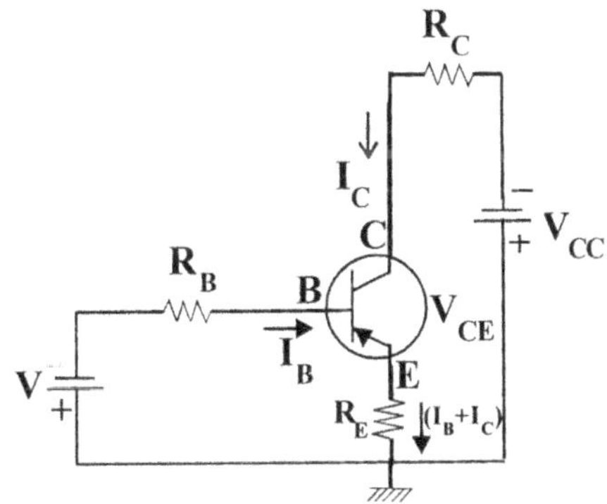

Fig 4.19   For Problem 4.9.

$$I_B = \frac{I_C}{\beta} = \frac{4}{50} = 80\mu A$$

$$R_e = \frac{V_{CC} - V_{CE} - I_C R_C}{(I_B + I_C)}$$

$$= \frac{16 - 8 - 4 \times 1.5}{4.08} = 0.49k$$

$$S = 12 = \frac{(\beta+1)1 + \dfrac{R_B}{R_e}}{1 + \beta + \dfrac{R_B}{R_e}}$$

or
$$\frac{R_B}{R_e} = 14.4$$

$\therefore$
$$R_B = 14.4 \times R_e = 7.05k.$$
$$V_{BN} = \text{Base to ground voltage}$$
$$= V_{BE} + I_E R_e$$

Therefore, I_E is negative.

$$V_{BN} = 0.2 + 4.08 \times 0.49 = 2.2V$$
$$V = V_{BN} + I_B R_b$$
$$= 2.2 + 0.08 \times 7.05 = 2.76V$$

$$V = \frac{R_2 . V_{CC}}{R_1 + R_2} \; ; \; R_B = \frac{R_1 R_2}{R_1 + R_2}$$

$\therefore$
$$\frac{V}{R_B} = \frac{V_{CC}}{R_1} = \frac{16}{R_1}$$

$\therefore$
$$R_1 = \frac{16 R_B}{V}$$

$$= \frac{16 \times 7.05}{2.76} = 41k$$

$$V = 2.76 \text{ V}$$

$$= \frac{R_2 \times V_{CC}}{R_1 + R_2} = \frac{R_2 \times 16}{41k + R_2}$$

$\therefore$
$$R_2 = 8.56k$$

4.16 STABILITY FACTOR S″ FOR SELF BIAS CIRCUIT

$$V - I_B \times R_B - V_{BE} - (I_B + I_C)R_e = 0 \qquad\qquad .... (\ 4.16\ )$$

or
$$V_{BE} = V - I_B \times (R_B + R_e) - I_C R_e \qquad\qquad .... (\ 4.17\ )$$

But
$$I_C = (1 + \beta) \times I_{CO} + \beta \times I_B \qquad\qquad .... (\ 4.18\ )$$

or
$$I_B = \frac{I_C - (1 + \beta)I_{CO}}{\beta} \qquad\qquad .... (\ 4.19\ )$$

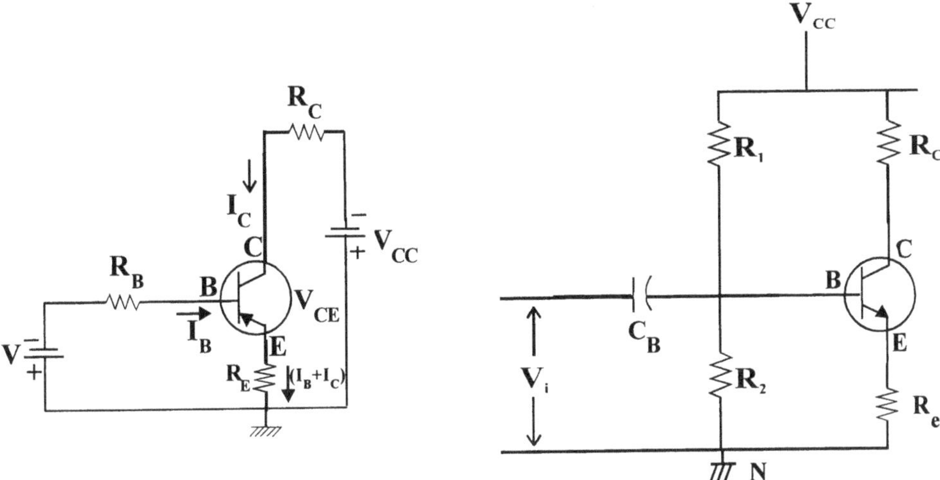

Fig 4.20 *Stability factor S".*   Fig 4.21 *Self bias circuit.*

Substituting equation (4.19) in (4.17),

$$V_{BE} = V + (R_B + R_e)\left(\frac{\beta+1}{\beta}\right) I_{CO} - \frac{R_B + R_e(1+\beta)}{\beta} I_C$$

$$= V - I_C\left\{\frac{(R_B + R_e)}{\beta} + R_e\right\} + (R_B + R_e)\frac{(\beta+1)}{\beta} I_{CO}$$

$$V_{BE} = V - I_C\left\{\frac{R_B + R_e(1+\beta)}{\beta}\right\} + (R_B + R_e)\left(\frac{\beta+1}{\beta}\right) I_{CO}$$

Let

$$(R_B + R_e)\left(\frac{\beta+1}{\beta}\right) \times I_{CO} = V'$$

$$\frac{\beta+1}{\beta} \simeq 1$$

$\therefore$

$$V' \simeq (R_b + R_e) I_{CO}$$

$$V_{BE} = V + V' - \frac{R_B + R_e(1+\beta)}{\beta} \times I_C$$

or

$$(-V_{BE} + V + V')$$

$$= \frac{R_B + R_e(1+\beta)}{\beta} \times I_C$$

$\therefore$

$$I_C = \frac{\beta(V + V' - V_{BE})}{R_B + R_e(1+\beta)}$$

V' can be assumed to be independent of β.

$\therefore$

$$S'' = \frac{\partial I_C}{\partial \beta} = \frac{I_C \cdot S}{\beta(1+\beta)} = S''$$

or
$$\frac{\Delta I_C}{\Delta \beta} = \frac{I_C . S}{\beta(1+\beta)} = S''$$

$$\Delta I_C = S'' \, \Delta\beta = \frac{I_C . S}{\beta(1+\beta)} \times \Delta\beta$$

$$\therefore \quad S'' = \frac{I_{C2} - I_{C1}}{\beta_2 - \beta_1} = \frac{\Delta I_C}{\Delta\beta}$$

$$I_{C1} = \frac{\beta_1 \left(V + V' - V_{BE}\right)}{R_B + R_e\left(1+\beta_1\right)} \qquad\qquad ....\ (\ 4.20\ )$$

$$I_{C2} = \frac{\beta_2 \left(V + V' - V_{BE}\right)}{R_B + R_e\left(1+\beta_2\right)} \qquad\qquad ....\ (\ 4.21\ )$$

$$\frac{I_{C2}}{I_{C1}} = \left\{\frac{\beta_2}{\beta_1}\right\} \frac{R_B + R_e\left(1+\beta_1\right)}{R_B + R_e\left(1+\beta_2\right)} \qquad\qquad ....\ (\ 4.22\ )$$

Subtracting 1 from both sides, and simplifying equation.

$$S'' = \frac{\Delta I_C}{\Delta \beta} = \frac{I_{C1} \, S_2}{\beta_1\left(1+\beta_2\right)}$$

Problem 4.10

A transistor type 2N335 is used in the self bias configuration may have any value of β between 36 and 90 at room temperature. Design a self bias circuit (Find Re, R_1 and R_2) subject to the following specification R_c = 4 KΩ, V_{CC} = 20V, normal bias point is to be at V_{CE} = 10V, I_C = 2mA and I_C should be in the range of 1.75 to 2.25 mA as β varies from 36 to 90. Neglect the variation of I_{CO}

Solution

$$I_C \gg I_B. \ \text{ So neglecting } I_B,$$

$$(R_C + R_e) = \frac{V_{CC} - V_{CE}}{I_C}$$

$$\therefore \quad V_{CC} - I_C . R_C - V_{CE} - I_C . R_E = 0$$

$$(R_C + R_e) = \frac{20 - 10}{2} = 5 \ K\Omega$$

$$R_C = 4 \ K\Omega$$

$$\therefore \quad R_e = (5 - 4) \ K\Omega = 1 \ K\Omega$$

$$\Delta I_C = I_{C2} - I_{C1} = 2.25 - 1.75$$
$$= 0.5 \ mA$$

$$\Delta \beta = \beta_2 - \beta_1$$
$$= 90 - 36 = 54$$

$$\therefore \qquad \frac{\Delta Ic}{\Delta \beta} = \frac{Ic_1 s_2}{\beta_1(1+\beta_2)} = \frac{52}{(1+90)}$$

$$= \frac{0.5}{54}$$

or $S_2 = 17.3$; S_2 is the Stability Factor (S).

$$\therefore \qquad S_2 = 17.3, \ R_e = 1 \ K\Omega, \ \beta_2 = 90,$$

$$S_2 = (1+\beta_2) \ 1 + \frac{1+\dfrac{R_B}{R_e}}{1+\beta_2+\dfrac{R_B}{R_e}}$$

$$\therefore \qquad R_B = 20.1 \ K\Omega$$

Neglecting the effect of I_{CO},

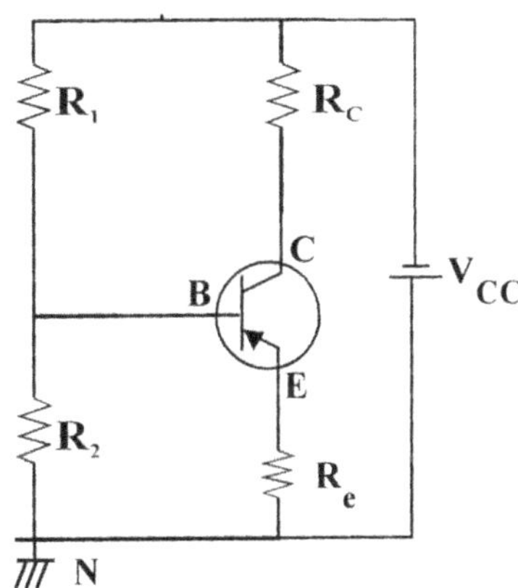

Fig 4.22 For Problem 4.10.

$$V = V_{BE} + \frac{R_B + R_e(1+\beta)}{\beta} \times I_C$$

$$= 0.6 + \left(\frac{20.1+37}{36}\right) 1.75 = 3.38V$$

$$R_1 = R_B. \frac{V_{CC}}{V} = 20.1 \times \frac{20}{3.38} = 119K$$

$$R_2 = \frac{R_1 V}{V_{CE} - V} = \frac{119 \times 3.38}{(20 - 3.38)} = 242K$$

Problem 4.11

A PNP transistor is used in self - biasing arrangement the circuit components are V_{cc} = 4.5V; R_2 = 2.7 K ; R_E = 0.27 K ; R_1 =27 K of β = 44. Find

    (a)   *Stability factor.*

    (b)   *Quiescent point*

    (c)   *Recalculate these values if the base - spreading resistance of 690 Ω is taken in account.*

Solution

The circuit and its Thevenin's equivalent are shown in Fig. 4.22 and Fig. 4.23.

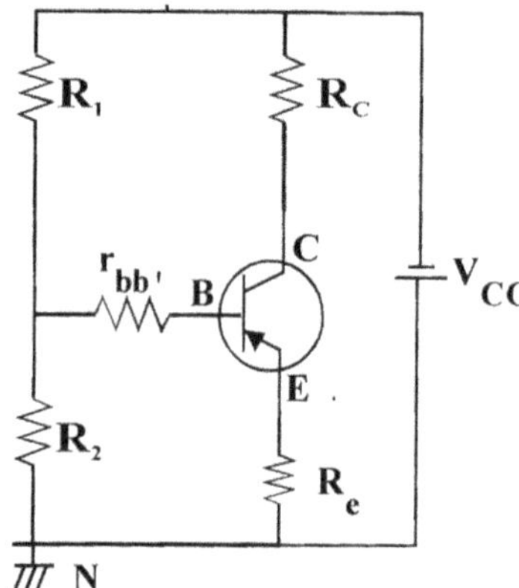

Fig 4.23 For Problem 4.11.

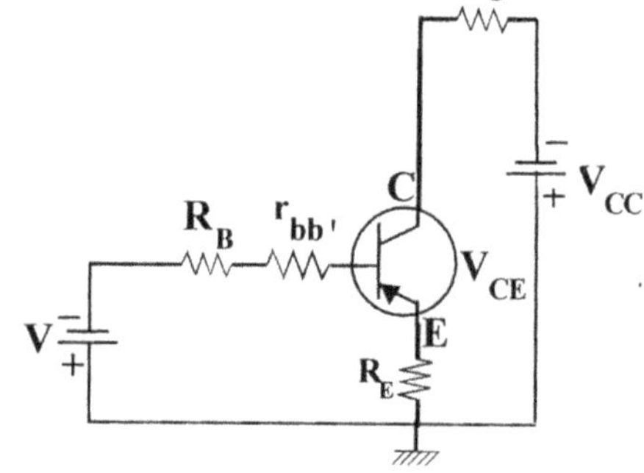

Fig 4.24 For Problem 4.11.

$$R_B = R_1 \parallel R_2 = \frac{R_1 R_2}{R_1 + R_2}$$

$$= \frac{2.7 \times 27}{29.7} = 2.46 \text{ k}\Omega$$

$$V_2 = V_{CC} \frac{R_2}{R_1 + R_2}$$

$$= +4.5 \times \frac{2.7}{29.7} = 0.41 \text{ V}$$

(a) $$S = \frac{1+\beta}{1+\beta \dfrac{R_e}{R_e + R_b}} = \frac{44+1}{1+44\dfrac{0.27}{2.73}} = \frac{45}{5.35} = 8.409$$

(b) In base circuit,

$$V_2 = I_E R_E + V_{BE} + R_B. I_B$$
$$= (1 + \beta) I_B . R_E + V_{BE} + R_B I_B$$
$$0.41 = (45) I_B (0.27) + 0.6 + (2.46) I_B.$$

$$\Rightarrow \quad I_B = \frac{(0.41 - 0.6)}{(45 \times 0.27 + 2.46)} = -0.013 \, mA$$

$$I_C = \beta I_B = 0.572 mA$$

In collector circuit.

$$V_{CC} = R_C I_C + V_{CE} + I_E R_E$$
$$= I_B (\beta . R_C + R_E + \beta R_E) + V_{CE}$$
$$\therefore \quad V_{CE} = 4.5 - 1.016 = 3.484 \, V$$

Quiescent Point : $V_{CE} = 3.484V$

$$I_C = 0.572 mA$$
$$I_B = 0.013 mA$$

(c) Taking $r_{bb'} = 690\Omega$;

$$S = \frac{45}{1 + 44 \dfrac{0.27}{3.42}} = 10.06$$

$$I_B = \frac{-0.19}{(45 \times 0.27) + 3.15} = -0.012 mA$$

$$I_C = -0.528 \, mA$$

$$V_{CE} = 4.5 - 0.938 = 3.562$$

Quiescent Point : $\quad V_{CE} = 3.562V$

$$I_C = 0.528 mA$$
$$I_B = 0.012 mA$$

These are the values taking $r_{bb'}$ into consideration.

Problem 4.12

For the given circuit, determine the stability factor S and Thermal Resistance θ_{Th}.

$$V_{CC} = 24V \qquad R_E = 270 \, \Omega$$
$$R_C = 10K \qquad \beta = 45$$
$$V_{CE} = 5V$$

Solution :

In Collector–Emitter Circuit,

$$Vcc = I_C R_C + V_{CE} + I_E R_E$$

$$= I_C R_C + V_{CE} + \frac{(1 + \beta)}{\beta} I_C . R_E$$

$$24 \text{ V} = 10\, I_C + 5 + \frac{46}{45} I_C\, (\, 270\,)$$

$$I_C = 1.849 \text{ mA.}$$

$$I_B = \frac{I_C}{\beta} = \frac{1.849}{45} = 41.09\ \mu A$$

(a)   In collector base circuit; $V_{CE} = I_B \times R + V_{BE}$

$$\therefore \qquad R = \frac{5 - 0.6}{41.09\mu} = 107.08 K\Omega$$

(b)   Stability factor $= S = \dfrac{1+\beta}{1+\beta\dfrac{R_C}{R_C + R}}$

$$= \frac{46}{1 + 45\dfrac{10}{117.08}}$$

$$= \frac{46}{1 + 3.844} = 9.496$$

$$\therefore \qquad R = 107.08 \text{ K } \Omega$$

$$S = 9.496$$

(c)   We know that $A_T = T_J = T_H = (\text{Th. Res})$

$$T_J - T_A = \theta_{Th} \times P_D.$$

Thermal resistance $\qquad = \dfrac{150 - 25}{125}$

$$= 1000^{\circ}C/W$$

SUMMARY

- In order that for a transistor, Emitter-Base junction is forward biased and collector-Base junction is reverse biased, biasing circuit must be employed. The three types of biasing circuits are

    1. Fixed bias or Base bias circuit.

    2. Collector to base bias

    3. Self bias or Emitter bias

- Self bias circuit is commonly used. It ensures that the operating point is in the centre of the active region. The operating point is defined by I_C, V_{CE} and I_B.

    Usually V_{CE} is chosen to be equal to $\dfrac{V_{CC}}{2}$.

- Stability factor is a function of I_{Co}, β, V_{BE}

$$S = \frac{\partial I_C}{\partial I_{co}}\bigg|_{\beta \text{ and } V_{BE} = \text{Constant}}$$

General Expression for,

$$S = \frac{1+\beta}{1-\beta\left(\dfrac{\partial I_B}{\partial I_C}\right)}$$

For a Self bias, circuit,

$$S = (1+\beta)\left[\frac{1+\dfrac{R_B}{R_e}}{1+\dfrac{R_B}{R_e}+\beta}\right]$$

- ◆ Quiescent point Q, (or operating point or biasing point) changes with tempera-ture. So to nullify for the effect of temperature, compensating techniques must be used. The circuits are :

    1. Diode Compensation for V_{BE}

    2. Diode Compensation for I_{C0}

    3. Thermistor and sensistor compensation.

- ◆ Thermistors have negative temperature coefficient. The materials used are oxides of Nickle, Cobalt and Manganese. Sensistors have positive temperature coefficient. They are very heavily doped semiconductors, resembling metallic thermal characteristics.

OBJECTIVE TYPE QUESTIONS

1. Operating point is also known as ....

2. Operating point depends upon the following BJT parameters ....

3. For a transistor to function as an amplifier, the operating point must be located at ....

4. For normal amplifier circuits, the operating point is chosen such that V_{CE} = ...V_{CC}

5. Fixed bias circuit is also known as ....

6. Self bias circuit is also known as ....

7. Stability factor S is defined as S = ...

8. S' is defined as S' = ...

9. Stability factor S'' is defined as S'' = ...

10. General expression for S = ...

11. Expression for S for self bias circuit is....................

12. For both Silicon and Germanium, V_{BE} or V_r cut in voltage decreases at the rate of ....

13. Thermistors have .... temperature coefficient of resistance

14. Semiconductor materials having positive temperature coefficient of R (PTCR) are called...................

15. Materials used for Thermistors are ....

16. Irrespective of the configuration of the Transistor (BTT) the relationship between the currents.......

17. For most configurations, the d.c. analysis of the circuit begins with ...... current of the BJT.

18. A BJT circuit is most stable, and least sensitive to temperature changes if the circuit has .... stability factor.

19. 'β' of a BJT is very sensitive to .... and V_{BE} .... at the rate of 2.5 mv for each 1° increase in temperature.

20. If a transistor (BJT) is in active region, V_{CE} will be about .... % of V_{CE}.

ESSAY TYPE QUESTIONS

1. Explain about the need for biasing in electronic circuits, what are the factors affecting the stability factor.

2. Draw the Fixed bias circuit and derive the expression for the stability factor S. What are the limitations of this circuit ?

3. Draw the Collector to Base bias circuit and derive the expression for the stability factors What are the limitations of this circuit ?

4. Draw the Self bias circuit and obtain the expression for the stability factor S. What are the advantages of this circuit ?

5. Compare all the three biasing circuits.

6. Explain the terms: Thermal Runaway and Thermal Resistance.

7. Explain the terms Bias Stabilization and Bias Compensation.

8. Draw the circuits and explain the principles of working of Diode Compensation for V_{BE} and I_{CO}.

MULTIPLE CHOICE QUESTIONS

1. When the input is symmetrical, to operate the BJT in active region, the quiescent point is chosen ....

    (a)   at the top edge of the load line

    (b)   at the bottom edge of the load line

    (c)   at the centre of the load line

    (d)   can be chosen any where on the load line

2. **AC load line is also known as ....**
   (a) dynamic load line
   (b) variable load line
   (c) quiescent load line
   (d) active load line

3. **For better stability of the amplifier circuit, value of stability factor 'S' must be**
   (a) large　　(b) as small as possible　　(c) 1　　(d) 0

4. **For fixed bias circuit, the expression for stability factor 'S' is**
   (a) $1 - \beta$　　(b) $1 + \beta^2$　　(c) $\dfrac{1+\beta}{1-\beta\frac{(\partial I_C)}{(\partial I_B)}}$　　(d) $1 + \beta$

5. **Voltage divider bias or universal bias circuit is also known as**
   (a) self bias circuit
   (b) collector bias circuit
   (c) collector to base bias circuit
   (d) Fixed bias circuit

6. **By definition, expression for stability factor S″ is**
   (a) $S'' = \left.\dfrac{\partial I_C}{\partial \beta}\right|_{\substack{V_{BE}=K \\ I_{CO}=K}}$
   (b) $S'' = \left.\dfrac{\partial \beta}{\partial I_C}\right|_{V_{BE},\,I_{CO}} = K$
   (c) $S'' = \dfrac{\partial I_\beta}{\partial I_C}$
   (d) $S'' = \dfrac{\partial I_\beta}{\partial \beta}$

7. **The units of stability factor S are**
   (a) ohms　　(b) constant　　(c) mhos　　(d) volt-amperes

8. **Compared to silicon BJT, thermal stability for germanium transistors is ...**
   (a) good　　(b) poor　　(c) same　　(d) can't be said

9. **Sensistors have ...... temperature coefficient**
   (a) –ve　　(b) + ve　　(c) zero　　(d) None of these

10. **The units of thermal resistance**
    (a) $^{\circ}C/\Omega$　　(b) $\Omega/^{\circ}C$　　(c) $\Omega/^{\circ}C$　　(d) $^{\circ}C/\Omega$.

CONCEPTUAL QUESTIONS (Interview Questions)

1.　　　Why transistor biasing is to be done?

Ans : In order that amplification is achieved, the BJT must be operated in the active region. To ensure thus, E-B junction of the transistor (BJT) must be forward biased and Collector - Base junction must be reverse biased. The operating point or quiescant point must be in the centre of the active region. To ensure thus, biasing must be done. Similarly to use the BJT as On - Off switch, the operating point must change between cut-off and saturation regions. So biasing is to be done.

5 Field Effect Transistors (FETs)

I.　INTRODUCTION

5.1　FIELD EFFECT TRANSISTOR (FET)

The Field Effect Transistor is a semiconductor device which depends for its operation on the control of current by an electric field. Hence, the name FET.

There are two types (FET) :

1. Junction Field Effect Transistor (JFET) or simply FET.
2. Insulated Gate Field Effect Transistor IGFET. It is also called as Metal Oxide Semiconductor (MOS) transistor or MOST or MOSFET.

The principle of operations of these devices, their characteristic are given in this chapter.

Advantages of FET over BJT (Transistor)

1. JFET is unipolar device. Its operation depends upon the flow of majority carriers only. Vacuum tube is another example of unipolar device. Because the current depends upon the flow of electrons emitted from cathode. Transistor is a bipolar device. So recombination noise is more.
2. JFET has high input resistance ($M\Omega$). BJT (Transistor) has less input resistance. So loading effect will be there.
3. It has good thermal stability.
4. Less noisy than a tube or transistor. Because JFET is unipolar, recombination noise is less.
5. It is relatively immune to radiation.
6. JFET can be used as a symmetrical bilateral switch.
7. By means of small charge stored or internal capacitance, it acts as a memory device.
8. It is simpler to fabricate and occupies less space in IC form. So packing density is high.

Disadvantage

1. Small gain - bandwidth product.

5.1.1　JFET

There are two types;

1. n - channel JFET.
2. p - channel JFET.

If n - type semiconductor bar is used it is n - channel JFET.

If p - type semiconductor bar is used it is p - channel JFET.

Ohmic contacts are made to the two ends of a semiconductors bar of n-type or p-type semiconductor. Current is caused to flow along the length of the bar, because of the voltage supply connected between the ends. If it is n - channel JFET, the current is due to electrons only and if it is p-channel JFET, the current is due to holes only.

The three lead of the device are **1.** Gate **2.** Drain **3.** Source. They are similar to **1.** Base **2.** Collector **3.** Emitter of BJT respectively.

5.1.2   n-Channel JFET

The arrow mark at the gate indicates the direction of current if the Gate source junction is forward biased.

On both sides of the n - type bar heavily doped (p^+) region of acceptor impurities have been formed by alloying on diffusion. These regions are called as Gates. Between the gate and source a voltage V_{GS} is applied in the direction to reverse bias the p-n junction.

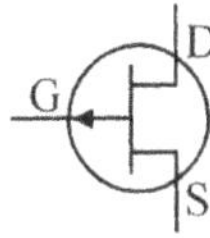

Fig 5.1 (n-channel JFET)

p-Channel JFET

The symbol for p-channel FET is given in Fig. 5.2. The semiconductor bar is p-type or source is p-type. Gate is heavily doped n-region. If G - S junction is forward biased, electrons from gate will travel towards source. Hence, the conventional current flows outside. So, the arrow mark is as shown in Fig. 5.2. indicating the direction of current, if G-S junction is forward biased.

The FET has three terminals, Source, drain and Gate.

Fig 5.2 (p-channel JFET).

Source

The source S is the terminal through which the majority carriers enter the bar (Fig. 5.3). Conventional current entering the bar at S is denoted as I_S.

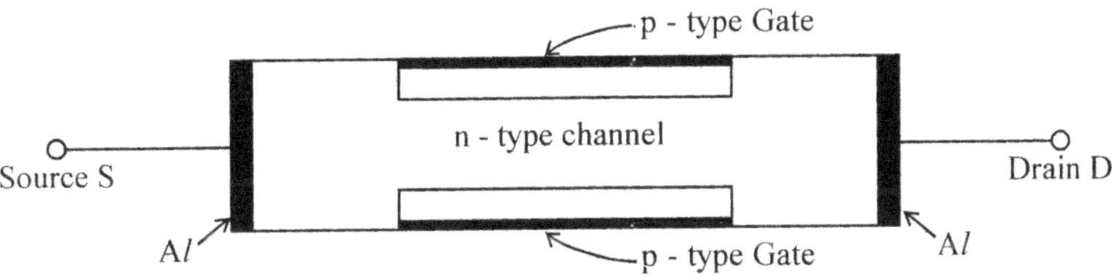

Fig 5.3 Structure of JFET.

Drain

The drain D is the terminal through which the majority carriers leave the bar. Current entering the bar at D is designated by I_D. V_{DS} is positive if D is more positive than S, source is forward biased.

Gate

In the case of n - channel FET, on both sides of the bar heavily doped (p+) region of acceptor impurities have been formed by diffusion or other techniques. The impurity regions form as Gate. Both these region (p+ regions) are joined and a lead is taken out, which is called as the gate lead of the FET. Between the gate and source, a voltage V_{GS} is applied so as to reverse bias the gate source p-n junction. *Because of this reason JFET has high input impedance. In BJT Emitter Base junction is forward biased. So it has less Z.* Current entering the bar at G is designated as I_G.

CHANNEL

The region between the two gate regions is the channel through which the majority carriers move from source to drain. By controlling the reverse bias voltage applied to the gate source junction the channel width and hence the current can be controlled.

II.   CONSTRUCTION

5.2   FET STRUCTURE

FET will have gate junction on both sides of the silicon bar (as shown in Fig. 5.4).

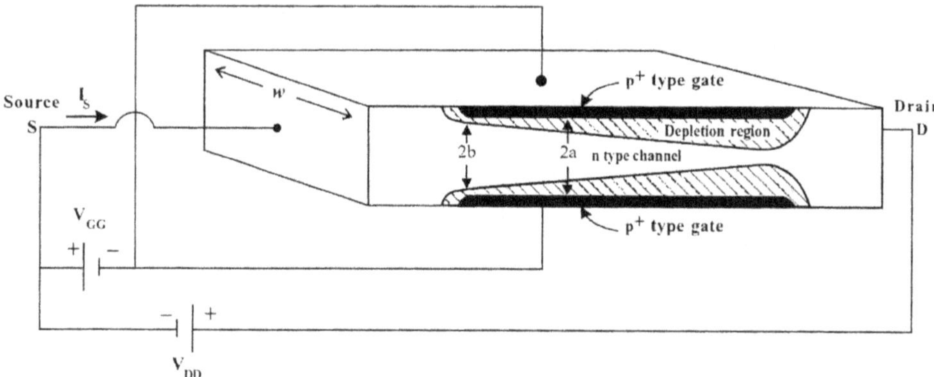

Fig 5.4   JFET structure.

But it is difficult to diffuse impurities from both sides of the wafer. So the normal structure that is adopted is, a p-type material is taken as substrate. Then n-type channel is epitaxially grown. A p-type gate is then diffused into the n-type channel. The region between the diffused p-type impurity and the source acts as the drain.

2b (x) =   Actual channel width, at any point x, the spacing between the depletion regions at any point x from source end because the spacing between the depletion regions is not uniform, it is less towards the drain and more towards the source (see Fig.5.5). 2b depends upon the value of V_{GS}.

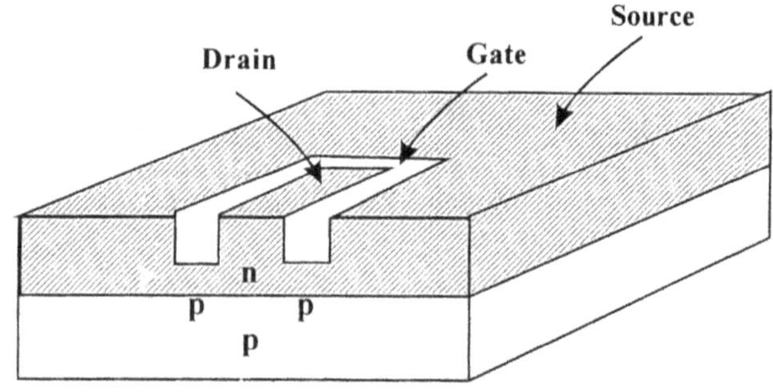

Fig 5.5   Structure

2a =   Channel width, the spacing between the doped regions of the gate from both sides. This is the channel width when $V_{GS} = 0$ (and the maximum value of channel width).

5.2.1 THE ON RESISTANCE r_{DS} (ON)

Suppose a small voltage V_{DS} is applied between drain and source, the resulting small drain current I_D will not have much effect on the channel profile. So the effective channel cross section A can be assumed to be constant, throughout its length.

$\therefore$ $A = 2b\,w$

Where, $2b$ = Channel width corresponding to zero I_D (Distance between the space charge regions)

w = Channel dimension perpendicular to b.

$\therefore$ Expressions for,

$$I_D = A \cdot e \cdot N_D \, \mu_n \, \varepsilon_x$$

$\because \qquad J = n\,e \cdot \mu \cdot \varepsilon \ (J = \sigma.\varepsilon. \ \sigma = ne.\mu)$

$$I = A \times J,$$

But $\qquad A = 2b\,(x) \cdot w$

Considering b along x-axis, since 'b' depends upon the voltage V_{GS}, $b(x)$ is the width at any point 'x'.

$$\therefore \qquad I_D = 2b\,(x)\,\omega.\ q.\ N_D \cdot \mu_n .\varepsilon_x$$

$$\varepsilon_n = \frac{V_{DS}}{L} \ ; \ L \text{ is the length of the channel}$$

(Electric field strength depends upon V_{DS}. Channel width 2b depends upon V_{GS}.)

$$\therefore \qquad I_D = 2b\,(x)\ w.\ q.\ N_D \cdot \mu_n \cdot \frac{V_{DS}}{L}$$

This expression is valid in the linear region only. This is the linear region, It behaves like a resistance, where ohmic resistance depends upon $V_{GS}.V_{DS} \, / \, I_D$ is called as the **ON drain resistance** or $r_{DS(ON)}$ ON because, JFET is conducting, in the ohmic regions when V_{DS} is small and channel area $A = 2a.w$,

$$r_{DS(on)} = \ L/2\ b(x)\ wq\ N_D\ \mu_n$$

$r_{DS(on)}$ will be few Ω to several 100 Ω.

Because, the mobility of electrons is much higher than that for holes, $r_{DS(on)}$ is small for n-channel FETs compared to p-channel FETs.

5.2.2 Pinch off region

The voltage V_{DS} at which, the drain current I_D tends to level off is called the **pinch off voltage**. When the value of V_{DS} is large, the electric field that appears along the x-axis, i.e., along the channel ε_x will also be more. When the value of I_D is more, the drain end of the gate i.e., the gate region near the drain is more reverse biased than the source end. Because, the drain is at reverse potential, for n-channel FET, source is n-type, drain is n-type and positive voltage is applied for the drain, gate is also reverse biased. Therefore, the drop across the channel adds to the reverse bias voltage of the gate. Therefore, channel narrows more near the drain region and less near the source region. So, the boundaries of the depletion region are not parallel when V_{DS} is large and hence ε_x is large.

As V_{DS} increases, ε_x and I_D increase, whereas the channel width $b(x)$ narrows ($\because$ depletion region increases). Therefore the current density $J = I_D \, / \, 2b(x) \, w$ increases. If complete pinch off were to take place $b = 0$. But this cannot happen, because if b were to be 0, $J \to \infty$ which cannot physically happen.

Mobility μ is a function of electric field intensity μ remains constant for low electric fields when $\varepsilon_x < 10^3 V/cm$.

$$\mu \ \alpha \ \frac{1}{\sqrt{\varepsilon_x}} \ \text{ when } \varepsilon_x \text{ is } 10^3 \text{ to } 10^4 \text{ V/cm.(for moderate fields).}$$

$$\mu \ \alpha \ \frac{1}{\varepsilon_x} \ \text{ for very high field strength} > 10^4 \text{ V/cm.}$$

$$I_D = 2bwe\ N_D\ \mu_n\ \varepsilon_x.$$

As V_{DS} increases, ε_x increases but μ_n decreases, b almost remains constant. Therefore I_D remains constant in the pinch off region.

Pinch off voltage (V_{PO} or V_P) is the voltage at which the drain current I_D levels off. When $V_{GS} = 0V$, if $V_{GS} = -1,V$, the voltage at which the current I_D levels off decreases, (This is not pinch off voltage). If V_{GS} is large -5, V_{DS} at which the current levels off may be zero. So the relationship between V_{DS}, V_{PD} and V_{GS} is

$$V_{DS} = V_{PD} + V_{GS} \qquad V_{DS} \text{ is positive n-channel FET.}$$

$V_{PO} = P$ indicates pinch off voltage, zero indicate the voltage when $V_{GS} = 0V$.

V_{DS} is the voltage at which the current I_D levels off.(See Fig. 5.6).

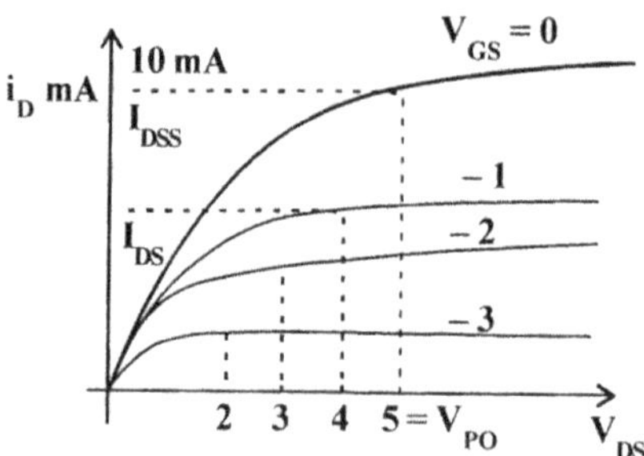

Fig 5.6   JFET drain characteristics.

5.2.3 EXPRESSION FOR PINCH OFF VOLTAGE, V_P

Net charge in an alloy junction, must be the same, semiconductor remains neutral.

$$\therefore \quad e.N_A\, w_P = e\, N_D \cdot w_n$$

If $N_A \gg N_D$, from the above eq, $w_p \ll w_n$. The relation between potential and charge density (ρ) is given by Poisson's equation.

$$\frac{d^2V}{dx^2} = \frac{e.N_D}{\epsilon}$$

We can neglect w_p and assume that the entire barrier potential V_B appears across the increased donor ions. $e.N_D$ = charge density P

$$\text{Integrating} \quad \frac{dv}{dx} = \frac{e.N_D}{\epsilon}.x$$

$$V = \frac{e.N_D}{\epsilon} \cdot \frac{x^2}{2}$$

At the boundary conditions, $x = w_n$

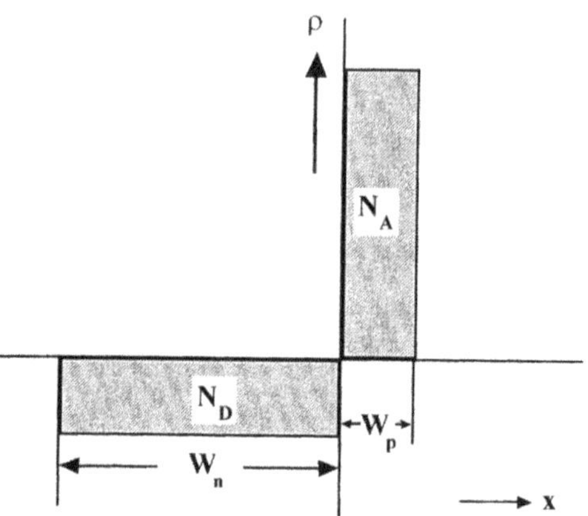

Fig 5.7   Charge density variation

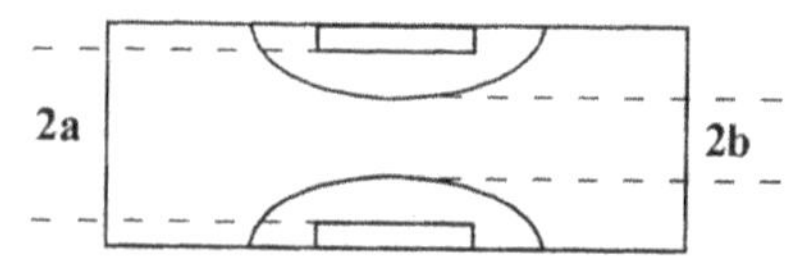

Fig 5.8 Structure

$$w_n = \text{depletion region width}$$

$$\therefore \quad V = \frac{e.N_D}{2 \in} \cdot w_n^2$$

$$\text{or} \quad w_n = \left\{ \frac{2 \in V}{e.N_D} \right\}^{\frac{1}{2}}$$

This is the expression for the penetration of depletion region into the channel. As the reverse bias potential between Gate and drain increased, the channel width narrows or the depletion region penetrates into the channel. Therefore the general expression for the spacecharge width $W_n(x) = w(n)$.

$$w(n) = \left\{ \frac{2 \in}{eN_D} \left\{ V_o - V(x) \right\} \right\}^{\frac{1}{2}}$$

$$V = \text{Voltage between drain and source } V_{DS}$$

$$w = \text{Penetration of depletion region into the channel,}$$

Where $V = V_0 - V(x)$; V_0 is the contact potential at x and $V_{(n)}$ is the applied potential across space charge region at x and is negative for an applied potential. This potential (between D and S) is not uniform throughout the channel. The potential is higher near the drain and decreases towards the source. The depletion region is more towards the drain and decrease towards the source.

$$\therefore \quad V = V_0 - V(x)$$

The actual potential is the difference of V_0 and $V(x)$.

$$\in = \text{Dielectric constant of channel material}$$

$$w(x) = a - b(x) = \left\{ \frac{2 \in}{eN_D} \left(V_0 - V(x) \right) \right\}^{\frac{1}{2}} \qquad .....(1)$$

$$a - b(x) = \text{Penetration } w(x) \text{ of depletion region into channel at a point}$$
$$x \text{ along the channel from one side of gate region only.}$$

$$2a = \text{Total width between the two gate diffusions.}$$

$$2b(x) = \text{Depletion region width between the two sides of the gate}$$
$$\text{regions.}$$

If $I_D = 0$, $b(x)$ is independent of x and it will be uniform throughout the channel and will be equal to 'b'. V_0 will be much smaller than $V(x)$.

If in equation (1), we substitute $b(n) = b = 0$, neglect V_0, and solve for V we obtain the pinch off voltage. Because at pinch off, the two depletion regions from both sides meet each other, therefore the spacing between them $b(x) = 0$.

Therefore, w at pinch off, V_0 neglected, $b = 0$,

$$w = \left\{ \frac{2 \in}{eN_D} V_p \right\}^{\frac{1}{2}}$$

$$\text{or} \quad w^2 = \frac{2 \in}{eN_D} \cdot V_P$$

5.3   FET OPERATION

If a p-n junction is reverse biased, the majority carriers will move away from the junction. That is holes on the p-side will move away from the junction leaving negative charge or negative ions on the p-side (because each atom is deprived of a hole or an electron fills the hole. So it becomes negative by charge). Similarly, electrons on the n-side will move away from the junction leaving positive ions near the junction. Thus, there exists space charge on both sides of the junction in a reverse biased p-n junction diode. So, the electric field intensity, the lines of force originate from the positive charge region to the negative charge region. This is the source of voltage drop across the junction. As the reverse bias across the junction

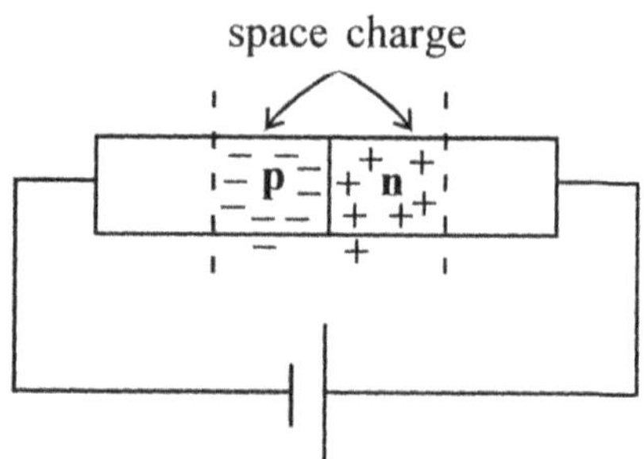

Fig 5.9 Space charge in p-n junction.

increases, the space charge region also increases, or the region of immobile uncovered charges increases (i.e., negative region on p-side and positive region on n-side increases as shown in Fig. 5.9). The conductivity of this region is usually zero or very small. Now in a FET, between gate and source, a reverse bias is applied. Therefore the channel width is controlled by the reverse bias applied between gate and source. Space charge region exists near the gate region on both sides. The space between them is the channel. If the reverse bias is increased, the channel width decreases. Therefore, for a fixed drain to source voltage the drain current will be a function of the reverse biasing voltage across the junction. Drain is at positive potential (for n-type FET). Therefore, electrons tend to move towards drain from the source (1) Because, source is at negative potential, they tend to move towards the drain. But because of the reverse bias applied to the gate, there is depletion region or negative charge region near the gate which restricts the number of electrons reaching the drain. Therefore the drain current also depends upon the reverse bias voltage across the gate junction. The term field effect is used to describe this device because the mechanism of current control is the effect of the extensions, with increase reverse bias of the field associated with region of uncovered charges.

The characteristics of n-channel FET between I_D, the drain current and V_{DS} the drain source voltage are as shown in Fig. 5.10, for different values of V_{GS}.

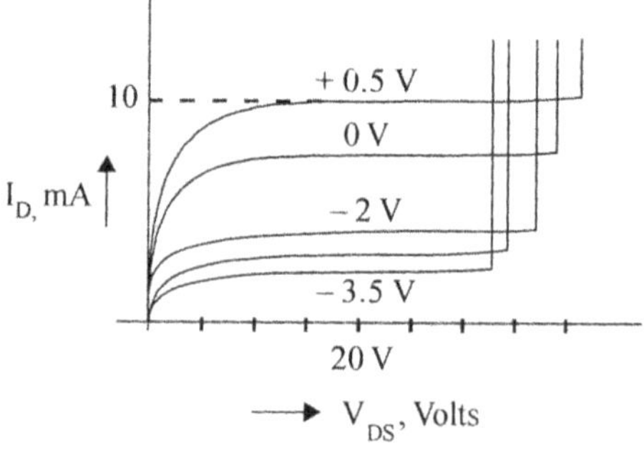

Fig 5.10 Drain charecteristics

To explain these characteristics, suppose $V_{GS} = 0$. When $V_{DS} = 0$, $I_D = 0$, because the channel is entirely open. When a small V_{DS} is applied (source is forward biased or negative voltage is applied to n-type source), the n-type bar acts as a simple semiconductor resistor and so I_D increases linearly with V_{DS}. With increasing current, the ohmic voltage drop between the source and the channel region reverse biases the junction and the conducting portion of the channel begins to constrict. Because, the source gets reverse biased or the negative potential at the n-type source reduces. Because of the ohmic drop, along the length of the channel it self, the constrictions is not uniform, but is more **pronounced** at distances farther from the source. i.e., the channel width is narrow at the **drain** and wide at the source. Finally, the current will remain constant at a particular value and the corresponding voltage at which the current begins to level off is called as the **pinch off voltage**. But the channel cannot be closedown completely and there by reducing the value of I_D to zero, because the required reverse bias will not be there.

Now if a gate voltage V_{GS} is applied in a direction to reverse bias the gate source junction, pinch off will occur for smaller values of (V_{DS}), and the maximum drain current will be smaller compared to when $V_{GS} = 0$. If V_{GS} is made $+ 0.5V$, the gate source junction is forward biased. But at this voltage, the gate current will be very small because for Si FET, $0.5V$ is just equal to or less than the cut in voltage. Therefore, the characteristic for $V_{GS} = + 0.5V$, I_D value will be comparatively larger, (compared to $V_{GS} = 0V$ or $- 0.5V$). Pinch off will occur early.

FET characteristics are similar to that of a pentode, in vacuum tubes.

The maximum voltage that can be applied between any two terminals of the FET is the lowest voltage that will cause avalanche breakdown, across the gate junction. From the FET characteristics it can be observed that, as the reverse bias voltage for the gate source junction is increased, avalanche breakdown occurs early or for a lower value of V_{DS}. This is because, the reverse bias gate source voltage adds to the drain voltage because drain though n -type (for n-channel FET), a positive voltage is applied to it. Therefore, the effective voltage across the gate junction is increased.

For n-channel FET, the gate is p-type, source and drain are n-type. The source should be forward-biased, so negative voltage is applied. Positive voltage is applied to the drain. **Gate source junction should be reverse biased,** and gate is p-type. Therefore,voltage or negative voltage is applied to the gate. Therefore, n-channel FET is exactly similar to a Vacuum Tube (Triode). Drain is similar to anode (at positive potential), source to cathode and gate to grid, (But the characteristics are similar to pentode).

For p-channel FET, gate is n-type, and positive voltage is applied, drain is at negative potential with respect to source.

Consider n-channel FET. The source and drain are n-type and p-type gate is diffused from both sides of the bar (See Fig. 5.11 below).

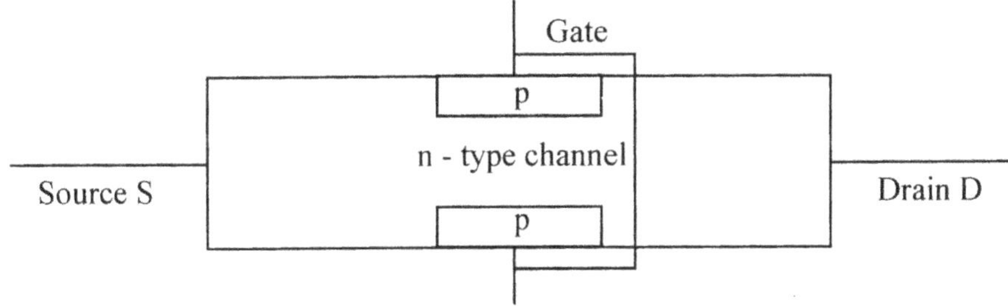

Fig 5.11 Structure of n-channel JFET.

Suppose source and gate are at ground potential and small positive voltage is applied to the drain. Source is n-type. So it is Forward biased. Because drain is at positive potential, electrons from the source will move towards the drain. Negligible current flows between source and gate or gate and drain, since these p-n junctions are reverse biased and so the current is due to minority carriers only. Because gate is heavily doped, the current between S and G or G and D can be neglected. The current flowing from source to drain I_D depends on the potential between source and drain, V_{DS}, resistance of the n-material in the channel between drain to source. This resistance is a function of the doping of the n-material and the channel width; length and thickness.

If V_{DS} is increased, the reverse bias voltage for the gate drain junction is increased, since, drain is n-type and positive voltage at drain is increased. This problem is similar to that of a reverse biased p-n junction diode.

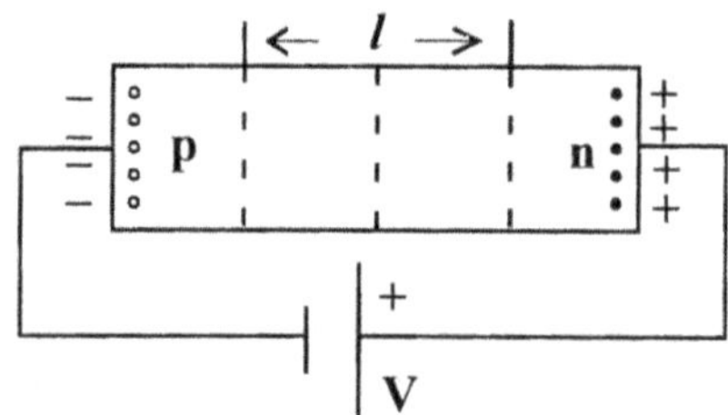

Fig 5.12 Reverse biased G - D junctions

Consider a p-n junction which is reverse biased. The holes on the p-side remain near the negative terminal and electrons on the n-side reach near the positive terminal as shown in Fig.5.12. There are no mobile charges near the junction. So we call this as the depletion region since it is depleted of mobile carriers or charges. As the reverse voltage is increased, the depletion region width 'l' increases.

This result is directly applicable for JFET. The depletion region extends more towards drain because points close to the drain are at higher positive voltage compared to points close to source. So the depletion region is not uniform $V_{DS} < V_{PO}$. But extends more towards drain than source. V_{P0} is the pinch off voltage.

As V_{DS} is increased, depletion region is increased (shaded portion). So channel width decreases and channel resistance increases. Therefore, the rate at which I_D increases reduces, eventhough there is positive potential for the electrons at the drain and hence they tend to move towards the drain and conventional current flows as shown by the arrow mark, in the Fig.5.13.

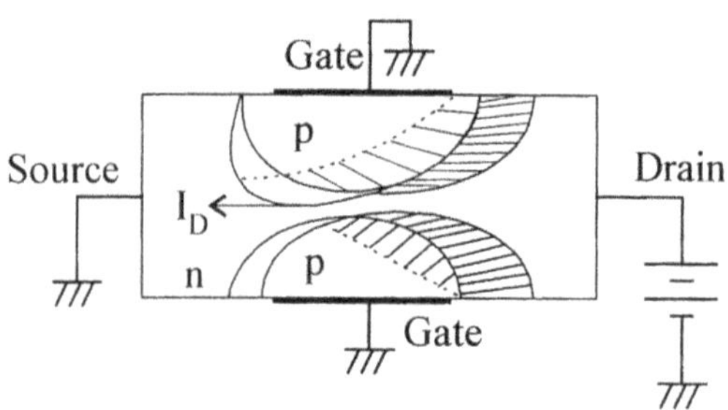

Fig 5.13 Spread of deplition regions

When $V_{DS} = V_{PO}$

When V_{DS} is further increased, the depletion region on each side of the channel join together as shown in Fig.5.14. ***The corresponding*** V_{DS} ***is called as*** V_{PD}, ***the pinch off voltage,*** because it pinches off the channel connection between drain and source.

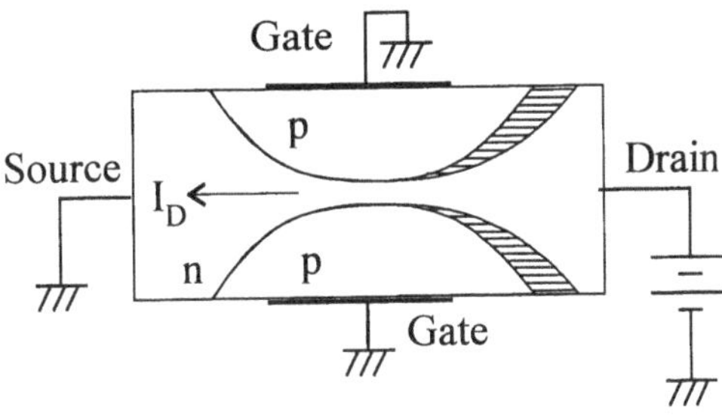

Fig. 5.14

When $V_{DS} > V_{PO}$

If V_{DS} is further increased, the depletion region thickens. So the resistivity of the channel increases. Because drain is at more positive potential, more electrons tend to move towards the drain. Hence I_D should increase. But, because channel resistance increases, I_D decreases. Therefore, the net result is I_D levels off, for any V_{DS} above V_{PO}.

$$I_D = \frac{V_{DS}}{r_{DS}}$$

In the linear region, r_{DS} is almost constant. So as V_{DS} is increased, I_D increases. When the two channels meet, as V_{DS} increases r_{DS} also increases. So I_D remains constant.(See Fig.5.15)

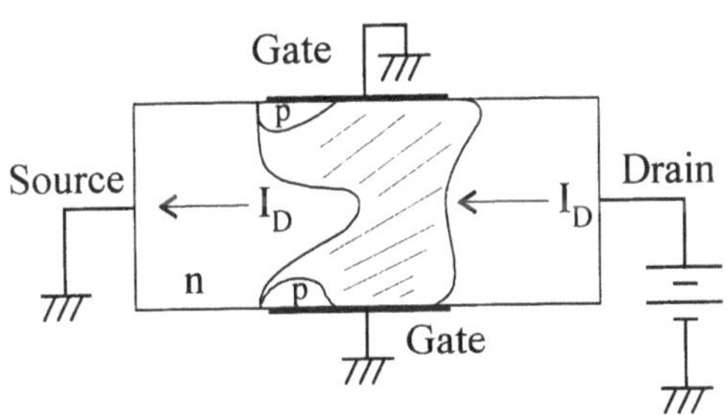

Fig 5.15 *Depletion region width variation in JFET.*

V.   V-I CHARACTERISTICS

5.4   JFET VOLT-AMPERE CHARACTERISTICS
Drian Characteristics

Suppose the applied voltage V_{DS} is small. The resulting small drain current I_D will not have appreciable effect on the channel profile. Therefore, the channel cross-section A can be assumed to be constant throughout, its length.

$$\therefore \qquad A = 2b.w,$$

Where 2b is the channel width corresponding to negligible drain current and w is channel dimension perpendicular to the 'b' direction.

$$\therefore \qquad I_D = Ae . N_D \, \mu_n \in$$
$$A = 2b . w$$

$$\epsilon = \frac{V_{DS}}{L}$$

$$= 2bwe \cdot N_D \mu_n \cdot \frac{V_{DS}}{L} \qquad \qquad .....(1)$$

Where L is the length of the channel. Eliminating 'b' which is unknown,

$$V_{GS} = \left(1 - \frac{b}{a}\right)^2 \cdot V_P$$

$$\left(1 - \frac{b}{a}\right) = \left(\frac{V_{GS}}{V_P}\right)^{1/2}$$

$$\frac{b}{a} = 1 - \left(\frac{V_{GS}}{V_P}\right)^{1/2}$$

$$b = a\left[1 - \left(\frac{V_{GS}}{V_P}\right)^{1/2}\right]$$

Substituting, this value in equation (1),

$$\boxed{I_D = \frac{2awe\,N_D\,\mu_n}{L}\left[1 - \left(\frac{V_{GS}}{V_P}\right)^{1/2}\right] V_{DS}}$$

This is the expression for I_D in terms of V_{GS}, V_P and V_{DS} because the value of b is not directly known.

$$V_P = \frac{e.N_D}{2\varepsilon} \cdot w^2$$

But, $w = a - b$ (x)

$b = 0$, at pinch off,

∴ $w = a =$ The spacing between the two gate dopings.

∴ $$\boxed{|V_P| = \frac{e.N_D}{2\,\epsilon} \cdot a^2}$$

V_{DS} controls the width of the depletion region $(a - b)$ because for a given V_{GS}, as V_{DS} increases, the reverse potential between drain and gate increase. Therefore, depletion region width increases.

EXPRESSION FOR V_{GS}

$$V = \frac{e.N_D x^2}{2\,\epsilon}$$

We can get the expression for V_{GS} if we replace x by (a - b) and V by V_{GS}.

∴ $$V_{GS} = \frac{e.N_D (a - b)^2}{2\,\epsilon}$$

But
$$|V_P| = \frac{e.N_D}{2\,\epsilon}.a^2$$

$\therefore$
$$\frac{e.N_D}{2\,\epsilon} = \frac{|V_p|}{a^2}$$

Substituting this in expression for V_{GS},

$$V_{GS} = \frac{V_p}{a^2}(a-b)^2$$

Channel width is controlled by V_{GS}. As V_{GS} increases, channel width reduces.)

Therefore, channel width is controlled by V_{GS} and depletion region width by V_{DS}.

$$\boxed{V_{GS} = \left(1 - \frac{b}{a}\right)^2 . V_p}$$

THE ON RESISTANCE r_{ds} (ON)

When V_{DS} is small, the FET behaves like an ohmic resistance whose value is determined by V_{GS}.

The ratio
$$\frac{V_{DS}}{I_D} = r_{ds(ON)}$$

$$I_D = 2bwe\,N_D\,\mu_n \; . \; \frac{V_{DS}}{L}$$

$\therefore$
$$\frac{V_{DS}}{I_D} = \frac{L}{2a\,we\,N_D\,\mu_n} = r_{ds(ON)}.$$

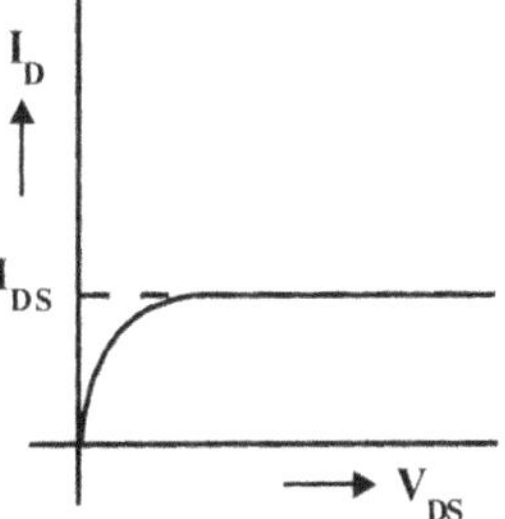

Fig 5.16 Drain characteristics

5.5 TRANSFER CHARACTERISTICS OF FET

$I_{DS} =$ The saturation value of the drain current or the value at which I_D remains constant.

$I_{DSS} =$ The saturation value of the drain current when gate is shorted to source that is $V_{GS} = 0$.

It is found that the transfer characteristics giving the relationship between I_{DS} and V_{GS} can be approximated by parabola.

$$I_{DS} = I_{DSS}\left(1 - \frac{V_{GS}}{V_p}\right)^2$$

Transfer characteristic follows the equation, I_D vs V_{GS}

$$I_D = (I_{DSS} + I_{GSS})\left(1 - \frac{V_{GS}}{V_p}\right)^2 - I_{GSS}$$

With negligible I_{GSS}, because I_{GS} is the leakage gate current between G and S when D is shorted, to S will be of the order of nano amperes, so it can be neglected.

$$\therefore \qquad I_D = I_{DSS}\left(1 - \frac{V_{GS}}{V_p}\right)^2$$

This V_p is $V_{GS(off)}$ and not V_{DS} because when $V_p = V_{GS(off)}, I_D = 0$.

I_{DSS} = The saturation value of drain current when gate is shorted to source or $V_{GS} = 0$.

The transfer characteristics can be derived from the drain characteristics.

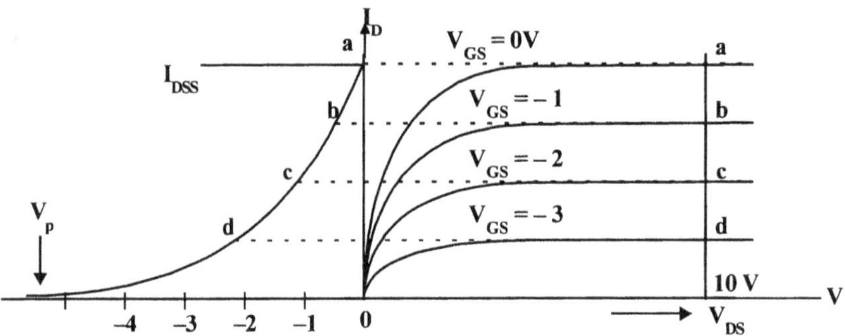

Fig 5.17 Drain and Gate characteristics of JFET.

To construct transfer characteristics, a constant value of V_{DS} is selected. Normally this is chosen in the saturation region, because the characteristics are flat. The intersection of the vertical line abcd gives a particular value of I_D for different values of V_{GS}. So the transfer characteristics between I_D and V_{GS} can be drawn.

If the end points V_p and I_{DSS} are known, a transfer characteristics for the device can be constructed.

Here V_p is again called the ***pinch off voltage*** or $V_{GS(off)}$ voltage. This is the voltage between gate and source for which I_D becomes zero. As the reverse potential between G and S increases, depletion region width increases. So channel width decreases. For some value of V_{GS} channel pinches off or I_D practically becomes zero (It cannot be exactly zero but few nano - amps). So this gate source voltage at which I_D becomes zero is also called as pinch off voltage or $V_{GS(off)}$ voltage. This pinch off voltage is different from V_{DS} voltage at which I_D levels off. Since in the latter case I_D is not zero, but channel width becomes zero and channel pinches off. This channel width is made zero either by controlling V_{DS} or V_{GS}. Hence there are two types of pinch of voltages. The specified pinch off voltage V_p for a given FET can be known if it is due to V_{DS} or V_{GS} from the polarity of the voltage V_p. For n-channel FET, V_{GS} is negative (since gate is p-type, G-S junction is reverse biased) V_{DS} is positive. If the specified V_p is –5V, (say) that it is due to $V_{GS(off)}$ because V_p is negative. If V_p is positive, it is due to V_{DS} at which I_{DSS} levels off.

5.5.1 FET Biasing for Zero Drift Current

As Temperature (T) increases, Mobility (μ) decreases at constant Electric Field (ε), therefore I_D increases. As T increases depletion region width decreases. So conductivity σ of the channel increases, I_D also increases.

I_D decreases by 0.7 % per degree centigrade.

I_D increase at a rate of 2.2 mV/^{0}C change in V_{GS}. Therefore change in V_{GS} = 2.2 mV/^{0}C.

∴   For zero drift current,

$$0.007\, |I_D| = 0.0022\, g_m$$

or
$$\frac{|I_D|}{g_m} = 0.314 \text{ V}$$

or
$$\boxed{|V_p| - |V_{GS}| = 0.63\text{V}}$$

$$I_{DS} = I_{DSS}\left(1 - \frac{V_{GS}}{V_P}\right)^2$$

$$g_m = g_{mo}\left(1 - \frac{V_{GS}}{V_p}\right)$$

$$g_{mo} = \frac{-2I_{DSS}}{V_p}$$

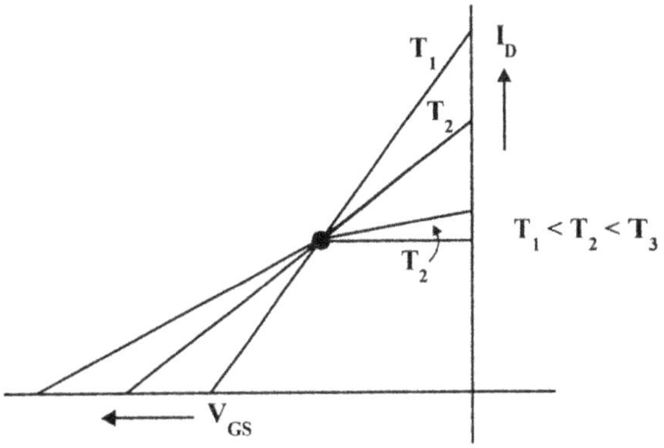

Fig 5.18 Biasing for zero drift current.

$T_1 < T_2 < T_3$

5.1 Problem

Design the bias circuit for zero drain current drift, for a FET having the following particulars :

$$V_p = -3V \; ; \; g_{mo} = 1.8 \text{ mA/V}$$
$$I_{DSS} = 1.75 \text{ mA, if } R_d = 5k\Omega$$

(a) Find I_D for zero drift current.

(b) V_{GS}

(c) R_S

(d) Voltage gain with R_s bypassed with a large capacitance,

Solution

For zero drift current,

$$|V_P| - |V_{GS}| = 0.63$$
$$V_p = -3V$$
$$V_{GS} = V_P - 0.63$$

(a)
$$I_D = I_{DSS}\left(1 - \frac{V_{GS}}{V_p}\right)^2 \simeq 0.08 \text{ mA}$$

(b) ∴
$$V_{GS} = -3 - 0.63 = -3.63 \text{ V}$$

(c)
$$R_S = \frac{-V_{GS}}{I_D} = \frac{3.63}{I_D} = \frac{3.63}{0.08 \text{ mA}} = 43 \text{ K}\Omega$$

(d)
$$g_m = g_{mo}\left|\frac{V_{GS} - V_p}{V_p}\right| = 0.378 \text{ mA/V}$$

$$A_v = g_m \cdot R_d = 1.89$$

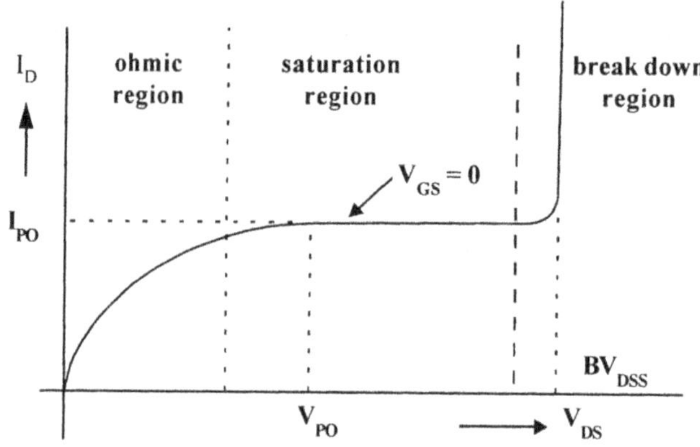

Fig 5.19 Drain characteristics.

The drain current I_D increases rapidly as V_{DS} increases, towards V_{PO}. Above V_{PO}. The current tends to level off at I_{PO} and then rises slowly. When V_{GS} = Breakdown voltage BV_{DSS}, breakdown (avalanche) occurs and the current rises rapidly. Current levels off because as V increases, resistance of the channel also rises. Therefore, I remains constant. As V

increases, R increases, therefore V/R = constant. The region before V_{PO} is called ***ohmic region*** because ohms law is obeyed (see Fig.5.41). Just as a p-n junction breakdown when reverse voltage is very high FET also break down if V_{DS} is high due to high electric field.

Suppose V_{DS} is fixed and V_{Gs} is varied. As V_{GS} is made negative, p-n junction is reverse biased and depletion region between gate and source increases. This decreases channel width and increases channel resistance. Hence I_D decreases. If gate voltage is made positive, depletion region decreases and hence I_D increases. The p-n junction, between gate and source becomes forward biased and current flows from gate to source. So n-type JFET is usually operated so that $V_{GS} = 0$ or negative.

V_{P0}	:	Pinch off voltage when $V_{GS} = 0$.
		0 indicate $V_{GS} = 0$.
I_{DSS} (I_{P0})	:	Saturation value of the drain current when gate is shorted to source ($V_{GS} = 0$).
I_{DSS}	:	Saturation value of the drain current for $V_{GS} \neq 0. -1, -2$ etc.
BV_{DSS}	:	Breakdown voltage V_{DS}, when gate is shorted to source.
$I_{D(off)}$	:	Drain current when the JFET is in OFF state.
IGSS	:	Gate cut-off current when Drain is shorted to source.
g_{mo}	:	Mutual conductance g_m when $V_{GS} = 0$.

The expression for the pinch off voltage $|V_P| = \dfrac{\in . N_D}{2\varepsilon} . a^2$

where N_D = Donor atom concentration

$\in$ = Permitivity $\in_o - \in_r$

a = Effective width of Silicon bar

5.2 Problem

For a p - channel Silicon FET, with a $= 2 \times 10^{-4}$ cm and channel resistivity $\rho = 10\,\Omega$ - cm. Find the pinch off voltage.

Solution

$$V_P = \frac{e.N_A}{2\in}.a^2 \quad \text{(for p - channel FET)}$$

For Si,

$$\in = 12\,e_o\,\mu_p = 500\ cm^2/v - sec.$$

$$\sigma = \frac{1}{\rho} \simeq p\mu_p \cdot e \simeq N_A\,\mu_p \cdot e$$

or

$$e.N_A = \frac{1}{\rho\mu_p}$$

$$= \frac{1}{10 \times 500}$$

or

$$e.N_A = 2 \times 10^{-4}$$

$\therefore$

$$V_p = \frac{2\times10^{-4}\left(2\times10^{-4}\right)^2}{2\times12\left(36\pi\times10^{11}\right)^{-1}} = 3.78V$$

5.5.2 CUT OFF

A FET is said to be cut off when the gate to source voltage $|V_{GS}|$ is greater than the pinch off voltage V_P. $|V_{GS}| > |V_p|$. Under these conditions, because for n - channel FET, gate is p - type and source is n - type, there will be some current called ***gate reverse current*** or ***gate cut off current***, designed as I_{GS}, this is when drain is shorted to source. There exists some drain current under these conditions also called $I_{D(off)}$.

I_D (off) and I_{GSS} will be in the range 0.1amps to few nano amps at 25^0C, and increase by a factor of 1000 at a temperature of 150^0C. I_{GSS} is due to the minority carries of the reverse biased gate. Source junction. $I_{D(off)}$ is due to the carriers reaching drain from source because channel resistivity will not become infinity. Because drain is at positive potential for n-channel FET, more electrons will reach the drain.

5.5.3 APPLICATIONS

1. JFETs make good digital and analog switches. off resistance is very high.

2. They are used in special purpose amplifiers such as very high input resistance amplifiers, buffers voltage follower etc.

3. It can be used as a voltage controlled resistance because resistivity of the channel varies with V_{DS}.

5.6 FET SMALL SIGNAL MODEL

The equivalent circuit for FET can be drawn exactly in the same manner as that for a vacuum tube. The drain current I_D is a function of gate voltage V_{GS} and drain voltage (V_{DS})

$$I_D = f(V_{GS} \cdot V_{DS}) \qquad\qquad .....(1)$$

Expanding (1) by Taylor's series,

$$\Delta I_D = \left.\frac{\partial I_D}{\partial V_{GS}}\right|_{V_{DS}} \Delta_{GS} + \left.\frac{\partial I_D}{\partial V_{DS}}\right|_{V_{GS}} \cdot \Delta V_{DS}$$

In the small signal notation the incremental values can be assumed to be a.c quantities.

$$\Delta I_D = I_d \; ; \; \Delta V_{GS} = V_{gs} \; ; \; \Delta V_{DS} = V_{ds}$$

$$I_d = g_m \cdot V_{gs} + \frac{1}{r_d} \cdot V_{ds}$$

where,
$$g_m = \left.\frac{\partial I_D}{\partial V_{GS}}\right|_{V_{DS}}$$

or
$$\left.\frac{\Delta I_D}{\Delta V_{GS}}\right|_{V_{DS}} = \left.\frac{I_d}{V_{gs}}\right|_{V_{DS}}$$

The ***mutual conductance*** or transconductance.

Sometimes it is designated as y_{f_s} or g_{f_s} the second subscript 's' indicating common source and y or g for forwards transadmittance or conductance respectively.

5.6.1 DRAIN RESISTANCE OR OUTPUT RESISTANCE r_D

$$r_D = \left.\frac{\partial V_{DS}}{\partial I_D}\right|_{V_{GS}} \simeq \left.\frac{\partial V_{DS}}{\partial I_D}\right|_{V_{GS}} = \left.\frac{V_{ds}}{I_d}\right|_{V_{GS}}$$

$$g_D = \frac{1}{r_D}$$

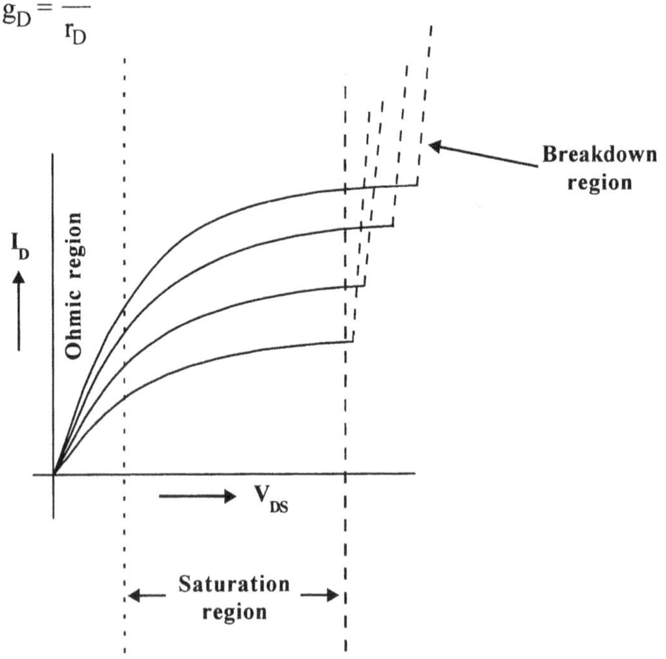

Fig 5.20 Drain characteristics of JFET.

Drain conductance = y_{DS} = output conductance for common source.

g_m of FET is analog to g_m of vacuum tube.

r_D of FET is analog to r_p of vacuum tube.

$$\mu \text{ for FET} = \left.\frac{\partial V_{DS}}{\partial V_{GS}}\right|_{I_D} = \left.\frac{\Delta V_{DS}}{\Delta V_{GS}}\right|_{I_D} = \left.\frac{V_{ds}}{V_{gs}}\right|_{I_D}$$

TYPICAL VALUES

$$g_m = 0.1 - 10 \text{ mA/V}$$
$$r_d = 0.1 - 1 \text{ M}\Omega$$
$$\mu = r_d \cdot g_m$$
$$C_{gs} = 1 - 10 \text{ pf}$$
$$C_{ds} = 0.1 - 10 \text{ pf}$$

5.6.2 EQUIVALENT CIRCUIT FOR FET

The expression for drain current I_d in the case of a FET is

$$I_d = g_m V_{gs} + \frac{1}{r_d} \cdot V_{ds}$$

This is similar to the expression I_p of a triode. So the equivalent circuit can be drawn as a current source in parallel with a resistance. Between gate and source the capacitance is C_{gs} and C_{gd} is the barrier capacitance between gate and drain. C_{ds} represents the drain to source capacitance of the channel. (See Fig. 5.21).

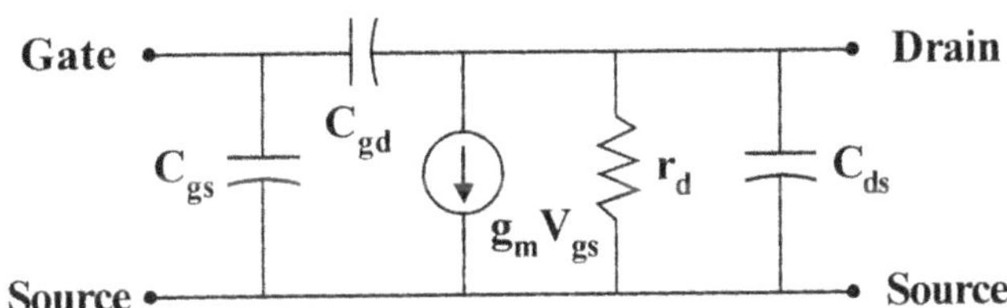

Fig 5.21   JFET equivalent circuit

This is high frequency equivalent circuit because we are considering capacitances. In low frequency circuit, X_C will be very large so C is open circuit and neglected.

FET is a voltage controlled device; by controlling the channel width by voltage, we are controlling the current. If we neglect the capacitances the equivalent circuit of FET can be drawn as given in Fig. 5.22. This is similar of a transistor *h-parameter* equivalent circuit.

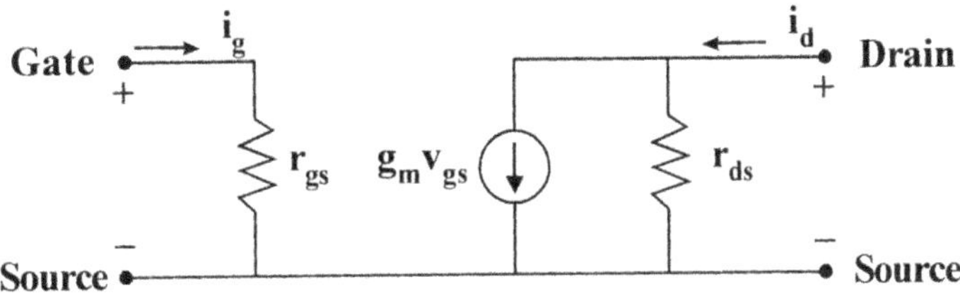

Fig 5.22   Simplified equivalent circuit

Arrow mark is downwards in the current source $g_m V_{gs}$ because the current (conventional current) is flowing from drain to source for n-channel FET (Because electrons are the majority carriers).

5.3 Problem

For a p-channel silicon FET, with $a = 2 \times 10^{-4}$ cm (effective channel width) and channel resistivity $\rho = 10\Omega - cm$. Find the pinch off voltage.

Solution

$$V_p = \frac{e.N_A}{2 \in} .a^2 \text{ for p-channel FET.}$$

For Si, $\qquad \in = 12 \in_0 \qquad\qquad \mu_p = 500 \text{ cm}^2/\text{v - sec.}$

$$\sigma = \frac{1}{\rho} = p\,\mu_p . e = N_A\, \mu_p . e$$

or $\qquad e.N_A = \dfrac{1}{\rho\mu_p} = \dfrac{1}{10 \times 500}$

or $\qquad e.N_A = 2 \times 10^{-4}$

$\therefore \qquad V_p = \dfrac{2 \times 10^{-4} \left(2 \times 10^{-4}\right)^2}{2 \times 12\left(36\pi \times 10^{11}\right)^{-1}} = 3.78 \text{ V}$

5.4   Problem

For an n-channel silicon FET, with $a = 3 \times 10^{-4}$ cm and $N_D = 10^{15}$ electrons/cm^3. Find (a) The pinch off voltage. (b) The channel half width for $V_{GS} = \dfrac{1}{2} V_P$ and $I_D = 0$.

Solution

(a) $\qquad V_p = \dfrac{e.N_D}{2 \in} .a^2$

$$= \frac{1.6 \times 10^{-19} \times 10^{15} \times \left(3 \times 10^{-4}\right)^2}{2 \times 12 \times \left(36\Pi \times 10^{11}\right)^{-1}}$$

$$= 6.8 V$$

(b) $\qquad b = ? \text{ for } \quad V_{GS} = \dfrac{V_P}{2},$

$$b = a\left[1 - \left(\frac{V_{GS}}{V_P}\right)^{\frac{1}{2}}\right]$$

$$= (3 \times 10^{-4}) \left(1 - \left(\frac{1}{2}\right)^{\frac{1}{2}}\right)$$

$$= 0.87 \times 10^{-4} \text{cm}$$

5.5 Problem

Show that for a JFET,

$$g_m = \frac{2}{|V_p|}\sqrt{I_{DSS}.I_{DS}}$$

Solution

$$I_{DS} = I_{DSS}\left(1 - \frac{V_{GS}}{V_p}\right)^2$$

$$g_m = \frac{\partial I_d}{\partial V_{GS}}$$

$$= -\frac{2I_{DSS}}{V_p}\left(1 - \frac{V_{GS}}{V_p}\right)$$

$$= g_m\left(1 - \frac{V_{GS}}{V_p}\right)$$

where g_{mo} is the transistor conductance g_m for $V_{GS} = 0$.

$$g_{mo} = -\frac{2I_{DSS}}{V_P}$$

$$\therefore \quad g_m = -\frac{2I_{DSS}}{V_P}\left(1 - \frac{V_{GS}}{V_P}\right)$$

But

$$I_{DS} = I_{DSS}\left(1 - \frac{V_{GS}}{V_P}\right)^2$$

or

$$\left(1 - \frac{V_{GS}}{V_P}\right) = \sqrt{\frac{I_{DS}}{I_{DSS}}}$$

$$\therefore \quad g_m = -\frac{2I_{DSS}}{V_p} \cdot \sqrt{\frac{I_{DS}}{I_{DSS}}}$$

$$= \frac{2}{|V_p|}\sqrt{I_{DSS}.I_{DS}}$$

5.6 Problem

Show that for small values of V_{GS}, compared with V_P, the drain current is $I_D \simeq I_{DSS} + g_{mo}V_{GS}$.

Solution

$$I_D \simeq I_{DSS} + g_{mo}V_{GS}$$

$$I_D = I_{DSS}\left(1 - \frac{V_{GS}}{V_P}\right)^2$$

$$= I_{DSS}\left(1 - \frac{2V_{GS}}{V_P} + \frac{V_{GS}^2}{V_P^2}\right)$$

$$\simeq I_{DSS}\left(1 - 2\frac{V_{GS}}{V_P}\right)V_{GS}$$

$$\simeq I_{DSS} - \left(\frac{2I_{DSS}}{V_P}\right)$$

$$I_D = I_{DSS} + g_{mo}\,V_{GS}\left[\because g_{mo} = \frac{-2I_{DSS}}{V_P}\right]$$

MOSFETs

5.7   FET TREE

The input Z of JFET is much higher compared to a junction transistor. But for MOSFETs, the input Z is much higher compared to JFETs. So these are very widely used in ICs and are fast replacing JFETs.

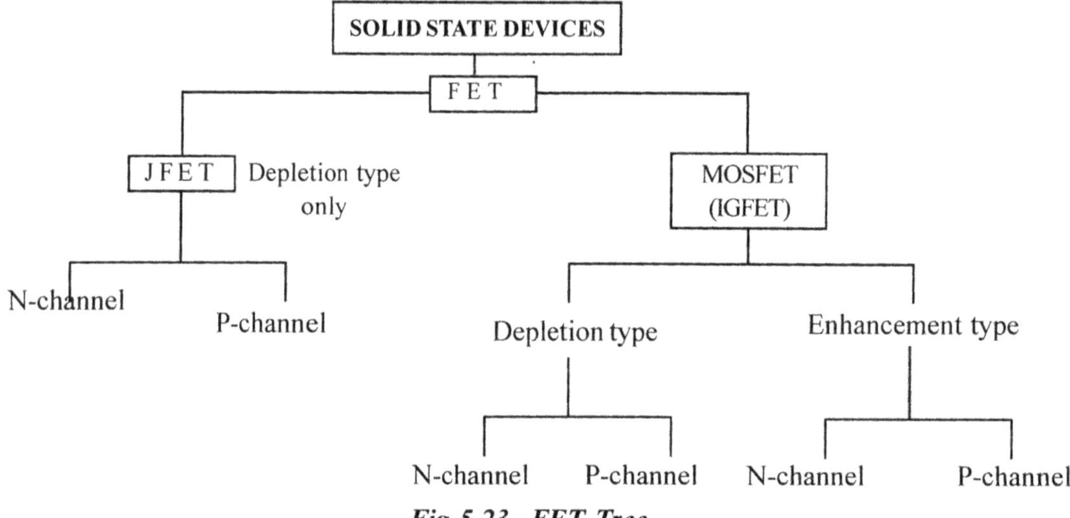

Fig 5.23   FET Tree.

There are two types of MOSFETs, enhancement mode and depletion mode. These devices, derive their name from Metal Oxide Semiconductor Field Effect Transistor (MOSFET) or MOS transistor (See Fig. 5.24). These are also known as IGFET (Insulated Gate Field Effect Transistor). The gate is a metal and is insulated from the semiconductor (source and drain are semiconductor type) by a thin oxide layer.

In FETs the gate source junctions is a p-n junction. (If the source is n-type, gate will be p-type). But in MOSFETs, there is no p-n junction between gate and channel, but rather a capacitor consisting of metal gate contact, a dielectric of SiO_2 and the semiconductor channel. [two conductors separated by dielectric]. It is this construction which accounts for the very large input resistance of 10^{10} to $10^{15}\Omega$ and is the major difference from the JFET.

The symbols for MOSFETs are,

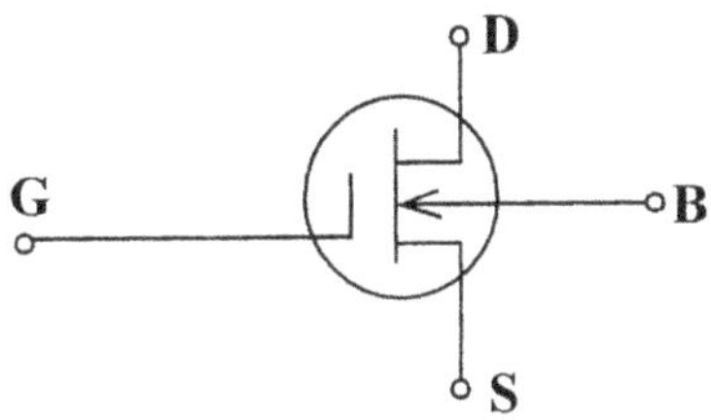

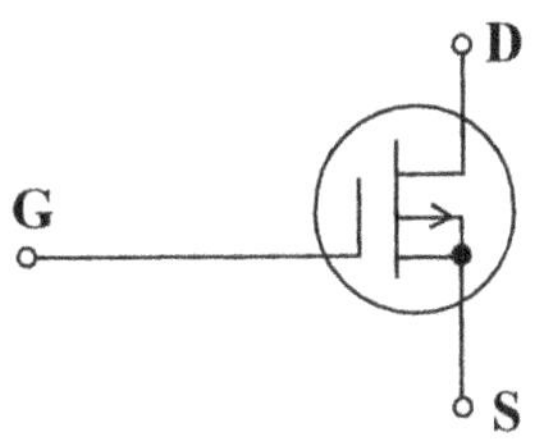

n-channel MOSFET, depletion type   *p-channel MOSFET, depletion type*
Fig 5.24   **Fig 5.25**

Some manufacturers internally connect the bulk to the source. But in some circuits, these two are to be separated (See Fig. 5.24). The symbol, if they are connected together is, given in Fig.5.25.

for p - channel, the arrow will point outwards.

Bulk is the substrate material taken.

Advantages of MOSFETs (over JFETs and other devices)

1. High package density $\sim 10^5$ components per square cm.
2. High fabrication yield. p - channel Enhancement mode devices require 1 diffusion and 4 photo masking steps.

   But for bipolar devices, 4 diffusions and $8 - 10$ photo masking steps are required.
3. Input impedance is very high $Z_n \sim 10^{14}\Omega$.
4. Inherent memory storage : charge in gate capacitor can be used to hold enhancement mode devices ON.
5. CMOS or NMOS reduces power dissipation, micropower operation. Hence, CMOS ICs are popular.
6. Can be used as passive or active element.

   Active : As a storage device or as an amplifier etc.

   Passive : As a resistance or voltage variable resistance.
7. Self Isolation : Electrical isolation occurs between MOSFETs in ICs since all p-n junctions are operated under zero or reverse bias.

Disadvantage

1. Slow speed switching.
2. Slower than bipolar devices.
3. Stray and gate capacitance limits speed.

Construction

5.7.1 Enhancement type MOSFET

n-channel is induced, between source and drain. So it is called as n-channel MOSFET.

For the MOSFET shown in Fig. 5.26, source and drain are n-type. Gate is A*l* (metal) in between oxide layer is there. The source and drain are separated by 1 mil. Suppose the p substrate is grounded, and a positive voltage is applied at the gate. Because of this an electric field will be directed perpendicularly through the oxide. This field will induce negative charges on the semiconductor side. The negative charge of electrons which are the minority carriers in the p - type substrate form an inversion layer. As the positive voltage on the gate increases, the induced negative charge in the semiconductor increases. The region below the SiO_2 oxide layer has n-type carriers. So the conductivity of the channel between source and drain increases and so current flows from source to drain through the induced channel. Thus, the drain current is enhanced by the positive gate voltage and such a device is called an ***enhancement type MOS***.

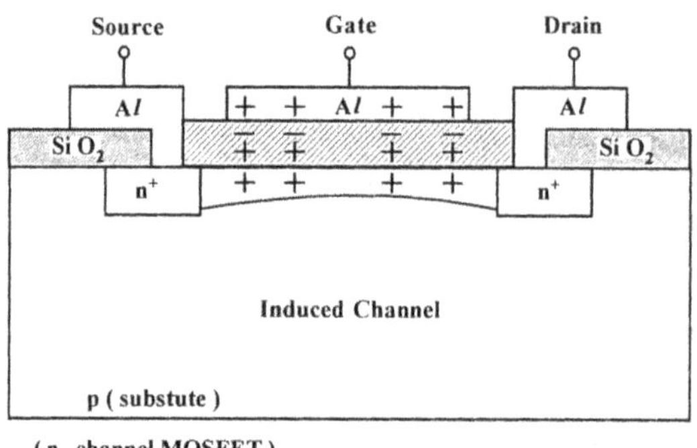

Fig 5.26   n-channel MOSFET

The volt-ampere drain characteristic of an n-channel enhancement mode MOSFET are as shown in Fig. 5.27.

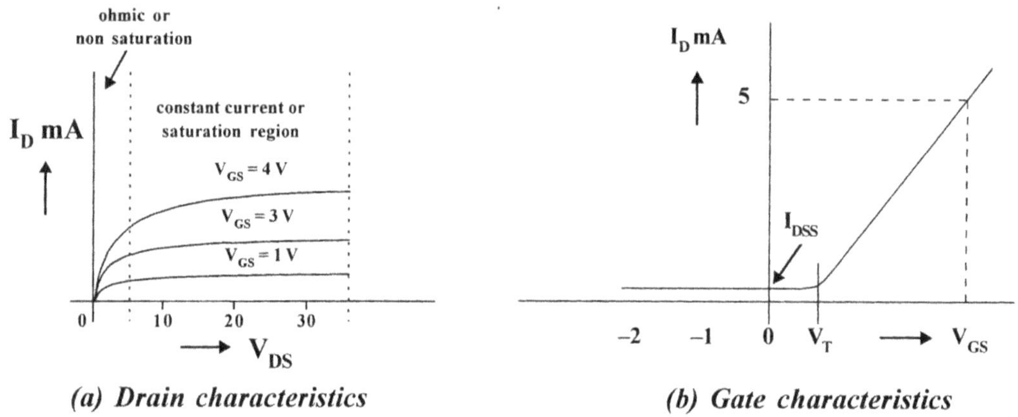

(a) Drain characteristics   *(b) Gate characteristics*

Fig 5.27

The current I_D for $V_{GS} \leq 0$ is very small being of the order of few nano amperes. As V_{GS} is made positive, the current I_D increases, slowly first and then much more rapidly with an increase in V_{GS} (Fig. 5.27(b)). Sometimes the manufacturers specify gate, source threshold voltage V_{GST} at which I_D reaches some defined small value $\sim 10\mu A$. For a voltage $< V_{GST}$, (V_{GS} threshold) I_D is very small. $I_{D(ON)}$ of MOSFET is the maximum value of I_D which remains constant for different values of V_{DS}.

In the ohmic region, the drain characteristic is given by

$$I_D = \frac{\mu C_0 . w}{2L} [2(V_{GS} - V_T) V_{DS} - V^2_{DS}]$$

where, μ = Majority carriers mobility.

C_0 = Gate capacitance per unit area.

L = Channel length.

w = Channel width perpendicular to C

The ohmic region in the I_D vs V_{DS} characteristic of FET or MOSFET is also known as *triode region*, because like in a triode I_D increases with V_{DS}. The constant current region is known as *pentode region* because the current remains constant with V_{DS}.

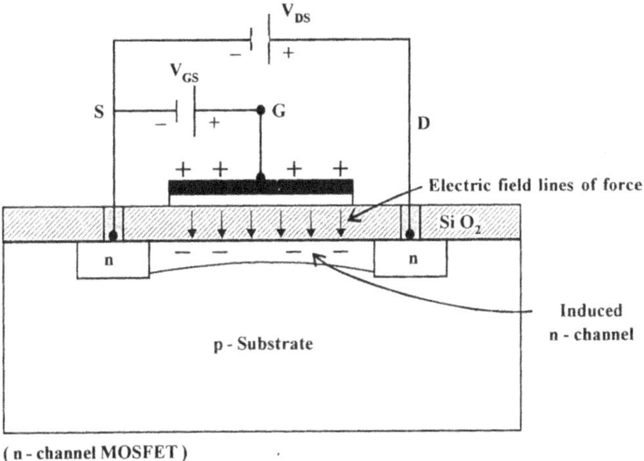

Fig 5.28 MOSFET structure

5.7.2 MOSFET OPERATION

Consider a MOSFET in which source and drain are of n-type. Suppose a negative potential is applied between gate and source. SiO_2 is an insulator. It is sandwiched between two conducting regions the gate (metal) and the S.C p-type substrate. Therefore, equivalent capacitor is formed, with SiO_2 as the dielectric whenever a positive charge is applied to one plate of a capacitor a corresponding negative charge is induced on the opposite plate by the action of the electric field with in the dielectric. A positive potential is applied to the gate. So a negative charge is induced on the opposite plate by the action of electric field within the dielectric, in the p - substrate. This charge results from the minority carriers (electrons) which are attracted towards the area below the gate, in the p - type substrate. The electrons in the p - substrate are attracted towards the lower region of SiO_2 layer because there is positive field acting from the gate through SiO_2 layer, because of the applied positive potential to the gate V_{GS}. As more number of electrons are attracted towards this region, the hole density in the p - substrate (below the SiO_2 layer between n - type source and drain only) decrease. This is true only in the relatively small region of the substrate directly below the gate. An n - type region now extends continuously from source to drain. The n-channel below the gate is said to be induced because it is produced by the process of electric inductor. If the positive gate potential is removed, the induced channel disappears.

If the gate voltage is further increased, greater number of negative charge carriers are attracted towards the induced channel. So as the carriers density increases, the effective channel resistance decreases, because there are large number of free carriers (electrons). So the resistance seen by the V_{DS} depends on the voltage applied to the gate. The higher the gate potential, the lower the channel resistance, and the higher drain current I_D. This process is referred to as enhancement because I_D increases, and the resulting MOSFET is called ***enhancement type MOSFET***. The resistance looking into the gate is high since the oxide is an insulator. The resistance will be of the order of $10^{15}\Omega$ and the capacitance value will also be large.

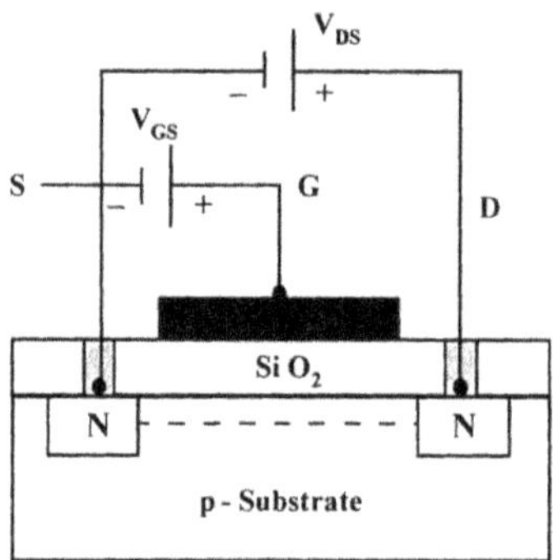

Fig 5.29 MOSFET biasing.

Depletion type MOSFET can also be constructed in the same way. These are also known as DE MOSFETs. Here as the gate voltage increases, the channel is depleted of carriers. So channel resistance increases.

Here the region below the gate is doped n - type. If negative voltage is applied to the gate, negative charge on the gate induces equal positive charge i.e., holes. These holes will recombine with the electrons of the n - channel between sources drain since channel resistivity increases. The channel is depleted of carriers. Therefore I_D decreases as V_{GS} increases. (negative voltage) If we apply positive voltage to V_{GS}, then this becomes enhancement type.

5.7.3 MOSFET CHARACTERISTICS

There are two types of MOS FETs,

>   1. Enhancement type
>   2. Depletion type.

Depletion type MOSFET can also be used as enhancement type. But enhancement type cannot be used as depletion type. So to distinguish these two, the name is given as depletion type MOSFET which can also be used in enhancement mode.

As V_{GS} is increased in positive values (0, +1, +2 etc) if I_D increases, then it is enhancement mode of operation because I_D is enhanced or increased.

As V_{GS} is increased in negative values (0, −1, −2, etc) if I_D decreases, then it is depletion mode of operation.

Consider n - channel MOSFET, depletion type. Fig. 5.52.

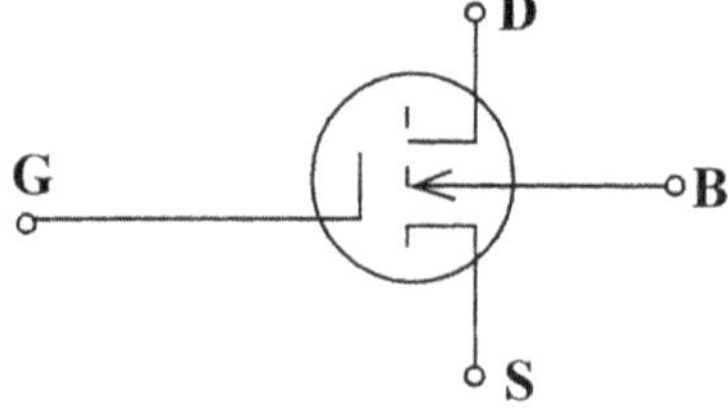

Fig 5.30 n - channel MOSFET.

DRAIN CHARACTERISTIC : I_D versus V_{DS}

When V_{GS} = OV, for a given V_{DS}, significant current flows just like in a JFET. When the gate is made negative (i.e., V_{GS} = –1V,) it is as if a negative voltage is applied for one plate of plate capacitor. So a positive charge will be induced below the gate in between the n type source and drain. Because it is semiconductor electrons and holes are induced in it below the gate. The channel between the source and drain will be depleted of majority carriers electrons, because these induced holes will recombine with the electrons. Hence, the free electron concentration in the channel between source and drain decreases or channel resistivity increases. Therefore current decreases as V_{GS} is made more negative i.e., –1, –2, etc. This is the depletion mode of operation, because the channel will be depleted of majority carriers as V_{GS} is made negative (for n-channel MOSFET).

In a FET, there is a p - n junction between gate and source. But in MOSFET, there is no such p - n junction. Therefore, positive voltage positive V_{GS} can also be applied between gate and source. Now a negative charge is induced in the channel, thereby increase free electron

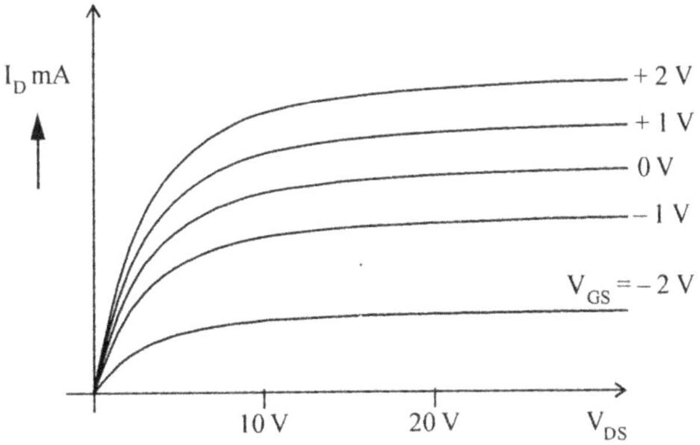

Fig 5.31 Drain characteristics (N MOS).

concentration. So channel conductivity increases and hence I_D increases. Thus, the current is enhanced. So, the device can be used both in enhancement mode and depletion mode.

In the transfer characteristic as V_{GS} increases, V_D increases. Similarly in the depletion mode as V_{GS} is increases in negative values, I_D decreases

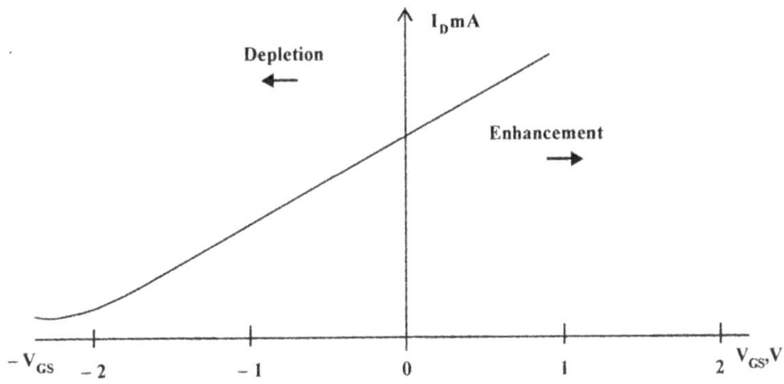

Fig 5.32 Gate characteristics

A JFET is a depletion type, because as V_{GS} is made positive (+1, +2 etc for n-channel FET) I_D increases. So it can be used in enhancement type. But there is no other type of JFET which is used only in enhancement mode and no depletion mode, hence no distinction is made.

5.7.4 MOSFET GATE PROTECTION

The SiO_2 layer of gate is extremely thin, it will be easily damaged by excessive voltage. If the gate is left open circuited, the electric field will be large enough (due to accumulation of charge) to cause punch through in the SiO_2 layer or the dielectric. To prevent this damage, some MOS devices are fabricated with zener diode between gate and substrate. When the potential at the gate is large, the zener will conduct and the potential at the gate will be limited to the zener breakdown voltage. When the potential at the gate is not large, the zener is open circuited and has no effect on the device.

If the body (bulk) of MOS transistor (MOSFET) is p - type Silicon, and if two 'n' regions separated by the channel length are diffused into the substrate to form source and drain n-channel enhancement device (designated as NMOS) is obtained. In NMOS, the induced mobile channel surface charges would be *electrons*.

Similarly, if we take n -type substrate and diffuse two p - regions separated by the channel length to form source and drain, PMOSFET will be formed. Here in the induced mobile channel surface charges would be *holes*.

5.7.5 COMPARISON OF p-CHANNEL AND n-CHANNEL MOSFETs

Initially there were some fabrication difficulties with n-channel MOSFETs. But in 1974, these difficulties were overcame and mass productions of n-channel MOSFETs began. Thus NMOSFETs have replaced PMOSs and PMOSFETs have almost become obsolete.

The hole mobility in Silicon at normal fields is 500 cm^2/v.sec. Electron mobility = 1300cm^2/v-Sec. Therefore p-channel ON resistance will be twice that of n-channel MOSFET ON resistance (ON resistance means the resistance of the device when I_D is maximum for a given V_{DS}) ON resistance depends upon 'μ' of carriers because $\sigma = \dfrac{1}{\rho} = ne\mu_n$ or $pe\mu_p$.

If the ON resistance of a p-channel device were to be reduced or to make equal to that of n - channel device, at the same values of I_D and V_{DS} etc, then the p-channel device must have more than twice the area of the n-channel device. Therefore n - channel devices will be smaller or packing density of n-channel devices is more ($R = \dfrac{\rho l}{A}$ .R is decreased by increase in A).

The second advantage of the NMOS devices is fast switching. The operating speed is limited by the internal RC time constant of the device. The capacitance is proportional to the junction cross sections.

The third advantage is NMOS devices are TTL compatible since the applied gate voltage and drain supply are positive for an n - channel enhancement MOS.

(Because in n-channel MOS, source and drain are n-type. So drain is made positive. For enhancement type the gate which is Al metal is made positive).

5.7.6 ADVANTAGES OF NMOS OVER PMOS

1. NMOS devices are fast switching devices since electron mobility is less than holes.

2. NMOS devices are TTL compatible since V_{GS} and V_D to be applied for NMOS devices are positive

3. Packing density of NMOS devices is more

4. The ON resistance is less because conductivity of NMOS devices is more since μ of electrons is greater than that of holes.

5.8 THE DEPLETION MOSFET

In enhancement type MOSFET, a channel is not diffused. It is of the same type (p-type or n-type) as the bulk or substrate. But if a channel is diffused between source and drain with the same type of impurity as used for the source and drain diffusion depletion type MOSFET will be formed.

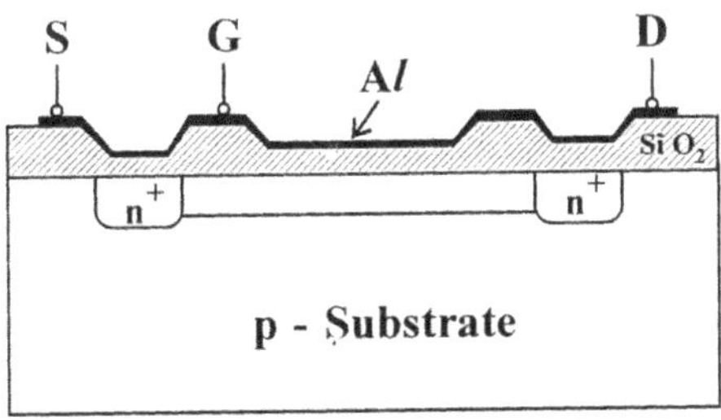

Fig 5.33 Depletion MOSFET.

The conductivity of the channel in the case of depletion type is much less compared to enhancement type. The characteristics of depletion type MOSFETs are exactly similar to that of JFET.

When V_{DS} is positive and $V_{GS} = 0$, large drain current denoted as I_{DSS} (Drain to source current saturation value) flows. If V_{GS} is made negative, positive charges are induced in the channel through SiO_2 of the gate capacitor. But the current in MOSFET is due to majority carriers. So the induced positive charges in the channel reduces the resultant current I_{DS}. As V_{GS} is made more negative, more positive charges are induced in the channel. Therefore its conductivity further decreases and hence I_D decreases as V_{GS} is made more negative (for n-channel depletion type). The current I_D decreases because, the electrons from the source recombine with the induced positive charges. So the number of electrons reaching the drain reduces and hence I_D decreases. So because of the recombination of the majority carriers, with the induced charges in the channel, the majority carriers will be depleted. Hence, this type of MOSFET is knows as *depletion type MOSFET*. JFET and depletion MOSFET have identical characteristics.

A MOSFET of depletion type can also be used as enhancement type. In the case of n-channel depletion type (source and drain are n-type), if we apply positive voltage to the gate-source junction, negative charges are induced in the channel. So, the majority carriers (electrons in the source) are more and hence I_D will be very large. Thus, depletion type MOSFET can also be used as enhancement type by applying positive voltage to the gate (for n-channel type).

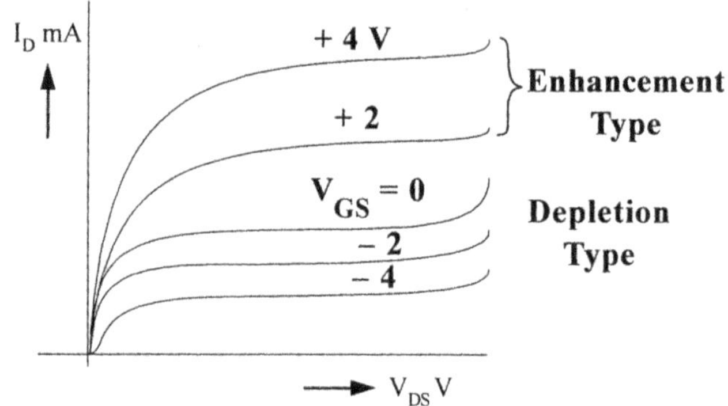

Fig 5.34 Drain characteristics of DMOSFET.

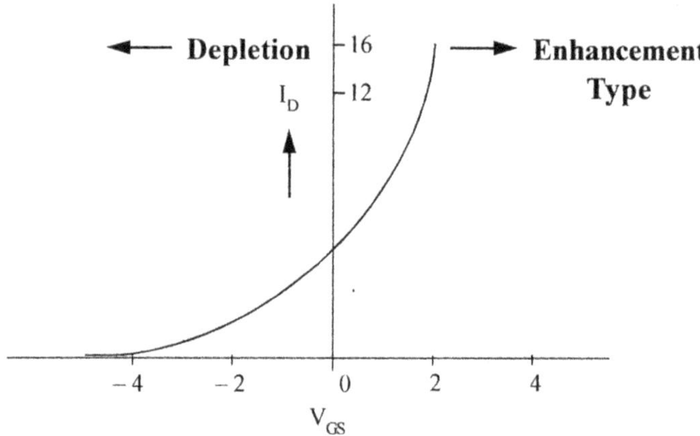

Fig 5.35 Gate characteristics of depletion type MOSFET.

Problem 5.6

A JFET is to be connected in a circuit with drain load resistor of 4.7 kΩ, and a supply voltage of $V_{DD} = 30V$. V_D is to be approximately 20V, and is to remain constant within $\pm$ 1V. Design a suitable self bias circuit.

Solution

$$V_D = V_{DD} - I_D \cdot R_L$$

$$I_D = \frac{V_{DD} - V_D}{R_L}$$

$$= \frac{30V - 20V}{4.7k\Omega} = 2.1 \text{ mA}$$

for V_D to be constant, it should be within $\pm$ 1V,

$$\Delta I_D = \frac{\pm V}{R_L}$$

$$= \frac{\pm 1V}{4.7k\Omega} \simeq \pm 0.2mA$$

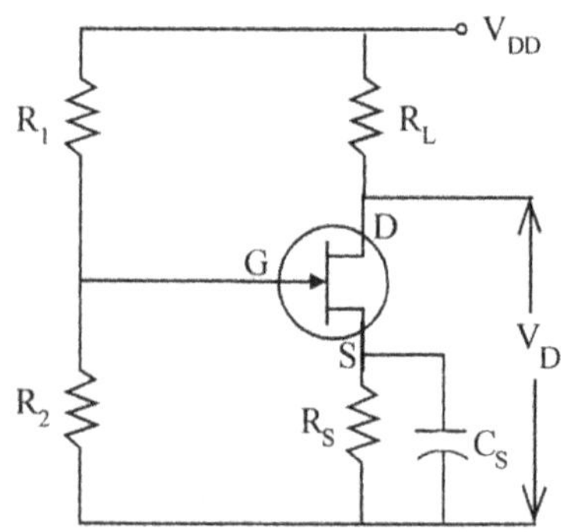

Fig 5.36 For Problem 5.20.

$$I_D = (2.1 \pm 0.2mA)$$
$$I_D(min) = (2.1 - 0.2) = 1.9mA$$
$$I_D(max) = (2.1 + 0.2) = 2.3mA$$

Indicate the points A and B on the maximum and minimum transfer characteristics of the FET. Join these two points and extend it till it cuts at point C.

The reciprocal of the slope of the line gives R_S

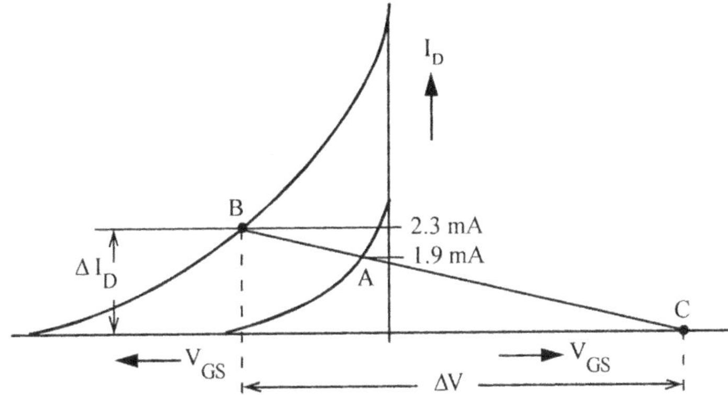

Fig 5.37 For Problem 5.20.

$$R_S = \frac{\Delta V}{\Delta I} = \frac{10V}{2.5mA} = 4k\Omega$$

The bias line intersects the horizontal axis at $V_G = 7V$

∴ An external bias of 7V is required.

$$V_G = \frac{R_2}{R_1 + R_2}.V_{DD}$$

∴
$$\frac{R_2}{R_1 + R_2} = \frac{V_G}{V_{DD}} = \frac{7V}{30v}$$

∴
$$\frac{R_2}{R_1} = \frac{7}{23}$$

R_2 and R_1 should be large to avoid overloading input signals. If $R_2 = 700 \, k\Omega$,
$$R_1 = 2.3 \, M\Omega$$

5.7 Problem

(a)   For the common source amplifier, calculate the value of the voltage gain, given
$$r_d = 100 \, k\Omega, \quad g_m = 300 \, \mu, \quad R_L = 10k\Omega, \quad R_o = 9.09k\Omega$$

Solution
$$A_V = \frac{-g_m.r_d.R_L}{rd + R_L} = \frac{-3000 \times 10^{-6} \times 100 \times 10^3 \times 10 \times 10^3}{\left(100 \times 10^3\right) + \left(10 \times 10^3\right)} = -27.3$$

(b)   If $C_{DS} = 3pf$, determine the output impedance at a signal frequency of 1MHz.

Solution

$$X_C = \frac{1}{2\pi f\, C_{DS}}$$

$$f = 1\ MHz,$$

$$X_C = \frac{1}{2\pi \times 1 \times 10^6 \times 3 \times 10^{-12}} = 53k\Omega$$

$$|Z_0| = \frac{R_0 . X_C}{\sqrt{R_0^2 + X_C^2}}$$

$$|Z_0| = \frac{9.09k\Omega \times 53k\Omega}{\sqrt{(9.09k\Omega)^2 + (53k\Omega)^2}} = 8.96k\Omega$$

5.9   CMOS STRUCTURE (COMPLEMENTARY MOS)

This is evolved because of circuits using both PMOS and NMOS devices. It is the most sophisticated technology. CMOS devices incorporate p-channel MOSFET and n-channel MOSFET. The advantages that we get with this complementary use of transistor are :

    1. Low power consumption.

    2. High speed.

But the disadvantages are

    1. Fabrication is more difficult. More number of oxidation and diffusion steps are involved.

    2. Fabrication density is less. Less number of devices per unit area, because the device chip area is more.

5.9.1 MNOS STRUCTURE

In this process, we will have Silicon semiconductor, SiO_2 over it and then layer of Silicon Nitride $(Si_3 N_4)$ and a metal layer. So we get the *Metal, Nitride*, $(Si_3 N_4)$ Oxide (SiO_2) and Semiconductor structure. Hence, the name MNOS. So the dielectrics between metal (Al) and semi-conductor (Si) is a sandwich of SiO_2 and $Si_3 N_4$, whereas in ordinary MOSFETs we have only SiO_2.

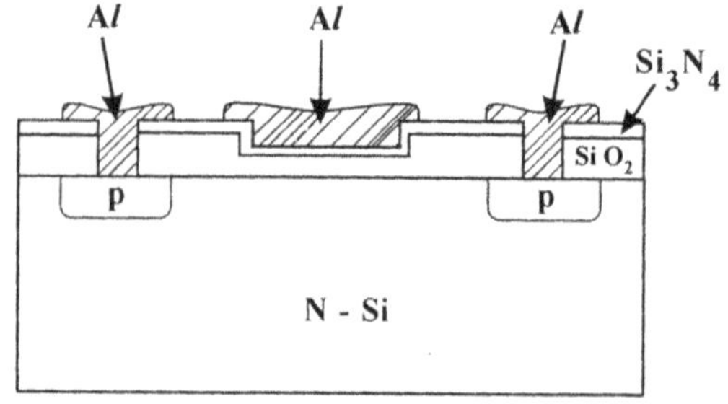

Fig 5.38 MNOS FET.

The advantages are :

1. Low threshold voltage V_T. (V_T is the voltage of the drain, (V_D) beyond which only the increase in drain current I_D will be significant)

2. Capacitance per unit area of the device is more compared to MOS structure. Because the dielectric constant will have a different value since because C is more, charge storage is more. It is used as a memory device.

5.9.2 SOS - MOS STRUCTURE

This is **Silicon On Saphire** MOS structure. The substrate used (Fig.5.39) is the silicon crystal grown on Saphire subtrate. For such a device, the parasitic capacitance will be very low.

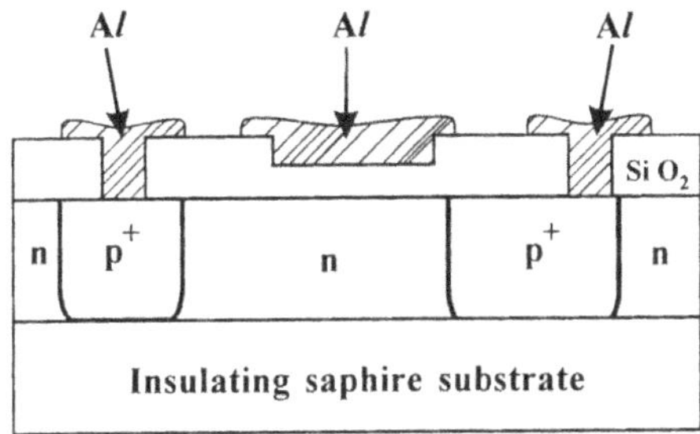

Fig 5.39 SOS – MOS Structure.

5.9.3 VOLTAGE BREAKDOWN IN JFETS

There is a p-n junction between gate and source and gate and drain. Just as in the case of a p-n junction diode if the voltage applied to the p-n junction of G and S, or G and D junction or D and S junctions increase, avalanche breakdown will occur.

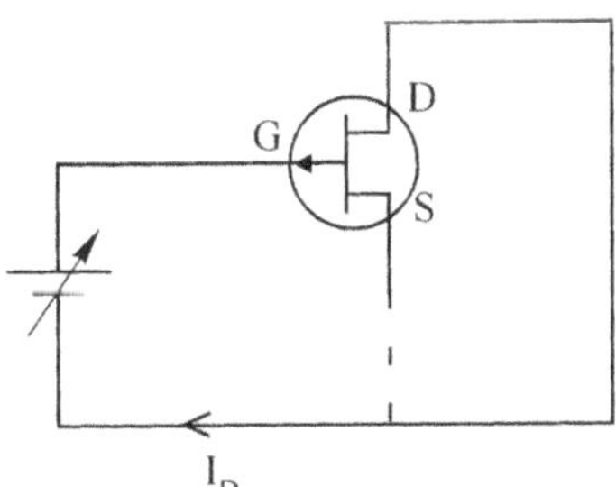

Fig 5.40   JFET circuit.

BV_{DGO} :

This is the value of V_{DG} that will cause junction breakdown, when the source is left open i.e., $I_S = 0$

BV_{GSS} :

This is the value of V_{GS} that will cause breakdown when drain is shorted to source. When drain is shorted to source, there will not be much change in the voltage which causes breakdown.

So, $BV_{GSS} = - BV_{DGO}$

For a p-channel FET, Gate is n-type, S and D are p - type. A positive voltage is applied to G and negative voltage to drain. Therefore with respect to D, V_{DG} is negative voltage. V_{GS} is positive voltage because G is at positive potential with respect to source. Since $BV_{GSS} = -BV_{DGO}$ gate junction breakdown can also result by the application of large V_{DS}.

$$V_{DS} = V_{DG} + V_{GS}$$

∴ More V_{DS} that can be applied to a FET = Max. V_{DG} + Max. V_{GS}

BV_{DS} X = Breakdown voltage V_{DS} for a given value of V_{GS}.

X indicates a specific value of V_{GS}

$$BV_{DSX} = BV_{DGO} + V_{GSX}$$

5.8 Problem

Determine the values of V_{GS}, I_D, and V_{DS} for the circuit shown in using the following data.

$$I_{DSS} = 5mA; \quad V_P = -5 \text{ V}; \quad R_S = 5k\Omega; \quad R_L = 2k\Omega; \quad V_{DD} = 10V$$

Solution

The gate is at DC ground potential, since no DC voltage is applied between gate and Source.

∴ $\qquad V_{GS} + I_S . R_S = 0$

But $\qquad I_S \simeq I_D, I_G$ is negligible

∴ $\qquad V_{GS} + I_D R_S = 0$

or $\qquad V_{GS} = - I_D . R_S$

$\qquad\qquad = -5000\ I_D.$

V_{GS} is unknown and I_D is unknown.

$$I_D = I_{DSS}\left(1 - \frac{V_{GS}}{V_P}\right)^2$$

$$= 5 \times 10^{-3}\left(1 - \frac{V_{GS}}{-5}\right)^2$$

But $\qquad I_D = - \dfrac{V_{GS}}{5000}$

$V_{GS} = (-5000)(5 \times 10^{-3})\,(1 + 0.2V_{GS})^2,$

$V_{GS}^2 + 11\ V_{GS} + 25 = 0$

$V_{GS} = -3.2V$ and $V_{GS} = -7.8V.$

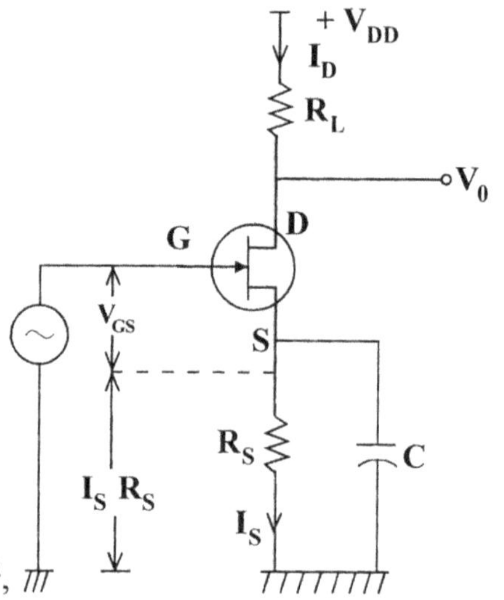

Fig 5.41 For Problem 5.22

The pinch off voltage V_P for which I_D becomes zero is given as 5V therefore V_{GS} cannot be 7.8 V. Therefore $V_{GS} = -3.2V$.

∴ $\qquad I_D = (5 \times 10^{-3})(1 - 0.64)^2 = 0.65 \text{ mA}$

$I_D . R_L = 2000\ I_D = 1.3V$

$I_D . R_S \simeq 5000\ I_D = 3.25V$

∴ $\qquad V_{DS} = 10 - 1.3 - 3.25 = 5.45V.$

The FET must be conducting.

If $V_{GS} = -7.8V$, the FET in cut off. Therefore $V_p = -5V$. Therefore V_{GS} is chosen as $-3.2V.$

5.10   SILICON CONTROLLED RECTIFIER

Silicon controlled rectifier (SCR) is a 4 layer p-n-p-n device. In one type of construction, n-type material is diffused into a silicon pnp pellet to form alloy junction. From the diffused n region, the cathode connection is taken. From the p region on the other side anode is formed. Gate is taken from the p - region into which n - type impurity is diffused. So in the symbol for SCR, the gate is shown close to the cathode, projecting into it. This is known as planar construction.

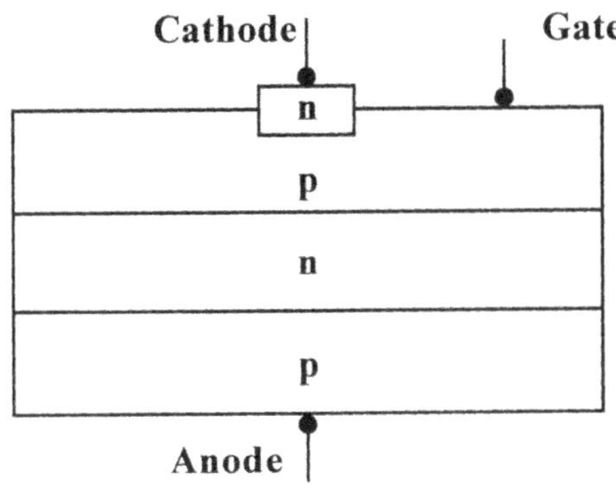

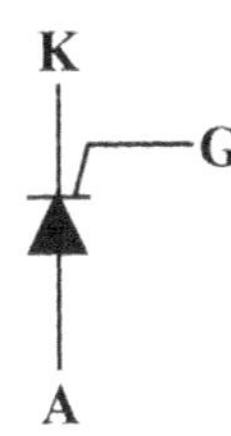

Fig 5.42 SCR structure.   *Fig 5.43 SCR symbol.*

5.10.1   ANNULAR TYPE OF CONSTRUCTION OF SCR

In this type of construction gate is central to the device, with the cathode surrounding it. This is also called as shorted emitter construction. The advantages of this type construction are

    (i) Fast turn on time

    (ii) Improved high temperature capabilities.

5.10.2   STATIC CHARACTERISTICS

The gate current I_G determines the necessary forward voltage to be applied between anode and cathode (Anode positive with reference to cathode) to cause conduction or to turn SCR 'ON'. So that current flows through the device.

As the gate current increases, the voltage to be applied between anode and cathode to turn the device ON decreases. If the gate is open, $I_G = 0$, the SCR is in the OFF state. This is also *known as forward blocking region.* The anode current that flows is only due to leakage. But if the voltage between anode and cathode is increased beyond a certain voltage, V_{BO} called as the forward breakover voltage, the SCR will turn ON and current will be limited by only the applied voltage V_{BO} called as the forward breakover voltage, the SCR will turn ON and current will be limited by only the applied voltage.

V_{BOO} : THE FORWARD BREAK OVER VOLTAGE

This is the voltage to be applied between anode and cathode to turn the SCR ON, when $I_G = 0$.

    B.O : Breakover O : Zero gate current.

As the gate current is increased, the forward break over voltage decreases. If the value of I_G is very large $\approx$ 50 mA, the SCR will turn ON immediately when same voltage is applied between anode and cathode.

Usually, some voltage is applied between the anode and cathode of the SCR, and it is turned ON by a pulse of current in the gate.

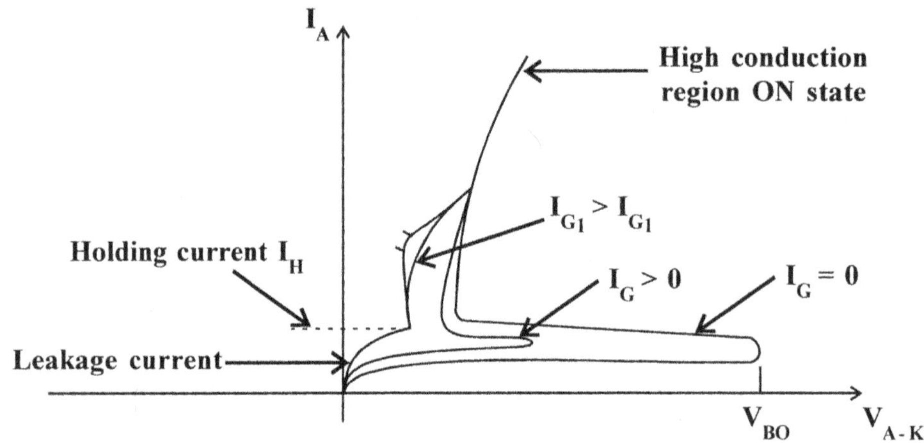

Fig 5.44 SCR Characteristics

HOLDING CURRENT I_H

Once SCR is ON, a minimum amount of current is required to flow to keep the device ON. If the current is lowered below I_H, by increasing external circuit resistance, the SCR will switch off.

Once the SCR is turned ON, the gate looses control over the device. The gate can not be used to switch off the device. If V_{A-K} is reduced to zero (as automatically happens in rectifying application) or if the current is reduced *below I_H, the device will turn off.*

I_H : Holding Current

Minimum Anode Current required to keep SCR in the conducting state. If I_A goes below I_H, SCR is turned off (To hold in conducting state).

LATCHING CURRENT I_L

Minimum value of I_A to be reached to keep SCR in the ON state even after the removal of gate trigger signal. (To latch on to the conducting state.)

5.10.3   ANALYSIS OF THE OPERATION OF SCR

The SCR can be regarded as two transistor connected together.

The SCR can be regarded as two transistor p-n-p and n-p-n. The Base of p-n-p transistor is connected to the collector of the n-p-n transistor. The base of n-p-n transistor is connected to the collector of p-n-p transistor. The emitter of n-p-n transistor is the cathode. The emitter of p-n-p transistor is the anode.

Suppose a positive voltage is applied to the anode (p type) since there is no base injection for both the transistors, the transistors are off. The only current is the leakage current. I_{CO1} and I_{CO2} of the two transistors. Therefore when SCR is not turned ON, the current that results is the leakage current ($I_{CO1} + I_{CO2}$).

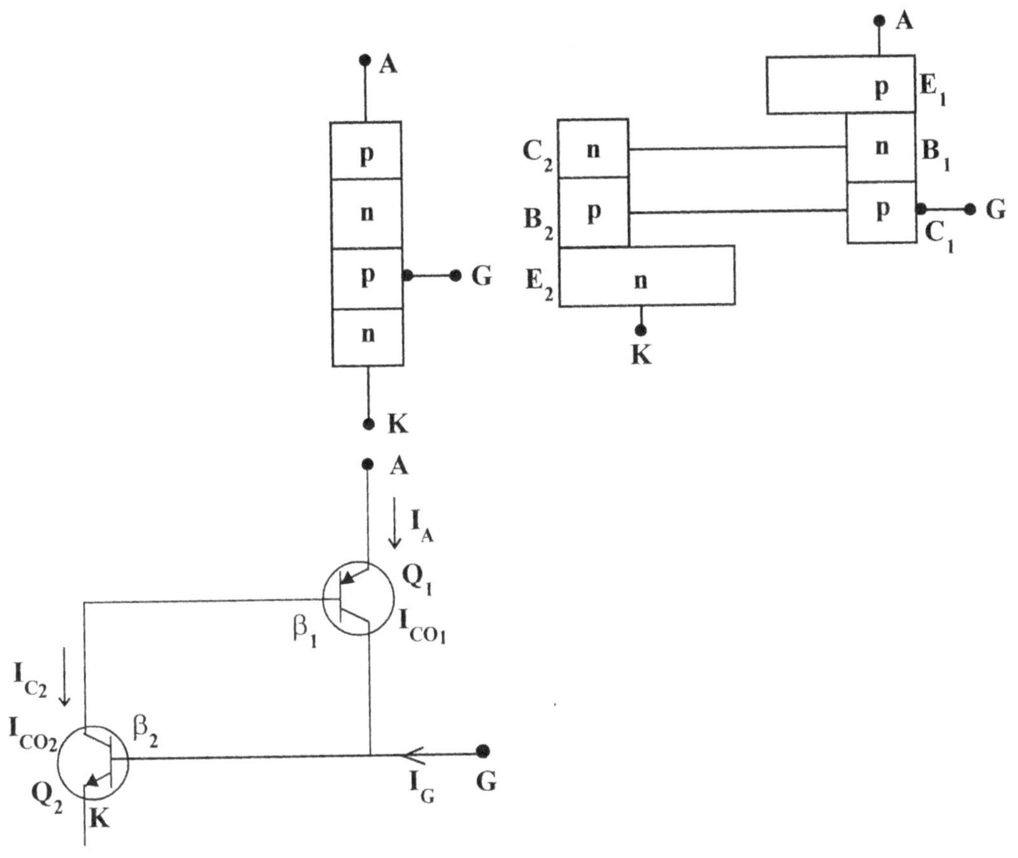

Fig 5.45 Two transistor analogy of SCR.

In a transistor, $I_C = \beta I_B + (\beta + 1) I_{CO}$ ....(1)

$\qquad\qquad\qquad = \beta I_B + \beta I_{CO} + I_{CO}$ ....(2)

$\qquad\qquad\qquad = \beta (I_B + I_{CO}) + I_{CO}$ ....(3)

CONSIDER TRANSISTOR Q_1

The base current for the transistor Q_1, I_{B1} is the collector current I_{C2} of Q_2.

$$I_C = \beta (I_B + I_{CO}) + I_{CO}$$

$$I_{B1} = I_{C2}$$

$\therefore\qquad I_{C1} = \beta_1 (I_{C2} + I_{CO1}) + I_{CO1}$ ....(4)

CONSIDER TRANSISTOR Q_2

The base current of transistor Q_2 is the collector current I_{C1} of Q_1.

$$I_{C2} = \beta_2 (I_{C1} + I_{CO2}) + I_{CO2}$$ ....(5)

Anode current is $I_{E2} = I_{B2} + I_{C2}$

But $I_{B2} = I_{C1}$

$\therefore$ $I_A = I_{C1} + I_{C2}$

I_A can be written in the forms,

$$\frac{(1+\beta_1)(1+\beta_2)(I_{CO1} + I_{CO2})}{1-\beta_1 \cdot \beta_2}$$

The two transistor will have wide base regions. So β will be small. The value of β depends upon the value of I_E. When I_E is very small, $\beta \simeq 0$.

Therefore when $\beta_1 = \beta_2 = 0$.

$$I_A = \frac{(I_{CO1} + I_{CO2})(1+0)(1+0)}{1-0.0}$$

$$\simeq I_{CO1} + I_{CO2}$$

When a negative voltage is applied from gate to cathode, to inject holes into the base of Q_1, emitter base junction of Q_1 is forward biased. So I_{C1} increases. But this is the base current for Q_2. So transistor Q_2 is tuned ON. This increases collector and emitter currents. The I_{C2} of Q_2 now becomes trigger current and so increase I_{C1}. So the action is one of internal regeneration feed back , until both the transistor are driven into saturation. Once the SCR is turned ON, removal of gate current will not stop conduction because already the transistor Q_2 is turned ON and it provides the base injection for Q_1.

SCR can be represented as a pnp and npn transistors connected together, as shown in Fig. 5.46.

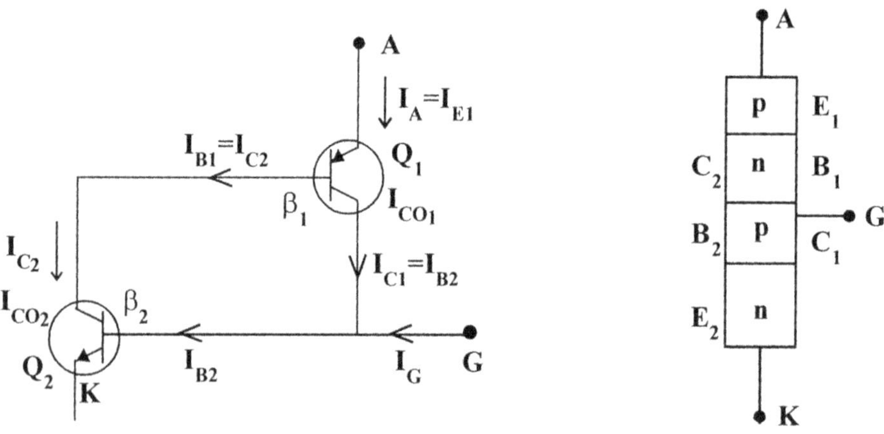

Fig 5.46 Two transistor analogy of SCR.

$$I_{B1} = I_{C2}$$
$$I_{C1} = I_{B2}$$

Anode current $= I_A = I_{E1}$

General expression for I_C in term of β, I_B and I_{CO} is

$$I_C = \beta I_B + (\beta + 1) I_{CO}$$

Since for transistor Q_1,

$$I_{C1} = \beta_1 I_{B1} + (\beta_1 + 1) I_{CO1} \qquad \qquad ....(6)$$

Since for transistor Q_1,

$$I_{C2} = \beta_2 I_{B2} + (\beta_2 + 1) I_{CO2}$$

But $\qquad I_{B2} = I_{C1}$

$\therefore \qquad I_{C2} = \beta_2 I_{C1} + (\beta_2 + 1) I_{Co2} \qquad \qquad ....(7)$

$\qquad I_A = I_{E1} = I_{C1} + I_{B1} \qquad \qquad$ But I_{C1} is given by eq. (6)

$$= \beta_1 \cdot I_{B1} + (\beta_1 + 1) I_{CO1} + I_{B1}$$

$\therefore \qquad I_{E1} = I_A = (\beta_1 + 1)(I_{B1} + I_{CO1}) \qquad \qquad ....(8)$

Since we must get an expression for I_{B1}.

Substitute eq. (6) in (7).

$\therefore \qquad I_{C2} = \beta_2 [\beta_1 I_{B1} + (\beta_1 + 1) I_{CO1}] + (\beta_2 + 1) I_{CO2}$

But $\qquad I_{C2} = I_{B1}$

$\therefore \qquad I_{B1} = \beta_2 \beta_1 I_{B1} + \beta_2 (1 + \beta_1) I_{CO1} + (\beta_2 + 1) I_{CO2}$

or $\qquad (1 - \beta_1 \beta_2) I_{B1} = \beta_2 (1 + \beta_1) I_{CO1} + (1 + \beta_2) I_{CO2}$

or $\qquad I_{B1} = \dfrac{\beta_2 (1 + \beta_1) I_{CO1} + (1 + \beta_2) I_{CO2}}{1 - \beta_1 \beta_2} \qquad ....(9)$

Substitute the value of I_{B1} from eq. (9) in (8).

$$\therefore \qquad I_{E1} = I_A = (1 + \beta_1)\left[\frac{\beta_2 (1 + \beta_1) I_{CO1} + (1 + \beta_2) I_{CO2}}{1 - \beta_1 \beta_2} + I_{CO1} \right]$$

$$= (1 + \beta_1) \left\{ \frac{\beta_2 I_{CO1} + \beta_1 \beta_2 I_{CO1} + I_{CO2} + \beta_2 I_{CO2} + I_{CO1} - \beta_1 \beta_2 I_{CO1}}{1 - \beta_1 \beta_2} \right\}$$

$$= (1 + \beta_1) \left\{ \frac{I_{CO1}(1 + \beta_2) + I_{CO2}(1 + \beta_2)}{1 - \beta_1 \beta_2} \right\}$$

$$\boxed{I_A = \frac{(1 + \beta_1)(1 + \beta_2)(I_{CO1} + I_{CO2})}{1 - \beta_1 \beta_2}}$$

5.10.4 Anode to Cathode Voltage - Current Characteristic

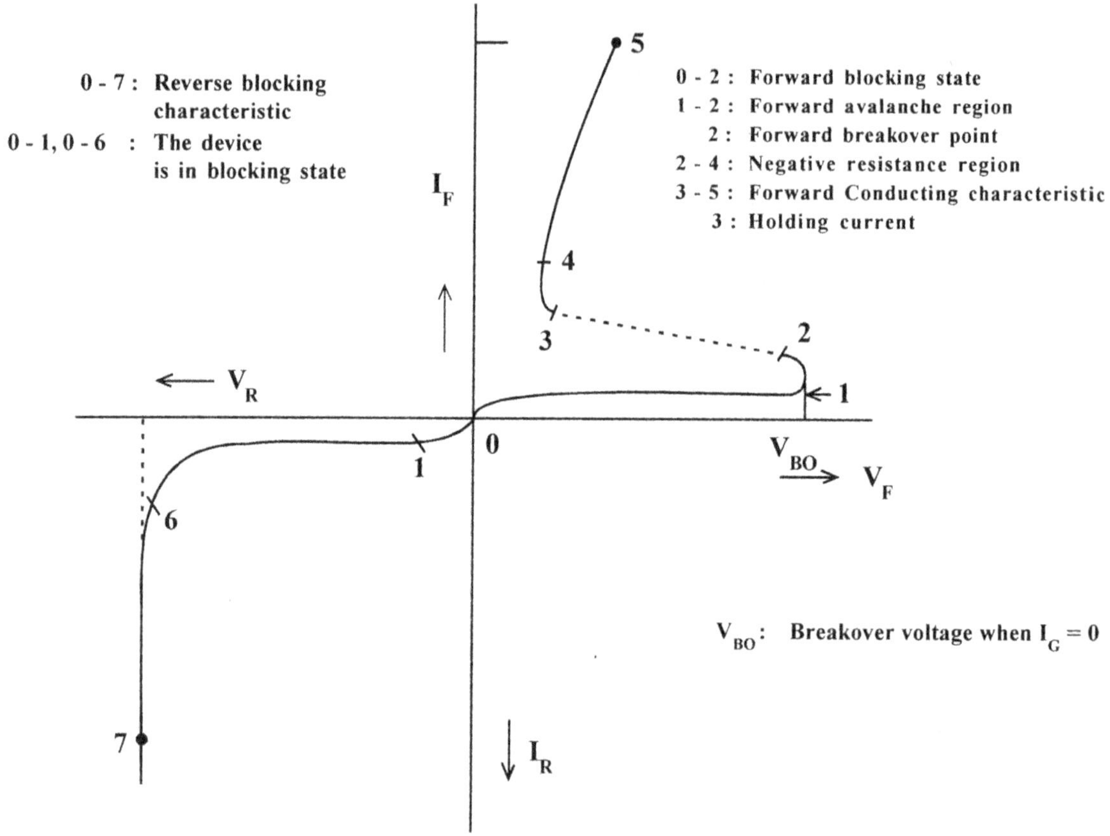

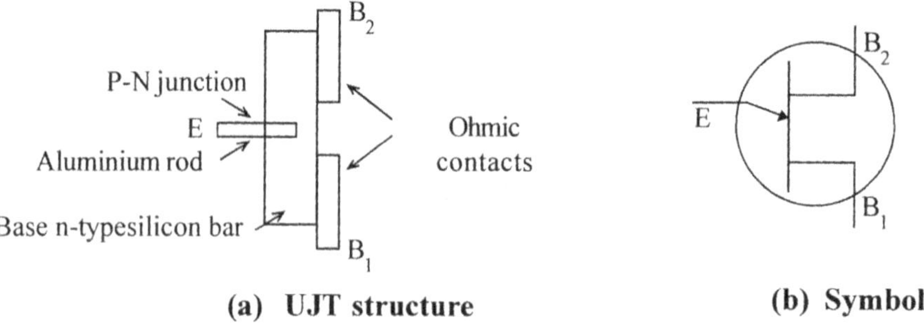

0 - 7 : Reverse blocking characteristic
0 - 1, 0 - 6 : The device is in blocking state
0 - 2 : Forward blocking state
1 - 2 : Forward avalanche region
2 : Forward breakover point
2 - 4 : Negative resistance region
3 - 5 : Forward Conducting characteristic
3 : Holding current
I_F
V_R
V_BO
V_F
I_R
V_BO : Breakover voltage when I_G = 0
0
1
2
3
4
5
6
7

Fig 5.47 SCR characteristics.

5.11 UNIJUNCTION TRANSISTOR (UJT)

This device has only one *p-n* junction and three leads like transistor. Hence it is called as unijunction transistor (UJT).

The construction of this device is as shown in Fig. 5.48.

B2
P-N junction
E
Aluminium rod
Base n-typesilicon bar
B1
Ohmic contacts
(a) UJT structure
E
B2
B1
(b) Symbol

Fig 5.48 (a) UJT structure (b) Symbol

A bar of high resistivity ***n-type silicon of typical dimensions*** $8 \times 10 \times 35$ mils called base B, is taken and two ohmic contacts are attached to it at the two ends to form the base leads B_2 and B_1. A 3 mil aluminium wire, called emitter E is alloyed to the base close to B_2 to form a *p-n* rectifying junction. This device was originally described in the literature as the ***double-base diode***. But now it is called as ***UJT*** ($\because$ it has only one *p-n* junction and two base leads).

5.11.1 UJT CONSTRUCTION AND OPERATION

In UJT, the alloy is formed by diffusing Alluminium into *n* type silicon close to the B_2 lead as shown in Fig. 5.49. The doping and construction is such that, when E - B_1, junction is forward biased, holes are injected into the *n*-silicon bar from the *p*-region towards the B_1 lead. Holes will not travel upwards because B_2 is at positive potential. The doping concentration of *p*-region is large. So there is sudden increase in the number of holes in the region close to B_1. To recombine with these holes, the free electrons from B_2 region will move towards the B_1 lead. Thus there is large increase in the number of holes and electrons in the region corresponding to R_{B_1}.

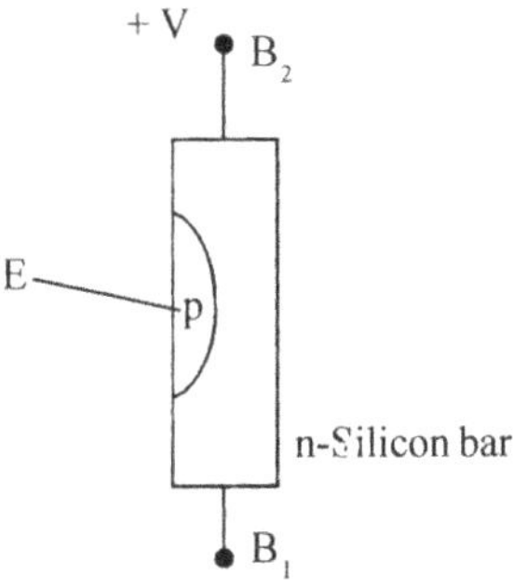

Fig 5.49   UJT structure.

$\therefore$ $\sigma = ne\,\mu_n + p\,e\,\mu_p$. *p* will be almost comparable with *n*. Because of increase in the values of *n* and *p*, σ increases, R_B, decreases. When the E-B, junction is forward biased, even after recombination of holes and electrons, still there will be large number of electrons and holes in the region close to B_1. So conductivity σ increases.

R_{B_1} and R_{B_2} have the same temperature coefficient. Therefore material is the same (*n* type silicon bar). With temperature, because of the increase in the number of free carriers, R_B decreases. The net effect is *n* decreases slightly with the temperature. The change is $\simeq 4\%$ for 100^0 C rise in temperature. $V_P = (\eta V_{BB} + V_V)$. V_V decreases with increase in temperature. Because threshold voltage decreases with increase in temperature. Therefore V_P decreases with temperature. Now R_2 has positive temperature coefficient. If R_2 is chosen such that it increases by the same amount as to decrease in R_{B_1} and η and V_V, V_P will remain constant. Thus, R_2 will provide temperature compensation.

UJT basically consists of a bar of *n*-type silicon with one *p*-type emitter providing *p-n* junction. The emitter will be close to the base two (B_2) than base one (B_1) (see Fig. 5.73). So UJT consists of emitter and two bases with the emitter close to the second base. The emitter is p-type and the two bases are *n*-type. The symbol for the voltage between emitter and base 1 (B_1 is *n*-type). Polarity is as shown in Fig. 5.50. Suppose V_{BB} is the voltage from B_2 to B_1. Now if $V_{BB} = 0$, and positive voltage V_E is applied. The resulting current I_E gives the emitter to base B_1 diode characteristic. With $V_{BB} = 10V$ or 20V, there is leakage current I_{EO} from B_2 to emitter.

($\because$ B_2 is *n*-type and E is *p*-type when B_2 is at a higher potential, the minority carries from B_2 and E will flow which results in leakage current). It takes $\simeq 7$ volts from E to B_1, to reduce this current to zero and then cause current to flow in the opposite direction, reaching peak point V_P. After this point I_E increases suddenly and V_E drops. This is the unstable resistance region. This lasts until the valley voltage is reached and the device saturates.

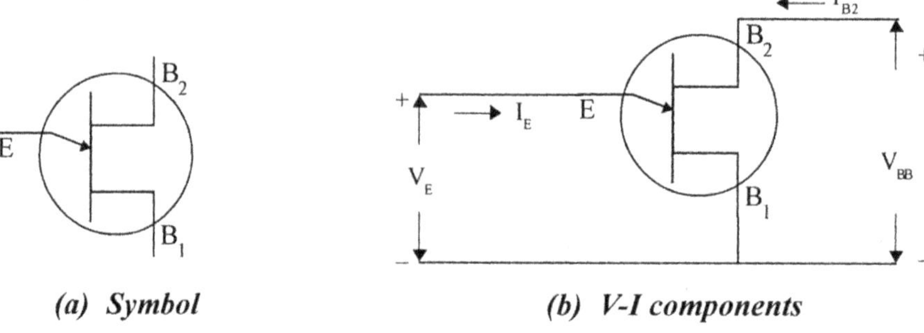

(a) Symbol   *(b) V-I components*

Fig 5.50

The equivalent circuit for UJT is as shown in Fig. 5.51. The resistance of the silicon bar is represented by two series resistors. R_{B_2} is the resistance of base two portion, R_{B_1} is variable resistance. Its value depends upon the bias voltage V_0. The p-n junction formed due to emitter is shown as a diode.

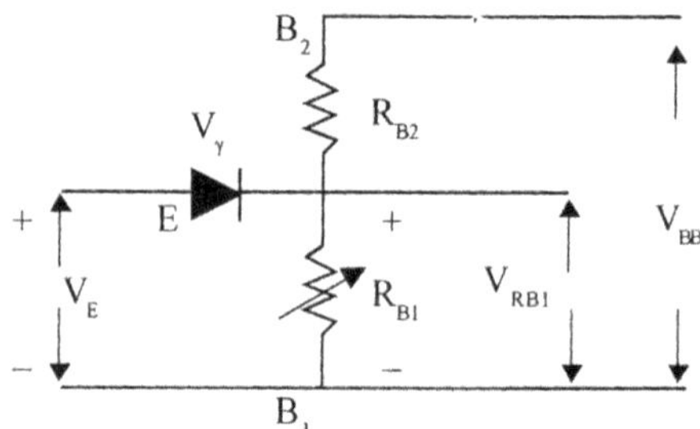

Fig 5.51 Equivalent circuit of UJT.

With no voltage applied to the UJT, $R_{BB} = R_{B_1} + R_{B_2}$. Its value varies from 4.7KΩ to 10KΩ for (2N1671). With emitter open, some voltage is applied at V_{BB}. It gets divided across R_{B_2} and R_{B_1}. (Fig. 5.51). The fraction of voltage appearing across R_{B_1} is

$$V_{RB1} = \eta V_{BB} = \frac{R_{B1}}{R_{B1} + R_{B2}} \times V_{BB}, \qquad \text{where,}$$

$$\eta = \frac{R_{B_1}}{R_{B_1} + R_{B_2}} = \textbf{Intrinsic stand off ratio.}$$

Its value lies between 0.5 and 0.82. Since no voltage is applied at the emitter, the diode is reverse biased, because the cathode voltage is ηV_{BB} and anode current is 0. Now as V_E is increased, from zero and when V_E is $> \eta V_{BB}$, the diode will be forward biased. So the resistivity between E and B_1 decreases since holes are injected from emitter into B_1. Therefore R_{B1} decreases or V_E decreases and I_E increases. So negative resistance region results. Peak voltage $V_P = \eta V_{BB} + V_\gamma$. $V_\gamma = 0.5V$.

5.11.2 V - I Characteristic of UJT

Suppose some voltage V_{BB} is applied between B_2 and B_1. If V_{BB} is small, $I_E = 0$. Then the silicon bar can be considered as an ohmic resistance R_{BB} between the leads B_2 and B_1. R_{BB} will be in the range 5 to 10 kΩ. $R_{BB} = (R_{B1} + R_{B2})$. When $I_E = 0$, the voltage across $R_{B1} = \eta V_{BB}$. Where,

$$\eta = \frac{R_{B1}}{R_{B1} + R_{B2}}.$$ η is called is *intrinsic stand off ratio* and its value lies between *0.5 and 0.75*.

If V_E is $< \eta V_{BB}$ the p-n junction diode or the E-B junction of UJT is reverse biased and the current I_E is negative. It is the reverse saturation current I_{E0} which is of the order of 10 R_B.

If V_{EE} is increased beyond ηV_{BB}, input diode becomes forward biased and becomes positive. So holes are injected from the emitter into the base region B_1. As V_{EE} increases, I_E increases. So the holes injected into B_1 region also increases ($\sigma = pe\,\mu_p$).

$\therefore$ As the hole concentration in the base region increases, its resistivity decreases. So the voltage across R_B, ηV_{BB} decreases. This is the voltage V_E between E and B_1. *So as I_E increases, V_E decreases and the device exhibits negative resistance region. (Fig. 5.52(b))*

When I_E is very large the current I_{B2} flowing from V_{BB} into B_2 load can be neglected. For current greater than the valley current, the resistance becomes positive.

$$V_P = \eta\, V_{BB} + V_V$$
$$V_P = \text{Peak Voltage}$$
$$V_V = \text{Valley Voltage}$$

For current greater than the valley current the resistance becomes positive. (Fig. 5.52.)

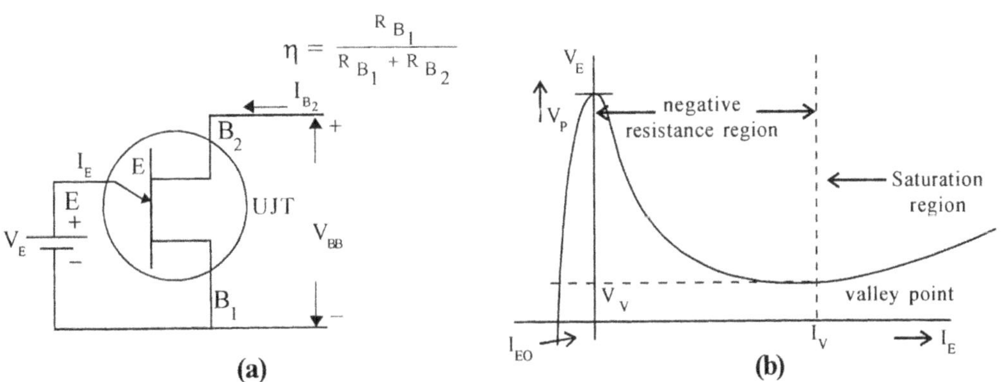

Fig 5.52 (a) UJT circuit (b) V-I Characterisitic of UJT

5.11.3 Applications of UJT

1. In Relaxation oscillator circuits to produce sawtooth waveforms.

2. In triggering SCR circuits.

3. In thyristor circuits.

5.12   LED'S

These are Light Emitting Diodes. On the application of voltage, when electron transition occurs from excited state to lower energy level, the difference in energy is released in the form of light. Thus light emission takes place. This property is exhibited by direct band gap semi conductor materials.

The symbol for LED and its V-I characteristics are shown in Fig.5.53. LEDs are always used in Forward Bias. Because when reverse biased energy transition and hence light emission will not take place.

Direct Band Gap Materials : Conduction band minimum energy and valence band maximum occur at the same value of K (wave vector) in the E – K diagram of semiconductors.

Indirect Band Gap Conduction band minimum and valence band maximum energy do not occur at the same value of K (wave vector) in the E – K diagram.

Photons have high energy and low momentum.

$$\lambda = \frac{hC}{E}; \quad h \cdot f = E$$

$$\boxed{\lambda = \frac{12{,}600}{E} \ A^0}$$

In the complete-electromagnetic spectrum visible spectrum extends in the range $\simeq 4000 - 7200 \ A^0$. Semiconducting materials suitable for light emitters should have energy gap E_G in the range $1.75 - 3.15$ eV.

Direct Band gap semiconductors used for LEDs : (Gallium Arsenide)Ga As, (Gallium Antimony) Ga Sb, (Arsenic) As, (Antimony) Sb, (Phosphorous) P.

Impurities added are Group II materials : (Zinc) Zn, (Magnesium) Mg, (Cadmium) Cd,

Group VI donars : Tellicum Te, Sulphur S.

Impurity concentration : $10^{17} - 10^{18}$ /cm^3. for donor atoms.

Impurity concentration for acceptor atoms : $10^{17} - 10^{19}$/cm^3.

LEDs are based on Injection Illuminisence. Due to injection of carriers light is emitted.

Gallium phosphide – Zinc oxide	LED	Red colour
Gallium phosphide – N	LED	Green colour
Silicon Carbide – SiC	LED	Yellow colour
Gallium phosphide, P, N	LED	Amber colour

5.13 PHOTO DIODES

Instead of photoconductive cells, if a reverse biased p-n junctions diode is used the current sensitivity to radiation can be increased enormously. The mechanism of current control through radiation is similar to that of a photo conductive cell. Photons create electron hole pairs on both sides of the junction. When no light is applied the current is the reverse saturation current due to minority carriers. When light falls on the device photo induced electrons and holes cross the junction and thus increasing the current through the device. For Photo diodes also dark current will be specified. This is the current through the device with no light falling on the device. It is typically 30 - 70 n A at $V_R = 30$ V.

The active diameter of these devices is only 0.1". They are mounted in standard To - 5 package with a window.

They can operate at frequencies of the order of 1 MHz. These are used in optical communication and in encoders.

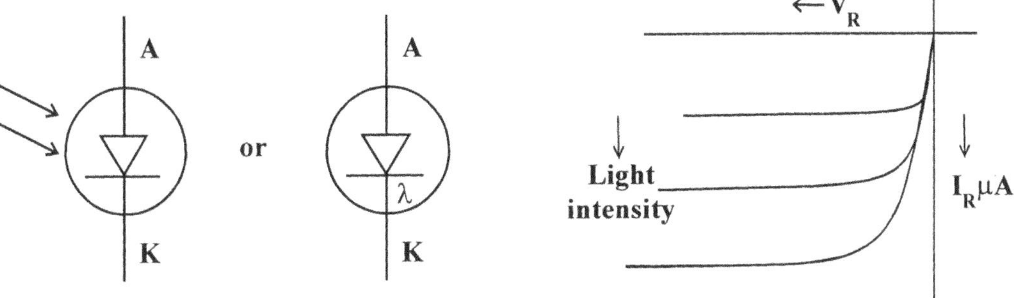

Fig 5.53 Symbols for photo diode.   Fig 5.54 Photo diode reverse characteristics.

5.14 PHOTO TRANSISTORS

In a photodiode, the current sensitivity is much larger compared to a photoconductive cell. But this can be further increased by taking advantage of the inherent current multiplication found in a transistor. Photo transistors have a lens, to focus the radiation or light on to the common base junction of the transistor. Photo induced current of the junction serves as the base current of the transistor (I_λ).

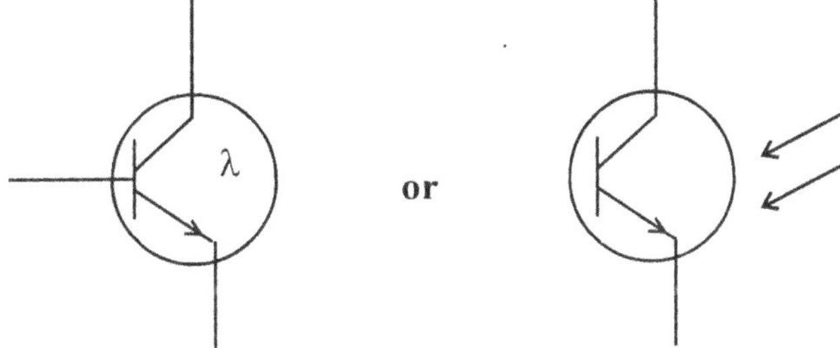

Fig. 5.55 Symbols for Photo transistors

Therefore collector current $I_C = (1 + h_{fe}) I_\lambda$

The base lead may be left floating or used to bias the transistor into some area of operation.

Symbol for photo transistor :

If the base lead in left open, the device is called photo duo diode.

Symbol is

It has two diodes pn and np. So it is called as photo duo diode.

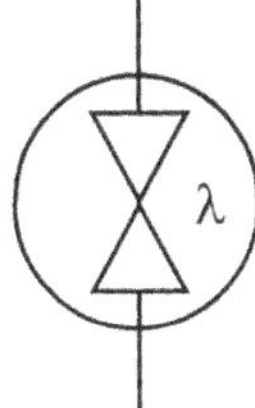

Fig. 5.56 Symbol for Photo duo diode.

SUMMARY

- JFET is UNIPOLAR Device (Unipolar only one type of carriers either holes or electrons)

- JFET device has Higher input resistance compared to BJT and Lower input resistance compared to MOSFET.

- The disadvantage of JFET amplifier circuits is Smaller Gain - Bandwidth product compared to BJT amplifier circuits.

- $r_{ds}\ (ON) = \dfrac{\partial V_{DS}}{\partial I_D}\bigg|_{V_{GS}\,=\,K}$

- $g_m = \dfrac{\partial I_D}{\partial V_{GS}}\bigg|_{V_{DS}\,=\,K}$

- $\mu = $ Amplification factor $= \dfrac{\partial V_{DS}}{\partial V_{GS}}\bigg|_{I_D\,=\,K}$

- Relation between μ, r_d and g_m for a JFET is $\mu = r_d \times g_m$

- Width of the depletion region W_n for n-channel JFET, interms of pinch off voltage V_P is $W = \left\{\dfrac{2\,\epsilon}{eN_D}\,V_P\right\}^{1/2}$

- Expression for I_{DS} in terms of I_{DSS}, V_{GS} and V_P is $I_D = I_{DSS}\left\{1 - \dfrac{V_{GS}}{V_P}\right\}^2$

- $I_{DSS} = $ Satuaration value of Drain current when gate is shorted to source.

- For zero drift current in the case of JFET, $0.007|I_D| = 0.0022\ g_m$

- For zero drift current, in the case of JFETs, $|V_P| - |V_{GS}| = 0.63V$

- Expression for g_m for JFET interms of g_{mO} is, $g_m = \left(1 - \dfrac{V_{GS}}{V_P}\right)$

- The two types of n-channel or p-channel MOSFETs are
    1. Depletion type
    2. Enchancement type

- In JFET terminology, B V_{DGO} parameter means the value of breakdown voltage V_{DG} at breakdown when the source is left open i.e., $I_S = 0$.

- The parameters B V_{GSS} stands for breakdown voltage V_{GS} when drain is shorted to source.

- At high frequencies, the expressions for input capacitance C_{in} of JFET shunting R_G is, $C_{in} = C_{gs} + [1 + g_m\ (R_L \parallel r_d)]\ C_{gd}$

- The purpose of swamping resistor r_s connected in series with source resistance R_S in common source (C.S) JFET amplifier is to reduce distortion.

- Simplified expression for voltage gian A_V in the case of common Drain JFET amplifier is $A_V = \dfrac{R_S}{R_S + \dfrac{1}{g_m}}$

- The distortion caused due to the non linear transfer curve of JFET device, in amplifier circuits is known as Square Law Distorion.

OBJECTIVE TYPE QUESTIONS

1. IGFET is the other name for .... device.
2. In JFET recombination noise is less because it is .... device.
3. The disadvantage of JFET amplifier circuit is ....
4. The D, G, S terminals of JFET are similar to .... terminals of BJT respectively.
5. The voltage V_{DS} at which I_D tends to level off, in JFET is called ....
6. The voltage V_{GS} at which I_D becomes zero in the transfer characteristic of JFET is called ....
7. The range of $r_{DS\,(ON)}$ for JFET is ....
8. For low electric fields E_x of the order of $10^3 V/cm$, $\mu\,\alpha$ ....
9. Expression for V_{GS} in terms of V_p, ...
10. Expression for I_{DX} in terms of I_{DSS} in I_{DS} ....
11. I_{DSS} in defined as ....
12. JFET can be used as .... resistor.
13. Relation between μ, r_d and g_m for JFET is ....
14. The square law device is ....
15. The MOSFET that can be used in both enhancement mode and depletion mode is ....

ESSAY TYPE QUESTIONS

1. Compare BJT, JFET and MOSFET devices in all respects.
2. Derive the expression for the width of Depletion region 'W' in the case of p-channel JFET.
3. Obtain the expression for the pinch off voltage V_p in the case of n-channel JFET
4. Deduce the condition for JFET biasing for zero drift current.
5. Draw the structure and explain the static Drain and Gate characteristics of n-channel JFET. Repeat the same for p-channel JFET.
6. Draw the structure of p-channel MOSFET and Qualitatively explain the static Drain and Gate characteristics of the device.
7. What are the applications of JFET and MOSFET devices?

8. Give the constructional features of UJT.

9. Qualitatively explain the static V-I characteristics of UJT.

10. What is the significance of negative resistance region? Explain how UJT exhibits this characteristics ?

MULTIPLE CHOICE QUESTIONS

1. Insulated Gate Field Effect Transistor (IGFET) is a ...

(a)   Normal JFET device   (b)   n-channel J FET device

(c)   p-channel JFET device   (d)   MOSFET device

2. Compared to BJT, JFET is

(a)   Less noisy   (b)   more noisy

(c)   Same noisy   (d)   can't be said

3. At pinch off voltage in a JFET device,

(a)   Channel width is zero and I_D becomes zero

(b)   Channel width is zero, but I_D is not zero

(c) V_{GS} becomes zero

(d)   None of these

4. For high eletric field strengths $\varepsilon > 10^4$ V/cm, mobility of carriers is proportional to

(a) ε_x (b) $\dfrac{1}{(\varepsilon_x)^2}$ (c) $\dfrac{1}{\varepsilon_x}$ (d)   None of these

5. The pinch off voltage Vp in the case of n-channel JFET is proportional to

(a) N_D^2 (b) $\dfrac{1}{N_D}$ (c) $\dfrac{1}{\sqrt{N_D}}$ (d) N_D

6. Expression for I_D in the case of JFET is,

(a) $I_{DSS}\left(1-\dfrac{V_{gs}}{V_p}\right)$ (b) $I_{DSS}\left(1-\dfrac{V_{gs}}{V_p}\right)^2$

(c) $I_{DSS}\left(1+\dfrac{V_{gs}}{V_p}\right)^2$ (d) $I_{DSS}\left(\dfrac{V_{gs}}{V_p}-1\right)^2$

7. The condition to be satisfied for zero drift current is,

(a) $0.007\,|I_D| = 0.0022$ gm   (b) $0.07\,|I_D| = 0.022$ gm

(c) $0.7\,|I_D| = 0.0022$ gm   (d) $0.007\,|I_D| = 0.022$ gm

8. The relation between μ, r_d and g_m for JFET is,

(a) $\mu = rd/gm$ (b) $rd = \mu^2 gm$ (c) $\mu = rd \cdot gm$ (d) $\mu = rd^2 \cdot gm$

Field Effect Transistors (FETs)

---

9. **Typical values of μ, r_d and g_m for JFET are,**

    (a) $\mu = 5$, rd = 100 Ω, gm = 100 m℧   (b) $\mu = 1000$, rd = 10 Ω, gm = 0.1 m℧

    (b) $\mu = 6$, rd = 1MΩ, gm = 10 K℧   (d) $\mu = 7$, rd = 1MΩ, gm = 0.5 mA/V

10. **The resistor which is connected in series with source resistance Rs to reduce distortion in JFET amplifier circuits is called**

    (a) Swamping resistor   (b) swinging resistor

    (c) bias resistor   (d) distortion control resistor

CONCEPTUAL QUESTIONS (INTERVIEW QUESTIONS)

1. What does the arrow mark in the symbols of pnp and npn transisters signify?

Ans: The arrow mark indicates the direction of emitter current flow when the E-B junction is forward biased.

For the P-N-P transistor emitter current flows towards the base, when E-B junction is forward biased.

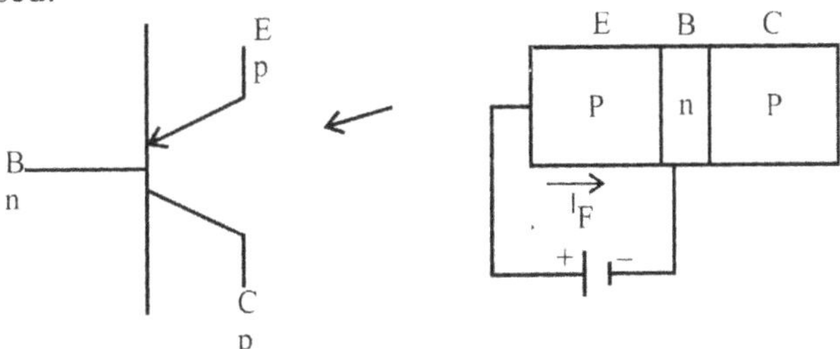

2. Can a BJT be used as a switch? If so how?

Ans: Yes, a transistor can be used as a switch. The device is to be biased between cut-off and saturation regions.

3. What is the concept of Amplification in BJT Circuit?

Ans: In the active region of the transistor, as I_B isincreased, I_C also increases. For a small increase in I_B in μA in the input side, there is large increase in I_C in mA, on the output side. This ratio of change in output to the change in input is large and is>1. So this is termed as amplification by the transistor.

4. Why the input resistance of a BJT is less than the input resistance of JFET?

Ans: On the input side of a BJT, the E-B junctio is to be Forward biased, since the BJT will conduct only when the input side is forward biased. As the forward resistance of a p-n junction is less, the input resistance of a BJT circuit will also be less. On the otherhand, in the case of JFET, the gate source junction is reverse biased, to control the channel width. As the reverse bias resistance is large, the input resistance of JFET is large.

5. Why the input resistance of MOSFET is greater than the input resistanse of JFET?

Ans: For MOSFET device, on the input side, Metal, Silicondioxide and Semiconductor structure is there. Silicondioxide is an insulator with very high resistance. The resistance of this MOS structure is much higher than the reverse resistance of a p-n junction as in the acse of jfet. So MOSFETs have much higher input resistance than JFETs.

Type Numbers of JFETs

2 N 4869   :   N - channel Silicon JFET

BF W10   :   N - channel Silicon JFET

BF W 11   :   N - channel Silicon JFET

2N5457
Case 29-04, Styles
TO-92(TO-226AA)

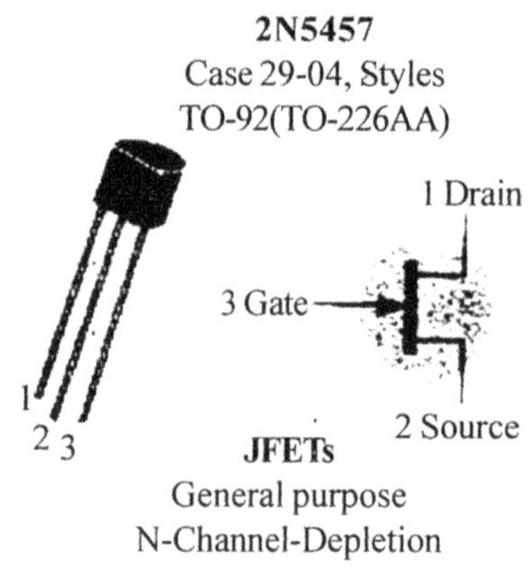

Fig. 5.59 Identification of JFET leads

2N4351
Case 20-03, Styles 2
TO-72(TO-206AF)

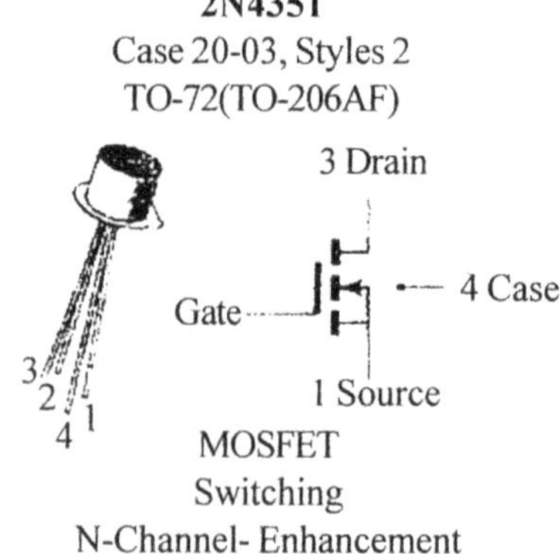

Fig. 5.60 Identification of MOSFET leads

Field Effect Transistors (FETs)

Specifications of JFETs (Typical Values)

Sl.No.	Parameter	Symbol	Typical Value	Units
1	Drain-Source voltage	V_{DS}	25	V
2.	Drain-Gate voltage	V_{DG}	25	V
3.	Reverse Gate - Source voltage	V_{GSR}	-25	V
4.	Gate Current	I_G	10	mA
5.	Junction temperature range	T_j	125	^oC
6.	Gate source Breakdown voltage	$V_{(BR)GSS}$	-25	V

Type No. 2N5457 n- channel JFET
Type No. 2 N 4351 n-channel Enhancement type MOSFET
MOSFET

Sl.No.	Parameter	Symbol	Typical Value	Units
1.	Drain-Source voltage	V_{DS}	25	V
2.	Drain-Gate voltage	V_{DG}	30	V
3.	Gate-source voltage	V_{GS}	30	V
4.	Drain current	I_D	30	mA
5.	Zero gate voltage drain current	I_{DSS}	10	nA
6.	Gate threshold voltage	$V_{GS(th)}$	1	V

UJT

Sl.No.	Parameter	Symbol	Typical Value	Units
1.	Power Dissipation	P_D	300	mw
2.	RMS Emitter current	I_E (rms)	50	mA
3.	Emitter reverse voltage	$V_{E\ R}$	30	V
4.	Inter base voltage	V_B are (inter)	30	V
5.	Intrinsic stand off ratio	η	0.56	-
6.	Peak Emitter current	I_p	0.14	μA
7.	Valley point current	I_V	4	mA
8.	Emitter reverse current	I_{EO}	1	μA

Specifications of Light Emitting diode (LED)

Sl.No.	Parameter	Symbol	Typical Value	Units
1.	Average forward Current	I_F	20	mA
2.	Power Dissipation	P D	120	mw
3.	Peak forward Current	i f (peak)	60	mA
4.	Axial luminous Intensity	I_V	1	mcd milli candelas
5.	Peak wavelength	λ Peak	600	nm
6.	Luminous efficiency	nv	140	lm/w

6 Amplifiers and Oscillators

6.1   HIGH INPUT RESISTANCE TRANSISTOR CIRCUITS

In some applications the amplifier circuit will have to have very high input impedance. Common Collector Amplifier circuit has high input impedance and low output impedance. But its $A_V < 1$. If the input impedance of the amplifier circuit is to be only 500 KΩ or less the Common Collector Configuration can be used. But if still higher input impedance is required a circuit shown in Fig.6.1 is used. This circuit is known as the **Darlington Connection** (named after Darlington) or **Darlington Pair Circuit**.

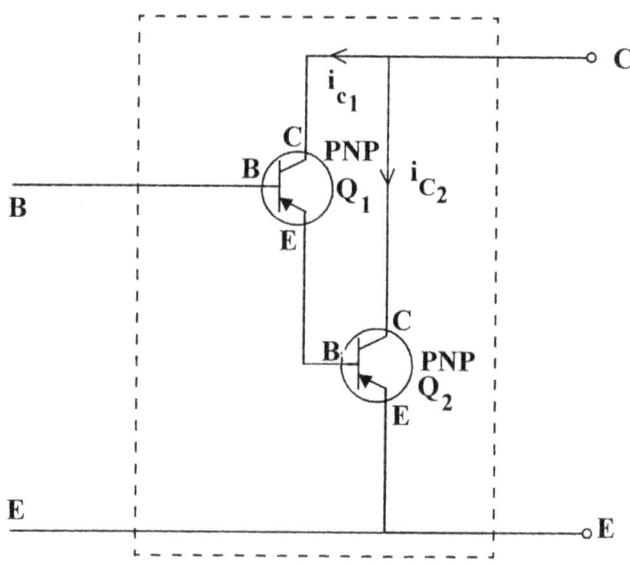

Fig 6.1   Darlington pair circuit.

In this circuit, the two transistors are in Common Collector Configuration. The output of the first transistor Q_1 (taken from the emitter of the Q_1) is the input to the second transistor Q_2 at the base. The input resistance of the second transistor constitutes the emitter load of the first transistor. So, Darlington Circuit is nothing but two transistors in Common Collector Configuration connected in series. The same circuit can be redrawn as AC equivalent circuit. So, DC is taken as ground shown in Fig.6.2.Hence, 'C' at ground potential. Collectors of transistors Q_1 and Q_2 are at ground potential. The AC equivalent Circuit is shown in Fig. 6.3.

There is no resistor connected between the emitter of Q_1 and ground i.e., Collector Point. So, we can assume that infinite resistance is connected between emitter and collector. For the analysis of the circuit, consider the equivalent circuit shown in Fig. 6.54 and we use Common Emitter **h-parameters**, h_{ie}, h_{re}, h_{oe} and h_{fe}.

For PNP transistor, I_c leaves the transistor, I_e enters the transistor and I_b leaves the transistor.

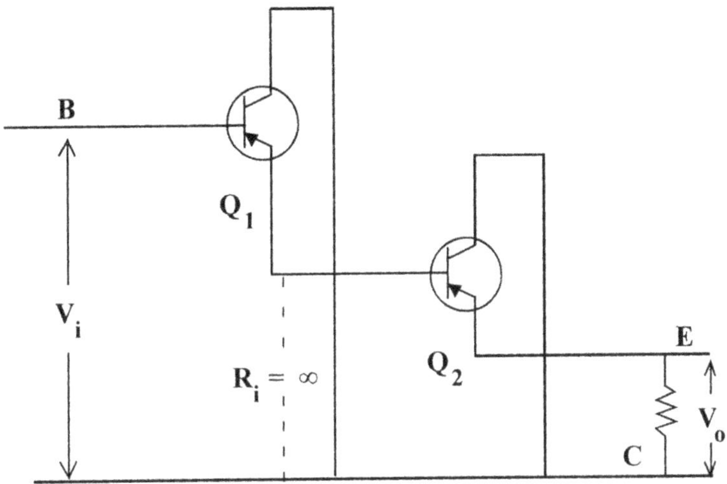

Fig 6.2 Equivalent circuit.

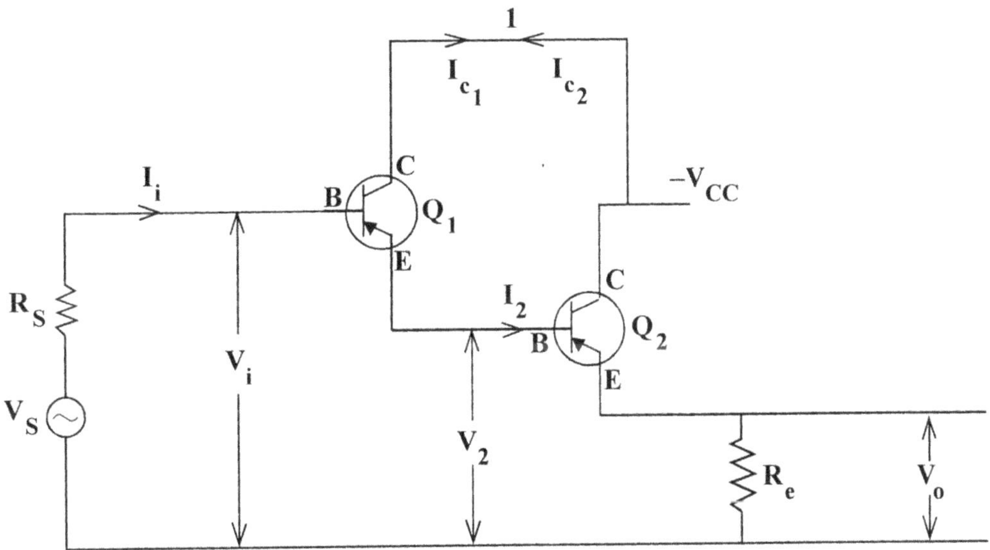

Fig 6.3 Darlington pair circuit.

6.1.1 Current Amplification For Darlington Pair

$$I_c = I_{c_1} + I_{c_2}.$$
$$I_{c_1} = I_{b_1} h_{fe};$$
$$I_{c_2} = I_{b_2} h_{fe}. \text{ (Assuming identical transistor and } h_{fe} \text{ is same)}$$

But
$$I_{b_2} = I_{e_1} \text{ (the emitter of } Q_1 \text{ is connected to the base of } Q_2)$$

$\therefore$
$$I_c = I_{b_1} h_{fe} + I_{e_1} h_{fe} \qquad \qquad .....(1)$$
$$I_{e_1} = I_{b_1} + I_{c_1} = I_{b_1} (1 + h_{fe}) \qquad \qquad .....(2)$$

$\therefore$
$$I_{c_1} = h_{fe} I_{b_1}$$

Substituting equation (2) in (1),

$$I_c = I_{b_1} h_{fe} + I_{b_1}(1 + h_{fe})h_{fe} = I_{b_1}(2h_{fe} + h_{fe}^2)$$

But $h_{fe}^2 \gg 2h_{fe}$

Since, h_{fe} is of the order of 100.

$$\therefore \qquad I_c = I_{b_1} h_{fe}^2$$

It means that we get very large current amplification $\left(A_1 = \dfrac{I_c}{I_{b_1}} \right)$ in the case of Darlington

Pair Circuit, it is of the order h_{fe}^2 i.e. $100^2 = 10,000$.

$$\therefore \qquad \boxed{A_1 = \dfrac{I_c}{I_{b_1}} \cong (h_{fe})^2}$$

6.1.2 Input Resistance (R_1)

Input resistance R_{i_2} of the transistor Q_2 (which is in Common Collector Configuration) in terms of *h- parameters* in Common Emitter Configuration is,

$$R_{i_2} = h_{ie} + (1 + h_{fe}) R_L$$

But $\qquad h_{ie} \ll h_{fe}R_L$, and $h_{fe} R_E \gg h_{ie}$ $A_{I_2} = \dfrac{I_0}{I_2} = (1 + h_{fe})$

Here R_L is R_e, since, output is taken across emitter resistance.

$$\therefore \qquad Ri_2 \cong (1 + h_{fe}) R_e$$

The input resistance Ri_1 of the transistor Q_1 is, since it is in Common Collector Configuration,

$$R_i = h_{ic} + h_{rc} A_I . R_L .$$

Expressing this in term of Common Emitter *h-parameters*,

$$hi_c \cong h_{ie}; h_{rc} \cong 1.$$

(For Common Collector Reverse Voltage Gain = 1) and R_L for transistor Q_1 is the input resistance of transistor Q_2.

$$\therefore \qquad R_{i_1} = h_{ie} + A_{I_1} R_{i2} . \quad R_{i2} \text{ is large,}$$

Therefore, $\quad h_{oe} . R_{i_2} \le 0.1$. and $A_I \# 1 + h_{fe}$

$$R_{i_1} \cong A_{I_1} R_{i2}$$
$$R_{i2} = (1 + h_{fe}) R_e$$

But the expression for Common Collector Configuration in terms of Common Emitter *h-parameters* is

$$A_I = \dfrac{1 + h_{fe}}{1 + h_{oe} . R_L}$$

Here, $R_L = R_{i2}$ and $R_{i2} = (1 + h_{fe}) R_e$.

$$\therefore \qquad A_{I1} = \dfrac{1 + h_{fe}}{1 + h_{oe}(1 + h_{fe})R_e}$$

$h_{oe} R_e$ will be < 0.1 and can be neglected.

$$\therefore \qquad h_{0e} \text{ value is of the order of } \mu \text{ mhos(micro mhos)}$$

$$\therefore \qquad A_{11} = \frac{1 + h_{fe}}{1 + h_{oe} h_{fe} R_e}$$

$$\therefore \qquad R_{i1} \simeq A_{11} \cdot R_{i2}$$

$$\boxed{R_i \simeq \frac{(1 + h_{fe})^2 R_e}{1 + h_{oe} h_{fe} R_e}}$$

This is a very high value. If we take typical values, of $R_e = 4K\Omega$, using *h-parameters*,

$$R_{i_2} = 205 \ K\Omega.$$
$$R_i = 1.73 \ M\Omega.$$
$$A_I = 427.$$

Therefore, Darlington Circuit has *very high input impedance* and very *large current gain* compared to Common Collector Configuration Circuit .

6.1.3 VOLTAGE GAIN

General expression for A_v for Common Collector in term of *h-parameters* is

$$A_v = 1 - \frac{h_{ie}}{R_i}; \ h_{ie} \cong h_{ic} \ \text{ or } A_{V1} = \frac{V_2}{V_i} = \left[\frac{1 - h_{ie}}{R_{i_1}} \right]$$

But $\qquad R_i \cong A_{11} \cdot R_2. \quad \therefore A_{V1} = \left(- \frac{h_{ie}}{A_{I_1} \ R_{i_2}} \right)$

$$\therefore \qquad A_{V2} = \frac{V_0}{V_2} = \left(1 - \frac{h_{ie}}{R_{i_2}} \right)$$

Therefore, over all Voltage Gain $A_v = A_{v_1} \times A_{v_2}$

$$= \left(1 - \frac{h_{ie}}{A_{I_1} \cdot R_{i_2}} \right) \left(1 - \frac{h_{ie}}{R_{i_2}} \right)$$

$$A_v \cong \left(1 - \frac{h_{ie}}{Ri_2} \right) \qquad \because A_{I_1} R_{i_2} \gg h_{ie} \quad \because A_{i_1} \text{ is } \gg 1$$

Therefore, A_v is always < 1.

6.1.4 OUTPUT RESISTANCE

The general expression for R_0 of a transistor in Common Collector Configuration in terms of Common Emitter *h-parameters* is,

$$R_0 = \frac{R_s + h_{ie}}{1 + h_{fe}}$$

$$\therefore \qquad R_{01} = \frac{R_s + h_{ie}}{1 + h_{fe}}$$

Now for the transistor Q_2, R_s is R_{01}.

$$\therefore \qquad R_{o2} = \frac{\dfrac{R_S + h_{ie}}{1 + h_{fe}} + h_{ie}}{1 + h_{fe}}$$

Therefore, R_{o2} is the output resistance of the Darlington Circuit.

$$\therefore \qquad R_{o2} = \frac{R_s + h_{ie}}{(1 + h_{fe})^2} + \frac{h_{ie}}{1 + h_{fe}}$$

This is a small value, since, $1 + h_{fe}$ is $\gg 1$.

Therefore, the characteristic of Darlington Circuit are

1. *Very High Input Resistance (of the order of $M\Omega$).*
2. *Very Large Current Gain (of the order of 10, 000).*
3. *Very Low Output Resistance (of the order of few Ω).*
4. *Voltage Gain, $A_v < 1$.*

Darlington Pairs are available in a single package with just three leads, like one transistor in integrated form.

Disadvantages :

We have assumed that the **h-parameters** of both the transistor are identical. But in practice it is difficult to make out. **h-parameters** depend upon the operating point of Q_1 and Q_2. Since the emitter current of transistor Q_1 is the base current for transistor Q_2, the value of $I_{c2} \gg I_{c1}$

1. *The quiescent or operating conditions of both the transistor will be different. h_{fe} value will be small for the transistor Q_1. $\because$ $h_{fe} = (I_c/I_b).I_{b_2}$ is less CDIL make CIL997 is a transistor of Darlington Pair Configuration with $h_{fe} = 1000$.*

2. *The second drawback is leakage current of the 1st transistor Q_1 which is amplified by the second transistor Q_2 ($\because$ $I_{e_1} = I_{b_2}$).*
   Hence overall leakage current is more. Leakage Current is the current that flows in the circuit with no external bias voltages applied

   (a) *The **h-parameters** for both the transistors will not be the same.*

   (b) *Leakage Current is more.*

   (b) *Darlington transistor pairs in single package are available with h_{fe} as high as 30,000*

6.2   BOOT STRAPPED DARLINGTON CIRCUIT

The maximum input resistance of a practical Darlington Circuit is only 2 MΩ. Higher input resistance cannot be achieved because of the biasing resistors R_1, R_2 etc. They come in parallel with R_i of

the transistors and thus reduce the value of R_i. The maximum value of R_i is only $\dfrac{1}{h_{ob}}$ since, h_{ob} is the resistance between base and collector. The input resistance can be increased greatly by boot strapping, the Darlington Circuit through the addition of C_o between the first collector C_1 and emitter B_2.

What is Boot Strapping ?

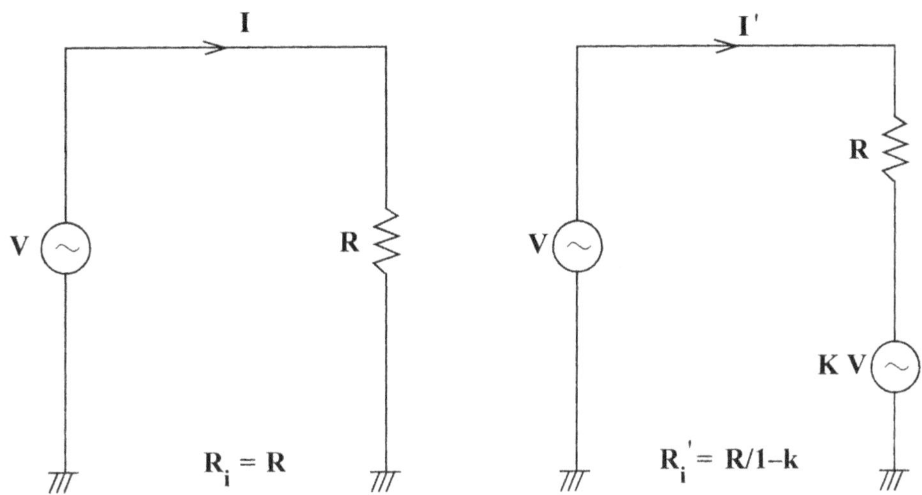

Fig 6.4 Equivalent circuit.						*Fig 6.5*

In Fig.6.10, V is an AC signal generator, supplying current I to R. Therefore, the input resistance of the circuit as seen by the generator is $R_i = \dfrac{V}{I}$ = R itself. Now suppose, the bottom end of R is not at ground potential but at higher potential i.e. another voltage source of KV (K< 1). is connected between the bottom end of R and ground. Now the input resistance of the circuit is (Fig.6.5).

$$R'_i = \frac{V}{I} \qquad\qquad I' = \frac{(V-KV)}{R}$$

or $$R_i = \frac{VR}{(1-K)} = \frac{R}{1-K}$$

I' can be increased by increasing V. When V increases KV also increases. K is constant. Therefore the potential at the two ends of R will increase by the same amount, K is less than 1, therefore $R_i > R$. Now if K = 1, there is no current flowing through R (So V = KV there is no potential difference). So the input resistance $R_i = \infty$. Both the top and bottom of the resistor terminals are at the same potential. This is called as the Boots Strapping method which increases the input resistance of a circuit. If the potential at one end of the resistance changes, the other end of R also moves through the same potential difference. It is as if R is pulling itself up by its boot straps. For c.c.amplifiers $A_v < 1 \cong 0.095$. So R_i can be made very large by this technique. $K = A_V \cong 1$. If we pull the boot with both the edges of the strap (wire) the boot lifts up. Here also, if the potential at one end of R is changed, the voltage at the other end also changes or the potential level of R_3 rises, as if it is being pulled up from both the ends.

For Common Collector Amplifier,

$$R_i = \frac{h_{ie}}{1-A_{v.}} ; \quad A_V \cong 1.$$

Therefore, R_i can be made large. $\because$ it is of the same form as

$$R_i = \frac{R}{1-K}$$

In the circuit shown in Fig. 6.6, capacitor C_o is connected between C_1 and E_2. If the input signal changes by V_o, then E_2 changes by $A_v.V_i$ (assuming the resistance of C_o is negligible)

Therefore, $\dfrac{1}{h_{ob}}$ is now effectively increased to $\dfrac{1}{h_{ob}(1-Av)} \cong 400M\Omega$

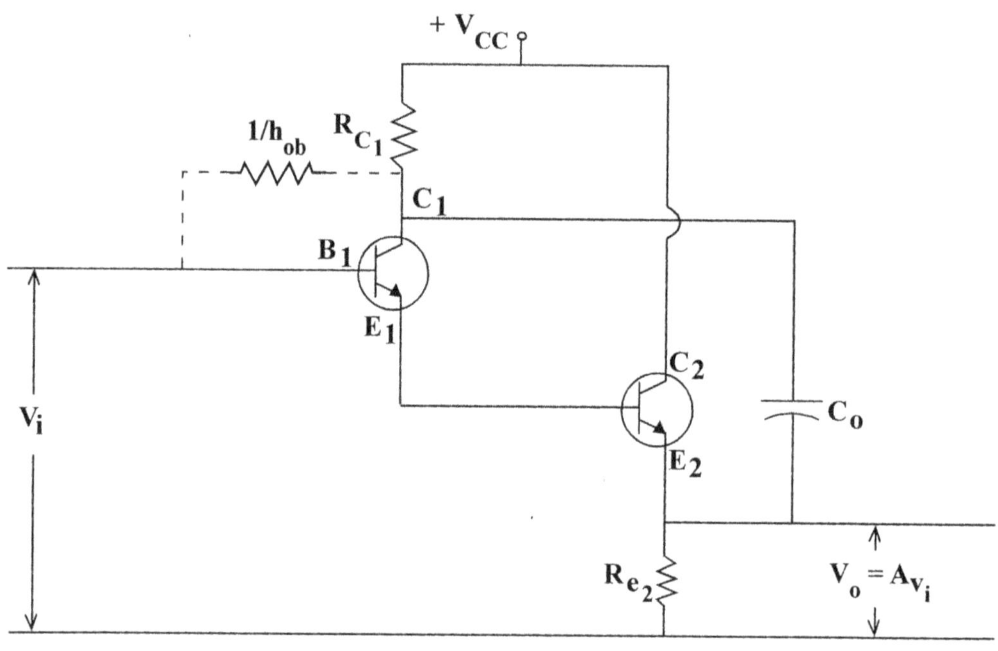

Fig 6.6 Boot strap circuit.

A.C. EQUIVALENT CIRCUIT :

The input resistance

$$R_i = \frac{V_i}{I_{b1}} \cong h_{fe_1} \, h_{fe_2} \, R_e$$

If we take h_{fe} as 50, $R_e = 4K\Omega$, we get R_i as 10MΩ. If a transistor with $h_{fe} = 100$ is taken, R_i will be much larger. The value of X_{C0} is chosen such that at the lower frequencies, under consideration X_{C0} is a virtual short circuit. If the collector C_1 changes by certain potential, E_2 also changes by the same amount. So C_1 and E_2 are boot strapped. There is $\dfrac{1}{h_{ob}}$ between B_1 and C_1.

$$\therefore \qquad R_{eff} = \frac{1}{h_{ob}(1-A_v)}$$

Direct short circuit is not done between C_1 and E_2. Since, DC condition will change, X_{C_0} is a short only for AC signals and not for DC.

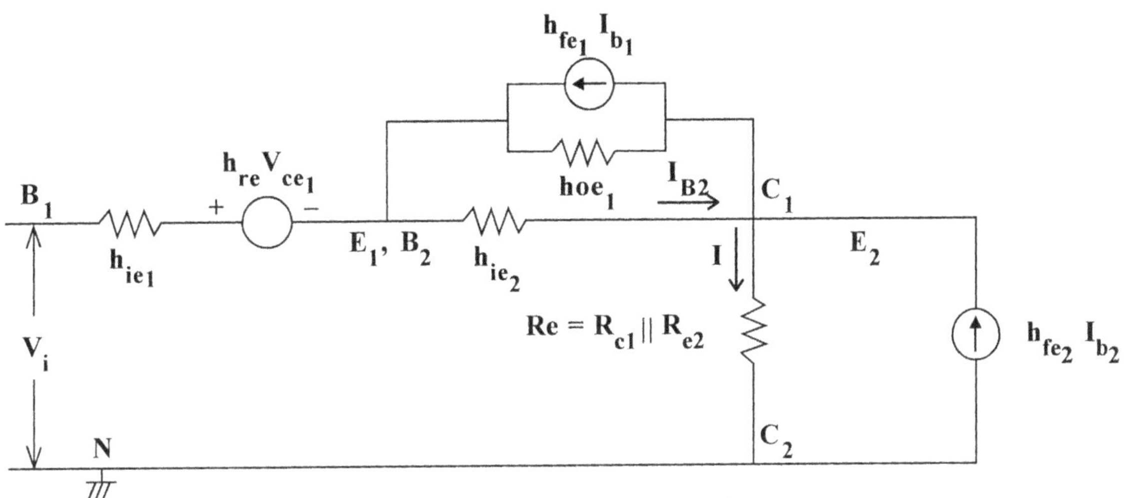

Fig 6.7 A.C. Equivalent circuit.

6.3   THE CASCODE TRANSISTOR CONFIGURATION

The circuit is shown in Fig. 6.8. This transistor configuration consists of a *Common Emitter Stage* in series with a *Common Base Stage*. The collector current of transistor Q_1 equals the emitter current of Q_2.

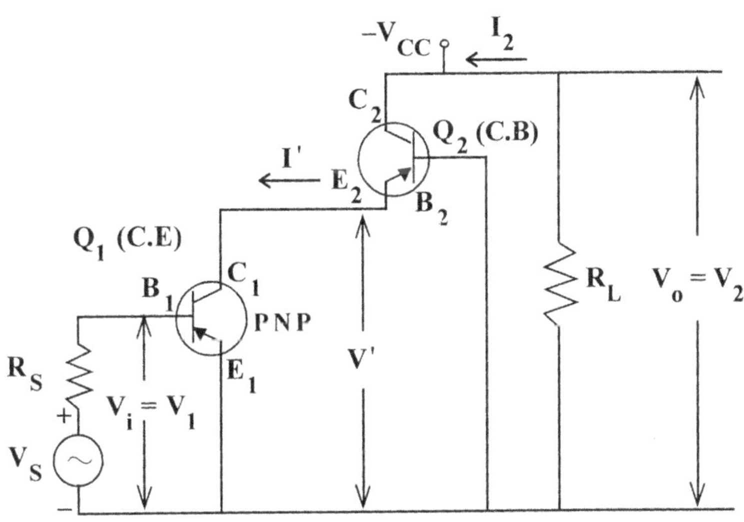

Fig 6.8 CASCODE Amplifier (C.E, C.B configuration).

The transistor Q_1 is in Common Emitter Configuration and transistor Q_2 is in Common Base Configuration. Let us consider the input impedance (h_{11}) etc., output admittance (h_{22}) i.e. the *h - parameters* of the entire circuit in terms of the *h - parameters* of the two transistors.

INPUT IMPEDANCE (h_{11})

$$h_{11} = \text{Input } Z = \left.\frac{V_1}{I_1}\right|_{V_2 = 0}$$

If V_2 is made equal to 0, the net impedance for the transistor Q_1 is only h_{ib_2}. But h_{ib}, for a transistor in common emitter configuration is very small $\cong 20$. We can conclude that the collector of Q_1 is effectively short circuited.

$$\therefore \qquad h_{11} \cong h_{ie}.$$

When $V_2 = 0$, C_2 is shorted. Therefore, $h_{i2} = h_{ib_2}$. But h_{ib_2} is very small. Therefore C_1 is virtually shorted to the ground.

$$\therefore \qquad \boxed{h_i = h_{ie}}$$

SHORT CIRCUIT CURRENT GAIN (h_{21})

$$h_{21} = \left.\frac{I_2}{I_1}\right|_{V_2 = 0}$$

$$h_{21} = \frac{I_2}{I_1} = \left.\frac{I'}{I_1} \times \frac{I_2}{I'}\right|_{V_2 = 0}$$

$$\frac{I'}{I_1} = h_{fe} \qquad \text{since, } I = I_{C_1}. \ I_1 = I_{B_1}$$

$$\frac{I_2}{I'} = -h_{fb} \quad \text{since, } I = I_B. \ I_2 = I_C.$$

$$\therefore \qquad h_{21} = -h_{fe}. \ h_{fb}.$$

$$h_{fe} \gg 1 . -h_{fb} \simeq 1, \quad \text{since } h_{fb} = \frac{I_C}{I_E}$$

$$\therefore \qquad \boxed{h_{21} \cong h_{fe}}$$

OUTPUT CONDUCTANCE (h_{22})

Output Conductance with input open circuited, for the entire circuit is,

$$h_{22} = \left.\frac{I_2}{V_2}\right|_{I_1 = 0}$$

when $I_1 = 0$, the output resistance of the transistor Q_1 is $\dfrac{1}{h_{oe}} \cong 40 \text{ K}\Omega$. (Since Q_1 is in Common Emitter configuration and h_{oe} is defined with $I_1 = 0$).

$$\therefore \qquad \frac{1}{h_{oe}} \cong 40 \text{K}\Omega$$

is the source resistance for Q_2. Q_2 is in Common Base Configuration. What is the value of R_o of the transistor Q_2 with $R_s \cong 40$k

It is $\cong 1/h_{ob}$ itself.

Since, h_{oe_1} is very large, we can say that $I'_1 = 0$. Or between E_2 and ground there is infinite impedance. Therefore, output conductance of the entire circuit is $h_{22} \cong h_{ob}$.

REVERSE VOLTAGE GAIN (h_{12})

$$h_{12} = \left.\frac{V_1}{V_2}\right|_{I_1 = 0}$$

$$= \left.\frac{V_1}{V'} \times \frac{V'}{V_2}\right|_{I_1 = 0}$$

$$\left.\frac{V_1}{V'}\right|_{I_i = 0} = h_{re}. \quad \left.\frac{V'}{V_2}\right| = h_{rb}$$

(Since, Q2 is in Common Base configuration)

$$\therefore \qquad h_{12} \cong h_{re}\, h_{rb}.$$

$$h_{re} \cong 10^{-4} \qquad h_{rb} = 10^{-4}. \quad \therefore\ h_{12}\ \text{is very small} .$$

$$\therefore \qquad h_i = h_{11} \cong h_{ie}. \qquad \text{Typical value} = 1.1\text{k}\Omega$$

$$h_f = h_{21} \cong h_{fe}. \qquad \text{Typical value} = 50$$

$$h_o = h_{22} \cong h_{ob}. \qquad \text{Typical value} = 0.49\ \mu\,\text{A/V}\ \ m$$

$$h_r = h_{12} \cong h_{re}\, h_{rb}. \text{ Typical value} = 7 \times 10^{-8}.$$

Therefore, for a CASCODE Transistor Configuration, its input Z is equal to that of a single Common Emitter Transistor (h_{ie}). Its Current Gain is equal to that of a single Common Base Transistor (h_{fe}). Its output resistance is equal to that of a single Common Base Transistor (h_{ob}). The reverse voltage gain is very very small, i.e., there is no link between V_i (input voltage) and V_2 (output voltage). In otherwords, there is negligible internal feedback in the case of cascode amplifier. A CASCODE Transistor Circuit acts like a single stage C.E. Transistor (Since h_{ie} and h_{fe} are same) with negligible internal feedback ($\therefore h_{re}$ is very small) and very small output conductance, ($\cong h_{ob}$) or large output resistance ($\cong 2\text{M}\Omega$ equal to that of a Common Base Stage). The above values are correct, if we make the assumption that hob $R_L < 0.1$ or R_L is < 200K. When the value of R_L is < 200 K. This will not affect the values of h_i, h_r, h_o, h_f of the CASCODE Transistor, since, the value of h_r is very very small.

CASCODE Amplifier will have

1. *Very Large Voltage Gain.*

2. *Large Current Gain (h_{fe}).*

3. *Very High Output Resistance.*

Problem 6.1

Find the voltage gains A_{v_s}, A_{v_1} and A_{v_2} of the amplifier shown in Fig. 6.9. Assume

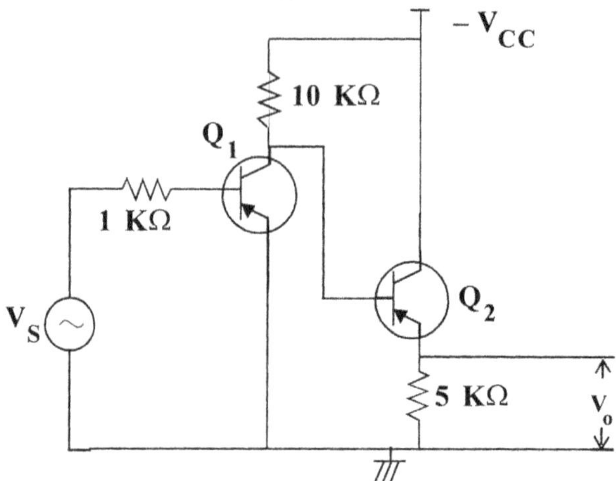

Fig 6.9 For Problem ---------

$$h_{ie} = 1K\Omega, \quad h_{re} = 10^{-4}, \quad h_{fe} = 50$$

and
$$h_{oe} = 10^{-8}\,A/V.$$

Solution

The second transistor Q_2 is in Common Collector Configuration Q_1 is in Common Emitter Configuration. It is convenient if we start with the II stage.

II Stage :

$$h_{oe}.\,R_{L2} = 10^{-8} \times 5 \times 10^3 = 5 \times 10^{-5} < 0.1.$$

Therefore, $h_{oe}\,R_{L2}$ is < 0.1, approximate analysis can be made. Rigorous Expressions of C.C. and C.B Configuration need not be used.

$$\therefore \qquad R_{i2} = h_{ie} + (1 + h_{fe})R_{L2}$$
$$= 1K\Omega + (1 + 50)\,5K\Omega = 256\ K\Omega.$$

Av_2 is the expression for Voltage Gain of the transistor in Common Collector Configuration in terms of Common Emitter *h-parameters* is,

$$A_{V2} = 1 - \frac{h_{ie}}{R_{i_2}}$$

$$= 1 - \frac{1k\Omega}{256k\Omega} = 1 - 0.0039 = 0.996$$

I Stage :

$$R_{L_1} = 10K \| R_{i_2} = 10k \| 256k\Omega = 9.36\ k\Omega.$$
$$h_{oe}.\,R_{L_1} < 0.1. \quad \therefore \ \text{Approximate equation can be used}$$
$$A_{I_1} = -50\,.\,(h_{fe})\,.$$
$$R_{i_1} = h_{ie}.\, = 1k\Omega.$$

$$A_{v_1} = -\,h_{fe}.\,\frac{R_{L_1}}{h_{ie}} = \frac{-50 \times 9.63k}{1k\Omega} = -481.5$$

$$= \left(A_I.\,\frac{R_{L1}}{R_i} \right)$$

Overall Voltage Gain $= A_{v_1}.A_{v_2} = -482 \times 0.996 = -480.$

$$\therefore \qquad A_{vs} = A_v \times \frac{R_S}{R_S + h_{oe}} = A_v \times \frac{1k\Omega}{2k} = -240.$$

6.4 THE JFET LOW FREQUENCY EQUIVALENT CIRCUITS

6.4.1 COMMON SOURCE AMPLIFIER

Capacitance and DC voltages are shorted. FET is replaced by its small signal model so the equivalent circuit is as shown in Fig. 6.10.

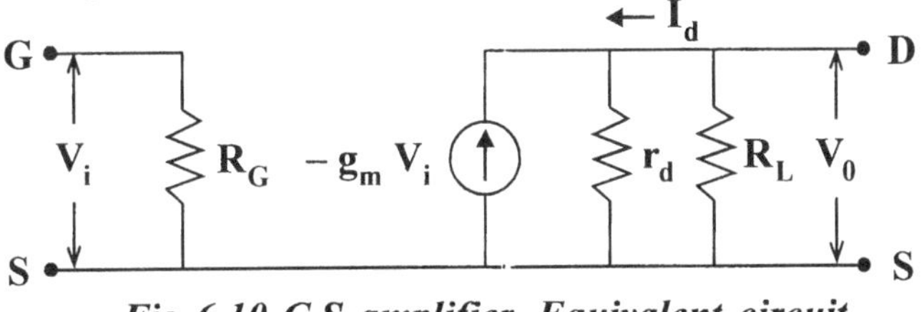

Fig 6.10 C.S amplifier. Equivalent circuit

I_d = AC drain current

r_d = drain resistance

$$V_0 = I_d \times (r_d \,||\, R_L)$$

$$= I_d \times \left[\frac{r_d.R_L}{r_d + R_L} \right]$$

But $\qquad I_d = -\,g_m\,V_i$ (The value of the current source)

$$V_0 = -\,g_m \cdot V_i \left\{ \frac{r_d \times R_L}{r_d + R_L} \right\}$$

$\therefore \qquad$ Voltage $= Av$

$$= \frac{V_0}{V_i}$$

$$= \frac{-g_m.r_d.R_L}{rd + R_L}$$

But $\qquad r_d \gg R_L$

$\therefore \qquad r_d + R_L \simeq r_d$

$$\therefore \qquad Av \simeq \frac{-g_m.r_d.R_L}{r_d}$$

$$Av \simeq -\,g_m \cdot R_L$$

For a FET, Y_{fs} = forward transfer admittance in C.S. configuration

$$= g_m$$

Y_{OS} = Out put admittance in C.S configuration

$$= \frac{1}{r_d}$$

At low frequencies, $Z_0 = r_d \parallel R_L$

$$\simeq R_L \qquad \because \; r_d \text{ is large.}$$

At high frequencies r_d and R_L are shunted by C_{ds}.

$$Z_0 = R_0 \parallel X_{ds} \qquad X_{ds} = \frac{1}{2\pi f C_{ds}}$$

At low frequencies, input impedance

$$Z_i = R_G \cdot \parallel R_{GS}$$

But R_{GS} is a very large value. Therefore $Z_i \simeq R_G$.

At high frequencies, the input capacitance shunting R_G becomes effective. The actual capacitance presented to an input signal is amplified by the Miller effect.

$\therefore$ $C_{in} = C_{gs} + (1 + Av) \cdot C_{gd}$

where A_v = circuit voltage gain $g_m \cdot (R_L \parallel rd)]$

                 $C_{in} = C_{gs} + [1 + g_m (R_L \parallel r_d)] C_{gd}$

6.4.2 COMMON DRAIN AMPLIFIERS (C.D)

This is also known as **source follower**. This is FET equivalent of emitter follower or common collectors transistor amplifier circuit.

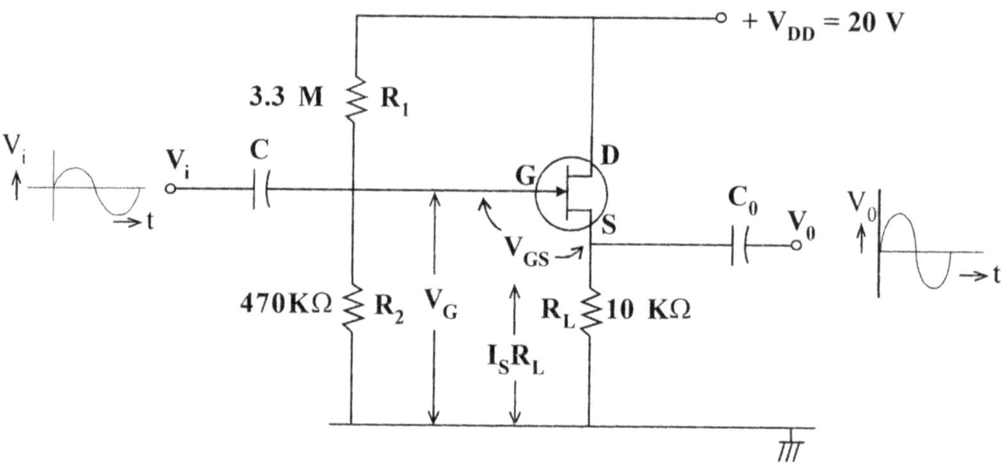

Fig 6.11 Common drain circuit.

V_G is not the same as V_{GS}. S is not at ground partial.

$$V_G = V_{GS} + V_{R_L}$$

$$= V_{GS} + I_D \cdot R_L$$

If the input increases by 1V, output also increases by IV. So, there is no phase-shift and voltage gain = 1.

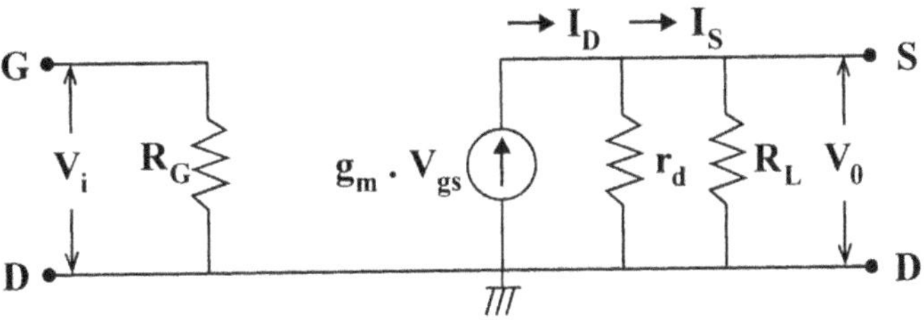

Fig 6.12 CD amplifier.

$$V_0 = I_D \times (r_d \parallel R_L)$$

$$= I_D \cdot \left\{ \frac{r_d \times R_L}{r_d + R_L} \right\}$$

$$I_D = g_m \cdot V_{gs}$$
$$V_{GS} = (V_i - V_0)$$
$$I_D = g_m (V_i - V_0) \qquad \text{from the circuit}$$

$$\therefore \qquad V_0 = g_m (V_i - V_0) \cdot \frac{r_d \times R_L}{r_d + R_L}$$

Solving for V_0

$$V_0 (r_d + R_L) = g_m \cdot V_i \cdot r_d \cdot R_L - g_m \cdot V_0 r_d \cdot R_L$$
$$V_0 (rd + R_L + g_m rd \cdot R_L) = g_m \cdot V_i \cdot r_d \cdot R_L$$

$$V_0 = g_m \cdot V_i \cdot \left\{ \frac{r_d . R_L}{r_d + R_L + g_m r_d . R_L} \right\}$$

$$\therefore \text{ Voltage gain} \quad A_v = \frac{V_d}{V_i}$$

$$\boxed{A_V = g_m \cdot \frac{r_d . R_L}{r_d + R_L + g_m r_d R_L}}$$

If $g_m \cdot r_d \cdot R_L \gg (r_d + R_L)$, $A_v \simeq 1$.

Z_0:

$$V_0 = g_m \cdot V_i \cdot \frac{r_d . R_L}{r_d + R_L + g_m r_d . R_L}$$

$g_m V_i$ is output current proportional to V_i, and,

$$Z_0 = \frac{(r_d . R_L)}{r_d + R_L + g_m r_d R_L}$$

$$\therefore \qquad Z_0 = \frac{r_d . R_L}{r_d + R_L (1 + g_m r_d)}$$

$$= \frac{[r_d / 1 + g_m r_d] R_L}{[(r_d / 1 + g_m r_d)] + R_L}$$

$$= R_L \parallel \frac{r_d}{1 + g_m r_d}$$

$$Z_0 \simeq R_L \parallel \frac{1}{g_m}$$

Z_i :

$$Z_i = R_1 \parallel R_2$$

C_{gd} is in parallel with Z_i

6.4.3 COMMON GATE FET AMPLIFIER CIRCUIT (C.G)

The circuit is shown in Fig. 6.13.

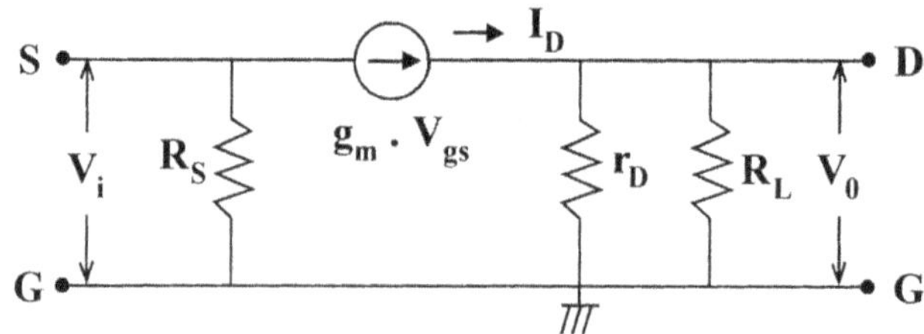

Fig 6.13 Common gate circuit.

Fig 6.14 C.G configuration.

$$V_0 = I_D \times (r_d \parallel R_L)$$

$$= I_D \cdot \left\{ \frac{r_d . R_L}{r_d + R_L} \right\}$$

$$I_D = g_m \cdot V_{gs} = g_m \cdot V_i$$

$$V_0 = g_m \cdot V_i \times \frac{r_d \cdot R_L}{r_d + R_L}$$

∴ voltage gain $= \boxed{A_v = \dfrac{V_0}{V_i} = \dfrac{g_m \cdot r_d \cdot R_L}{r_d + R_L}}$

$$Z_0 = R_2 \parallel r_d \simeq R_L$$

$$I_D = g_m \cdot V_{gs} = g_m \cdot V_i$$

∴ $$Z_i = \frac{V_i}{I_D} = \frac{1}{g_m}.$$

This is the input impedance of the device.

The circuit input impedance is, $\dfrac{1}{g_m} \parallel R_S$.

6.5 COMPARISON OF FET AND BJT CHARACTERISTICS

From construction point of view triode is similar to BJT (Transistor).

Cathode E ;   B → Grid; Anode → collector.

But from characteristics point of view pentode is similar to a BJT.

1. In FET and pentode current is due to one type of carriers only. In pentode it is only due to electrons. But in BJT it is due to both electron and holes.

2. In FET pinch off occurs and current remains constant. But no such channel closing in Transistor and Pentode.

3. As T increases, I_{CO} increases, so I in a BJT increases. But in FET, as temperature increases, 'μ' decreases. So I_D decreases. In pentode, also as temperature increases, I increases. But it is less sensitive compared to a transistor or FET.

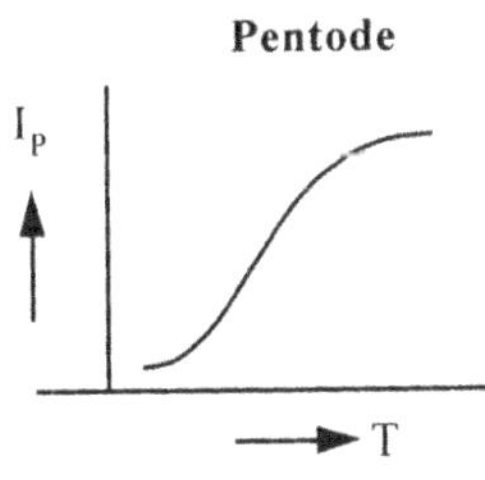

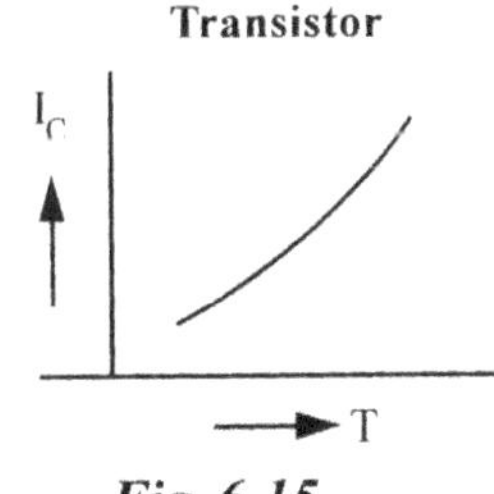

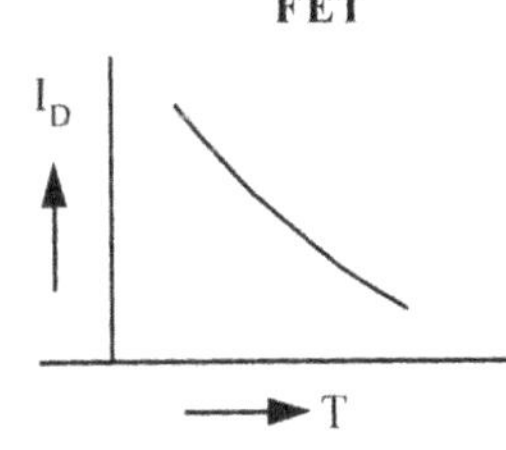

Fig 6.15

4. As I_B decreases I_C decreases in a transistor.
   As V_{GS} decreases, I_D decreases in FET.
   As $V_{control\ grid}$ increases I_p decreases in pentode.

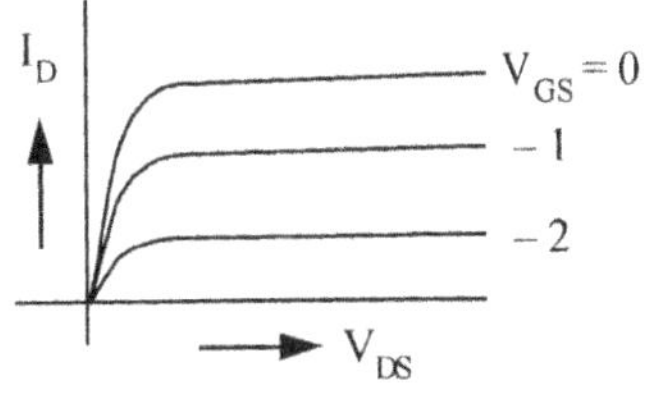

Fig 6.16

5. FET and pentode are voltage dependent devices. Transistor is current dependent.

6. r_D of FET is very large. $M\Omega$, r_C of transistor is less (in $k\Omega$).

7. Breakdown occurs in FET and BJT for small voltages 50, 60V etc. But in pentode it is much higher.

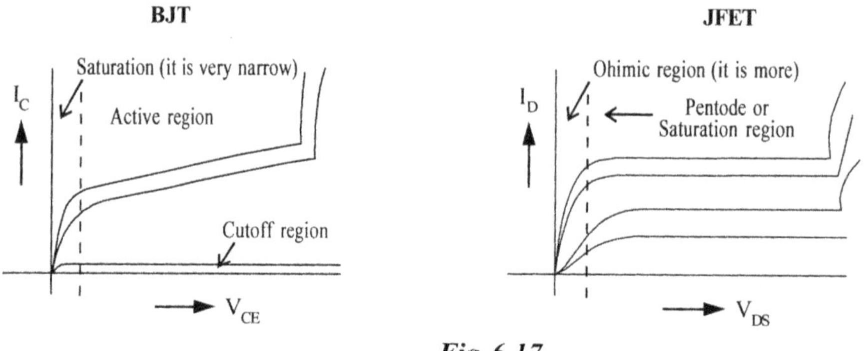

Fig 6.17

6.6   R. C. COUPLED AMPLIFIER

The circuit is as shown in Fig. 6.18 In the A.C. equivalent circuit, the collector output of the transistor (BJT) is coupled to the output point V_0 through a resistor (R) and capacitor (C). R_L and C_0 are the coupling elements. So it is called R-C coupled amplifier.

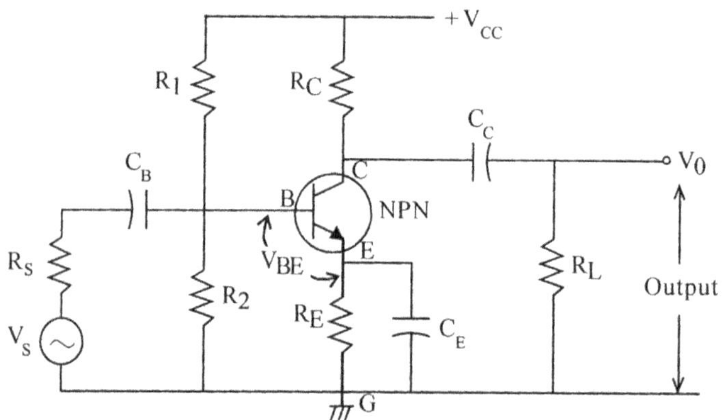

Fig 6.18   R.C coupled amplifier circuit.

R_1, R_2 and R_C, R_E are biasing resistors. It is a self bias circuit. The operating point 'Q' is located at the centre of the active region of the transistor characteristics. V_{CE} will be about $V_{CC}/2$, typically 5 to 7 V. V_{BE} must be 0.5 V for silicon transistor. I_B will be about 100

μA and I_C 5 mA. These are the proper bias conditions. R_S is the source resistance of the input A.C source. If an external resistance is connected on the input side, it is to limit the input A.C. voltage so that the BJT is not damaged due to excess input current. C_B is the blocking capacitor. It blocks D.C components present in the A.C source, so that the D.C biasing provided is not changed. C_C is the coupling capacitor, to couple the amplified A.C signal at the collector of the transistor to the output point V_0 R_E is emitter resistor to provide biasing, so that $V_{BG} - V_{EG}$ = V_{BE} = 0.5 to 0.6 V for silicon transistor. R_C is collector resistor to limit the collector current from V_{CC}, to protect the transistor. C_E is emitter bypass capacitor. It by passes A.C signal, so that A.C signal does not pass through R_E, C_E is chosen such that $X_{CE} = \dfrac{1}{10} R_E$, at the lower cut off frequency f_1, X_{CE} will be any way smaller than R_E.

The A.C equivalent circuit is as shown in Fig. 6.19.

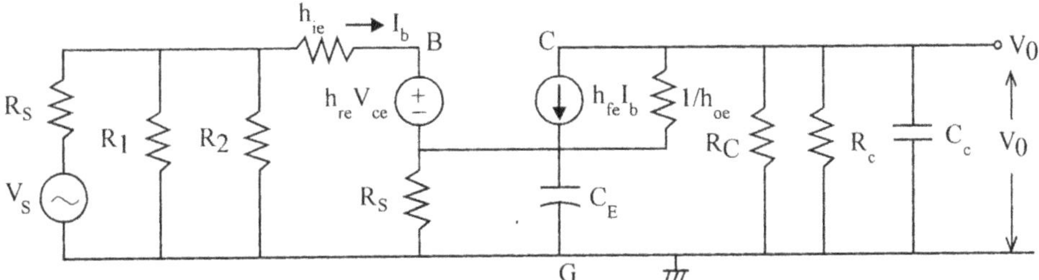

Fig 6.19 A.C equivalent circuit.

Frequency Response is a plot between frequency and voltage gain. Voltage gain is measured in decibels.

$$1 \text{ decibel} = \frac{1}{10} \text{ bel.} = 10 \log \frac{P_O}{P_i}$$

$$P_0 = \frac{V_0^2}{R} \; ; \; P_i = \frac{V_i^2}{R} \; ;$$

$$\therefore \qquad \text{Gain} = 10 \log \frac{V_0^2/R}{V_i^2/R} = 20 \log \left(\frac{V_0}{V_i}\right) db$$

As the frequency of the input signal varies, gain of the amplifier also varies, because of the variation of the reactance of capacitors with frequency. At high frequencies, capacitance associated with BJT also must be considered.

f_1 or f_L is called Lower cut off frequency or Lower half power point or Lower 3db frequency $(f_2 - f_1)$ is called Bandwidth.

f_1 and f_2 are chosen such that the gain at these frequencies is $\dfrac{1}{\sqrt{2}}$ or 0.707 of the maximum

gain value. At f_1 and f_2, output power is half of the maximum value. Because at these frequencies

voltage is $\dfrac{V_m}{\sqrt{2}}$. So output power is $\dfrac{V^2}{R} = \left(\dfrac{V_m}{\sqrt{2}}\right)^2 \bigg| R = \dfrac{V_m^2}{2R}$. $P_{max} = \dfrac{V_m^2}{R}$. So f_1 and f_2 are

called as half power points. At f_1 and f_2, $P_0 = \dfrac{P_{max}}{2}$. In decibels it is $20\,\log\left(\dfrac{1}{2}\right) \simeq 3db$. f_0 f_1

and f_2 are also called as 3 db points.

FREQUENCY RESPONSE

The graph of voltage gain versus frequency is as shown in Fig.6.20.

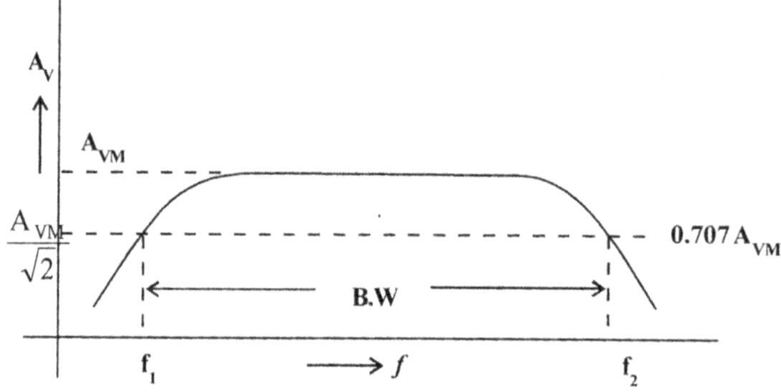

Fig 6.20   Frequency response.

    Initially as frequency increases, capacitance reactance due to C_b, $\alpha_{Cb} = \dfrac{1}{2\pi f C_b}$ decreases.
So the drop across it also decreases and hence net input at the base of transistor increases and

so V_0 increases. Hence $A_V = \dfrac{V_0}{V_i}$ also increases.

    In the mid frequency range, X_{Cb} is almost a ***short circuit,*** since its reactance decreases

with increasing frequency. In the low frequency range, $X_{CC} = \dfrac{1}{2\pi f C_C}$ is open circuit (in the
A.C equivalent circuit).

    In the mid frequency range, the capacitance reactances of both C_b and C_c are negligible.
So the net change with frequency, remains constant as shown in the figure.

    As the frequency is further increased, capacitance reactance due to C_C decreases. Hence
V_0 decreases because, V_0 is taken across C_c. So as frequency increases, V_0 decreases and
hence A_V decreases. So the shape of the curve is as shown in the Fig. 6.70.

$$\text{Expression } A_V \text{ (M.F)} = \dfrac{-hf_e R_L}{R_S + h_{ie}}$$

$$f_1 \simeq \frac{1}{2\pi C_b R_c}$$

$$f_2 \simeq \frac{1}{\pi C_c R_C}$$

PHASE RESPONSE

Phase shift between V_0 and V_i varies as shown in Fig. 6.21.

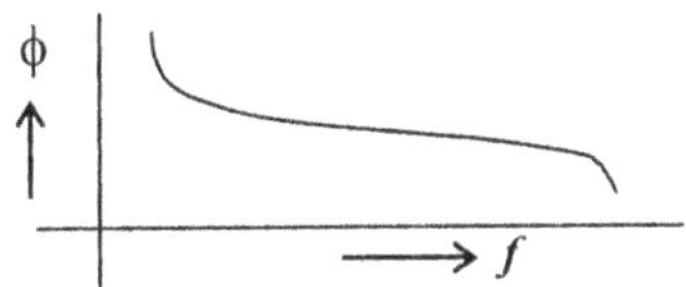

Fig 6.21   Phase response.

6.7   CONCEPT OF f_α, f_β AND f_T

In order to obtain some idea of a transistor's high frequency capability and what transistor to choose for a given application, we examine how transistor's CE short-circuit forward-current gain varies with frequency. When $R_L = 0$, the approximate high-frequency equivalent circuit is drawn below.

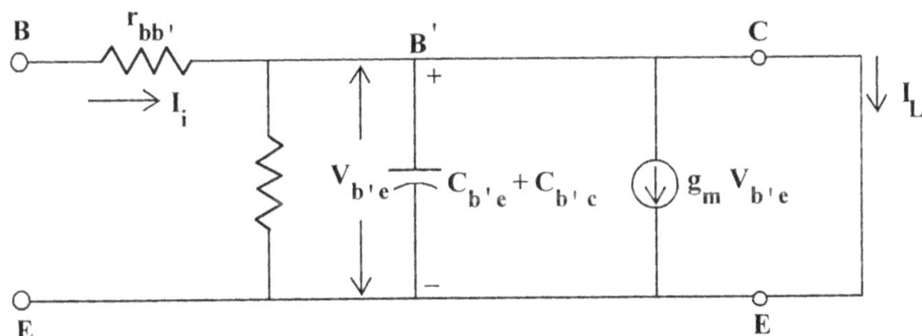

Fig 6.22   Approximate high-frequency circuit.

The current gain,
$$A_1 = \frac{I_L}{I_i} = \frac{-g_m V_{b'e}}{I_i}$$

$$I_i = V_{b'e}\, g_{b'e} + V_{b'e}\, [j\omega\, (C_{be} + C_{b'c})]$$

$$A_1 = \frac{-g_m V_{b'e}}{V_{b'e}\left[g_{b'e} + j\omega(C_{b'e} + C_{b'c})\right]} = \frac{-g_m}{g_{b'e} + j\omega(C_{b'e} + C_{b'c})}$$

$$\because g_{b'e} = \frac{g_m}{h_{fe}}$$

$$A_1 = \frac{-g_m h_{fe}}{g_m + j2\pi f(C_{b'e} + C_{b'c})h_{fe}}$$

$$= \frac{-h_{fe}}{\left[(j + j2\pi f \ (C_{b'e} + C_{b'c}) \dfrac{h_{fe}}{g_m} \right]} = \frac{-h_{fe}}{1 + j2\pi f \ (C_{b'e} + C_{b'c})}$$

$$\left[\because r_{b'e} \ \frac{h_{fe}}{g_m} \right]$$

$$A_I = \frac{-h_{fe}}{1 + j\left(\dfrac{f}{f_\beta} \right)}$$

Where
$$f_\beta = \frac{1}{2\pi r_{b'e}(C_{b'e} + C_{b'c})}$$

h_{fe} = CE small signal short-circuit current gain.

β *Cut-off Frequency 'f_β' :*

The 'β' cut-off frequency f_β, also referred as f_{hfe}, is the CE short-circuit small-signal forward-current transfer ratio cutoff frequency. It is the frequency at which a transistors CE short-circuit current gain drops 0.707 from its value at low frequency. Hence, f_β represents the maximum attainable bandwidth for current gain of a CE amplifier with a given transistor.

α *Cut-off Frequency 'f_α' :*

A transistor used in the CB connection has a much higher 3-dB frequency. The expression for the current gain by considering approximate high frequency circuit of the CB with output shorted is given by

$$A_I = \frac{I_L}{I_i} = \frac{-h_{fb}}{1 + j\left(\dfrac{f}{f_\alpha} \right)}$$

Where
$$f_\alpha = \frac{1}{2\pi r_{b'e}(1 + h_{fb})C_{b'e}} \approx \frac{h_{fe}}{2\pi \ r_{b'e} C_{b'e}}$$

$$\boxed{\ f_\alpha = \frac{h_{fe}f_\beta(C_{b'e} + C_{b'c})}{C_{b'e}}\ }$$

f_α is the alpha cutoff frequency at which the CB short-circuit small-signal forward current transfer ratio drops 0.707 from its 1KHZ value.

Gain Bandwidth Product :

Although f_a and f_b are useful indications of the high-frequency capability of a transistor, are even more important characteristics is f_T.

The f_T is defined as the frequency at which the short-circuit CE current gain has a magnitude of unity. The 'f_T' is lies in between f_β and f_α.

$$|A_I| = \frac{h_{fe}}{\sqrt{1+\left(f/f_\beta\right)^2}}$$

at $\qquad f = f_T, \; |A_I| = 1, \; \left(\dfrac{f_T}{f_\beta}\right)^2 \gg 1,$

we obtain $\qquad f_T / f_\beta \approx h_{fe}$

$$\boxed{f_T \approx h_{fe} \, f_\beta}$$

$\therefore$ f_T is the product of the low frequency gain, h_{fe}, and CE bandwidth f_β. f_T is also the product of the low frequency gain, h_{fb}, and the CB bandwidth, f_α . i.e $f_T = h_{fB} \cdot f_\alpha$. The value of f_T ranges from 1MHz for audio transistors upto 1GHZ for high frequency transistors.

The following Fig. 6.23 indicates how short-circuit current gains for CE and CB connections vary with frequency.

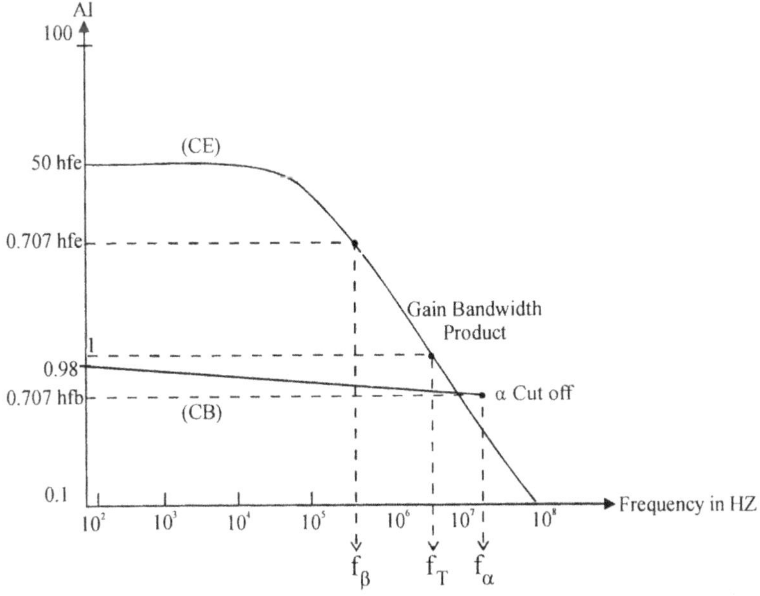

Fig 6.23

6.8   FEEDBACK AMPLIFIERS

Any system whether it is electrical, mechanical, hydraulic or pneumatic may be considered to have at least one input and one output. If the system is to perform smoothly, we must be able to measure or control output. If the input is 10mV, gain of the amplifier is 100, output will be 1V. If the input deviates to 9mV or 11mV, output will be 0.9 V or 1.1V. So there is no control over the output. But by introducing feedback between the output and input, there can be control over the output. If the input is increased, it can be made to increase by having a link between the output and input. By providing feedback, the input can be made to depend on output.

One example is, the temperature of a furnace. Suppose, inside the furnance, the temperature should be limited to 1000^0C. If power is supplied on, continuously, the furnace may get over heated. Therefore we must have a thermocouple, which can measure the temperature. When the output of the thermocouple, reaches a value corresponding to 1000^0C, a relay should operate which will switch off the power supply to the furnace. Then after sometime, the temperature may come down below 1000^0c. Then again another relay should operate, to switch on the power. Thus the thermocouple and relay system provides the feedback between input and output.

Another example is traffic light. If the timings of red, green and yellow lights are fixed, on one side of the road even if very few vehicles are there, the green light will be on for sometime. On the other hand, if the traffic is very heavy on the other side of the road, still if the green lamp glows for the same period, the traffic will not be cleared. So in the ideal case, the timings of the red and green lamps must be proportional to the traffic on the road. If a traffic policeman is placed, he provides the feedback.

Another example is, our human mind and eyes. If we go to a library, our eyes will search for the book which we need and indicates to the mind. We take the book, which we need. If our eyes are closed, we can't choose the book we need. So eyes will provide the feedback.

Basic definitions

Ideally an amplifier should reproduce the input signal, with change in magnitude and with or without change in phase. But some of the short comings of the amplifier circuit are

1. Change in the value of the gain due to variation in supplying voltage, temperature or due to components.
2. Distortion in wave-form due to non linearities in the operating characters of the amplifying device.
3. The amplifier may introduce noise (undesired signals)

   The above drawbacks can be minimizing if we introduce feedback

6.9   CLASSIFICATION OF AMPLIFIERS

Amplifiers can be classified broadly as,

1. Voltage amplifiers.
2. Current amplifiers.
3. Transconductance amplifiers.
4. Transresistance amplifiers.

This classification is with respect to the input and output impedances relative to the load and source impedances.

6.9.1 VOLTAGE AMPLIFIER

This circuit is a 2-port network and it represents an amplifier (see in Fig 6.24). Suppose $R_i \gg R_S$, drop across R_S is very small.

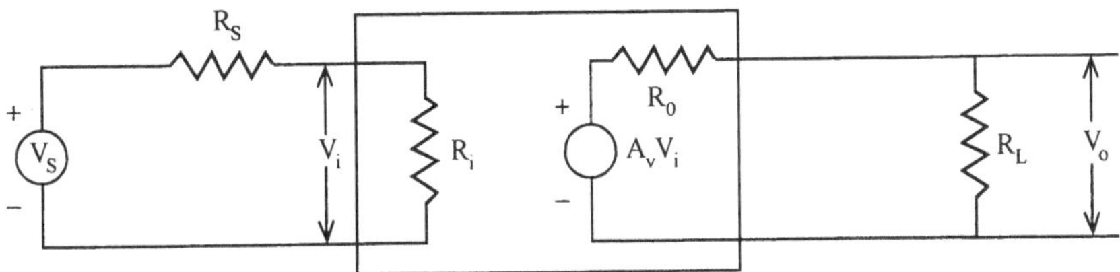

Fig 6.24 Equivalent circuit of voltage amplifiers.

$$\therefore \qquad V_i \simeq V_S.$$

Similarly, if $R_L \gg R_O$, $V_O \simeq A_V \cdot V_i$.

But $V_i \sim V_S$.

$$\therefore \qquad V_O \simeq A_V \cdot V_S.$$

$\therefore$ Output voltage is proportional to input voltage.

The constant of proportionality A_V doesn't depend on the impedances. (Source or load). Such a circuit is called as *Voltage Amplifier*.

Therefore, for ideal voltage amplifier

$$R_i = \infty.$$
$$R_O = 0.$$
$$A_V = \frac{V_o}{V_i}$$

with $\qquad R_L = \infty.$

A_V represents the open circuit voltage gain. For ideal voltage amplifier, output voltage is proportional to input voltage and the constant of proportionality is independent of R_S or R_L.

6.9.2 CURRENT AMPLIFIER

An ideal current amplifier is one which gives output current proportional to input current and the proportionality factor is independent of R_S and R_L.

The equivalent circuit of current amplifier is shown in Fig.6.25.

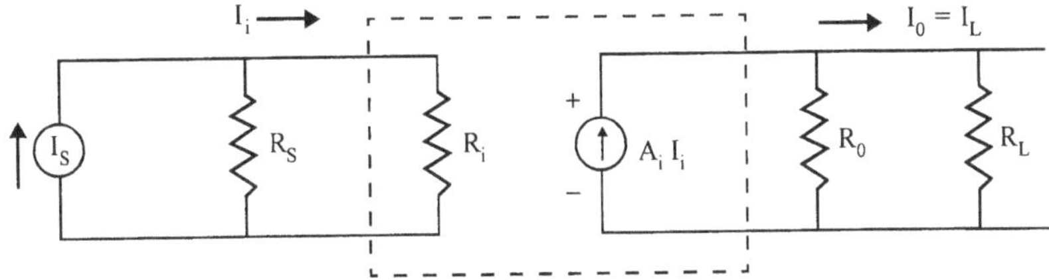

Fig 6.25 Current amplifier.

For ideal Current Amplifier,

$$R_i = 0$$
$$R_O = \infty.$$

If
$$R_i = 0, I_S \sim I_i.$$

$\therefore$
$$R_O = \infty,$$
$$I_L = I_O = A_i I_i = A_i I_S.$$

$\therefore$
$$A_I = \frac{I_L}{I_i} \text{ with } R_L = 0.$$

$\therefore$ A_I represents the short circuit current amplification.

6.9.3 TRANSCONDUCTANCE AMPLIFIER

Ideal Transconductance amplifier supplies output current which is proportional to input voltage independently of the magnitude of R_S and R_L.

Ideal Transconductance amplifier will have

$$R_i = \infty.$$
$$R_O = \infty.$$

In the equivalent circuit, on the input side, it is the Thevenins' equivalent circuit. A voltage source comes in series with resistance. On the output side, it is Norton's equivalent circuit with a current source in parallel with resistance. (See Fig. 6.26).

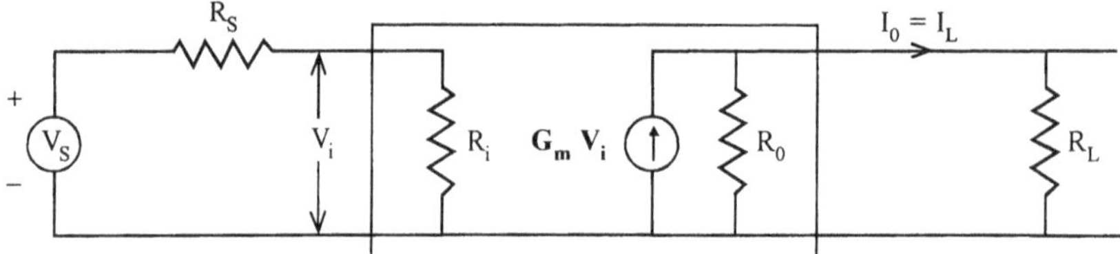

Fig 6.26 Equivalent circuit of Transconductance amplifier.

6.9.4 TRANS RESISTANCE AMPLIFIER

It gives output voltage V_0 proportional to I_s, independent of $R_s \propto R_L$. For *ideal amplifiers*

$$R_i = 0, R_0 = 0$$

Equivalent circuit

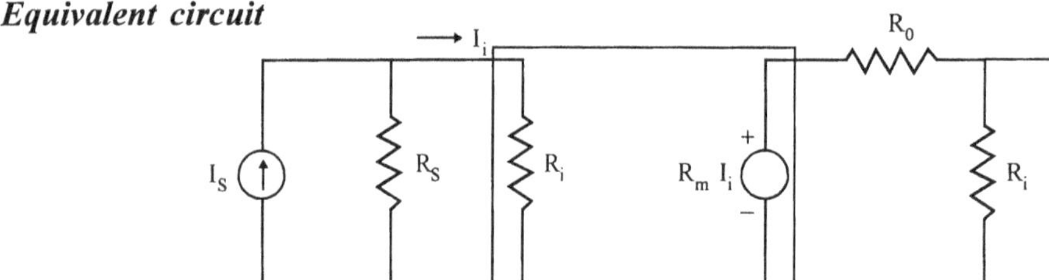

Fig 6.27 Trans resistance amplifier.

Norton equivalent circuit on the R_i input side.

Thevenins' equivalent circuit on the output side.

6.10 FEEDBACK CONCEPT

A sampling network samples the output voltage or current and this signal is applied to the input through a feedback two port network. The block diagram representation is as shown in Fig. 6.28.

GENERALIZED BLOCK SCHEMATIC

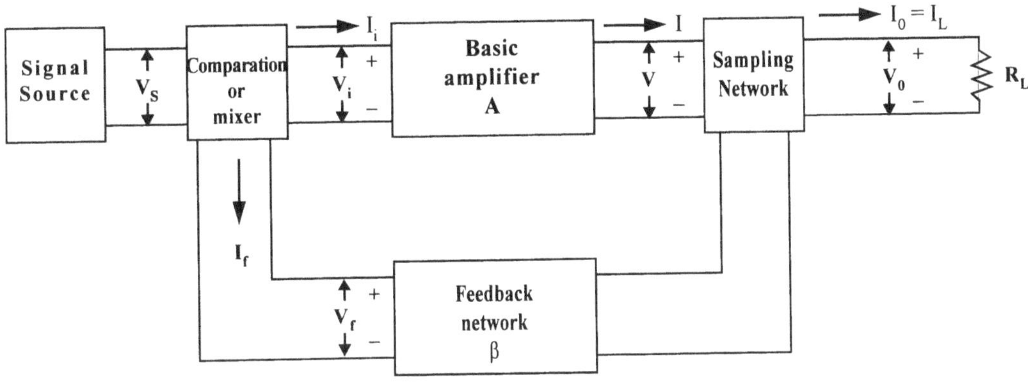

Fig 6.28 Block diagram of feedback network.

Signal Source

It can be a voltage source V_s or a current source I_s

FEEDBACK NETWORK

It is a passive two port network. It may contain resistors, capacitors or inductors. But usually a resistance is used as the feedback element. Here the output current is sampled and feedback. The feedback network is connected in series with the output. This is called as *Current Sampling or Loop Sampling*.

A voltage feedback is distinguished in this way from current feedback. For voltage feedback, the feedback element (resistor) will be in parallel with the output. For current feedback the element will be in series.

COMPARATOR OR MIXER NETWORK

This is usually a differential amplifier. It has two inputs and gives a single output which is the difference of the two inputs.

$$V = \text{Output voltage of the basic amplifier } before\ sampling \text{ [see the block diagram of feedback]}$$

$$V_i = \text{Input voltage to the basic amplifier}$$

$$A_V = \text{Voltage amplification} = V/V_i$$

$$A_I = \text{Current amplification} = I/I_i$$

$$G_M = \text{Transconductance of basic amplifier} = I/V_i$$

$$R_M = \text{Transresistance} = V/I_i.$$

All these four quantities, A_v, A_I, G_M and R_M represent the transfer gains (Though G_M and R_M are not actually gains) of the basic amplifier without feedback. So the symbol 'A' is used to represent these quantities.

A_f is used to represent the ratio of the output to the input *with feedback*. This is called as the transfer gain of the amplifier with feedback.

$$\therefore \qquad A_{Vf} = \frac{V_o}{V_s}$$

$$A_{If} = \frac{I_o}{I_s}$$

$$G_{Mf} = \frac{I_o}{V_s}$$

$$R_{Mf} = \frac{V_o}{I_s}$$

Feedback amplifiers are classified as shown below.

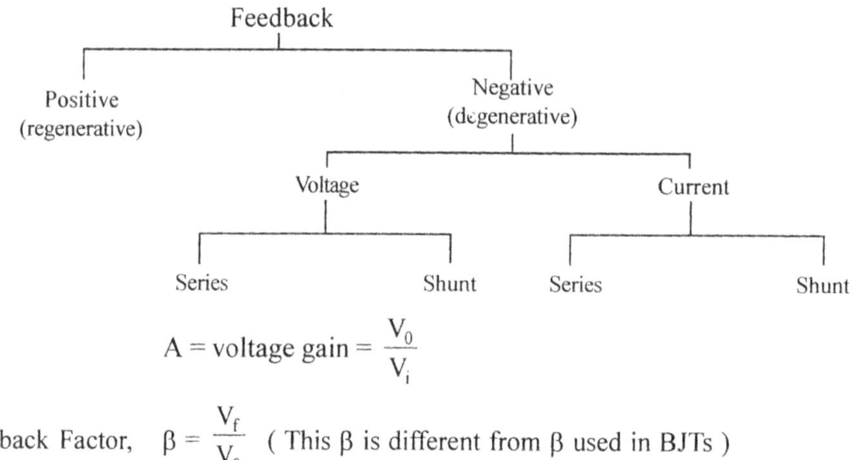

$$A = \text{voltage gain} = \frac{V_0}{V_i}$$

Feedback Factor, $\quad \beta = \dfrac{V_f}{V_0} \quad$ (This β is different from β used in BJTs)

$$(\beta)\,(A)\,(-1) = -\beta A$$

Loop gain / Return Ratio

-1 is due to phase-shift of 180^0 between input and output in Common Emitter Amplifier. Since $\text{Sin}(180^0) = -1$.

Return difference $D = 1-$ loop gain negative sign is because it is the difference

$$D = 1 - (-\beta A)$$

$$\boxed{D = 1 + \beta.A.}$$

Types of Feedback

How to determine the type of feedback ? Whether current or voltage ? If the feedback signal is proportional to voltage, it is *Voltage Feedback.*

If the feedback signal is proportional to current, it is *Current Feedback.*

Conditions to be satisfied

1. Input signal is transmitted to the output through amplifier A and not through feedback network β.
2. The feedback signal is transmitted to the input through feedback network and not through amplifier.
3. The reverse transmission factor β is independent of R_s and R_L.

6.11 TYPES OF FEEDBACK

Feedback means a portion of the output of the amplifier circuit is sent back or given back or feedback at the input terminals. By this mechanism the characteristics of the amplifier circuit can be changed. Hence feedback is employed in circuits.

There are two types of feedback.

1. Positive Feedback
2. Negative Feedback

Negative feedback is also called as ***degenerative feedback***. Because in negative feedback, the feedback signal opposes the input signal. So it is called as degenerative feedback. But there are many advantages with negative feedback.

Advantages of Negative Feedback

1. Input impedance can be increased.
2. Output impedance can be decreased.
3. Transfer gain A_f can be stabilized against variations in ***h-parameter*** of the transistor with temperature etc.

   i.e. stability is improved.
4. Bandwidth is increased.
5. Linearity of operation is improved.
6. Distortion is reduced.
7. Noise reduces.

6.12 EFFECT OF NEGATIVE FEEDBACK ON TRANSFER GAIN

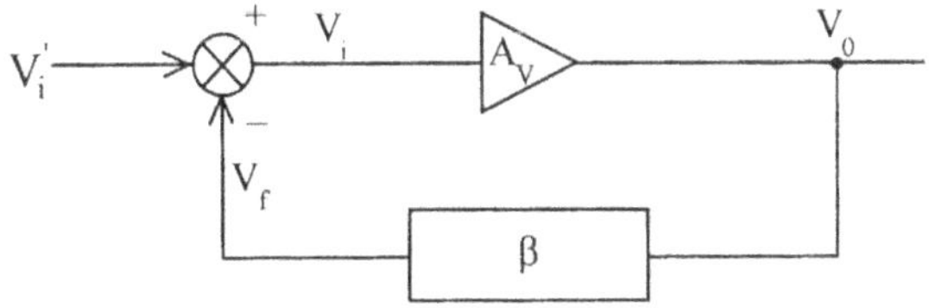

Fig 6.29 Block schematic for negative feedback.

$$A_v = \frac{V_0}{V_i}$$

$$A_v' = V_0 / V_i'$$

$$V_i' = V_i - \beta V_0$$

$$V_0 = A_v (V_i - \beta V_0)$$
$$V_0 = A_v . V_i - A_v . \beta V_0$$
$$V_0 (1 + \beta A_v) = A_v V_i$$

$$\therefore \quad A_{vf} = \frac{V_0}{V_i} = \frac{A_v}{1 + \beta A_v}$$

6.12.1 REDUCTION IN GAIN

$$A'_v = \frac{A_v}{1 - \beta A_v}$$

(for positive feedback)

A_v = Voltage gain without feedback. (Open loop gain).

If the feedback is negative, β is negative.

$$\therefore \quad A'_v = \frac{A_v}{1 - (-\beta . A_v)}$$

For negative feedback $A'_v = \dfrac{A_v}{1 + \beta A_v}$

Denominator is > 1. $\therefore$ $A'_v < A_v$

$\therefore$ There is reduction in gain.

6.12.2 INCREASE IN BANDWIDTH

If, f_H is upper cutoff frequency.

f_L is lower cut off frequency.

f is any frequency.

Expression for A_v (voltage gain at any frequency f) is,

$$A_v = \frac{A_v (\text{mid})}{1 + j . \dfrac{f}{f_H}} \qquad \qquad \qquad \dots\dots (1)$$

A_v (mid) = Mid frequency gain

$$A_v = \frac{A_v (\text{mid})}{1 + \beta A_v} \quad \text{for negative feedback.} \qquad \dots\dots(2)$$

Substituting equation (1) in (2) for A_v,

$$A_v = \frac{A_v (\text{mid})/1 + j . \dfrac{f}{f_H}}{1 + \beta \left[\dfrac{A_v \text{mid}}{1 + jf / f_H} \right]} \qquad \text{(for negative Feedback, } \beta \text{ is } V_e)$$

Simplifying,

$$A_v = \dfrac{\dfrac{A_V\,(\text{mid})}{f_H + j.f}\left[1 + j\dfrac{f}{f_H}\right]}{1 + j.f + \beta\,[A_V\ \text{mid}]}$$

$$= \dfrac{A_V\,(\text{mid})}{f_H + j.f + \beta.A_{Vmid}\cdot f_H}$$

$$A_v{}' = \dfrac{A_v(\text{mid})/1 + \beta_v A_v(\text{mid})}{1 + \dfrac{jf}{f_H[1 + \beta_v A_v(\text{mid})]}} \qquad\qquad \dots\dots (3)$$

Equation (3) can be written as,

$$A_v{}' = \dfrac{A_V'\,(\text{mid})}{1 + \dfrac{jf}{f_H'}}$$

Where

$$A_v{}'_{(\text{mid})} = \dfrac{A_v(\text{mid})}{1 + \beta_v A_v(\text{mid})}$$

and

$$f_H{}' = f_H\,(1 + \beta_v\,A_{v\,(\text{mid})})$$

$\because$ β is negative for negative feedback, $f_H{}' > f_H$.

$\therefore$ ***Negative feedback, increases bandwidth.***

Similarly,

$$f_L{}' = \dfrac{f_L}{1 + \beta_v A_{v(\text{mid})}} \qquad\qquad A_v = \dfrac{A_{v(\text{mid})}}{1 - j\left(\dfrac{f_L}{f}\right)}$$

or

$$A_v{}' = \dfrac{A_{V\,(\text{mid})}/1 + \beta A_{V(\text{mid})}}{1 + j\dfrac{f_L}{f(1 + \beta.A_V)}}$$

$$A_v{}' = \dfrac{A_V}{1 + j\dfrac{f_L'}{f}}$$

6.12.3 Reduction in Distortion

Suppose, the amplifier, in addition to voltage amplification is also producing distortion D.

$$V_0 = A_v\,.\,V_i + D.$$

Where,

$$V_i = V_i{}' - \beta_v.\,V_0 \quad \text{(for negative feedback)}$$

$$V_i{}' = V_i - \beta\,V_0 \quad \text{and } \beta \text{ is negative for negative feedback,}$$

$\therefore$

$$V_o{}' = A_v\,[V_i{}' - \beta_v V_o] + D$$

$$V_o \left[1 + A_V\, \beta_V \right] = V_i'.A_V + D$$

$$\therefore \qquad V_0 = \frac{V_i'.A_v}{(1 + A_v.\beta_v)} + \frac{D}{(1 + \beta_V A_V)}$$

For negative feedback, β is negative therefore denominator is > 1

$\therefore$ The distortion in the output is reduced.

$$= \frac{D}{1 + \beta_v A_v} \text{ is } < D$$

The physical explanation is, suppose the input is pure sinusoidal wave. There is distortion in the output as shown, in Fig. 7.7.

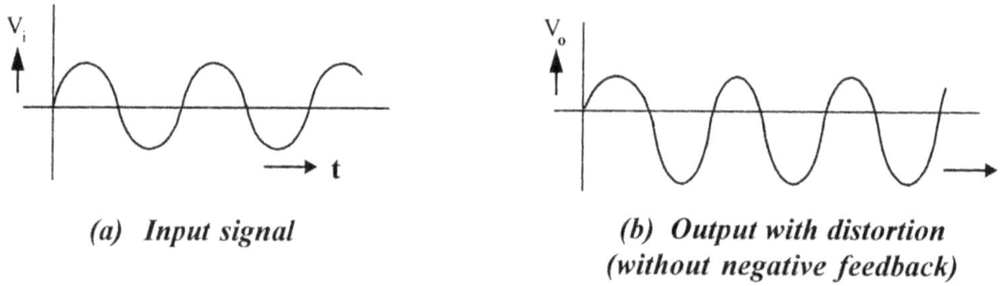

(a) Input signal   *(b) Output with distortion*

(without negative feedback)

Fig 6.30

Now, if a part of the distorted output is fed back to the input, so as to oppose the input, the feedback signal and input will be out of phase.

So the new input V_i will have distortion introduced init, because of mixing of distorted V_f with pure V_i. So the distortion in the output will be reduced because, the distortion introduced in the input cancels the distortion produced by the amplifier. Because these two distortions are out of phase. The feedback signal cancels the distortion produced by the amplifier, Therefore these two are out of phase.

6.12.4   FEEDBACK TO IMPROVE SENSITIVITY

Suppose an amplifier of gain A_1 is required. Build an amplifier of gain $A2 = DA_1$ in which D is large. Feedback is now introduced to divide the gain by D. Sensitivity is improved by the same factor D, because both gain and instability are divided by D. The stability will be improved by the same factor.

6.12.5   FREQUENCY DISTORTION

$$A_f = \frac{A}{1 + \beta A}$$

If the feedback network does not contain reactive elements. The overall gain is not a function of frequency. So frequency duration is less. If β depends upon frequency, with negative feedback, Q factor will be high.

6.12.6 BAND WIDTH

It increases with negative Feedback (as shown in Fig.7.2)

$$f_1^1 < f_1, \; f_2^1 > f_2$$

$$BW' = (f_1' - f_2') : BW = (f_2 - f_1)$$

$$f_1' = \frac{f}{1+\beta A_m} \quad f_2' = f_2 \, (1 + \beta A_m) \qquad BW' > BW$$

$$BW = f_2 - f_1 \simeq f_2$$

$$(BW)_f = f_2' - f_1 \sim f_2'$$

$$\therefore \quad (BW)_f = (1 + \beta A_m) \, BW$$

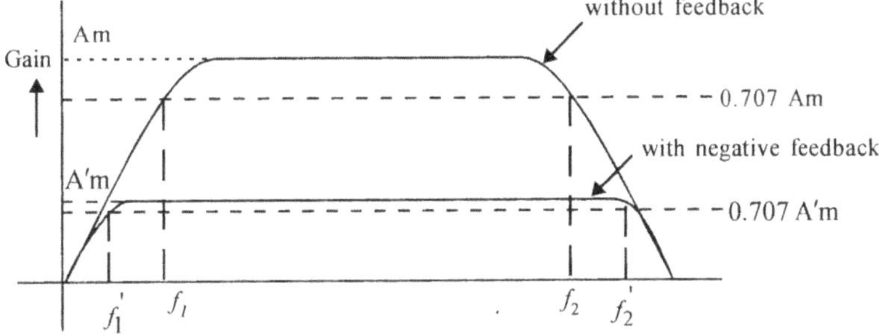

Fig 6.31 Frequency Response with and without negative feedback.

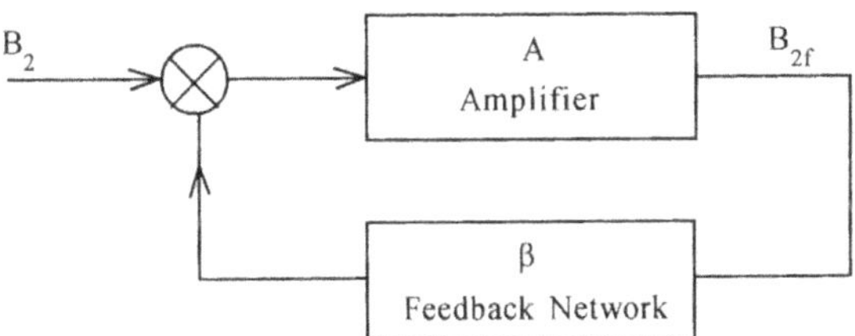

Fig 6.32 Feedback network.

6.12.7 SENSITIVITY OF TRANSISTOR GAIN

Due to aging, temperature effect etc., on circuit capacitance, transistor or FET, stability of the amplifier will be affected

Fractional change in amplification with feedback divided by the fractional change without feedback is called ***Sensitivity*** of Transistor.

$$= \frac{\left|\dfrac{dA_f}{A_f}\right|}{\left|\dfrac{dA}{A}\right|}$$

$$A_f = \frac{A}{1+\beta A}$$

Differentiating, $\quad \dfrac{dA_f}{A_f} = \dfrac{1}{|1+\beta A|}\left|\dfrac{dA}{A}\right|$

$\therefore$ Sensitivity, $\quad = \dfrac{1}{1+\beta A}$

Reciprocal of sensitivity is **Densitivity** $\boxed{D = (1+\beta A).}$

6.12.8 REDUCTION OF NONLINEAR DISTORTION

Suppose the input signal contains second harmonic and its value is B_2 before feedback. Because of feedback. B_{2f} appears at the output. So positive βB_{2f} is fed to the input. It is amplified to $-A\beta B_{2f}$.

$\therefore$ Output with two terms $B_2 - A\beta B_2$ f.

$\therefore \qquad\qquad B_2 - A\beta B_2 f = B_{2f}$

or $\qquad\qquad B_{2f} = \dfrac{B_2}{1+\beta A} \quad B_{2f} < B_2$

So it is reduced.

6.12.9 REDUCTION OF NOISE

Let N be noise constant without feedback and N_F with feedback. N_F is fed to the input and its value is βN_F. It is amplified to $-\beta A N_F$.

$\therefore \qquad\qquad N_F = N - \beta A N_F$

$\qquad\qquad N_F(1+\beta A) = N$

as $\qquad\qquad N_F = \dfrac{N}{1+\beta A}$

$\therefore \qquad\qquad N_F < N$. Noise is reduced with negative feedback.

6.13 TRANSFER GAIN WITH FEEDBACK

Consider the generalized feedback amplifier shown in Fig.6.33.

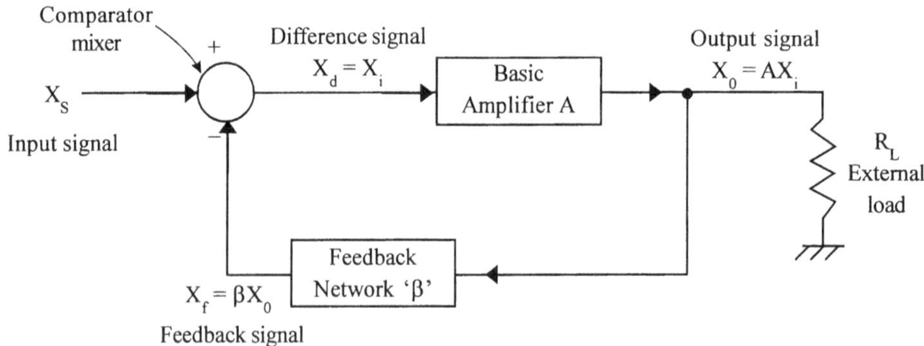

Fig 6.33

The basic amplifier shown may be a voltage amplifier or current amplifier, transconductance or transresistance amplifier.

The four different types of feedback amplifiers are,

1. Voltage series feedback.
2. Voltage shunt feedback.
3. Current series feedback.
4. Current shunt feedback.

X_s = Input signal.

X_0 = Output signal.

X_f = feedback signal.

X_d = Difference signal. [Difference between the input signal and the feedback signal].

β = Reverse transmission factor or feedback factor = X_f / X_o.

The feedback network can be a simple resistor. The mixer can be a difference amplifier. The output of the mixer is the difference between the input signal and the feedback signal.

$$X_d = X_s - X_f = X_i.$$

X_d is also called as error or comparison signal.

$$\beta = \frac{X_f}{X_o}$$

It is often a positive or negative real number.

This β should not be confused with the β of a transistor $\dfrac{I_C}{I_B}$

Transfer gain $\qquad A = \dfrac{X_o}{X_i} = \dfrac{X_o}{X_d}$

$$X_i = X_d \qquad\qquad\qquad .....(1)$$

Gain with feedback $A_f = \dfrac{X_o}{X_s}$

But $\qquad\qquad X_d = X_s - X_f = X_i \qquad\qquad .....(2)$

$$\beta = \frac{X_f}{X_o} \qquad\qquad\qquad .....(3)$$

From (2), $\quad X_d = X_s - X_f.$

Substitute (2) in (1).

$\therefore \qquad\qquad A = \dfrac{X_0}{X_s - X_f}$

Dividing Numerator and Denominator by X_s,

$$A = \frac{X_o/X_s}{1 - \dfrac{X_f}{X_s}} = \frac{X_o/X_s}{1 - \dfrac{X_f}{X_o} \cdot \dfrac{X_o}{X_s}}.$$

We know that $X_o / X_s = A_f;\ \dfrac{X_f}{X_o} = \beta.$

$$\therefore \qquad A = \frac{A_f}{\left(1 - \beta.A_f\right)}$$

or $\qquad A - \beta.\ A_f.\ A = A_f.$

$$A = A_f\,(\,1 + \beta A)$$

$$\therefore \qquad \boxed{A_f = \frac{A}{1 + \beta A}}$$

A_f = gain with feedback.

A = transfer gain without feedback.

If $|A_f| < |A|$ the feedback is called as negative or degenerative, feedback

If $|A_f| > |A|$ the feedback is called as positive or regenerative, feedback

6.13.1   LOOP GAIN

In the block diagram of the feedback amplifier, the signal X_d which is the output of the comparator passes through the amplifier with gain A. So it is multiplied by A. Then it passes through a feedback network and hence gets multiplied by β, and in the comparator it gets multiplied by -1. In the process we have started from the input and after passing through the amplifier and feedback network, completed the loop. So the total product $A\beta$.

In order that **series feedback** is most effective, the circuit should be driven from a **constant voltage** source whose internal resistance R_s is small compared to R_i of the amplifier. If R_s is very large, compared with R_i, V_i will be modified not by V_f but because of the drop across R_s itself. So effect of V_f will not be there. Therefore for series feedback, the voltage source should have less resistance. (See Fig.6.34).

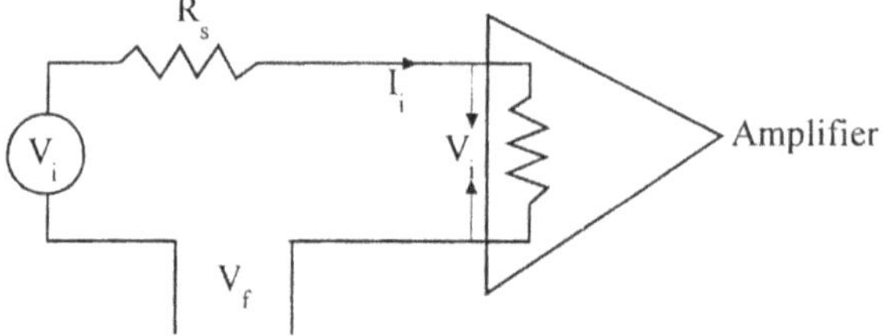

Fig 6.34   Series feedback.

In order that **shunt feedback** is most effective, the amplifier should be driven from a constant current source whose resistance R_s is very high ($R_s \gg R_i$).

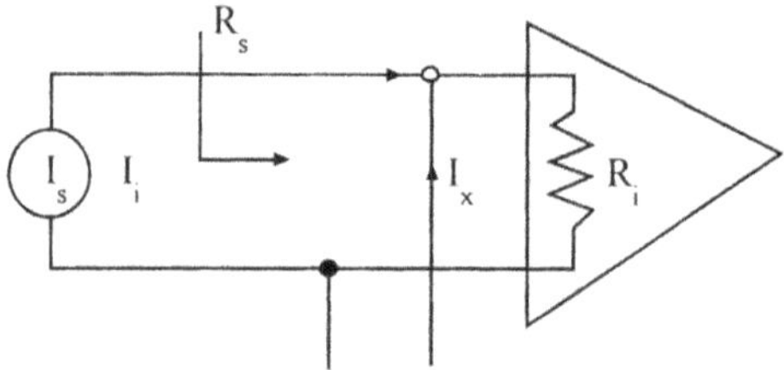

Fig 6.35 Current shunt feedback.

If the resistance of the source is very small, the feedback current will pass through the source and not through R_i. So the change in I_i will be nominal. Therefore the source resistance should be large and hence a current source should be used.

If the feedback current is same as the output current, then it is series derived feedback.

When the feedback is shunt derived, output voltage is simultaneously present across R_L and across the input to the feedback. So in this case V_f is proportional to V_o.

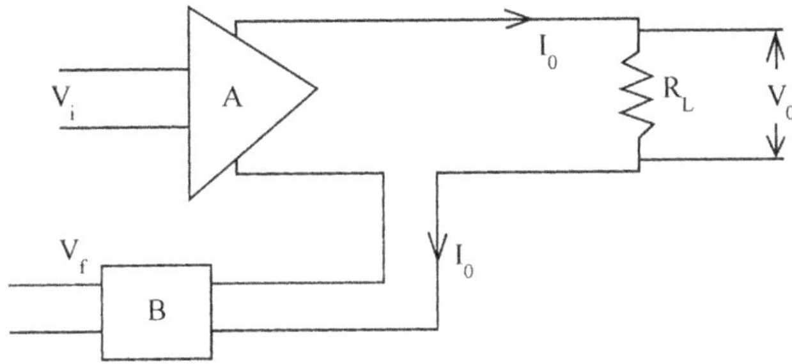

Fig. 6.36 Series derived feedback.

An amplifier with shunt derived negative Feedback increases the output resistance. When the feedback is series derived, I_o remains constant, so R_o increase.

Similarly if the feedback signal is *shunt fed* it reduces the input resistance.

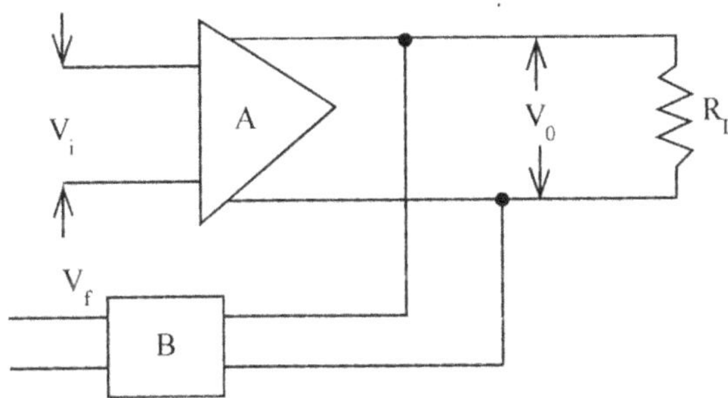

Fig. 6.37 (a) Shunt derived feedback.

If it is *series fed*, it increases the output resistance. Therefore I_o remains constant, even through V_o increase, so R_0 increases.

Return Ratio

βA = Product of feedback factor β and amplification factor A is called as **Return Ratio**.

Return Difference (D)

The difference between unity (1) and return ratio is called as **Return difference**.

$$D = 1 - (-\beta A) = 1 + \beta A.$$

Amount of feedback introduced is expressed in decibels

$$= N \log \left(\frac{A'}{A}\right) \quad N = 20 \log \left|\frac{A_f}{A}\right| = 20 \log \left|\frac{1}{1 + A\beta}\right|$$

If feedback is negative N will be negative because $A' < A$.

6.14 CLASSIFACTION OF FEEDBACK AMPLIFIERS

There are four types of feedback,

    1. Voltage series feedback.

    2. Voltage shunt feedback.

    3. Current shunt feedback.

    4. Current series feedback.

Voltage Series Feedback

Feedback signal is taken across R_L proportional to V_o. So it is voltage feedback. V_f is coming in series with V_i So it is Voltage series feedback.(See Fig.6.38).

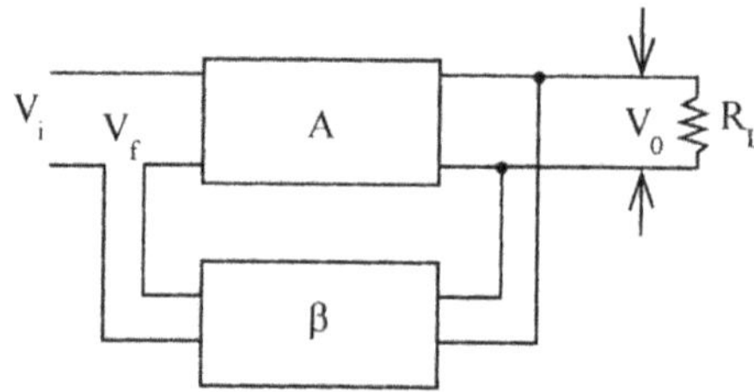

Fig 6.38   Schematic for voltage series feedback.

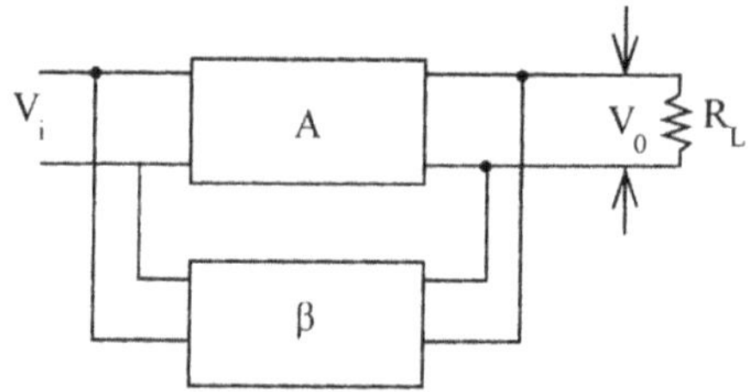

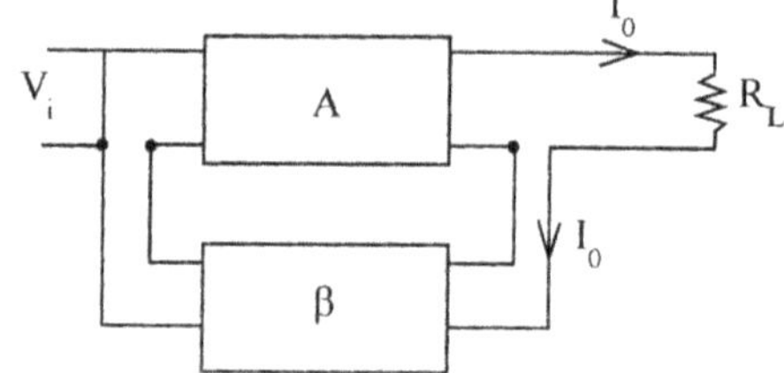

(a)   Voltage shunt Feedback   *(b)   Current Shunt Feedback*

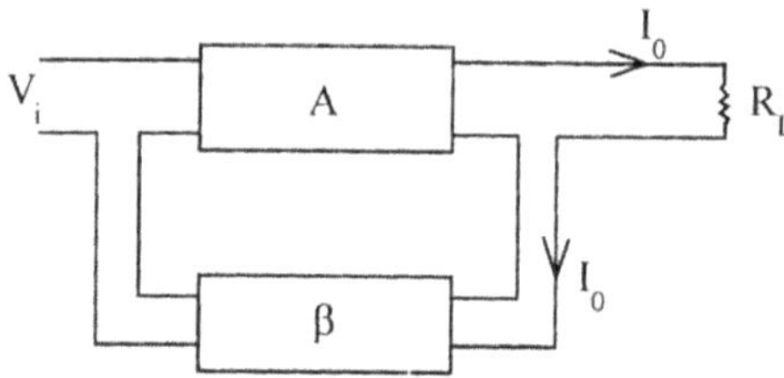

(c)   Current Series Feedback

Fig 6.39

Improvement of Stability with Feedback

Stability means, the stability of the voltage gain. The voltage gain must have a stable value, with frequency. Let the change in A_v is represented by S.

$$\frac{dA_v}{A_v} = S\left(\frac{dA'_v}{A'_v}\right) = S = \frac{\left(\dfrac{d A'_v}{A'_v}\right)}{\left(\dfrac{d A_v}{A_v}\right)}$$

$$A'_v = \frac{V_0}{V_S} \text{ (for negative feedback)}$$

$$A'_v = \frac{A_v}{1 + A_v \beta_1}$$

Differentiating with respect to A_v, $\dfrac{dA_v}{A_v} = \dfrac{\left(1 + \beta.A_v\right) - A_v\left(+\beta\right)}{\left(1 + \beta.A_v\right)^2} = \dfrac{1}{\left(1 + A_v\beta\right)^2}$

Dividing by A_v on both sides, of $\dfrac{dA'_v}{dA_v} = \dfrac{1}{\left(1 + \beta A_v\right)^2}$

$$\frac{dA'_v}{A_v}\frac{1}{dA_v} = \frac{1}{\left(1 + \beta A_v\right)} \cdot \frac{1}{A'_v}$$

But $\qquad A'_v = \dfrac{A'_v}{1 + \beta A_v} \text{ and } \dfrac{dA'_v}{A'_v} = S' \ \dfrac{dA_v}{A_v} = S$

$$S = \frac{dA'_v}{dA_v} = \frac{1}{\left(1 + \beta A_v\right)^2}$$

$$\boxed{S = \frac{dA_v}{\left(1 + \beta A_v\right)}}$$

For negative Feedback, β is negative $\therefore$ denominator > 1.

$\therefore \qquad\qquad S' < S$

i.e., variation in A_v, or % change in A_v is less with -negative feedback.

$\therefore$ *Stability* is good.

6.15 EFFECT OF FEEDBACK ON INPUT RESISTANCE

6.15.1 INPUT RESISTANCE WITH SHUNT FEEDBACK

With feedback $\qquad R_{if} = \dfrac{V_i}{I_s}$

$$I_f = \beta_i\, I_o. \qquad \left[\because \beta_i = \frac{I_f}{I_o}\right]$$

$$I_s = I_i + \beta\, I_o$$

$$R_i' = \frac{V_i}{I_i + \beta_i I_O}$$

$$I_o = A_i \cdot I_i$$

$$R_i' = \frac{V_i}{I_i + \beta_i A_i I_i}$$

$$R_i = \frac{V_i}{I_i(1 + \beta_i A_i)}$$

$$\boxed{R_i' = \frac{R_i}{(1 + \beta_i A_i)}}$$

$\therefore$ *Input resistance decreases with shunt feedback.*

If the feedback signal is taken across R_L, it is α V_o or so it is *Voltage feedback.*

If the feedback signal is taken in series with the output terminals, feedback signal is proportional to I_o. So it is *current feedback*.

If the feedback signal is in series with the input, it is *series feedback*.

If the feedback signal is in shunt with the input, it is *shunt feedback*.

EXPRESSION FOR R_I WITH CURRENT SHUNT FEEDBACK

A_i = Shunt circuit current gain of the BJT

A_I = practical current gain I_0 / I_i

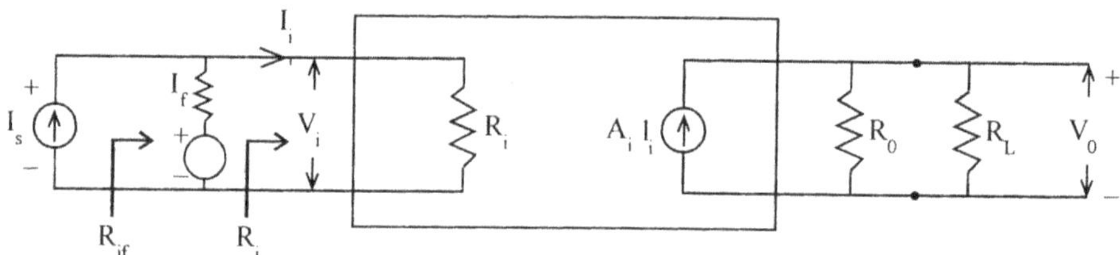

Fig. 6.40 Current shunt feedback.

A_i represents the Shunt circuit current gain taking R_s into accounts,

$$I_S = I_i + I_f = I_i + \beta I_0$$

and

$$I_0 = \frac{A_i R_o I_i}{R_o + R_L} = A_I I_i \qquad \therefore \quad \text{Let} \quad A_I = \frac{A_i R_o}{R_o + R_L}$$

$$A_I = \frac{I_o}{I_i} = \frac{A_i R_o}{R_o + R_i} = (I_i + I_f)$$

$$I_S = I_i + \beta \cdot I_0 = I_i \left(1 + \beta \cdot \frac{I_o}{I_i}\right) = I_i (1 + \beta \cdot A_I)$$

$$R_{if} = V_i / I_s \qquad R_i = V_i / I_i$$

$$\boxed{R_{if} = \frac{V_i}{(1 + \beta A_I)I_i} = \frac{R_i}{1 + \beta \cdot A_I}}$$

for shunt feedback the input resistance decreases.

6.15.2 Input Impedance with Series Feedback

$$V_i' = V_i + \beta V_0 \text{ (in general case)}.$$

For negative feedback, β is negative.

$$V_i = \frac{V_i^1}{(1+\beta A)}$$

$$\frac{V_i}{I_i} = \frac{V_i'}{I_i(1+\beta A)}$$

In general, R_i increases,

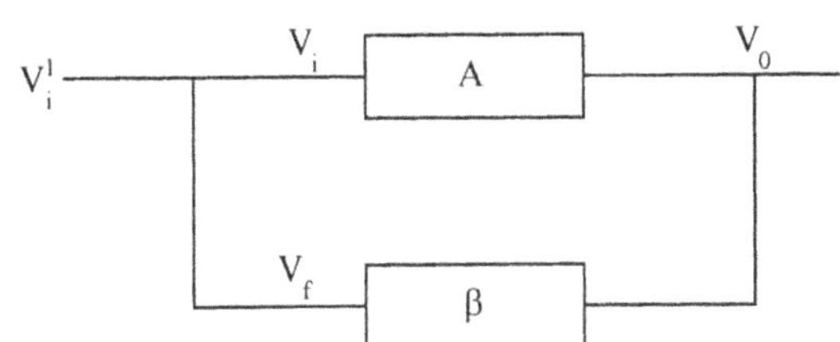

Fig 6.41 Feedback network.

$$V_i' = V_i + \beta V_0$$

$$V_i' = V_i + \beta.A. V_i \qquad \because \quad V_0 = A. V_i$$

$$V_i' = V_i (1+\beta A)$$

But
$$V_i' = I_i R_i$$

$\therefore$
$$V_i' = (1+\beta A) R_i I_i$$

$$\frac{V_i'}{I_i} = \text{Input Z seen by the source} = R_i (1+\beta A)$$

$\therefore$
$$R_{if} = R_i (1+\beta A)$$

Expression for R_i with Voltage Series Feedback

In this circuit A_v represents the open circuit voltage gain taking R_s into account. (see Fig.7.18).

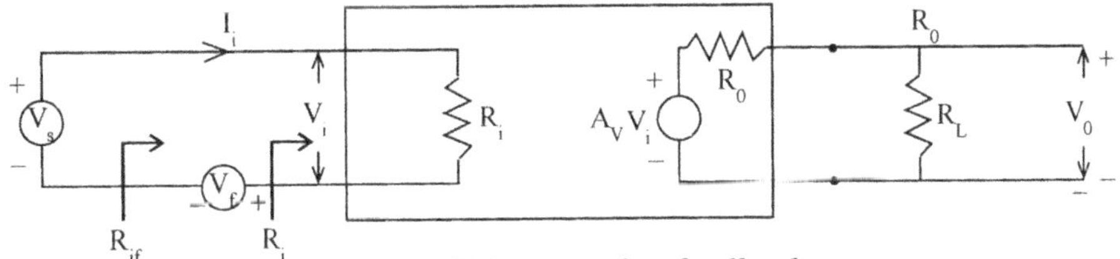

Fig. 6.42 Voltage series feedback.

$$V_i = V_S - V_f$$

or
$$V_s = V_i + V_f$$

$$R_0' = R_{if} = V_S/I_i.$$

$$V_f = I_i R_i + V_f = I_i R_i + \beta V_0$$

$$V_0 = \frac{A_v V_i R_L}{R_0 + R_L} = (A_V I_i R_i)$$

Let,
$$\frac{A_v.R_L}{R_0 + R_L} = A_V, \quad V_i = I_i. R_i.$$

$$A_V = \frac{V_o}{V_i} = \frac{A_v R_L}{R_o + R_L}$$

$$V_S = I_i R_i + \beta \cdot V_0$$

$$\boxed{\frac{V_s}{I_i} = R_i + \frac{\beta.V_o}{I_i} = R_i\left(1 + \frac{\beta.V_o}{I_i \times R_i}\right)}$$

$$R_{if} = R_i\left(1 + \beta.\frac{V_o}{V_i}\right) = R_i\left(1 + \beta.A_V\right)$$

A_V = voltage gain, without feedback.

for series feedback input resistance increases.

6.16 EFFECT OF NEGATIVE FEEDBACK ON R_O

Voltage feedback (series or shunt) R_o decreases.

Current feedback (series or shunt) R_o increases.

Series feedback (voltage or current) R_i increases.

Shunt feedback (voltage or current) R_i decreases.

OUTPUT RESISTANCE

Negative feedback tends to decrease the input resistances. Feeding the voltage back to the input in a degenerative manner is to cause lesser increase in V_0. Hence the output voltage tends to remain constant as R_L changes because output resistance with feedback $R_{of} \ll R_L$.

Negative feedback, which samples the output current will tend to hold the output current constant. Hence an output current source is created ($R_{0f} \gg R_L$). So this type of connection increases output resistance.

For voltage sampling $R_{of} < R_0$. For current feedback $R_{of} > R_0$.

6.16.1   VOLTAGE SERIES FEEDBACK

Expression for R_{0f} looking into output terminals with R_L disconnected,

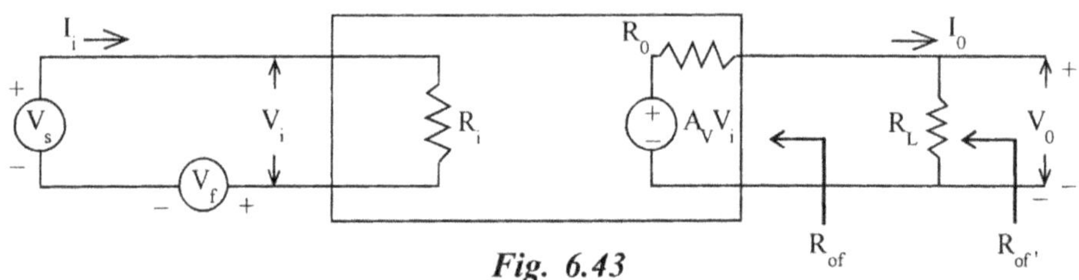

Fig. 6.43

R_0 is determined by impressing voltage 'V' at the output terminals or messing 'I', with input R_{of}' terminals shorted.

Disconnect R_L. To find R_{of}, remove external signal (set $V_s = 0$, or $I_s = 0$)

Let $R_L = \infty$.

Impress a voltage V across the output terminals and calculate the current I delivered by V. Then, $R_{0f} = V|I$.

$$I = \frac{V_0 - A_V V_i}{R_0} = \frac{V_0 + \beta A_v V}{R_o}$$

$\because$ $\qquad$ V_0 = output voltage.

$\qquad$ $V_i = -\beta_V$

Because with $\qquad$ $V_S = 0$, $V_i = -V_f = -\beta_V$

Hence, $\qquad$ $R_{of} = \dfrac{V_0}{I} = \dfrac{R_o}{1 + \beta A_v}$

This expression is excluding R_L. If we consider R_L also R_{of} is in parallel with R_L.

$$R_{of}' = \frac{R_{of} \cdot R_L}{R_{of} + R_L} \qquad \text{Substitute the l value of } R_{of}$$

$$R_{of}' = \frac{\dfrac{R_o}{1 + \beta A_V} \times R_L}{\dfrac{R_0}{1 + \beta A_V} + R_L} = \frac{R_o R_L}{R_o + R_L + \beta \cdot A_v R_L}$$

6.16.2 CURRENT SHUNT FEEDBACK

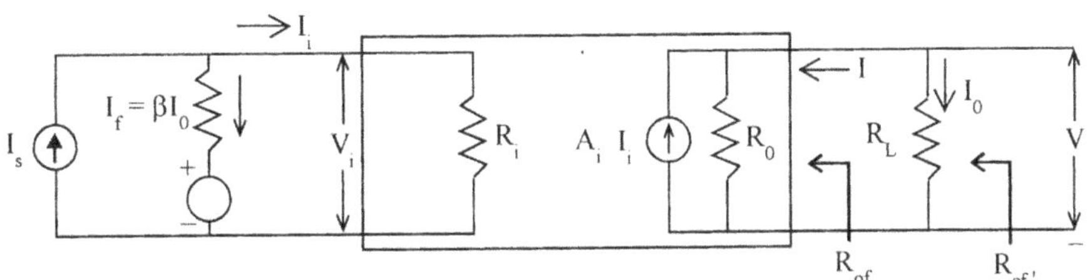

Fig. 6.44 *Block schematic for current shunt feedback amplifier.*

$\therefore$ $\qquad$ $I_0 = -I$

$\qquad$ A_i = Shunt circuit current gain

$\qquad$ A_I = Practical current gain $\left(\dfrac{I_0}{I_i} \right)$

$$I = \frac{V}{R_o} - A_i I_i$$

With, $\qquad$ $I_S = 0$, $I_i = -I_f = -\beta I = +\beta I$,

$$I = \frac{V}{R_o} - \beta A_i I \quad \text{or} \quad I(1 + \beta A_i) = \frac{V}{R_o}$$

$$R_{of} = \frac{V}{I} = R_o(1 + \beta A_i)$$

$\qquad$ A_i = short circuit current gain.

R_{of}' includes R_L as part of the amplifier.

$$R'_{of} = \frac{R_{of} \cdot R_L}{R_{of} + R_L}$$

$$= \frac{R_o(1 + \beta A_i)R_L}{R_o(1 + \beta A_i) + R_L}$$

$$= \frac{R_o R_L}{R_o + R_L} \times \frac{1 + \beta A_i}{1 + \frac{\beta A_i R_o}{R_o + R_L}}$$

$$A_I = \frac{I_o}{I_i} = \frac{A_i R_o}{R_o + R_L}$$

$$R'_o = \frac{R_o \cdot R_L}{R_o + R_L}$$

$\therefore \qquad R'_{of} = R'_o \dfrac{1 + \beta A_i}{1 + \beta A_I}$

If, $\qquad R_L = \infty,$

$\qquad A_I = 0, \ R'_o = R_0$

$\therefore \qquad \boxed{R'_{of} = R_0(1 + \beta A_i)}$

Problem 6.2

The following information is available for the generalized feedback network. Open loop voltage amplification $(A_v) = -100$. Input voltage to the system $(V_i') = 1\,\text{mV}$. Determine the closed loop voltage amplification, the output voltage, feedback voltage, input voltage to the amplifier, and type of feed back for (a) $\beta = 0.01$, (b) $\beta = -0.005$ (c) $\beta = 0$ (d) $\beta = 0.01$.

Solution

$$A'_v = \text{closed loop voltage amplification} = \frac{A_v}{1 - \beta_v A_v}$$

Sign must be considered,

$$A'_v = \frac{-100}{1 - 0.01(-100)} = -50.$$

$\because$ It is positive feedback, A'_v is less negative $\quad \therefore$ there is increase in gain.

V_o Output voltage $= V_i A_v$

$$= -50 \times 10^{-3}\,\text{mV}$$

$$= -50\,\text{mV}.$$

V_x Feedback voltage $= \beta V_o = 0.01(-50 \times 10^{-3})$

$$= -0.5\,\text{m V}.$$

$V_i = V_i' + \beta_v \cdot V_o$

$$= 10^{-3} + (-0.5 \times 10^{-3})$$

$$= 0.5\,\text{m V}.$$

This is negative feedback because, $V_i < V_i'$.

$$\therefore \qquad |A_V'| < |A_v|.$$

Problem 6.3

In the above problem determine the % variation in A_v^1 resulting from 100 % increase in A_v when $\beta_v = 0.01$.

When $A_v = -100$, $A_v^1 = -50$.

Solution

If A_v increases by 100 %, Then new value of $A_v = -200$.

$$A_V' = \frac{A_v}{1 - \beta_v A_v}$$

$$A_V' = \frac{A_v}{1 - \beta_v A_v}$$

Now, $\qquad A_V' = \dfrac{-200}{1 - 0.01(-200)} = -66.7.$

Change in A_V' is $-66.7 - 50 = 16.7$

$$\therefore \qquad \text{Variation} = \frac{16.7}{50} \times 100 = 33.3\%$$

Problem 6.4

An amplifier with open loop voltage gain $A_v = 1000 \pm 100$ is available. It is necessary to have an amplifier where voltage gain varies by not more than $\pm\, 0.1\,\%$

   (a) Find the reverse transmission factor β of the feedback network used

   (b) Find the gain with feedback.

Solution

(a) $\qquad \dfrac{dA_f}{A_f} = \dfrac{1}{1 + \beta A} \cdot \dfrac{dA}{A}$

or $\qquad \dfrac{0.1}{100} = \dfrac{1}{(1 + \beta A)} \cdot \dfrac{100}{1000}$

$\therefore \qquad (1 + \beta A) = 100$ or $\beta A = (100 - 1) = 99.$

Hence, $\qquad \beta = \dfrac{99}{1000} = 0.099$

(b) $\qquad A_f = \dfrac{A}{1 + \beta A} = \dfrac{1000}{1 + 99} = 10$

Problem 6.5

A gain variation of $+\,10\,\%$ is expected for an amplifier with closed loop gain of 100. How can this variation be reduced to $+\,1\,\%$

Solution

$$S' = \frac{S}{1-\beta A_v}.$$

for positive feedback,

$$S' = \frac{S}{1-\beta A_v}$$

$$S' = 0.01, \quad S = 0.1$$

for negative Feedback, $(1+\beta A_v) = \dfrac{0.1}{0.01} = 10 = \dfrac{S}{S'}$

$$A'_V = 100 = \frac{A_v}{1+\beta A_v} = \frac{A_v}{10}$$

$$\therefore \quad A'_V = 100 \times 10 = 1000.$$

$$1 + \beta A_v = 10 ; \qquad \therefore \quad \beta.A_v = 10 - 1 = 9$$

$$\therefore \quad \beta = \frac{9}{A_v} = \frac{9}{1000} = 0.009$$

$\therefore$ By providing negative feedback, with $\beta = 0.009$, we can improve the stability to 1%.

Problem 6.6

An amplifier with $A_v = -\,500$, produces 5% harmonic distortion at full output. What value of β is required to reduce the distortion to 0.1% ? What is the overall gain?

Solution

$$D' = \frac{D}{1+\beta A_v} \quad \text{(for negative feedback)}$$

$$D = 5; \quad D' = 0.1$$

$$N' = \frac{N}{\left(1+\beta A_v\right)}$$

$$0.1 = \frac{5}{1+\beta\,[500]}$$

or $\qquad\qquad \beta = \dfrac{4.9}{50} = 0.098.$

$$A'_V = \frac{-A_v}{1+\beta A_v} = -\frac{500}{50} = -10.$$

$$\frac{D'}{D} = \frac{A_v'}{A_v}.$$

$$\therefore \qquad A_v' = A_v\left(\frac{0.1}{5}\right) = -10$$

Problem 6.7

An amplifier has voltage gain of 10,000 with ground plate supply of 150 V and voltage gain of 8000 at reduced rate supply of 130 V. On application of negative feedback. The voltage gain at normal plate supply is reduced by a factor of 80. Calculate the 1) voltage gain of the amplifier with feedback for two values of plate supply voltage. 2) Percentage reduction in voltage gain with reduction in plate supply voltage for both condition, with and without feedback.

Solution

$$A = \frac{V_o}{V_s}. \text{ When } V_s = 150 \text{ V and } A = 10{,}000.$$

$$\therefore \qquad V_o = 10{,}000 \times 150 = 15 \times 10^5 \text{ V. (at normal supply voltages)}$$

$$\beta = \frac{V_f}{V_o}. \quad A_{fb} = \frac{A}{1+\beta A}$$

But
$$A_{fb} = \frac{10{,}000}{80}$$

$\because$ it is reduced by a factor of 80 with feedback. $\left(\frac{1}{80}^{Th} \text{ of } 10{,}000\right)$

$$\therefore \qquad \frac{10{,}000}{80} = \frac{10{,}000}{1+\beta \times 10000}$$

$$1 + \beta(10{,}000) = 80$$

$$\therefore \qquad \beta = +\frac{79}{10^4}$$

$$= 0.0079$$

V_o when $V_s = 130$ V is gain $\times V_s = 8000 \times 130 = \mathbf{104 \times 10^4}$ **V**

Voltage gain of amplifier with feedback when $V_s = 150$V,

$$A_{fb_1} = \frac{10{,}000}{80} = 125.$$

$$V_s = 130 \text{ V}.$$

$$A_{fb_2} = \frac{8000}{1+0.0079 \times 8000} = 124.7$$

$$\% \text{ stability of gain without feedback} = \frac{10{,}000 - 8000}{10{,}000} \times 100 = 20\%$$

$$\% \text{ stability of gain with feedback} = \frac{125 - 124.7}{125} \times 100 = 0.24\%$$

$\therefore$ With negative feedback stability is improved.

6.17   ANALYSIS OF FEEDBACK AMPLIFIERS

When an amplifier circuit is given, separate the basic amplifier block and the feedback network. Determine the gain of the basic amplifier A. Determine the feedback factor β. Knowing A and β of the feedback amplifier, the characteristics of the amplifiers R_i, R_o, noise figure, A_f etc., can be determined .

Complete analysis of the feedback amplifier is done by the following steps :

Step I : First identify whether the feedback signal X_f is a voltage feedback signal or current feedback signal. If the feedback signal X_f is applied in shunt with the external signal, *it is shunt feedback*. If the feedback signal is applied in series, it is *series feedback*.

Then determine whether the sampled signal X_o is a voltage signal or current signal. If the sampled signal X_o is taken between the output node and ground, it is voltage feedback.

If the sampled signal is taken from the output loop, it is current feedback.

Step II :

2. Draw the basic amplifier without feedback

3. Replace the active device (BJT or FET) by proper model (hybrid $-\pi$ equivalent circuit or *h-parameter* model)

4. Indicate V_f and V_0 in the circuit.

5. Calculate $\beta = V_f / V_0$

6. Calculate 'A' of basic amplifier.

7. From A and β, calculate $A_f = \dfrac{A}{1+\beta A}$; R_{if} and R_{of}.

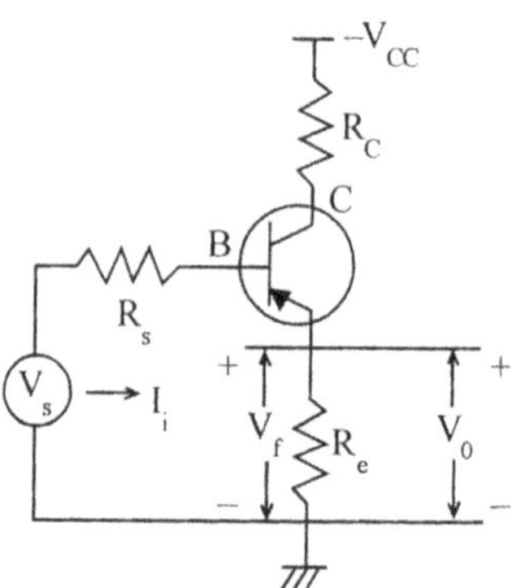

Fig. 6.45 Circuit for voltage series feedback.

6.17.1   Voltage series Feedback

Amplifier circuit is shown in Fig. 6.46.

This is emitter follower circuit because output is taken across R_e. The feedback signal is also across R_e. So $V_f = V_o$. because feedback signal is proportional to output voltage. If V_o increases V_f also increase and if V_0, V_f also decreases because V_f a V_o. So it is

Voltage feedback. Now the drop across R_e, i .e . V_f opposes the input voltage. It reverse biases the feedback in V_f coming in series with V_{BE} and it opposes it .So it is negative ***series voltage feedback***.

In the current series feedback, V_f is the voltage across R_e but, output is taken across R_c or R_L and not Re. So in that case $V_f \propto I_o$ or I_c or not V_o . But in this case, it is emitter follower circuit. Output is taken across R_e and that itself is the feedback signal V_f.

Let us draw the base amplifier without feedback.

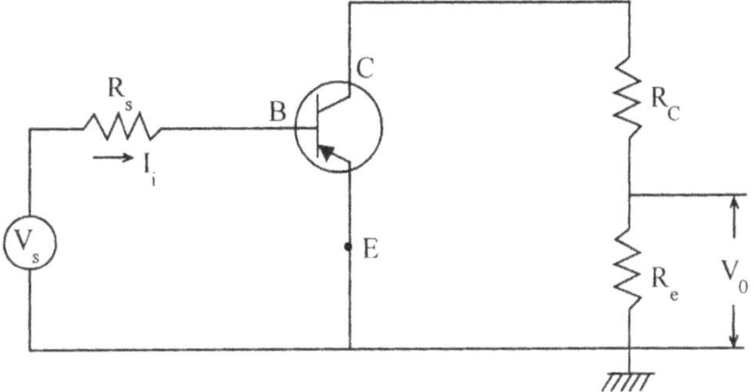

Fig. 6.46 Simplified circuit.

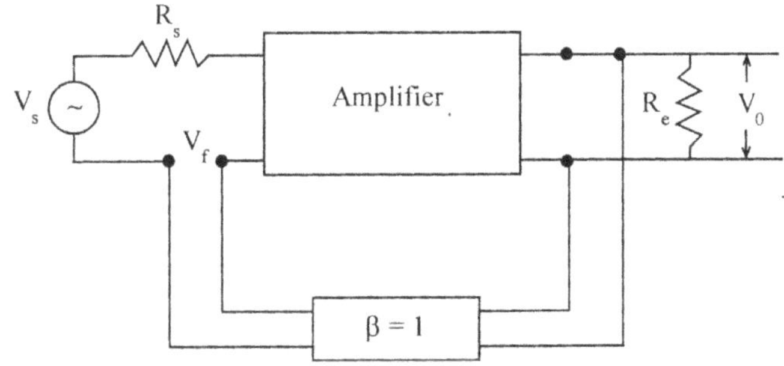

Fig. 6.47 Block schematic.

EQUIVALENT CIRCUIT

Now replace the transistor by its low frequency h-parameter equivalent circuit.

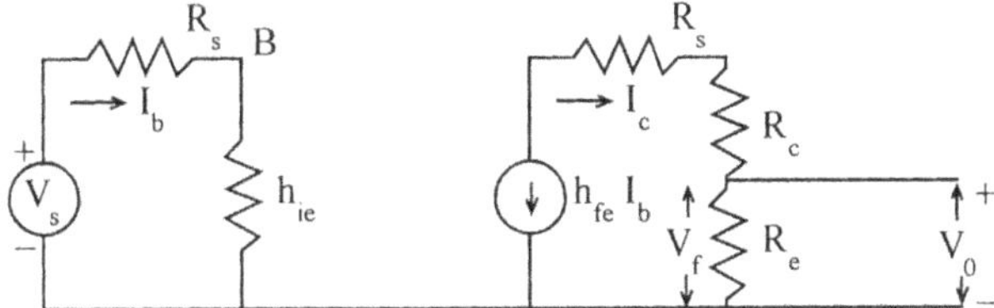

Fig. 6.48 Block Schematic

$$V_o = V_f$$

$$\therefore \quad \boxed{\beta = \frac{V_o}{V_f} = 1.}$$

V_S is considered as part of the amplifier. So $V_i = V_S$.

$$A_v = \frac{V_o}{V_i} \; . \; V_o = h_{fe}. \, I_b.R_e$$

$$I_C \simeq I_e \; ; \; I_C = h_{fe} . \, I_b \quad (\because \text{It is voltage source, } R_s \text{ in services with } h_{ie})$$

$$V_i = (R_s + h_{ie}) \, I_i$$

$$I_i = I_b$$

$$A_v = \frac{h_{fe} \, I_b \, R_e}{I_b(R_s + h_{ie})}$$

$\therefore$
$$A_v = \frac{h_{fe}.R_e}{R_s + h_{ie}} = \text{Voltage gain without feedback}$$

A'_V : Voltage gain with feedback.

$$A'_V = \frac{A_v}{1 + \beta A_v}$$

$$\beta = 1$$

$$A_V = \frac{h_{fe}.R_e}{R_s + h_{ie}}$$

$$= \frac{\dfrac{h_{fe}. \, R_e}{R_S + h_{ie}}}{1 + \dfrac{1.h_{fe}. \, R_e}{R_S + h_{ie}}}$$

$$A'_V = \frac{h_{fe}.R_e}{R_s + h_{ie} + h_{fe}R_e}$$

$$h_{fe} \, R_e \gg (R_s + h_{ie})$$

$\therefore$
$A_{Vf} \cong 1.$ or <1, which is true for emitter follower.

R'_i = Input resistance without feedback is $R_s + h_{ie}$.

R'_i = Input resistance with feedback

$R'_i = R_i (1 + \beta \, A_v)$ (voltage Feedback increase input resistance)

$\beta = 1,$

$$A'_V = \frac{h_{fe}.R_e}{R_s + h_{ie}}$$

$\therefore$
$$R'_i = (R_s + h_{ie}) \left(1 + \frac{h_{fe}.R_e.1}{R_s + h_{ie}}\right)$$

$$R_i' = \frac{(R_s + h_{ie})(R_s + h_{ie} + h_{fe}R_e)}{(R_s + h_{ie})}$$

$$\boxed{R_i' = R_s + h_{ie} + h_{fe}R_e}$$

R_i' : Output resistance with feedback

Output resistance of the circuit is R_e. Because output taken across R_e and not R_c and ground. Load resistance is also R_e.

∴ Considering load resistance,

$$R_0' = \frac{R_0'}{1 + \beta A_v}$$

$$R_0' = \frac{R_e}{1 + \dfrac{1 \cdot h_{fe} \cdot R_e}{R_s + h_{ie}}}$$

Problem 6.8

Calculate A_{Vf}, R_{of} and R_{if} for the voltage series feedback for the circuit shown in Fig. 6.49.

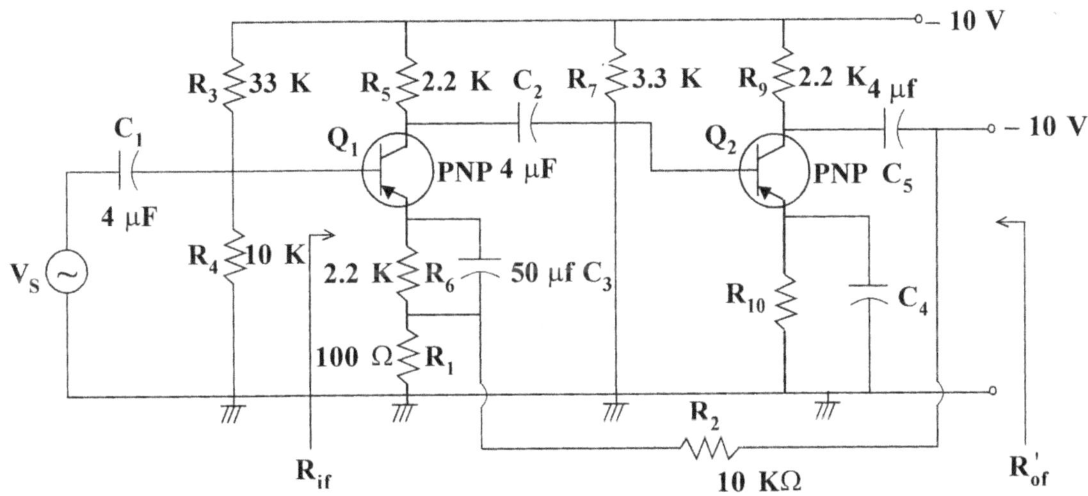

Fig. 6.49 For Problem 6.7.

Assume, $R_s = 0$, $h_{fe} = 50$,

$h_{ie} = 1.1\ kh_{re}$

$= h_{oe} = 0$ and identical transistor.

The second collector is connected to the first emitter through the voltage divider $R_1\ R_2$. Capacitor C_1, C_2, C_5 are DC blocking capacitors. C_3 and C_4 are bypass capacitors for the emitter resistors. All these capacitances represent negligible reactance at high frequencies.

Solution

Voltage gain without feedback $A_v = A_{v_1} \times A_{v_2}$.

Load resistance, R'_{L1}

R_5 in parallel with R_7 ($\because$ at high frequency X_{C2} is negligible), in parallel with

$$R_8, \text{ and } h_{ie2} \ R'_{L1} = 2.2 \parallel 33 \parallel 4.3 \parallel 1.1 \ k\Omega = 980\Omega.$$

$$R'_L = R_5 \parallel R_7 \parallel R_8 \parallel h_{ie2}$$

Effective load resistance R'_{L2} of transistor Q_2 is its collector resistance R_9 in parallel with

$$(R_1 + R_2); \ R'_{L2} = R_9 \parallel (R_1 + R_2)$$

$\therefore \qquad R'_{L2} = 2.2 \ k\Omega \parallel^{le} \text{ with } 10.1 k\Omega \simeq 2 \ k\Omega$

Effective emitter impedance R_e of Q_1 is R_1 parallel with R_6.

$$R_{e1} = R_1 \parallel^{le} R_6 \qquad \because \ C_3 \text{ is short for AC}$$
$$R_e = 0.1 \parallel^{le} \text{ with } 2.2 \ k\Omega = 0.098 k\Omega = 98\Omega$$

The voltage gain A_{V_1} of Q_1 for a common emitter transistor with emitter resistance

is $\qquad A_{V_1} = \dfrac{V_1}{V_i} = \dfrac{-h_{fe}R'_L}{h_{ie} + (1 + h_{fe})R_e}$

For Q_1 emitter is not at GROUND potential. So for A_V, this formula must be used.

$$= \frac{-50 \times 0.980}{1.1 + (51) \times 0.098} \simeq -7.78$$

Voltage gain A_{v2} of transistor Q_2,

$$A_{v2} = A_1 \cdot \frac{R_L}{R_i} = -\frac{h_{fe}R_L}{h_{ie}}$$

$$A_{v2} = -\frac{h_{fe} \cdot R_{L2}}{h_{ie}} = -\frac{50 \times 2}{1.1} = -91$$

For Q_2 emitter is bypassed. So $R_e = 0$. $\therefore$ For A_v this formula is used.

Voltage gain A_v of the two stages is cascade without feedback

$$A_v = \frac{V_o}{V_i} = A_{v1} \times A_{v2} = 7.78 \times 91 \simeq 728$$

$$\beta = \frac{R_1}{R_1 + R_2} = \frac{100}{100 + 10,000} = \frac{100}{10100} = \frac{1}{101} \simeq 0.01$$

$$A_v \times \beta = 728 \times 0.01 = 7.28$$

$$D = 1 + \beta A_v = 8.28 \ .$$

$$A_{vf} = \frac{A}{1 + \beta A} = \frac{A_v}{D} = \frac{728}{8.28} \simeq 90$$

$$A_{vf} = \frac{A}{1 + \beta A} = \frac{A_v}{D} = \frac{728}{8.28} \simeq 90$$

Input resistance without external feedback

$$R_i = h_{ie} + (1 + h_{fe})\, R_e = 1.1 + (1 + 50)\, 0.098 \simeq 6.1 \text{K } \Omega$$

$$R_i' = R_{if} = R_i(1 + \beta A) = 6.1\,(1 + 7.28) = 50.5 \approx 51$$

Output resistance without feedback $R' = R'_{L_2} = 2 \text{ k}\Omega$

Output resistance without feedback $R_{0f}' = \dfrac{R_o}{1 + \beta A} = \dfrac{2}{8.28} = 0.24 \text{ K } \Omega$

Equivalent Circuit

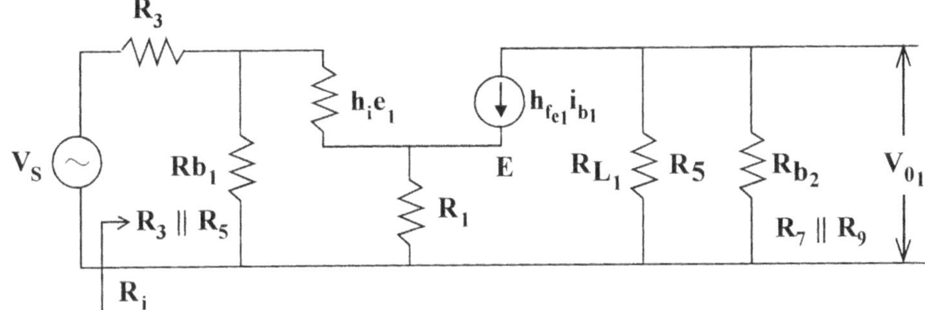

Fig. 6.50 For Problem 6.7.

BJT will not behave like a fixed resistor with h_{ie}, value . So circuit analysis is not simply parallel or series combination.

$$A_V = \frac{V_o}{V_i} = \frac{R_L}{R_s + r_{bb'}}$$

$$R_o = \frac{V_o}{R_L} = \frac{V_i}{R_s + r_{bb}'}$$

$$A_I = \frac{I_o}{I_i} = \frac{R_s}{R_s + r_{bb}'}$$

6.17.2 Current Shunt Feedback

The circuit is shown in Fig. 7.27.

Amplifier circuit parameters in terms of transistor h-parameters.

$$h_{ie} = R_i$$
$$h_{fe} = \beta$$
$$h_{oe} = R_0$$

The circuit shows two transistors Q_1 and Q_2 in cascade(See Fig.7.30). Feedback is provided from the emitter of Q_2 to the base of Q_1. This is negative feedback because, V_{i2} the input voltage to Q_2 is >> V_{i1}. V_{i2} is out of phase with V_{i1}. V_{i2} >> V_{i1} because Q_1 is in Common emitter configuration. A_V is large. Also, V_{i2} is 180^0 out of phase with V_{i1}. Q_2 is emitter follower because emitter is not at ground potential. Voltage is taken across R_e. [This voltage follows the collector voltage. So it is emitter follower]. So $A_v < 1. \simeq 0.99$

$\therefore$ V_{e2} is slightly less than V_{i2}, and there is no phase shift. [because emitter follows action]

$\therefore$ V_{e2} is in phase with V_{i2}.

    V_{i2} is out of phase with V_{i1}.

$\therefore$ V_{e2} is out of phase with V_{i1} (180^0). So it is negative feedback

    $V_{i2} \gg V_{i1}$. $V_{e2} \simeq V_{i2}$.

    $V_{e2} \gg V_{i1}$.

If the input signal V_s increases, I_s' the input current from source also increases. If I_s' increases,

I_f also increases ($\because$ V_{e2} increases as I_s' increases)

$I_i = I_s' - I_f$. (I_i is the base current for the transistor Q_1. So it is negative feedback)

This is current shunt feedback because,

$$I_f = \frac{V_{e_2} - V_i}{R'}$$

But $V_{i1} \ll V_{e2}$.

$$I_f \simeq + \frac{V_{e_2}}{R'} .$$

I_0 = collector current of Q_2. $V_{e2} = (I_0 - I_f) R_e$

  $\simeq$ emitter current of Q_2.

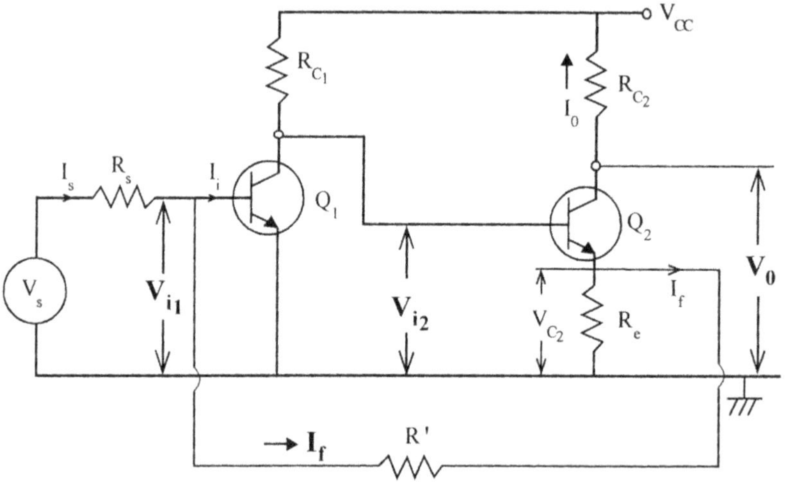

Fig. 6.51 Current shunt feedback.

Dividing by R'.

$$\frac{V_{e2}}{R'} = I_f = \frac{(I_0 - I_f)R_e}{R'} .$$

$\therefore$

$$I_f = + \frac{(I_0 - I_f)R_e}{R'} ;$$

$$I_f + \frac{I_f . R_e}{R'} = \frac{I_0 . R_e}{R'}$$

$$I_f \left(1 + \frac{R_e}{R'}\right) = \frac{I_0 . R_e}{R'}$$

$$I_f = \frac{R' . I_0 . R_e}{(R'+R_e)R'} = \frac{R_e}{(R'+R_e)} I_0$$

$$\boxed{\beta = \frac{I_f}{I_0} = \frac{R_e}{R'+R_e}}$$

$$I_f \; \alpha \; I_0$$

$\therefore$ This is current feedback

A'_I : Current gain with feedback.

$$A'_1 = \frac{I_0}{I_s'}$$

$$I'_S = I_i + I_f \text{ which is small } \mu A \; I_i = I_b$$

$\therefore$ $$I'_S \simeq I_f$$

$\therefore$ $$A'_1 = \frac{I_0}{I_f} = \frac{1}{\beta} \qquad\qquad \because \; \beta = \frac{I_f}{I_0}$$

$\therefore$ $$\boxed{A'_1 = \frac{R'+R_e}{R_e}}$$

A'_V : Voltage gain with feedback.

$$\frac{V_o}{V_s} \text{ (with feedback)}$$

$$V_0 = I_0 . R_{c2}$$
$$V_s = I_s . R_s .$$

$$\frac{V_o}{V_s} = \frac{I_0 . R_{c2}}{I_s R_s} \qquad\qquad I_S \simeq I_f$$

$$A'_V \simeq \frac{I_0}{I_f} . \frac{R_{c2}}{R_s} \qquad\qquad \frac{i_0}{I_f} = \frac{1}{\beta}$$

$$= \frac{1}{\beta} . \frac{R_{c2}}{R_s} \qquad\qquad \beta = \frac{R_e}{R'+R_e}$$

$$\boxed{A'_V = \frac{R'+R_e}{R_e} . \frac{R_{c2}}{R_s}}$$

We shall determine R_i', R_0' taking numerical values. The actual circuit, without feedback, considering R' can be drawn as shown in Fig. 6.52 To represent R' on the input side, with output terminals open circuited ($\because$ it is emitter follower configuration, output is taken across emitter and ground). When E_2 is left open (To make $I_0 = 0$). R' is in series with R_{e2} and thus total resistance is between base B_1, and ground. Hence on the input side, R' is in series with R_{e2}.

To find the output resistance, the base of Q_1 is shorted to ground because one terminal of R' is connected to (input shorted looking the high output terminals) B_1 is at ground. R' or R_{e2} are in parallel.

$\therefore$ The circuit is as shown below, (Fig. 6.52)

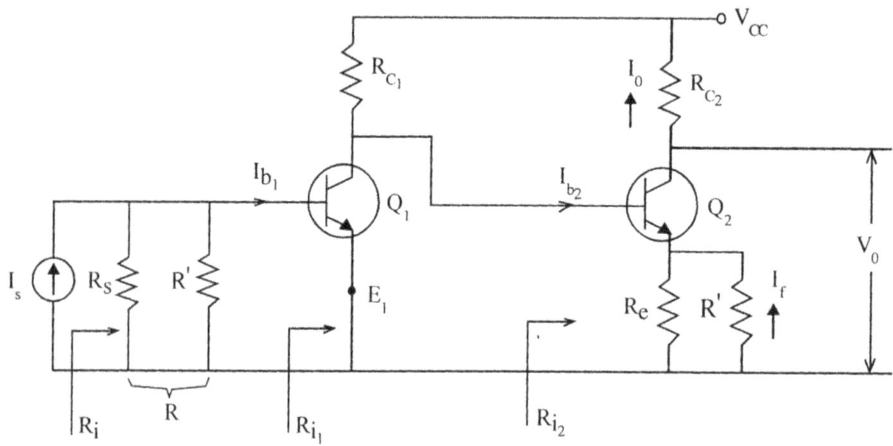

Fig. 6.52 Current shunt feedback without R'.

Because it is shunt feedback, we have shown this as a current source with R_s in parallel with I_s.

Problem 6.9

To find A_v', R_i', R_0'. A_i' for the circuit shown in Fig. 7.31.

$$R_{c1} = 3 \text{ k}\Omega, \qquad R_{c2} = 500\Omega; \quad R_{e2} = 50\Omega$$
$$R' = R_s = 1.2 \text{ k}\Omega \quad h_{fe} = 50.$$
$$h_{ie} = 1.1 \text{ k}\Omega \qquad h_{re} = h_{re} = 0.$$

Solution

In this problem, output is taken at the collector 2 and not emitter 2. So the formulae derived earlier can be used directly.

$$A_I = \frac{-I_{c_2}}{I_s} = \frac{-I_{c2}}{I_{b2}} \cdot \frac{I_{b2}}{I_{c1}} \cdot \frac{I_{c1}}{I_{b1}} \cdot \frac{I_{b1}}{I_s}$$

(Multiplying and dividing by I_{b2}, I_{c_1} and I_{b_1})

$$\frac{-I_{c_2}}{I_{b_2}} = -h_{fe} = -50; \text{ (Emitter follower)}$$

$$\frac{I_{c_1}}{I_{b_1}} = + h_{fe} = + 50 \text{ (common emitter configuration)}$$

$$I_{b_2} \# I_{c_1},$$

$\because$ The current gets divided in the ratio of the resistances, current gets divided depending upon R_{c_1} and R_{i_2} of transistor Q_2.

$$\frac{I_{b2}}{I_{cl}} = \frac{-R_{c_1}}{R_{c_1} + R_{i_2}}$$

$$= \frac{-3}{3 + 3.55} = -0.457.$$

$$R_{i2} = h_{ie} + (1 + h_{fe})(R_{e_2} \| R').$$

(For emitter follower configuration this is the input resistance).

$$= 1.1 + (50 + 1)\left[\frac{0.05 \times 1.20}{1.25}\right] \quad (R_{e\,2} = 0.05 \text{ k }\Omega)$$

$$= 3.55 \text{ k }\Omega$$

Let
$$R = R_S \text{ in parallel with } (R' + R_{e2})$$

$$= \frac{1.2 \times 1.25}{1.2 + 1.25} = 0.612 \text{ k }\Omega$$

$$= \frac{I_{b_1}}{I_S} = \frac{R}{R + h_{ie}}$$

$$= \frac{0.61}{0.61 + 1.1} = 0.358 \ \Omega$$

$\therefore$
$$A_1 = (-50)(-0.457)(50)(0.358)$$

$$= + 406$$

$$\beta = \frac{R_{e_2}}{R_{e_2} + R'}$$

$$= \frac{50}{1,250} = 0.04.$$

$$(1 + \beta A_i) = 1 + (0.040)(406) = 17.2$$

$$A_i' = \frac{A_1}{1 + \beta A_i} = \frac{406}{17.2} = 23.6$$

$$A_V' = \frac{V_o}{V_s} = \frac{-I_{c_2} . R_{c_2}}{I_s . R_s} = \frac{A_1' . R_{c_2}}{R_s}$$

$$= \frac{(23.6)(0.5)}{1.2} = 9.83$$

$$\left[A_V' \text{ is also} = \frac{R_{c_2}}{\beta R_s} = \frac{0.5}{(0.040)(1.2)} = 10.4 \right]$$

R_i = Input resistance without feedback

= R in parallel with h_{ie}

$$= \frac{(0.61)(1.1)}{1.71} = 0.394 \text{ k}\Omega$$

$$R_i' = \frac{R_i}{1 + \beta A_i} = \frac{394}{17.2} = 23.0\Omega$$

R_{oL} = Output resistance considering R_L,

$$= R_o \text{ in parallel with } R_{c2} = R_{c2} \quad \because R_o \text{ is large } \frac{1}{h_{ie}} = \alpha.$$

R_L' = with feedback considering R_L,

$$= \frac{R_{0L}(1 + \beta A_i)}{1 + \beta A_i} = R_0' = R_{c2} = 500$$

When we represent R on the input side and output side and calculate the value of A_I, it is not the current gain with feedback. Because, R' is represented on the input side leaving E_2 terminal open and on the output side shorting B_1 to ground (to make I_o and $I_s = 0$). So thus A_I will not be the same if R' is actually connected between E_2 and B_1. We are taking into account the effect of R' and not the feedback effect. In practice for shunt or series feedback, the signal generator will act as a current source or voltage source. Therefore it is capable of supplying current or voltage required .In theory we assume ideal voltage and current sources. Therefore for shunt feedback we must have a current sources irrespective of input voltage. The current source will supply sufficient current to drive the resistance.

$$R_i = (1+9) \text{ 20 k}\Omega = 200 \text{ k}\Omega$$

R_s is negligible compared to 200 kΩ.

$$R_0 = \frac{r_0}{1 + \beta A_o} = \frac{20\Omega}{10} = 2\Omega$$

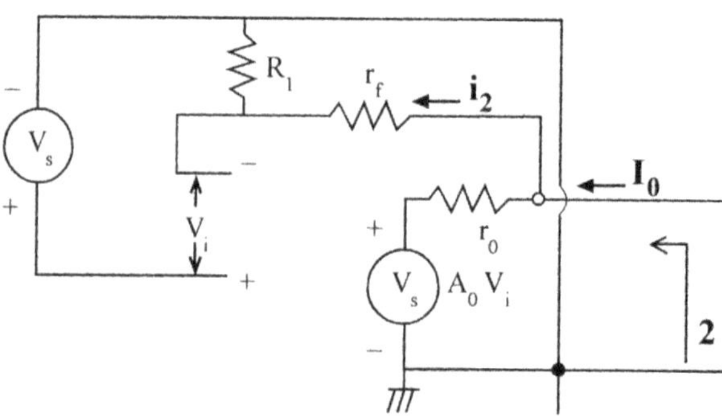

Fig. 6.53 Equivalent circuit.

6.17.3 CURRENT SERIES FEEDBACK

INPUT RESISTANCE (R_i')

$$V_i = V_i' - V_f$$

$$R_i' = \frac{V_i'}{I_i} = \frac{V_i + V_f}{I_i} = \frac{V_i}{I_i} + \frac{V_f}{I_i} \qquad \therefore \qquad V_i' = V_i + V_f .$$

$$\frac{V_i}{I_i} = R_i;$$

$$V_f = I_e R_e \; ; \; I_i = I_e - I_c \qquad [\because \text{negative current series feedback}] \; (I_b)$$

$$\therefore \qquad R_i' = R_i + \frac{I_e R_e}{I_e - I_c}$$

$$R_i' = R_i + \frac{R_e}{1 - \dfrac{I_c}{I_e}}$$

Dividing II term by I_e.

$$\frac{I_C}{I_e} = \alpha = \frac{\beta}{1 + \beta}$$

$$\frac{I_C}{I_e} = \frac{I_C}{I_C + I_b}$$

$$= \frac{\dfrac{I_c \mid I_b}{\dfrac{I_c}{I_b} + 1}} = \frac{h_{fe}}{1 + h_{fe}} .$$

$$\therefore \qquad R_i' = R_i + \frac{R_e}{1 - \dfrac{h_{fe}}{1 + h_{fe}}}$$

$$R_i' = R_i + (1 + h_{fe}).R_e$$

But $\qquad R_i = h_{ie}$ for the transistor

$$\boxed{\therefore \; R_i' = h_{ie} + (1 + h_{fe}) \, R_e}$$

If we consider the bias resistor also, R_1 and R_2 will come in parallel with R_i'

$$= R_i \parallel R_1 \parallel R_2$$

VOLTAGE GAIN (A_V')

$$A_v = \frac{A_i.R_L}{R_i} \qquad [\text{This is the general expression for } A_V \text{ interms of } A_i]$$

$$\therefore \qquad A_V' = \frac{A_i.R_L}{R_i'}$$

[With feedback, A_i will not change, R_L will not change, but R_i will be R'_i]

$$A_i \simeq - h_{fe}$$

$$\therefore \quad A_v' = \frac{-h_{fe}.R_L}{h_{ie} + (1 + h_{fe})R_e}$$

$$\therefore \quad A_v' < A_v$$

OUTPUT RESISTANCE (R_0')

$$R \text{ without feedback} \simeq \frac{1 + h_{fe}}{h_{oe}}$$

This will be very large.

Taking into effect, the feedback,

$$R_0' = R_0 \parallel R_L$$

$$\frac{1}{R_0} = h_{oe} - \frac{h_{fe}h_{re}}{h_{ie} + R_s}$$

ACTUAL EXPRESSION FOR β THE FEEDBACK FACTOR :

With negative feedback, $R_i' = R_i (1 + \beta. A_v)$

$$A_v = \frac{-h_{fe}.R_L}{R_i} ;$$

$$\therefore \quad R_i' = R_i \left[1 - \frac{h_{fe}.R_L}{R_i}.\beta \right]$$

$$= R_i - h_{fe}. R_L. \beta \qquad\qquad .....(1)$$

The expression we get for

$$R_i' = R_i + (1 + h_{fe}).R_e \qquad\qquad .....(2)$$

Comparing (1) and (2), we find that

$$\beta = \frac{(1 + h_{fe})}{h_{fe}} . \frac{R_e}{R_L}$$

This is the actual expression for β

$$\frac{1 + h_{fe}}{h_{fe}} \simeq 1.$$

$$\therefore \quad \beta \simeq \frac{R_e}{R_L}$$

6.17.3   CURRENT SERIES FEEDBACK (TRANS CONDUCTANCE AMPLIFIER)

Consider the circuit shown in Fig. 7.30 output is taken between collector and ground. The drop across R_e is the feedback signal V_f. The sampled signal is the load current I_o and not V_0. This is current series feedback. Because V_f is in series with V_i. It is current series feedback.

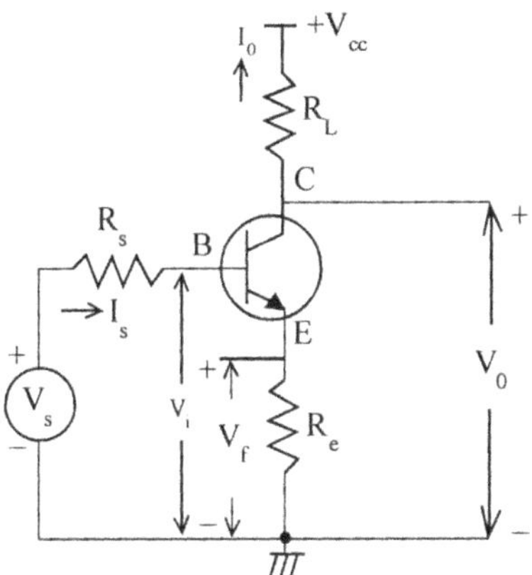

Fig. 6.54 Current series feedback.

$$V_f = I_e\, R_e.$$
$$I_e \simeq I_c = I_o.$$
$$\therefore\quad V_f = I_o\, R_e .$$

$\because$ R_e is constant,

$$V_f \,\alpha\, I_o.$$

[β must be independent of R_s or R_L]

$$\beta = \frac{V_f}{V_o} = \frac{-I_o.R_e}{I_o.R_L} = -\frac{R_e}{R_L}$$

The basic amplifier circuit without feedback is shown in Fig.6.55 below,

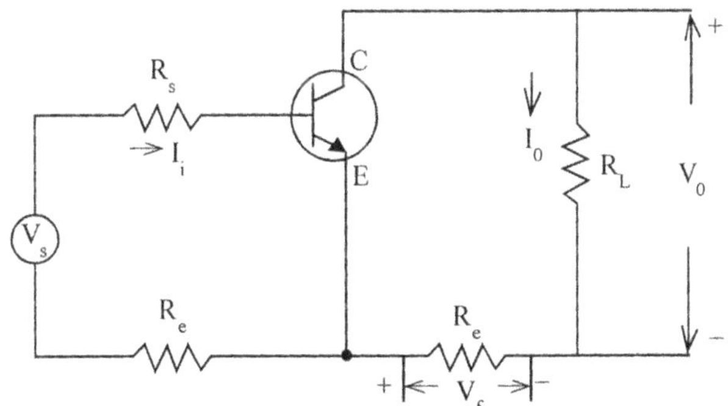

Fig.6.55 Circuit without feedback.

The equivalent circuit when the active device is replaced by *h-parameter* circuit is shown in Fig.6.56.

This has to be considered as Transconductance Amplifier, Since, β is taken as $\dfrac{V_f}{V_o}$. It depends, on R_L, because for this circuit, G_M is considered.

                                                     Basic Electronics Engineering

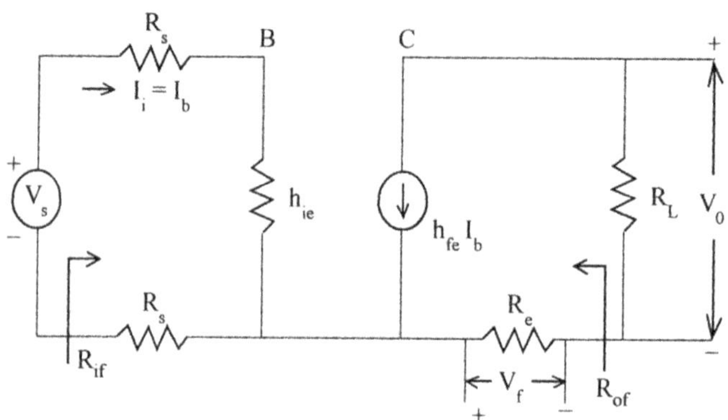

Fig. 6.56 h - parameter equivalent circuit.

$$\beta = \frac{V_f}{I_o} = \frac{-I_o.R_e}{I_o} = -R_e.$$

($\because$ the sampled signal is I_o and not V_o.)

$$G_M = \frac{I_o}{V_i} \quad I_0 = -h_{fe}.I_b \text{ (Input current gain)}$$

Since,
$$V_i = I_b (R_s + h_{ie} + R_e)$$

$\therefore$
$$G_m = \frac{-h_{fe}.I_b}{I_b(R_s + h_{ie} + R_e)}$$

$$= \frac{-h_{fe}}{R_s + h_{ie} + R_e}$$

$$D = 1 + \beta.G_M = 1 + \frac{h_{fe}.R_e}{R_s + h_{ie} + R_e}$$

$$= \frac{R_s + h_{ie} + (1 + h_{fe})R_e}{R_s + h_{ie} + R_e}$$

$$G_{Mf} = \frac{G_M}{D} = \frac{-h_{fe}}{R_s + h_{ie} + (1 + h_{fe})R_e}$$

Voltage gain
$$A_{vf} = \frac{I_o.R_L}{V_s} = \frac{I_o}{V_s} = G_{Mf}$$

$\therefore$
$$A_{vf} = G_{Mf}.R_L$$

$$= \frac{-h_{fe}.R_L}{R_s + h_{ie} + (1 + h_{fe})R_e}$$

$$(1 + h_{fe}) R_e \gg R_s + h_{ie}.$$

Since h_{fe} is a large quantity.

$\therefore$ Denominator $\simeq (1 + h_{fe}) R_e.$

$$h_{fe} \gg 1.$$

$\therefore$ Denominator $\simeq h_{fe} \cdot R_e$.

$\therefore$

$$\boxed{A_{vf} = \frac{-h_{fe}R_L}{h_{fe} \cdot R_e} = -\frac{R_L}{R_e}}$$

$R_i = R_s + h_{ie} + R_e$ (without feedback).

$R_{if} = R_i\, D = R_s + h_{ie} + (1 + h_{fe})\, R_e$.

6.17.4 Voltage Shunt Feedback (Trans Resistance Amplifier)

This is *voltage shunt feedback* because, the feedback current through R_B is proportional to the output voltage or it is in shunt with the input. So it is voltage shunt feedback. [We are not interested whether voltage is fed or current is fed. But the feedback signal is proportional to output voltage. *So it is voltage feedback*]

The circuit can be written as, shown in Fig. 6.57.

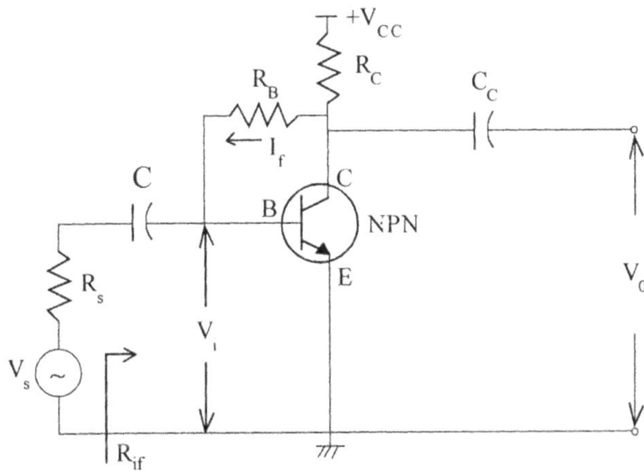

Fig. 6.57 Circuit for voltage shunt feedback.

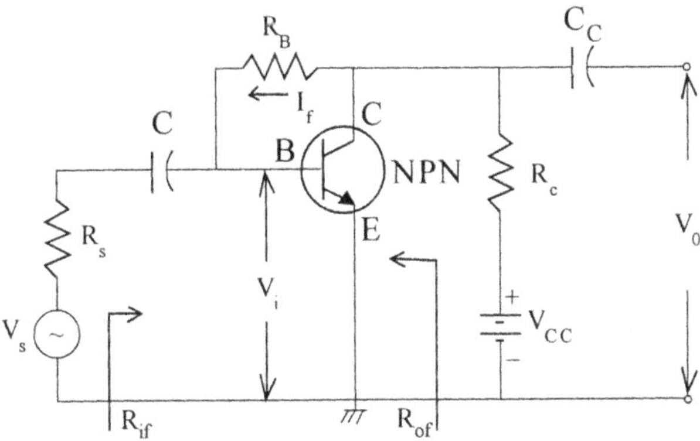

Fig. 6.58 Redrawn circuit for voltage shunt feedback.

The low frequency **h-parameter** equivalent circuit is, shown in Fig. 6.59.

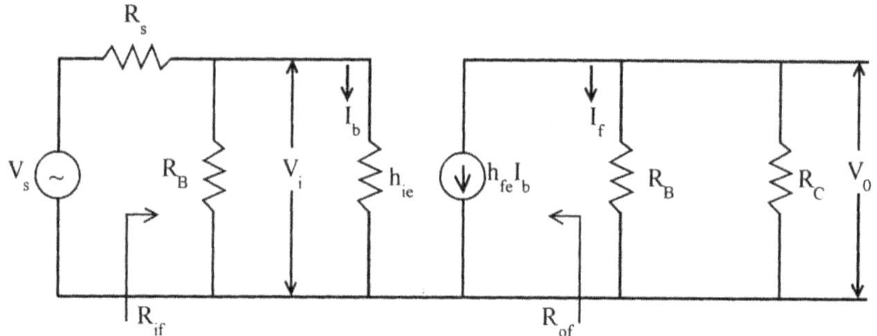

Fig. 6.59 h-parameter equivalent circuit.

$$V_0 >> V_i \qquad \because \text{Amplification is there.}$$

$$I_f \simeq \frac{-V_0}{R_B},$$

$$\therefore \qquad I_f \simeq \frac{-V_0}{R_B};$$

$$\therefore \qquad \beta = \frac{-I_f}{V_0}$$

$$\beta = \frac{-1}{R_B}$$

$$A_v = \frac{V_0}{V_i} = \frac{-h_{fe} \cdot R_L'}{h_{ie}} \text{ without feedback} \left(\frac{A_I R_L}{R_v} \right)$$

$$R_L' = R_B \parallel R_C.$$

$$A_v{}^1 = \frac{V_0}{V_s} \text{ (with feedback)}$$

$$\boxed{A_v{}^1 = A_v \cdot \frac{R_i'}{R_s + R_i'}}$$

R_i' = Input resistance with feedback

$$A_{vf} = \frac{V_0}{V_s}$$

$$= \frac{V_0}{I_s \cdot R_s} \simeq \frac{1}{\beta R_s} = \frac{-R_B}{R_s}$$

$\mathbf{R_i'}$:

$$I_s = I_f + I_b, \quad I_b \text{ is negligible}$$

$$\therefore \quad I_s \simeq I_f \quad \text{or} \quad \beta = \frac{-1}{R_B}.$$

$$R_i' = h_{ie} \text{ in parallel with } \frac{R_B}{1 - A_v}$$

Transresistance $\quad R_M = \dfrac{V_o}{I_s} = \dfrac{-I_o . R_c}{I_s} = \dfrac{-h_{fe} . I_b . R_c}{I_s}$

$$R_c' = R_c \parallel R_B \, ; \quad I_s = \frac{\left(R + h_{ie}\right)}{R} I_b. \qquad \text{Where } R = R_s \parallel R_B .$$

$$\therefore \quad R_M = \frac{-h_{fe} . R_c' . R}{\left(R + h_{ie}\right)} \, ;$$

$$I_b = \frac{-h_{fe} . R_e' . R}{\left(R + h_{ie}\right)}$$

$$R_M' = \frac{R_M}{1 + \beta R_M}.$$

$$R_i = \frac{R \times h_{ie}}{R + h_{ie}}$$

$$R_i' = \frac{R_i}{1 + \beta R_M}$$

$$R_0' = \frac{R_B}{R_s} \times \frac{R_s + h_{ie}}{h_{fe}}$$

Problem 6.10

For the circuit shown, $R_c = 4 \text{ k}\Omega$, $R_B = 40 \text{ k}$, $R_s = 10 \text{ k}$, $h_{ie} = 1.1 \text{ k}\Omega$, $h_{fe} = 50$, $h_{re} = h_{oe} = 0$. Find A_{vf}, R_{if}.

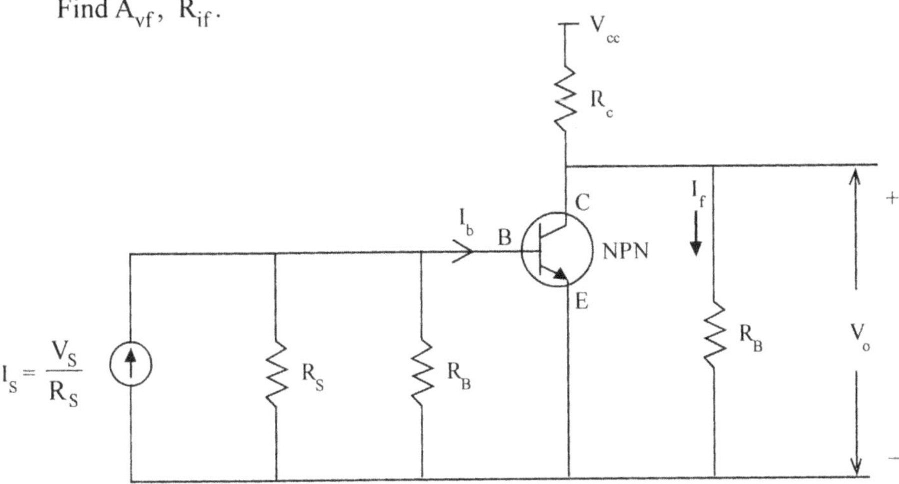

Fig. 6.70 For Problem 6.10.

Solution

$$R'_C = R_c \,||^{le}\, R_B = \frac{4 \times 40}{4 + 40} = 3.64 \text{ k}\Omega$$

$$R = R_s \,||^{le}\, R_B = \frac{10 \times 40}{10 + 40} = 8 \text{ k}\Omega$$

$$\text{Transresistance } R_M = \frac{V_o}{I_s} = \frac{-I_C.R_c'}{I_s} = \frac{-h_{fe}.I_b.R_c'}{I_s} \qquad \because \quad I_c = h_{fe} \cdot I_b$$

$$I_s = \frac{(R + h_{ie})}{R} I_b; \qquad \therefore \quad R_M = \frac{-h_{fe}.I_b\, R_C' \cdot R}{(R + h_{ie}).I_b}$$

$$R_M = -160 \text{ k}$$

$$\therefore \qquad \beta = -\frac{-1}{R_B} = -0.025 \text{ mA|V}$$

$$R'_M = \frac{R_M}{1 + \beta R_M} = \frac{-160}{5.00} = -32.0 \text{ k}\Omega$$

$$A'_V = \frac{V_o}{V_s} = \frac{V_o'}{I_s.R_s}$$

$$= \frac{V_o'}{I_s} = R'_M$$

$$\therefore \qquad A'_V = \frac{R_M'}{R_s} = -\frac{32}{10} = -3.2 \,.$$

$$R_i = \frac{R \times h_{ie}}{R + h_{ie}} = \frac{8 \times 1.1}{8 + 1.1} = 0.968 \text{ k}\Omega;$$

$$R'_i = \frac{R_i}{1 + \beta R_M} = \frac{968}{5.0} = 193\Omega$$

6.18 OSCILLATORS

Oscillator is a source of AC voltage or current. We get A.C.output from the oscillator circuit. In alternators (AC generators) the thermal energy is converted to electric energy at 50Hz.

In the oscillator circuits that we are describing now, the electric energy in the form of DC is converted into electric energy in the form of AC. 'Invertors' in electrical engineering convert DC to AC, but there, only output power is the criterian and not the actual shape of the wave form.

An amplifier is different from oscillator in the sense that an amplifier requires some A.C. input which will be amplified. But an oscillator doesn't need any external AC signal. This is shown in Fig. 6.71 below :

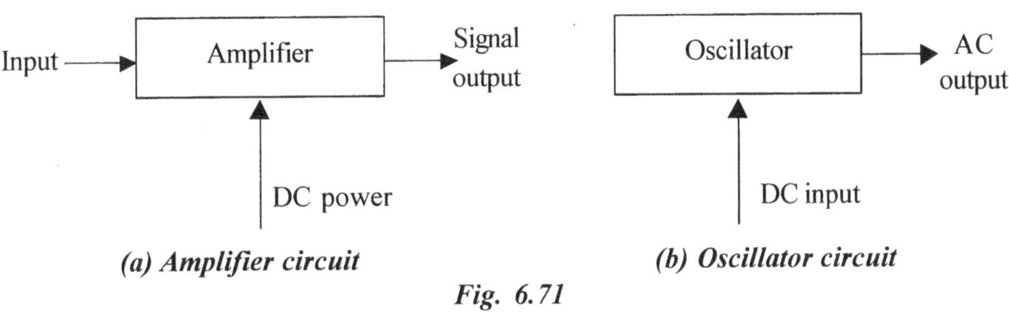

(a) Amplifier circuit   **(b) Oscillator circuit**

Fig. 6.71

For an amplifier, the additional power due to amplification is derived from the DC bias supply. So an amplifier effectively converts DC to AC. But it needs AC input. Without AC input, there is no AC output. In the oscillator circuits also DC power is converted to AC. But there is no AC input signal. So the difference between amplifier and oscillator is in amplifiers circuits, the DC power conversion to AC is controlled by the AC input signal. But in oscillators, it is not so.

There are two types of oscillators circuits :

1. Harmonic Oscillators

2. Relaxation Oscillators.

Harmonic Oscillators produce sine waves. Relaxation Oscillators produce sawtooth and square waves etc. Oscillator circuits employ both active and passive devices. Active devices convert the DC power to AC. Passive components determine the frequency of oscillators.

6.18.1 PERFORMANCE MEASURES OF OSCILLATOR CIRCUITS :

1. *Stability :* This is determined by the passive components. R,C and L determine frequency of oscillations. If R changes with T, f changes so stability is affected. Capacitors should be of high quantity with low leakage. So silver mica and ceramic capacitors are widely used.

2. *Amplitude stability :* To get large output voltage, amplification is to be done.

3. *Output Power :* Class A, B and C operations can be done. Class C gives largest output power but harmonics are more.

    Class A gives less output power but hormonics are low.

4. *Harmonics :* Undesirable frequency components are harmonics. An elementary sinusoidal oscillator circuit is shown in Fig.6.72.

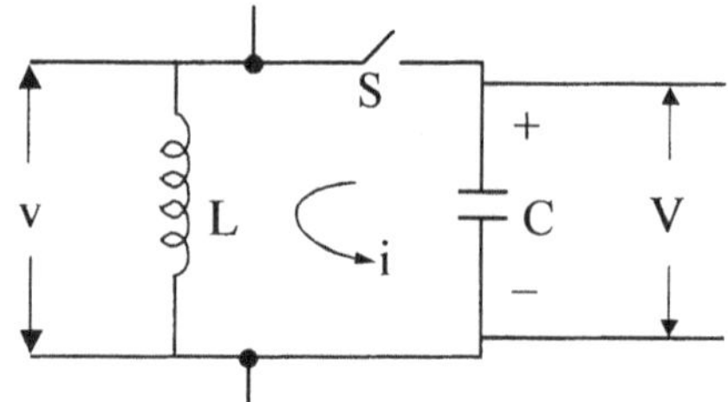

Fig. 6.72 LC Tank circuit.

L and C are reactive elements. They can store energy. The capacitor stores energy whenever there is voltage across its plates. Inductor stores energy in its magnetic field whenever current flows. Both C and L are lossless, ideal devices. So quality factor is infinity ∞. Energy is introduced into the circuit by charging capacitor to 'V' Volts. If switch is open, C cannot discharges because there is no path for discharge current to flow.

Suppose at $t = t_0$, switch 'S' is closed. Then current flows. Voltage across L will be V. At $t = t_0$, the Voltage across C is V volts. When switch is closed, current flows. So the charge across Capacitor C decreases. Voltage across C decreases, as shown in the waveform. So as the energy stored in Capacitor decreases, the energy stored in inductor L increases, because current is flowing through L Thus total energy in the circuit remains the same as before. When V across C becomes 0, current through the inductor is maximum. When the energy in C is 0, energy in L is maximum. Then the current in L starts charging C in the opposite directions. So at $t = t_1$ current in L is maximum and for $t > t_1$, the current starts charging C in the opposite direction. So V across C becomes negative as shown in the wavefrom. Thus we get sinusoidal oscillations from LC circuit.

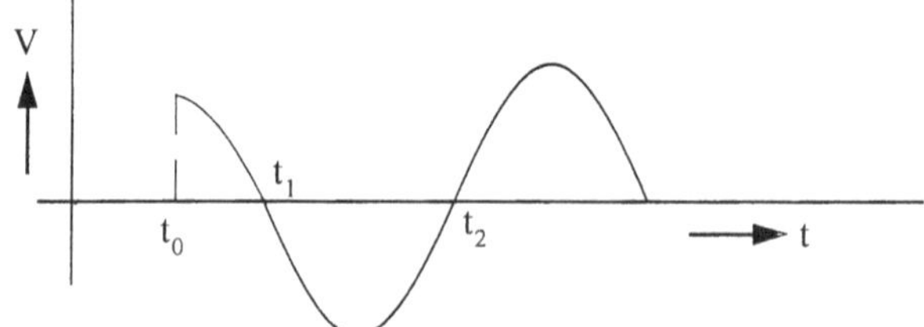

Fig. 6.73 Waveform.

Thus we are getting sinusoidal variations without giving any one input. What we have done is depositing some charge on C, so that the circuit operates on its own. But why should we get sinusoidal wave and not triangular or square wave ? Sinusoidal function is the only function that satisfies the conditions governing the exchange of the energy in the circuit.

But the above circuit is not a practical circuit. Because we have to take output from the circuit; i.e. energy has to be extracted, from the circuit. As we draw energy from the circuit, the energy stored in C and L decreases. So output also decreases or the voltage across C and L decreases so we get damped oscillation as shown below.

Fig. 6.74 Damped oscillations.

The energy lost by the elements must be replenished, so that oscillations are obtained continuously. If a negative resistance R is connected in the circuit, it replenishes, whatever energy that is lost in the circuit. Certain devices like tunnel diode, UJT, etc., exhibit -ve resistance. The energy supplied by the -ve resistance to the circuit actually comes from the DC bias supply.

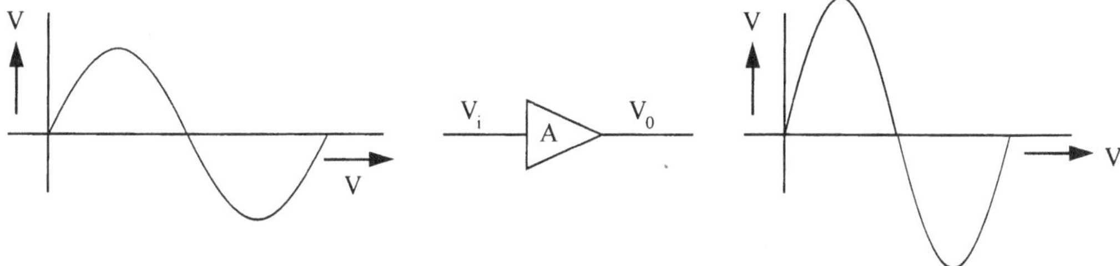

Fig. 6.75 Amplifier response.

Another method of producing sinusoidal oscillations is :

Suppose we have an amplifier with gain A_V and phase shift 180^0, $A_V \gg 1$.

If we connect V_0 through a feedback network to V_i as shown in Fig.6.76 so that after feedback, the feedback signal $V_X = V_i$, and also, the feedback network provides 180^0, phase shift, after feedback the feedback signal V_X will be removed. Thus, without any input we get sinusoidal output.

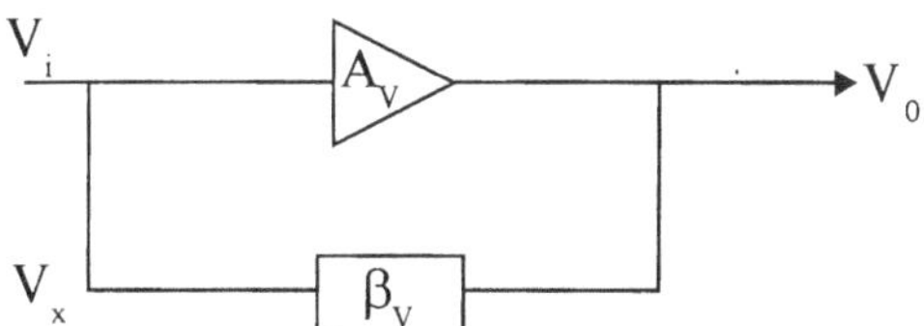

Fig. 6.76 Feedback network.

V_i is provided from the feedback signal of V_0 itself. To get initially Vo, the noise signal of transistor after switching itself is sufficient. Thus without external AC input we get sinusoidal oscillators.

$$\beta_v = \frac{V_x}{V_o}.$$

$$A_v = \frac{V_o}{V_i}.$$

$$V_x = V_i$$

$$\therefore \quad \beta . A_v = \frac{V_x}{V_o} \times \frac{V_o}{V_i}$$

$$= \frac{V_x}{V_i} \quad \text{or} \quad V_x = V_i$$

$$\therefore \quad \beta A_V = 1$$

Total phase shift $= 360^0$ $(180 + 180)$. Therefore, to get sustained oscillations,

1. The loop gain must be unit 1.
2. Total Loop phase shift must be 0^0 or 360^0. (Amplifier circuit produces 180^0 phase shift and feedback network another 180^0.

6.19 SINUSOIDAL OSCILLATORS

Figure 6.77 shows an amplifier, a feedback network and input mixing circuit not yet connected to form a closed loop. The amplifier provides an output signal X_0, as a consequence of the signal X_i applied directly to the amplifier input terminal. Output of feedback network

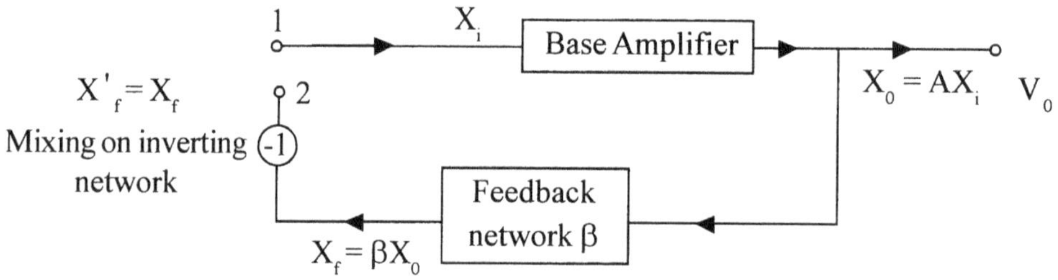

Fig. 6.77 Block schematic.

is $X_f = \beta X_0 = A\,\beta X_i$ and the output of the mixing circuit,

is $X'_f = -X_f = -A\beta X_i$

Loop gain, $= \dfrac{X'_f}{X_i} = \dfrac{-A\beta X_i}{X_i} = -\beta A.$

If X'_f were to be identically equal to X_i, input signal, $-\beta A$ should be $= 1$, so the output will be X_0, even if the input source is removed. The condition that $-\beta A = 1$ means, the loop gain must be 1.

6.20 BARKHAUSEN CRITERION

We make an asssumption that the circuit operates only in the linear region, and the amplifier feedback network contains reactive elements. For a sinusoidal wave form, if $X_i = X'_f$, the amplitude,

phase and frequency of X_i and X'_f be identical. *The frequency of a sinusoidal oscillator is determined by the condition that loop gain, Phase shift is zero at that frequency.*

For oscillator circuits positive feedback must be there i.e., V_f must be in phase with V_i to get added to V_i . When active device BJT or FET gives 180^0 phase shift, the feedback network must produce another 180^0 phase shift so that net phase shift is 0^0 or 360^0 and V_f is in phase with V_i to make it positive feedback.

Oscillations will not be sustained if, at the oscillator frequency the magnitude of the product of the transfer gain of the amplifier and of β are less than unity.

The conditions $-A\beta = 1$ is called ***Barkhausen criterion***

i.e. $|\beta A| = 1$ and phase of $-A\beta = 0.$

But
$$A_f = \frac{A}{1+\beta A}. \text{ If } \beta A = -1, \text{ than } A_f \to \infty.$$

which implies that, there exists an output voltage even in the absence of an externally applied voltage.

A 1-V signal appearing initially at the input terminals will, after a trip around the loop and back to the input terminals, appear there, with an amplitude larger than 1V. This larger voltage will then reappear as a still larger voltage and so on. So if $|\beta A|$ is larger than1, the amplitude of oscillations will continue to increase without limit. But practically , to limit the increase of amplitude of oscillations, nonlinearity ability of the circuit will be set in. Though a circuit has been designed for $|\beta A| = 1$, since circuit components and transistor change characteristics with age, temperature, voltage etc. $|\beta A|$ will become larger or smaller than 1. If $\beta A < 1$, the oscillations will stop. If $\beta A > 1$, the amplitude practically will increase. So, to achieve a sinusoidal osciallation practically $|\beta A| > 1$ to 5%, so that with incidental variations in transistor circuit parameters, $|\beta A|$ shall not fall below unity.

The type of noise in electronic circuits and the causes are :

1. ***Johnson noise or Thermal Noise :*** Due to temperature.

2. ***Schottky noise or Shot noise :*** Because of variation is concentration in the Semiconductor Devices.

6.21 R – C PHASE-SHIFT OSCILLATOR (USING JFET)

This is voltage series feedback. FET amplifier is followed by three cascaded arrangements of a capacitor C, resistor R. The output of the last RC combination is returned to the gate. This forms the feedback connection. The FET amplifier shifts the phase of voltage appearing at the gate by 180^0. The RC network shifts the phase by additional amount. At some frequency, the phase-shift introduced by this network will be exactly 180^0. The total phase-shift at this frequency, from the gate around the circuit and back to the gate is $+180 - 180 = 0^0$. At this particular frequency, the circuit will oscillate. (Fig. 6.78).

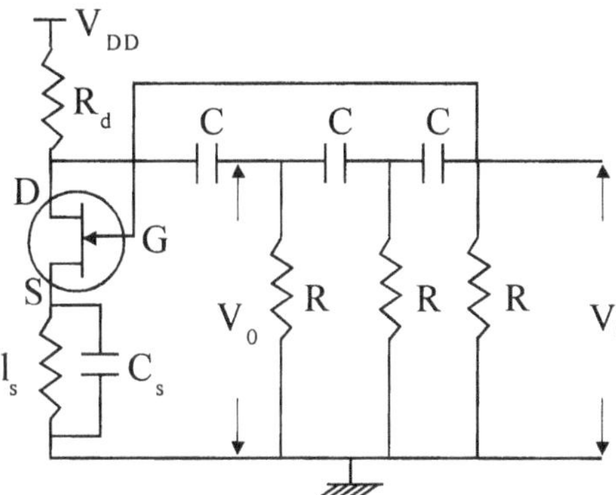

Fig. 6.78 JFET R – C phaseshift oscillator circuit.

$\therefore$ At that frequency, $V_f' = -V_f$ $\because$ 180^0 out of phase.

The equivalent circuit is, show below in Fig. 6.79.

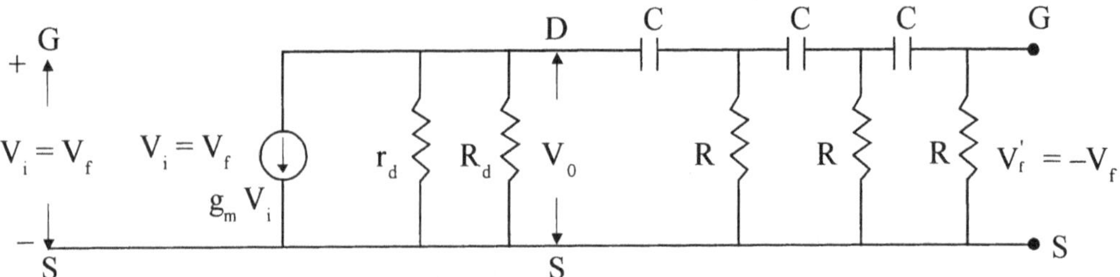

Fig. 6.79 Equivalent circuit.

The transformation of RC net work is $\dfrac{V_f}{V_o} = -\beta$; $\therefore$ $\beta = \dfrac{V_f}{V_0}$ and $V_f' = -V_f$

$\therefore$ $-\beta = \dfrac{V_f'}{V_o} = \dfrac{1}{1 - 5\alpha^2 - j(6\alpha - \alpha^3)}$.

where $\alpha = \dfrac{1}{\omega RC}$.

The phase-shift of $\dfrac{V_f'}{V_o}$ is 180^0, for $\alpha^2 = 6$ or

β must be real. Therefore $j(6\alpha - \alpha^3) = 0$.

$\therefore$ $f = \dfrac{1}{2\pi RC\sqrt{6}}$. $\because$ $6 = \dfrac{1}{\omega^2 R^2 C^2}$.

$$\omega^2 = \dfrac{1}{6R^2C^2}.$$

$\therefore$ $$\boxed{f = \dfrac{1}{2\pi RC\sqrt{6}}.}$$

when $\alpha^2 = 6, -\beta = \dfrac{1}{1 - 5 \times 6 - 5(6\alpha - 6\alpha)} = \dfrac{-1}{29}$

or $$\boxed{\beta = \dfrac{1}{29}.}$$

In order that $|\beta A|$ is not les than unity, A should be atleast $|29|$. So select F E T whose μ is atleast 29.

6.21.1 TO FIND THE β OF THE RC PHASE-SHIFT NETWORK (JFET)

Each RC network introduces a phase-shift of 60^0. Therefore, total phase-shift = 180^0. (See Fig.6.80).

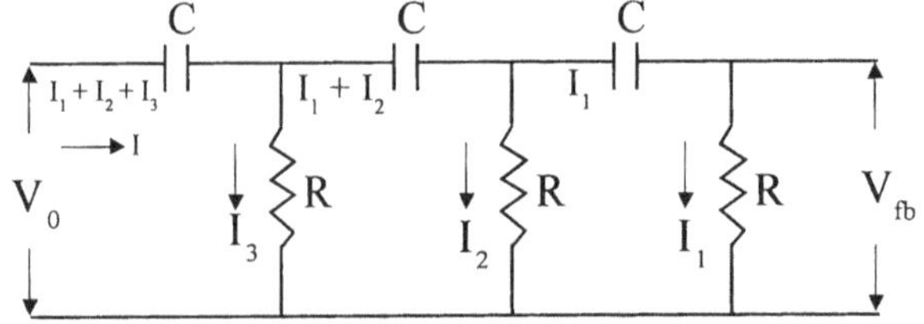

Fig. 6.80 R C phase shift network.

$$V_{fb} = I_L \cdot R \qquad \qquad .....(1)$$

$$I_2 \cdot R = \frac{I_1}{j\omega C} + I_1 \cdot R \qquad \qquad .....(2)$$

$$I_3 \cdot R = \frac{(I_2 + I_1)}{j\omega C} + I_2 \cdot R \qquad \qquad .....(3)$$

$$V_o = \frac{(I_1 + I_2 + I_3)}{j\omega C} + I_3 \cdot R \qquad \qquad .....(a)$$

But,
$$I_3 \cdot R = \frac{(I_2 + I_1)}{j\omega C} + I_2 \cdot R$$

$$I_2 \cdot R = \frac{I_1}{j\omega C} + I_1 \cdot R$$

$$I_1 R = V_{FB}$$

$$\therefore \quad I_3 \cdot R = \frac{(I_2 + I_1)}{j\omega c} + \frac{I_1}{j\omega c} + V_{fb} \cdot$$

By substituting 1, 2, and 3 in (a) we get,

$$V_o = \frac{(I_1 + I_2 + I_3)}{j\omega c} + \frac{(I_2 + I_1)}{j\omega c} + \frac{I_1}{j\omega c} + V_{fb} \cdot$$

$$V_o = \frac{3I_1 + 2I_2 + I_3}{j\omega c} + V_{fb} \cdot \qquad \qquad .....(b)$$

To eliminate I_1, I_2, and I_3,

$$I_1 = \frac{V_{fb}}{R} \qquad \qquad .....(c)$$

$$I_2 = \frac{I_1}{j\omega cR} + I_1 = \frac{V_{fb}}{j\omega cR^2} + \frac{V_{fb}}{R}$$

$$I_2 = V_{fb} \left(\frac{1}{R} + \frac{1}{J\omega CR^2} \right) \qquad \qquad .....(d)$$

$$I_3 = \frac{(I_2 + I_1)}{j\omega CR} + I_2$$

$$I_3 = \frac{V_{fb}}{j\omega cR} \left(\frac{2}{R} + \frac{1}{j\omega cR^2} \right) + V_{fb} \left(\frac{1}{R} + \frac{1}{j\omega CR^2} \right) \qquad \qquad .....(e)$$

Substitute c, d, and e equation in (b) and simplify

$$V_o = \frac{3V_{fb}}{R(j\omega c)} + \frac{2V_{fb}}{j\omega c} \left(\frac{1}{R} + \frac{1}{j\omega cR^2} \right) + \frac{V_{fb}}{(j\omega c)^2 R} \left(\frac{2}{R} + \frac{1}{j\omega cR^2} \right)$$

$$+ \frac{V_{fb}}{j\omega c} \left(\frac{1}{R} + \frac{1}{j\omega cR^2} \right) + V_{fb}$$

$$\frac{V_o}{V_{fb}} = \frac{1}{\beta} = 1 - \frac{5}{\omega^2 c^2 R^2} + j\left(\frac{1}{\omega^3 C^3 R^3} - \frac{6}{\omega CR}\right)$$

Let, $$\alpha = \frac{1}{\omega CR}$$

∴ $$\frac{1}{\beta} = 1 - 5\alpha^2 + j(\alpha^3 - 6\alpha)$$

or $$\beta = \frac{1}{1 - 5\alpha^2 - j(6\alpha - \alpha^3)}$$
$$\beta A = 1$$

But β is a complex number. Therefore equating imaginary part to zero.

$$6\alpha - \alpha^3 = 0.$$

$$6\alpha = \alpha^3. \quad \text{or} \quad \alpha = \sqrt{6}.$$

But $$\alpha = \frac{1}{\omega CR}.$$

∴ $$\frac{1}{\omega CR}. = \sqrt{6}.$$

or $$\omega = \frac{1}{\sqrt{6}CR}$$

or $$\boxed{f = \frac{1}{2\pi CR\sqrt{6}}}$$

The gain A, corresponding to this frequency will be 29.

6.22 TRANSISTOR RC PHASE-SHIFT OSCILLATOR

For transistor circuit, voltage shunt feedback is employed, because the input impedance of transistor is small. If voltage series feedback is employed, the resistance of feedback network will be shunted by the low 'R' of the transistor. (Fig. 6.81(a) and Fig.6.81(b)).

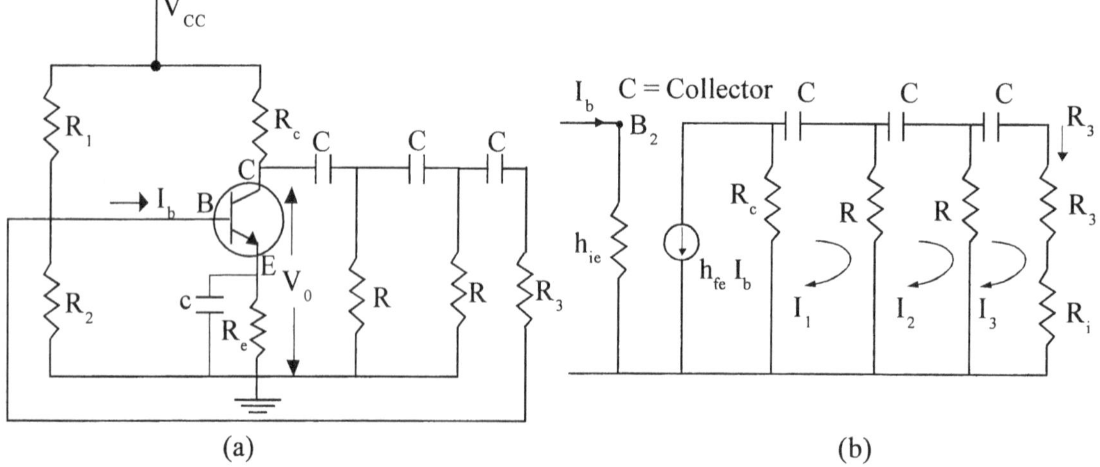

(a)

(b)

Fig. 6.81 Transistor phase shift oscillator.

The value of $R_3 = R - R_i$, where $R_i \simeq h_{ie}$, the input resistance of transistor. $\therefore$ The three RC sections of the phase-shifting network are identical R_1 R_2 and R_e are biasing resistors.

The feedback current $x_f = - I_3$ Input current $x_i = I_b$ (Negative sign is because it is negative feedback)

$\therefore$ Loop current gain $= \dfrac{-x_f}{x_i} = \dfrac{I_3}{I_b}$

By writing K V L for the three nodes, $\dfrac{I_3}{I_b}$ can be found out.

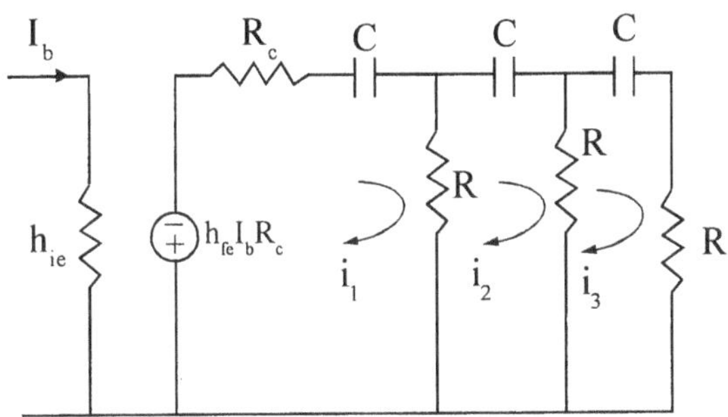

Fig. 6.82 R - C Equivalent circuit.

Loop 1 : $(R_C + R - \dfrac{j}{\omega c})\, I_1 - R\, I_2 = - h_{fe}\, I_b\, R_C$ $\hspace{2cm}$ .....(1)

Loop 2 : $-RI_1 + (2R - \dfrac{j}{\omega c})\, I_2 - RI_3 = 0$ $\hspace{2cm}$ .....(2)

Loop 3 : $-RI_2 + (2R - \dfrac{j}{\omega c})\, I_3 = 0$ $\hspace{2cm}$ .....(3)

By taking 'R' as common, the modified given by equations are

$$\left(\dfrac{R_C}{R} + 1 - \dfrac{j}{\omega RC}\right) I_1 - I_2 = -h_{fe}\, I_b\, \dfrac{R_C}{R} \hspace{2cm} .....(4)$$

$$-I_1 + \left(2 - \dfrac{j}{\omega RC}\right) I_2 - I_3 = 0 \hspace{2cm} .....(5)$$

$$-I_2 + \left(2 - \dfrac{j}{\omega RC}\right) I_3 = 0 \hspace{2cm} .....(6)$$

Let $\dfrac{R_C}{R} = K$ and $\alpha = \dfrac{1}{\omega RC}$, the equations are

$$(1 + K - j\alpha)\, I_1 - I_2 = - h_{fe}\, I_b = 0 \hspace{2cm} .....(7)$$
$$-I_1 + (2 - j\alpha)\, I_2 - I_3 = 0 \hspace{2cm} .....(8)$$
$$-I_2 + (2 - j\alpha)\, I_3 = 0 \hspace{2cm} .....(9)$$

$$\begin{bmatrix} 1+k-j\alpha & -1 & 0 \\ -1 & 2-j\alpha & -1 \\ 0 & -1 & 2-j\alpha \end{bmatrix} \begin{bmatrix} I_1 \\ I_2 \\ I_3 \end{bmatrix} = \begin{bmatrix} -h_{fe}I_b k \\ 0 \\ 0 \end{bmatrix}$$

$$I_3 = \frac{1}{\Delta} \begin{vmatrix} 1-k-j\alpha & -1 & -h_{fe}I_b K \\ -1 & 2-j\alpha & 0 \\ 0 & -1 & 0 \end{vmatrix} = \frac{1}{\Delta} \left[-h_{fe}I_b K \right]$$

$$\Delta = [1 + K - j\alpha] \, [(2 - j\alpha)^2 - 1] + [-(2 - j\alpha)]$$
$$= [1 + K - j\alpha] \, [4 - \alpha^2 - j4\alpha - 1] - 2 + j\alpha$$
$$= -j\alpha^3 - j6\alpha - j4K\alpha - 5\alpha^2 - \alpha^2 K + 3K + 1$$
$$\Delta = -(5 + K)\,\alpha^2 + 3K + 1 + j\,[\alpha^3 - (6 + 4K)\,\alpha]$$

$$\therefore \; I_3 / I_b = \frac{-h_{fe}K}{-(5+K)\alpha^2 + 3K + 1 + j\,[\alpha^3 - (6+4K)\alpha]} \qquad \qquad .....(10)$$

The Barkhausen condition that the loop gain $\dfrac{I_3}{I_b}$ phase shift must equal zero. The phase shift equals zero provided imaginary part is zero.

Hence $\qquad \qquad \alpha^3 = (6 + 4K)\,\alpha$

$\qquad \qquad \qquad \quad \alpha^2 = 6 + 4K$

$\qquad \qquad \qquad \quad \alpha = \sqrt{6 + 4K}$

Since $\qquad \qquad \alpha = \dfrac{1}{\omega RC}, \qquad \qquad \omega = \dfrac{1}{RC\sqrt{6 + 4K}}$

$\therefore$ The frequency of oscillation f is given by

$$f = \frac{1}{2\pi RC\sqrt{6 + 4K}} \qquad \qquad .....(11)$$

For mantaining the oscillations at the above frequency, $|I_3 / I_b| > 1$

$$\left| \frac{I_3}{I_b} \right| = \left| \frac{-h_{fe}K}{-(5+K)(6+4K) + 3K + 1} \right| > 1$$

$$\left| -h_{fe}K \right| > \left| -(5+K)(6+4K) + 3K + 1 \right|$$

$$\left| -h_{fe}K \right| > \left| -4K^2 - 23K - 29 \right|$$

$$h_{fe}\,K > 4K^2 + 23K + 29$$

$$h_{fe} > 4K + 23 + \frac{29}{K} \qquad \qquad .....(12)$$

To determine the minimum value of h_{fe}, the optimum value of K should be determined by differentiating h_{fe} w.r.t K and equate it to zero.

$$\frac{dh_{fe}}{dk} > \frac{d}{dk}\left(4K + 23 + \frac{29}{K}\right)$$

$$0 > 4 - \frac{29}{K^2}$$

$$\therefore K^2 < \frac{29}{4}, \qquad \therefore \mathbf{K < 2.7} \qquad\qquad .....(13)$$

Substitute the optimum $K = 2.7$ in eq. (12)

$$h_{fe} > 4 \times 2.7 + 23 + \frac{29}{2.7}$$

$$\therefore h_{fe} > 44.5 \qquad\qquad .....(14)$$

The BJT with a small-signal common-emitter short-circuit gain h_{fe} lessthan 44.5 cannot be used in this phase shift oscillator.

In R–C phase shift oscillator circuit, each RC network produces 60^0 phase shift. Thus 3 sections produce the required 180^0 phase shift. If there are 4-sections, each sections must produce 45^0 phase shift. But more number of components are to be used. If only 2 sections are these, 90^0 phase shift must be produced, which is not possible for practical R-C network.

6.23   A GENERAL FORM OF LC OSCILLATOR CIRCUIT

Many Oscillator Circuits fall in to the general form as shown in Fig.6.83 (a) and (b).

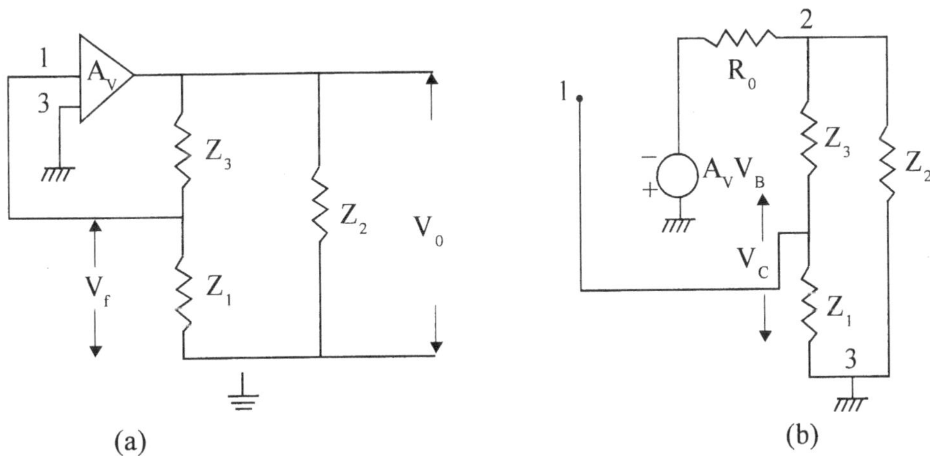

(a)   (b)

Fig. 6.83 General form of oscillator circuit

The active device can be FET, transistor or operational amplifier, Fig.8.13(b) shows the equivalent circuit using an amplifier with negative gain A_v and output resistance R_0. This is Voltage Series Feedback.

6.24   LOOP GAIN

$$\beta = -\frac{V_f}{V_o}$$

$$\therefore \quad \beta = \frac{-V_f}{V_0} = -\frac{Z_1}{Z_1 + Z_3}.$$

The load impedance, $Z_L = (Z_3$ in series with $Z_1)$ parallel with Z_L.

R_0 is the output resistance.(See Fig. 6.84).

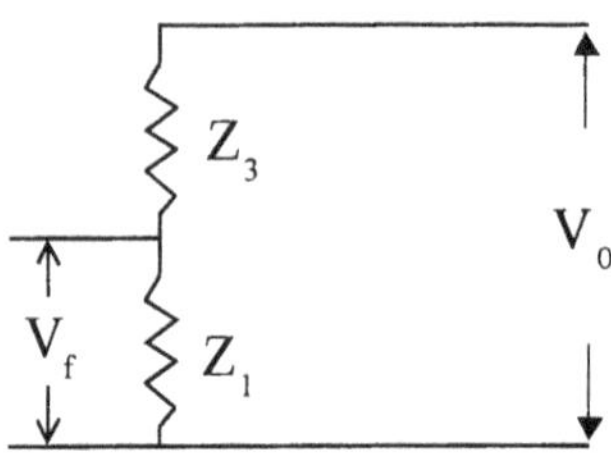

Fig. 6.84 Potential divider network.

Gain without feedback $A = \dfrac{-A_V.Z_L}{Z_L + R_O}$

$$\therefore \quad -\beta\,A = \frac{-A_V.Z_1 Z_L}{\left(Z_L + R_O\right)\left(Z_1 + Z_3\right)}; \quad Z_L = \frac{Z_2(Z_1 + Z_3)}{Z_2 + Z_1 + Z_3}$$

$$= \frac{-A_V.Z_2(Z_1 + Z_3)Z_1}{(Z_1 + Z_2 + Z_3)\dfrac{(Z_2(Z_1 + Z_3))}{(Z_1 + Z_2 + Z_3)} + R_O)(Z_1 + Z_3)}$$

$$-\beta\,A = \frac{-A_v.Z_1\,Z_2}{R_O(Z_1 + Z_2 + Z_3) + Z_2(Z_1 + Z_3)}$$

If the impedances are pure reactances, then $Z_1 = j\,x_1, \quad Z_2 = jx_2; \quad Z_3 = jx_3$

$$\therefore \quad -A\,\beta = \frac{A_v X_1\,X_2(j)^2}{jR_o(X_1 + X_2 + X_3) - X_2(X_1 + X_3)}$$

If $\beta A = 1$, or for zero phase-shift, imaginary part must be zero.

$$jR_o\,(X_1 + X_2 + X_3) = 0.$$

or $\qquad X_1 + X_2 + X_3 = 0$

$$-A\beta = \frac{A_v X_1\,X_2}{-X_2(X_1 + X_3)} = -\frac{A_v\,X_1}{X_1 + X_3}$$

But $\quad X_1 + X_2 + X_3 = 0 \quad \therefore \; X_1 + X_3 = -X_2$

$$\therefore \qquad -A\beta = \frac{A_V X_1}{X_2}.$$

$\therefore \; -A\beta$ must be positive, and at least unity in magnitude. Than X_1 and X_2 must have the same sign.

So if X_1 and X_2 are capacitive, X_3 should be inductive and vice versa.

If X_1 and X_2 are capacitors, the circuit is called *Colpitts Oscillator (Fig. 6.85(a))*

If X_1 and X_2 are inductors, the circuit is called *Hartley Oscillators (Fig. 6.85(b))*

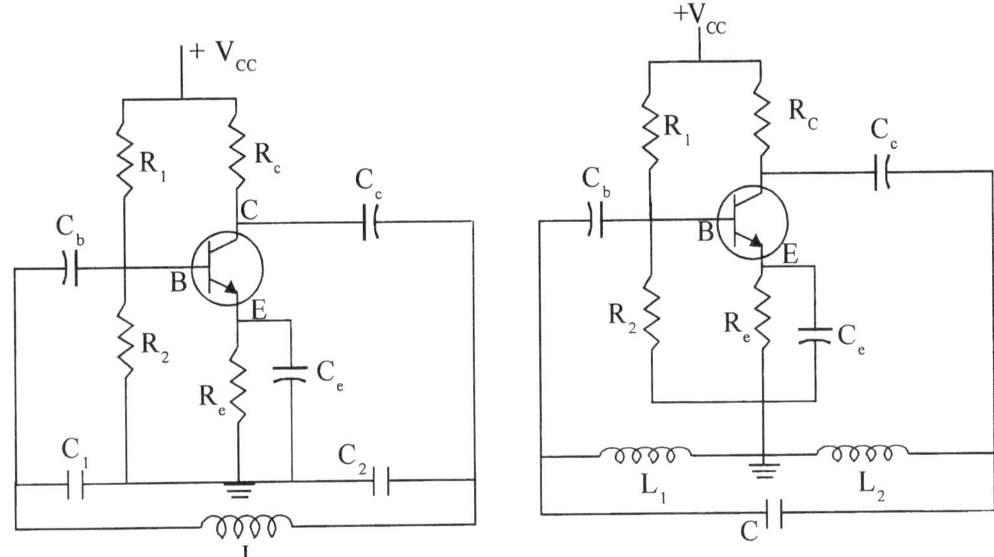

(a) Colpitts oscillator   *(b) Hartley oscillator circuit*

Fig. 6.85

6.24.1 FOR HARTLEY OSCILLATOR

Condition for oscillations, $X_1 + X_2 + X_3 = 0$.

where
$$X_1 = j\omega L_1 \quad X_2 = j\omega L_2; \quad X_3 = +\frac{1}{j\omega C}$$

$\therefore$
$$j\omega L_1 + j\omega L_2 + \frac{1}{j\omega C} = 0 \quad \text{or} \quad \omega L_1 + \omega L_2 = \frac{1}{\omega C}$$

$$\omega^2(L_1 + L_2) = \frac{1}{C}; \quad \frac{1}{\sqrt{(L_1 + L_2)C}} = \omega$$

$$\boxed{f = \frac{1}{2\pi\sqrt{(L_1 + L_2)C_3}}}$$

6.24.2 FOR COLPITTS OSCILLATOR

$$X_1 = -\frac{j}{\omega C_1}; \quad X_2 = \frac{-j}{\omega C_2}; \quad X_3 = j\omega L.$$

$$X_1 + X_2 + X_3 = 0;$$

$$\frac{-j}{\omega C_1} - \frac{-j}{\omega C_2} + j\omega L = 0 \quad \text{or}$$

$$\omega L = \frac{1}{\omega C_1} + \frac{1}{\omega C_2}$$

$$\omega L = \frac{\omega C_2 + \omega C_1}{\omega C_1\, C_2} = \frac{C_2 + C_1}{\omega C_1\, C_2} \quad \text{or} \quad \omega = \left(\frac{1}{\sqrt{L_3}} \sqrt{\frac{C_2 + C_1}{C_1\, C_2}} \right)$$

$$\boxed{f = \frac{1}{2\pi\sqrt{L\, C_T}}}$$

where
$$C_T = \frac{C_1\, C_2}{C_1 + C_2}$$

6.25 WIEN BRIDGE OSCILLATOR

In this circuit, a balanced bridge is used as the feedback network. The active element is an operational amplifier. It employs lead-lag Network. Frequency f_0 can be varied in the ratio of 10 : 1 compared to 3 : 1 in other oscillator circuits.

External voltage V_0' is applied betweeen 3 and 4, as shown in Fig.6.86.

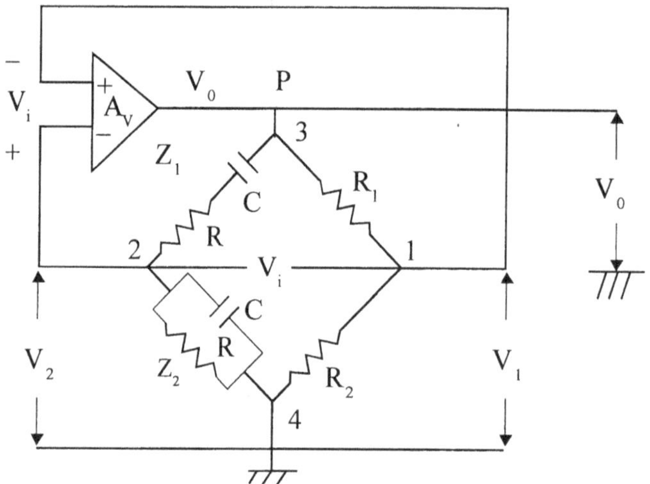

Fig. 6.86 Wien bridge oscillator circuit.

To find loop gain, $-\beta A$, (–sign because phase-shift feedback)

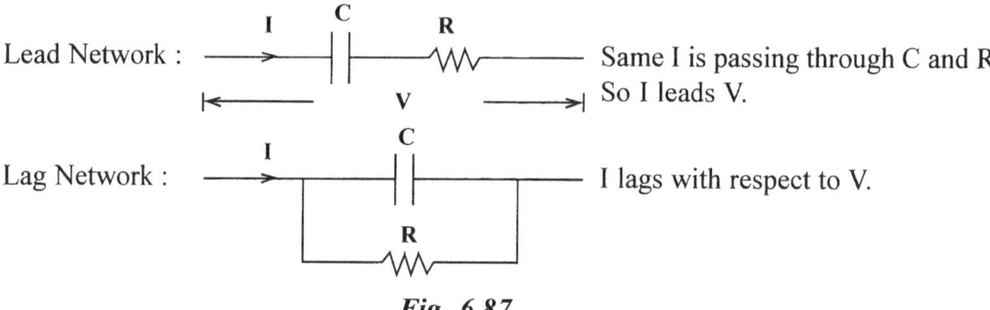

Fig. 6.87.

Frequency variation of 10 : 1 is possible in Wein Bridge compared to 3 : 1 in other oscillator circuits.

$$V_0 = A_v V_i; \quad V_i \text{ is } (V_2 - V_1); \quad V_i = V_f.$$

Loop gain
$$= \frac{V_0}{V_0{'}} = \frac{A_v \cdot V_i}{V_O{'}} = -\beta A. \qquad \qquad \dots\dots(1)$$

V_1 and V_2 are auxillary voltages. $V_i = V_2 - V_1$.

$$-\beta = \frac{V_i}{V_0{'}} \quad \because V_f = V_i \quad \therefore A = A_v.$$

$$-\beta = \frac{V_i}{V_0{'}} = \frac{V_2 - V_1}{V_0{'}} = \left[\frac{Z_2}{Z_1 + Z_2} - \frac{R_L}{R_1 + R_2} \right]$$

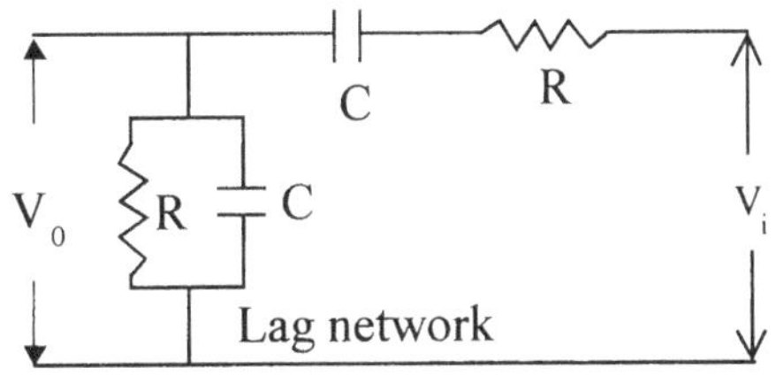

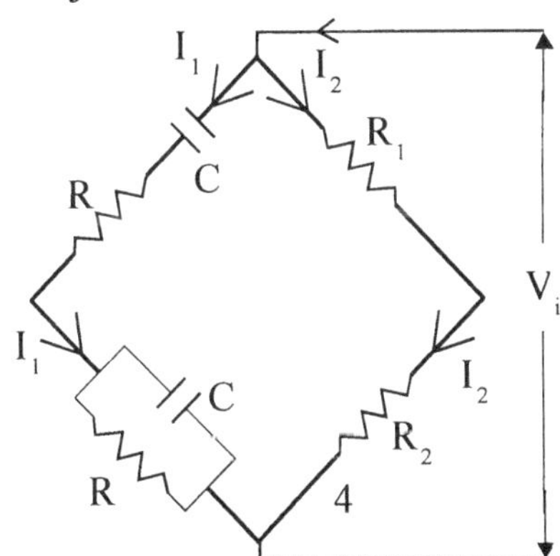

Fig. 6.88

6.26   EXPRESSION FOR *f*

Fig. 6.89   Wien Bridge oscillator circuit.

$$I_1 = \frac{\left(R + \dfrac{1}{j\omega c} \right)}{I_1 \left(\dfrac{R}{1 + j\omega cR} \right)} = \frac{I_2 R_1}{I_2 R_L}$$

$$R_1 = \frac{\left(R_2 R + \dfrac{R_2}{j\omega C} \right)(1 + j\omega CR)}{R} \; ;$$

$$R_1 R = \left(R_2 R + \frac{R_2}{jwC} \right)(1 + jwCR)$$

$$R_2 \cdot R + \frac{R_2}{j\omega C} = \left(\frac{R_1 R}{1 + j\omega CR} \right)$$

$$[(R\,R_2)\,(j\omega C) + R_2]\,R_2\,j\omega CR - \omega^2\,C^2\,R^2\,R_2 = R_1\,R\,j\omega C$$

$$(R_1\,R_2)\,(j\omega C) + R_2 + R_2\,j\omega CR - \omega^2 c^2 R^2 R_2 = R_1\,R\,j\omega C$$

Equating imaginary parts,

$$R_1\,R_2\,\omega C + R_2\,\omega CR = R_1\,R\omega C$$

Equating real parts,

$$A = 1 + \frac{R_1}{R_2} + \frac{C_1}{C_2}, \quad R_2 = \omega^2\,C^2\,R^2\,R_2$$

$$\therefore \qquad \omega^2 = \frac{1}{C^2 R^2} \quad \text{or} \quad \boxed{f = \frac{1}{2\pi RC}}$$

$$R_1 = \frac{R_2 \cdot R}{R} + \frac{R_2}{j\omega CR} + j\omega CRR_2 + R_2$$

$$2R_2 = R_1 \quad \text{or} \quad \boxed{\frac{R_2}{R_1 + R_2} = \frac{1}{3}} \quad \text{So the minimum gain of the amplifier must be 3.}$$

$$\frac{-R_2}{\omega CR} = \omega CRR_2$$

$$\omega^2 = \frac{1}{(RC)^2} \; ; \quad \text{or} \quad \boxed{f = \frac{1}{2\pi RC}}$$

This is the frequency at which the circuit oscillates. Continuous variation of frequency is accomplished by using the capacitors 'C'.

6.27　Thermistor

Conductivity of Germanium and Si increases with temperature. A semiconductor when used in this way, taking advantage of this property is called a ***Thermistor***. Ex : NiO, Mn_2O_3 etc. These are used for temperature compensation in oscillator circuits.

6.28　Sensistor

A heavily doped semiconductor can exhibit a positive temperature coefficient of resistance because under heavy doping, semiconductor acquires the properties of a metal. So R increases because mobility decreases. Such a device is called ***Sensistor***. These are also used for temperature compensation like thermistors.

6.29　Amplitude Stabilization

The amplitude of oscillations can be stabilized by replacing R_2 with a senistor. If β is fixed, as A increases, amplitude of oscillations increases. If a senistor is introduced, as its 'R' changes with temperature, it changes β, so that βA is always constant, (when A changes).

By equating the imaginary part to zero we get $f = \dfrac{1}{2\pi\,RC\sqrt{6+4K}}$ where $K = \dfrac{R_C}{R}$

Forward Current Gain $\boxed{h_{fe} = 4k + 23 + \dfrac{29}{K}.}$

Practically RC phase-shift oscillators can be used from several hertz to several hundred kilohertzs. In the mega hertz range, tuned LC circuits are more advantageous. Frequency of oscillators can be changed by changing R and C. Amplitude of oscillations will not vary if any C is varied, because X_C varies, but the imaginary part will be zero. Phase-shift oscillator is operated in class A in order to keep distortion minimum.

6.30 Applications

Elevation levelling systems, Burglar detection: frequency of oscillators change as a result of local disturbance. The change in frequency causes further electronic action or alarm signal.

6.31 Resonant Circuit Oscillators

Initial transient due to switching will initiate electrical signals. These are feedback to the transistor as input. The input is amplified and obtained as output. Oscillations are sustained. Output is coupled to the base of transistor through L_1. LL_1 is a transformer. R represents the resistance in series with the winding in order to account for the losses in the transformer shown in Fig. 6.90. If

'r' is very small, then at $\omega = \dfrac{1}{\sqrt{LC}}$ the Z is purely resistive. Then voltage drop across inductor is

180^0 out of phase with the applied input voltage to the FET. If the direction of winding of the secondary connected to the gate is such that it introduces another 180^0 phase-shift, the total phase-shift is zero.

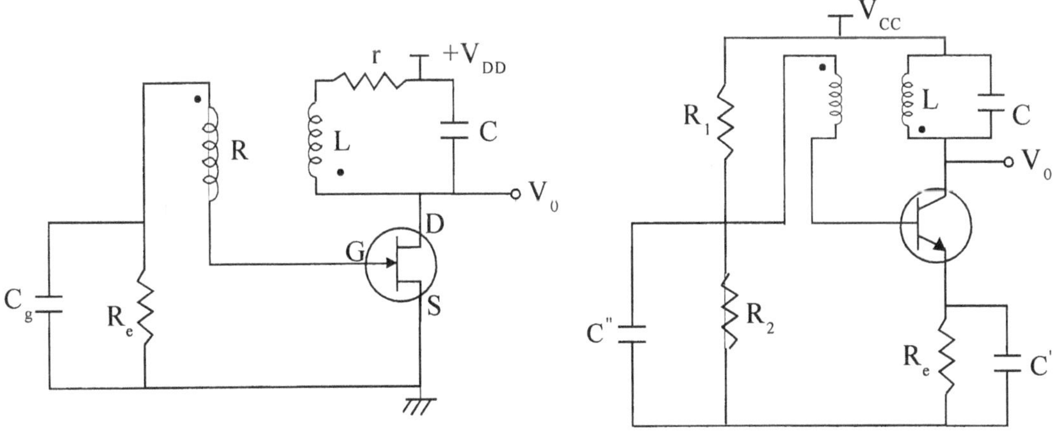

Fig. 6.90 Resonant circuit oscillator.

The ratio of the amplitude of secondary to the primary voltage is M/L where M is the inductance.

$$i.e., \quad \frac{V_f}{V_o} = \frac{M}{L} = \beta$$

Voltage gain $A_v = -\mu$

$$\beta\, A_v = -1$$

$\therefore$ $\quad\quad -\mu\beta = -1;$

$$\mu = \frac{1}{\beta}$$

$\therefore$ $\quad\quad \mu = \frac{L}{M}$

$\therefore$ $\quad\quad \mu = L/M$

If we consider resistance 'r' also, $\omega^2 = \dfrac{1}{LC}\left(1 + \dfrac{r}{r_d}\right)$

$$\boxed{g_m = \frac{\mu r C}{\mu M - L}.}$$

6.32   CRYSTAL OSCILLATORS

When certain solid materials are deformed, they generate within them, an electric charge. This effect is reversible in that, if a charge is applied, the material will mechanically deform in response. This is called *Piezoelectric effect.*

Naturally available materials :   1. Quartz   2. Rochelle salt.

Synthetic materials :   1. Lithium sulphate 2. Ammonium-di-hydrogen phosphate, PZT (Lead Zirconate Titanate), $BaTiO_3$ (Barium Titanate).

If the crystal is properly mounted, deformations take place within the crystal, and an electro mechanical system is formed which will vibrate when properly excited. The resonant frequency and Q depend upon crystal dimensions etc. With these, frequencies from few KHz to MHz and Q in the range from 1000s to 100,000 can be obtained. Since Q is high, and for Quartz, the characteristics are extremely stable, with respect to time, temperature etc., very stable oscillators can be designed. The frequency stability will be $\pm\, 0.001\%$. It is same as $\pm\, 10$ parts per million (10 ppm).

The electrical equivalent circuit of a crystal is as shown in Fig.6.91. L,C,R are analogous to mass, spring constant, and viscous damping factor of the mechanical system.

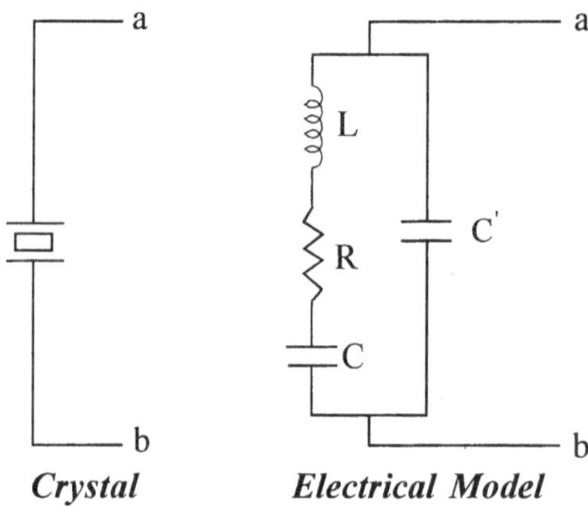

Crystal   *Electrical Model*

Fig. 6.91.

Values for a 90 kHz crystal are L = 137 H, C = 0.023pF; R = 15kΩ corresponding to Q = 5,500. The dimension of a crystal will be 30 × 4 ×1.5 mm. C′ is the electostatic capacitance between electrodes with the crystal as a dielectric. C′ = 3.5 pF and is larger than C.

When the crystal slabs are cut in proper directions, with regard to the crystal axis, a potential difference exists between the faces of the crystal slab when pressure is brought to bear on them. And if the slab is placed in an electrostatic field, the slab undergoes deformation. (Fig.6.92). If the electric field is an alternating one, with a frequency which sets the slab into mechanical resonance, the slab will physically vibrate vigorously. Such a crystal can be employed to maintain an oscillation of great frequency stability. When the L.C. circuit in the plate circuit is tuned close to the crystal resonant frequency, steady oscillations will be established. These are maintained by C whose oscillations value is small. By placing crystal between the gate and source, of the FET amplifier, and feeding back a small A.C. voltage from the output, to keep crystal vibrating, the circuit becomes an oscillator with precise stability. Accuracy ≃ 0.01% '*f*' range is 0.1 to 20 MHz. By keeping crystal in an oven, accuracy can be improved to 0.001%.

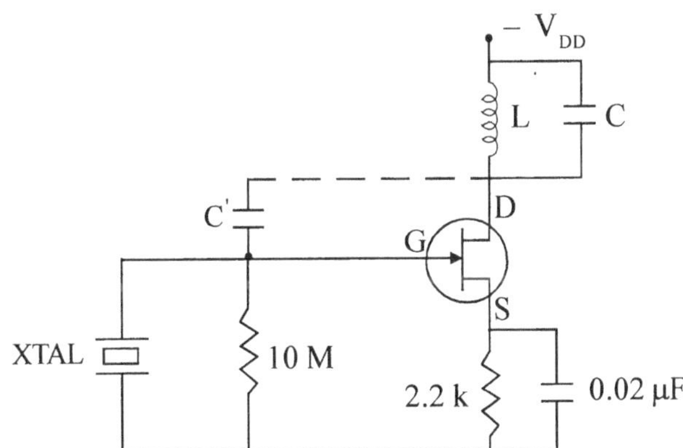

Fig. 6.92 Crystal oscillator circuit.

6.33   FREQUENCY STABILITY

It is a measure of the ability of the circuit to maintain exactly the same frequency for which it is designed over a long time interval. But actually in many circuits the '*f*' will not remain fixed but it drifts from the designed frequency continuously. This is because of variations of number of parameters, circuit components, transistor parameters, supply voltages temperature, stray capacitances etc. In order to increase the stability the factors which effect the '*f*' largely should be taken care of. If '*f*' depends only on R and C high precision R and C should be employed. Also temperature compensating elements are to be employed.

Effect of temperature on inductors and capacitors amounts to more than 10 parts per million per degree change.

Crystal will have mass, elasticity and damping. Crystal will have very high mass to elastic ratio $\left(\dfrac{L}{C}\right)$ and to a high ratio of mass to damping (high Q). Its value is of the order of 10,000 to 30,000.

The crystal is coupled with external C, and the LC circuit oscillates. In the oscillator circuits, instead of external L and C, crystal is connected.i.e. crystal L and C are being used. So 'f' is fixed by crystal itself. 'f' will not vary with temperature. To change 'f' another crystal is used.

6.34   FREQUENCY OF OSCILLATIONS FOR PARALLEL RESONANCE CIRCUIT

$$Z = \frac{(R - j\omega C) \times \dfrac{1}{j\omega C}}{(R + j\omega C) + \dfrac{1}{j\omega C}} = \frac{R + j\omega C}{(1 - \omega^2 LC) + j\omega RC}$$

$$Y_1 = \frac{1}{R + j\omega C} = \frac{R - j\omega C}{R^2 + \omega^2 C^2} \quad Y_2 = j\omega C$$

Total admittance $\quad Y = Y_1 + Y_2 \quad = \dfrac{R}{R^2 + \omega^2 C^2} - j\left(\dfrac{\omega C}{R^2 + \omega^2 C^2} - \omega C\right)$

At resonance, imaginary part is zero $\qquad \therefore \quad R \to \infty$.

$$\therefore \qquad = \frac{\omega_2}{R^2 + \omega^2 C^2} = \omega C \quad \text{or} \quad \frac{L}{C} = R^2 + \omega^2 C^2$$

$$\omega^2 C^2 = \frac{L}{C} - R^2 \quad \text{or} \quad \omega^2 = \frac{1}{LC} - \frac{R^2}{L^2}$$

$$\text{or} \qquad f = \frac{1}{2\pi}\sqrt{\frac{1}{LC} - \frac{R^2}{L^2}}$$

This is the expression for frequency of oscillations.

Amplitude of oscillator will be very small (v is Less) for series resonance circuit. So parallel tuned circuits are employed.

6.35 1-MHz FET CRYSTAL OSCILLATOR CIRCUIT

In the basic configuration of oscillator circuit,

$$Z_1 + Z_2 + Z_3 = 0.$$

Z_1 is crystal Z_2 is L and C combination.

Z_3 is C_{gd}. The frequency of oscillator essentially depends up on crystal only as shown

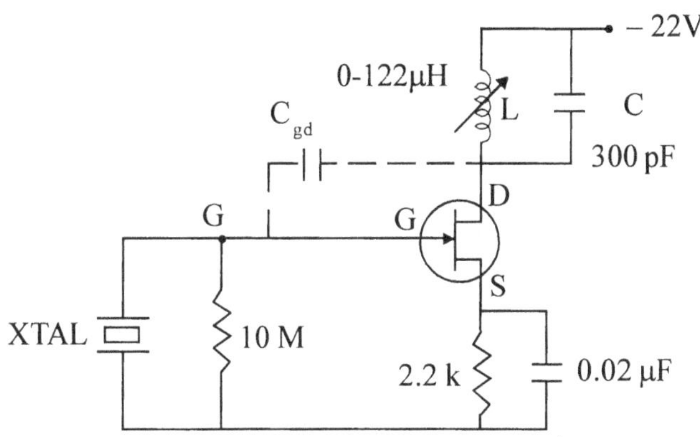

Fig. 6.93 *Crystal oscillator circuit.*

in Fig. 6.93. Z_2 and Z_3 values are insignificant compared to Z_1 of the crystal. Therefore the stability is high.

When electrical input is given, mechanical vibrations are set in the crystal, due to piezoelectric effect. But the crystal is being considered as an inductor capacitor combination only. So it acts like a $|Z|$ R_0 L, C combination. Frequency of oscillations depend only on crystal. Other L and Cs are insignificant, as their values are less compared to L, C and R of the crystal.

SUMMARY

♦ Hybrid Parameters are h - parameters, so called because the units of the four parameters are different. These are used to represent the equivalent circuit of a transistor.

♦ The h - parameters and their typical values, in Common Emitter Configuration are,

$$h_{ie} = 500 \ \Omega; \quad h_{re} = 1.5 \times 10^{-4}; \quad h_{oe} = 6 \ \mu\Omega; \quad h_{fe} = 250;$$

♦ The second subscript denotes the configuration. similarly b and c represent Common Base and Common Collector Configurations.

♦ The Transistor Amplifier Circuit Analysis can be done, using h-parameters, replacing the transistor by h-parameter equivalent.

♦ Expressions for Voltage Gain A_V, Current Gain A_I, Input Resistance R_i, Output Resistance R_o etc., can be derived in terms of h-parameters.

♦ $r_{bb'}$ is called the base spread resistance. It is the resistance between the fiction base terminal B′ and the outside base terminal B.

♦ C.E. Amplifier circuit characteristics are :

(a) Low to moderate R_i (300 Ω – 5 kΩ)

(b) Moderately high R_0 (100 kΩ – 800 kΩ)

(c) Large current amplification

(d) Large voltage amplification

(e) Large power gain

(f) 180^0 phase shift between V_i and V_0.

♦ For C.B. amplifier :

Low R_i. High R_0, $A_i < 1$, $A_v > 1$, No phase inversion, Moderate A_p

♦ For C.C. amplifier,

High R_i, Low R_0, Large A_I, $A_V < 1$, Low A_p

♦ In Darlington pair circuit, the two BJTs are in C.C. configuration. The characteristics are :

High R_i, Large A_I, Very low R_0, $A_v < 1$

♦ In CASCODE amplifier configuration, one BJT in C.E. configuration is in series with another BJT amplifier in C.B. configuration. Its characteristics are,

Very large A_v, Large A_I and thigh R_0.

♦ In the case of R-C coupled amplifier, the output is coupled to the load through R and C, hence the name. The lower cut-off frequency f_1 is also called lower 3-db point or lower half power point. The upper cut-off frequency f_2 is also called upper 3-db point or upper half power frequency. In the low frequency range and high frequency range, the gain decreases. In the mid frequency range, the gain remains constant, $(f_2 - f_1)$ is called Band width.

♦ An Amplifier circuit is to provide voltage gain or current gain or both in the form of power gain. But the other desirable characteristics of the amplifier circuits are high Z_i, Low Z_o, Large B.W, low distortion, low noise and high stability. To achieve these characteristics a part of output signal is feedback and coupled to input, to oppose in phase with (V_i). So it is negative feedback. Though gain reduces due to negative feedback, it is employed in amplifier circuits to get the other advantages.

♦ Feedback factor $\beta = V_f / V_o$. The product βA is called returned ratio. $(1 + \beta A)$ is called return difference D or desensitivity factor.

♦ The different types of feedback are   (i) voltage series   (ii) current series   (iii) voltage shunt and   (iv) current shunt.

♦ With voltage feedback (series or shunt) output resistance R_0 decreases.

♦ With current feedback (series or shunt) R_0 increases.

♦ With series feedback (current or voltage) R_i increases.

♦ With shunt feedback (current or voltage) R_i decreases.

♦ With negative feedback, distortion, noise, gain reduce by a factor $(1 + \beta A)$. Bandwidth, f_2, stability improve by $(1 + \beta A)$.

♦ If the feedback signal is proportional to voltage, it is voltage feedback. If the feedback signal is porportional to current, it is current feedback. If the feedback signal V_f is coming in series with input signal V_i, it is series feedback. If V_f is in parallel with V_i it is shunt feedback.

♦ If the feedback signal V_f is out of phase with V_i, opposing it is negative feedback. If V_f is in phase with V_i, adding to it or aiding V_i, it is positive feedback.

♦ Oscillators generate A.C. output without external A.C input by using Noise A.C signal generated in switching. D.C power from V_{CC} or V_{DD} is converted to A.C. power. The range of frequency signals generated by the circuit can be high and output power is small.

♦ For sustained oscillations, the conditions known as Barkhausen criterion to be satisfied are (i) $|\beta A| \geq 1$ (ii) Total loop phase shift must be 0^0 or 360^0 (or phase shift of feedback network must be 180^0 when the amplifying device produces another 180^0).

♦ In the case of JFET, R - C phase shift oscillator circuit, $f_0 = \dfrac{1}{2\pi RC \sqrt{6}}$

♦ In the case BJT, R - C phase shift oscillator circuit, $f_0 = \dfrac{1}{2\pi RC \sqrt{6 + 4k}}$ where

$K = \dfrac{R_C}{R}$

♦ For Hartley oscillator circuit which employs two inductors and one capacitor in the feedback network is, $f_0 = \dfrac{1}{2\pi \sqrt{(L_1 + L_2)C_3}}$

♦   For Colpitts oscillator circuit with two capacitors and one inductor in the feedback

$$\text{network,} \; f_0 = \frac{1}{2\pi} \frac{1}{\sqrt{L_3}} \sqrt{\frac{C_2 + C_1}{C_1 C_2}}$$

♦   For Wein Bridge oscillator circuit, the minimum gain of amplifier must be 3. It can produce variations in f_0 in the ratio of 10 : 1 compared to a variation of 3 : 1 in other types of oscillator circuits.

♦   Thermistors with (NTCR) and sensistor with positive temperature coefficient of resistance (PTCR) are used for frequency stability in oscillator circuits.

♦   The specification parameters of oscillator circuits are (i) Amplitude stability (ii) Frequency stability (iii) Frequency range (iv) Distortion in output waveform etc.

♦   Crystal oscillators produce highly stable output waveform in the high frequency range of MHz also.

OBJECTIVE TYPE QUESTIONS

1.   The units of h-parameters are ....

2.   h-parameters are named as hybrid parameters because ....:.......

3.   The general equations governing h-parameters are

$$V_1 = \text{..............................}$$
$$I_2 = \text{..............................}$$

4.   The parameter h_{re} is defined as $h_{re} = $ ....

5.   h-parameters are valid in the .... frequency range.

6.   Typical values of h-parameters in Common Emitter Configuration are ....

7.   The units of the parameter h_{rc} are ....

8.   Conversion Efficiency of an amplifier circuit is ....

9.   Expression for current gain A_I in terms of h_{fe} and h_{re} are $A_I = $ ....

10. In Common Collector Configuration, the values of $h_{rc} \simeq$ ....

11. In the case of transistor in Common Emitter Configuration, as R_L increases, R_i ....

12. Current Gain A_I of BJT in Common Emitter Configuration is high when R_L is ....

13. Power Gain of Common Emitter Transistor amplifier is ....

14. Current Gain A_I in Common Base Configuration is ....

15. Among the three transistor amplifier configurations, large output resistance is in .... configuration.

16. Highest current gain, under identical conditions is obtained in .... transistor amplifier configuraiton.

17. C.C Configuration is also known as .... circuit.
18. Interms of h_{fe}, current gain in Darlington Pair circuit is approximately ....
19. The disadvantage of Darlington pair circuit is ....
20. Compared to Common Emitter Configuration R_i of Darlington piar circuit is ....
21. In CASCODE amplifier, the transistors are in .... configuration.
22. The salient featuers of CASCODE Amplifier are ....
23. The disadvantage of negative feedback is _______________ .
24. The expression for sensitivity of an amplifier with negative feedback is
     _______________ .
25. The expression for Desentivity of negative feedback amplifier is D =
     _______________ .
26. The relation between bandwidth of an amplifier without feedback and with negative
     feedback is _______________ .
27. Relation between upper cut-off frequency f_2' with negative feedback and f_2 without
     negative feedback is f_2' = _______________ .
28. Negative feedback is also called as _______________ .
29. For Ideal transconductance amplifier R_i = _______________ .
30. For practical transresistance amplifier, it is desirable that R_i is _______________
     and R_0 is _______________ .
31. Voltage sampling is also known as _______________ .
32. Characteristics of ideal voltage amplifier are : _______________ .
33. Desirable characteristics of practical current amplifier are = _______________ .
34. βA in feedback amplifier circuits is called _______________ .
35. With voltage feedback, (series or shunt), output resistance R_0 of an amplifier
     _______________ .
36. With series feedback, (voltage or current), input resistance R_i of an amplifier
     _______________ .
37. Expression for reverse transmission factor or feedback factor β =
     _______________ .
38. Identify the type of feedback.

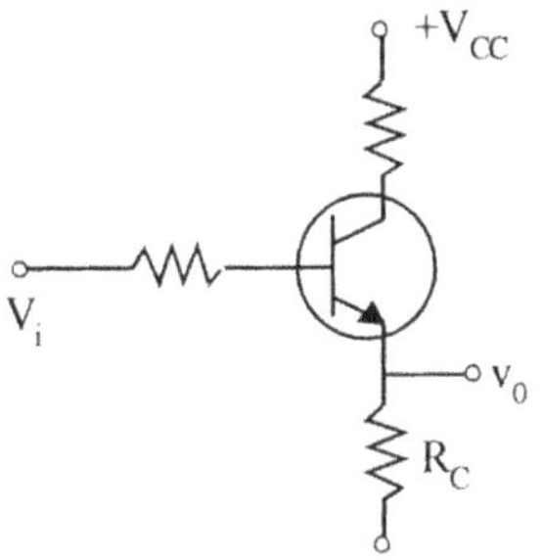

(i) 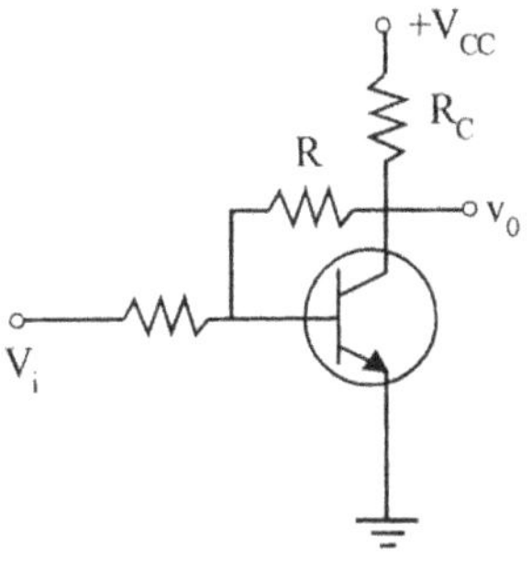   (ii)

39. In the case of Voltage - series feedback, expression for output impedance with feed back z_{of} is ....

40. In the case of Voltage shunt feedback amplifier, expression for input impedance with feedback is Z_{if} = ...

41. In the case of current series feedback amplifier, expression for output impedance with feedback z_{of} is ....

42. Expression for z_{if} with current shunt feedback is ....

43. The difference between an amplifier circuit and oscillator circuit is ___________________.

44. A.C. output signal is generated by the oscillator circuits without external A.C input, by amplifying _______________ signal.

45. Oscillator circuits employ _______________ type of feedback.

46. One condition for sustained oscillations is total loop phase shift must be _______________ degrees.

47. As per Berkauhsen criteria for sustained oscillations, $|\beta A|$ must be _______________.

48. The range of frequencies over which R - C phase shift oscillator circuit is used is ___________________.

49. In the Mega Hertzs frequency range, the type of oscillator circuit used is ___________________.

50. The range of frequency over which Wein Bridge oscillator used is _______________.

51. In the feedback network if two inductors and one capacitor elements are used, the oscillator circuit is _______________.

52. The oscillator circuit which employs two capacitors and one inductor in the feedback network, is _______________ oscillator circuit.

53. Naturally occuring materials used for in crystal oscillator circuits, exhibiting piezoelectric effect are 1. _______________ 2. _______________.

54. Synthetic materials which exhibit piezoelectric effect are,

 1. _______________ 2. _______________

 3. _______________ 4. _______________

55. The ratio of frequency variation possible with Wein Bridge Oscillator circuit is _______________ is compared to the ratio of _______________ with others oscillator circuits.

56. Thermistors have temperature coefficient and the materials used are _______________.

57. Typical values of L, C, R and Q of a crystal used in oscillator circuits are

 L = _______________ C = _______________ R = _______________ Q = _______________.

58.   Expression for frequency of Osillations in the case of Phase-shaft oscillator circuit is f_o = ... and the value of B must be at least ....

59.   Expression for frequency of oscillations in the case of Wien Bridge Oscillator is f_o = ...

60.   In the case of Colpitts oscillator f_o = ...

61.   For Hartley oscillator, f_o = ...

62.   Frequency of oscillations, in the case of uJT oscillator circuit is $fo\simeq$ ....

ESSAY TYPE QUESTIONS

1.   Write the general equations in terms of h-parameters for a BJT in Common Base Amplifiers configuration and define the h-parameters.

2.   Convert the h-parameters in Common Base Configuration to Common Emitter Configuration, deriving the necessary equations.

3.   Compare the transistor (BJT) amplifiers circuits in the three configurations with the help of h-parameters values.

4.   Draw the h-parameter equivalent circuits for Transistor amplifiers in the three configurations.

5.   With the help of necessary equations, discuss the variations of A_V, A_I, R_i R_o, A_P with R_S and R_L in Common Emitter Configuration.

6.   Discuss the Transistor Amplifier characteristics in Common Base Configuration and their variation with R_S and R_L with the help of equations.

7.   Compare the characteristics of Transistor Amplifiers in the three configurations.

8.   Draw the circuit for Darlington piar and derive the expressions for A_I, A_V, R_i and R_o.

9.   Draw the circuit for CASCODE Amplifier. Explain its working, obtaining overall values of the circuit for h_i, h_f, h_o and h_r.

10.   Explain the concept of feedback as applied to electronic amplifier circuits. What are the advantages and disadvantages of positive and negative feedback ?

11.   With the help of a general block schematic diagram explain the term feedback.

12.   What type of feedback is used in electronic amplifiers ? What are the advantages of this type of feedback ? Prove each one mathematically.

13.   Define the terms Return Ratio, Return Difference feedback factor, closed loop voltage gain and open loop voltage gain. Why negative feedback is used in electronic amplifiers eventhough closed loop voltage gain decreases with this type of feedback ?

14. Give the equivalent circuits, and characteristics of ideal and practical amplifiers of the following types (i) Voltage amplifier, (ii) Current amplifiers, (i i i) Transresistance amplifier, (iv) Transconductance amplifier.

15. Derive the expression for the input resistance with feedback R_{if} (or R_i') and output resistance with feedback R_{0f} (or R_i') in the case of

    (a)   Voltage series feedback amplifier.

    (b)   Voltage shunt feedback amplifier.

    (c)   Current series feedback amplifier.

    (d)   Current shunt feedback amplifier.

16. In which type of amplifier the input impedance increases and the output impedance decreases with negative impedance ? Prove the same drawing equivalent circuit.

17. Draw the circuit for Voltage series amplifier and justify the type of feedback. Derive the expressions for A_V', β, R_i' and R_0' for the circuit.

18. Draw the circuit for Current series amplifier and justify the type of feedback. Derive the expressions for A_V', β, R_i' and R_0' for the circuit.

19. Draw the circuit for Voltage shunt amplifier and justify the type of feedback. Derive the expressions for A_V', β, R_i' and R_0' for the circuit.

20. Draw the circuit for Current shunt amplifier and justify the type of feedback. Derive the expressions for A_V', β, R_i' and R_0' for the circuit.

21. Explain the basic principle of generation of oscillations in LC tank circuits. What are the considerations to be made in the case of practical L.C. Oscillator Circuits ?

22. Deduce the Barkausen Criterion for the generation of sustained oscillations. How are the oscillations initiated?

23. Draw the circuit and explain the princple of operaiton of R.C.phase-shift oscillator circuit. What is the frequency range of generation of oscillations ? Derive the expression for the frequency of oscillations.

24. Derive the expression for the frequency of Hartely oscillators.

25. Derive the expression for the frequency of Colpitt Oscillators.

26. Derive the expression for the frequency of Wein Bridge Oscillators.

27. Derive the expression for the frequency of Crystal Oscilators.

28. Explain how better frequency stability is obtained in crystal oscillator ?

29. Draw the equivalent circuit for a crystal and explain how oscillations can be generated in electronic circuits, using crystals.

30. Why three identical R-C sections are used in R-C phase-shift oscillator circuits? Consider the other possible combinations and limitations.

MULTIPLE CHOICE QUESTIONS

1. The h-parameters have
   (a)   Same units for all parameters   (b)   different units for all parameters
   (c)   do not have units   (d)   dimension less

2. The expression defining the h-parameter h_{22} is,

   (a) $h_{22} = \left.\dfrac{I_1}{I_2}\right|_{I_1=0}$ (b) $h_{22} = \left.\dfrac{I_2}{I_1}\right|_{I_1=0}$

   (c) $h_{22} = 12.11$ (d) $h_{22} = \left.\dfrac{I_2}{V_2}\right|_{I_1=0}$

3. h - Parameters are valid over a ...... frequency range
   (a)   R.F.   (b)   For DC only
   (c)   Audio frequency range   (d)   upto 1 M Hz

4. By definition the expression for h_{oe} is,

   (a) $\left.\dfrac{\partial I_C}{\partial I_B}\right|_{I_E=K}$ (b) $\left.\dfrac{\partial I_C}{\partial V_C}\right|_{I_B=k}$ (c) $\left.\dfrac{\partial V_C}{\partial I_C}\right|_{I_B=k}$ (d) $\left.\dfrac{\partial I_B}{\partial V_B}\right|_{I_C=k}$

5. Typical value of h_{fb} is
   (a)   −0.98   (b)   −101   (c)   + 2.3   (d)   26.6 Ω

6. The expression for h_{ie} interms of h_{fb} and h_{ib} is, $h_{ie} \simeq$

   (a) $h_{ie} \simeq \dfrac{h_{fb}}{1-h_{fb}}$ (b) $\dfrac{h_{fb}}{1+h_{fb}}$ (c) $\dfrac{h_{ib}}{1+h_{fb}}$ (d) $\dfrac{h_{ib}}{1-h_{fb}}$

7. The ratio of AC signal power delivered to the load to the DC input power to the active device as a percentage is called
   (a)   conversion   (b)   Rectification η (c)   power η (d)   utilisation factor

8. The general expression for A_I interms of h_f h_o and Z_L is ... $A_I =$

   (a) $-\dfrac{h_o}{1+h_f Z_L}$ (b) $-\dfrac{h_f}{1+h_o Z_L}$ (c) $\dfrac{h_o}{1-h_f Z_L}$ (d) $\dfrac{h_f}{1-h_o Z_L}$

9. Expression for Avs in terms of A_s, R_s, Z_L, Z_i, Z_o is .....

   (a) $\dfrac{A_I}{R_s + Z_i}$ (b) $\dfrac{A_I \cdot Z_L}{R_s - Z_i}$ (c) $\dfrac{Z_L}{R_s + Z_i}$ (d) $\dfrac{A_I Z_L}{R_s + Z_i}$

10. **In the case of BJT, the resistance between fictious base terminal B' and outside base terminal B is, called**
    (a)   Base resistance   (b)   Base drive resistance
    (c)   Base spread resistance   (d)   Base fictious resistance

11. **In large signal analysis of amplifiers ....**
    (a)   The swing of the input signal is over a wide range around the operating point.
    (b)   Operating point swimgs over large range
    (c)   stability factor is large
    (d)   power dissipation is large

12. **The units of the parameter hoe ...**
    (a)   (b) Ω (c)   constant   (d) V-A

13. **Typical value of hre is**
    (a) 10^{-4} V/A   (b) 10^{-4} (c) 10^{-6} A/V   (d) 10^4

14. **CASCODE transistor amplifier configuration consists of**
    (a)   CE-CE amplifier stages   (b)   CE-CC stages
    (c)   CE-CB stages   (d)   CB-CC stages

15. **For CASCODE amplifier, on the input side, the amplifier stage is in ....**
    (a)   C.B configuration   (b)   C.C. configuration
    (c)   emitor follower   (d)   C.E. configuration

16. **CASCODE amplifier characteristics are**
    (a)   Low voltage gain, large current are
    (b)   Low voltage gain, low current gain
    (c)   large voltage gain, large current gain, high output resistance
    (d)   Large current gain, Large voltage gain Low putput resistance

17. **Characteristics of Darlington circuit are typical values**
    (a) $Ri = M\Omega$, $A_I = 10$ k, $R_o = M\Omega$, $Av = 100$
    (b) $Ri = M\Omega$, $A_I = 10$ k, $R_o = 10\Omega$, $Av < 1$
    (c) $Ri = 10\Omega$, $A_I = 1$, $R_o = 1.0\Omega$, $Av > 1$
    (d) $Ri = M\Omega$, $A_I = 1$, $Ro = 10$ Ω, $Av < 1$

18. **The disadvantage of Darlington pair circuit is ...**
    (a)   low current gain   (b)   low output resistance
    (c)   leakage current is more   (d)   high input resistance

19. **Compared to C.C. configuration, Darlington pair circuit has**
    (a)   low current gain   (b)   low voltage gain
    (c)   large current gain   (d)   large p output resistance

20. In the case of Darlington circuit, A_I is approximately

(a) $2 h_{fe}$ (b) $4 h_{fe}$ (c) h_{fe} (d) $(h_{fe})^2$

21. Loop sampling is also known as

(a)   Voltage sampling   (b)   Current sampling

(c)   Power sampling   (d)   Node sampling

22. Positive feedback is also known as

(a)   regenerative feedback   (b)   degenerative feedback

(c)   Loop feedback   (d)   return feedback

23. Equation for feedback factor 'β' is ....

(a) $\dfrac{V_o}{V_f}$ (b) $\dfrac{V_o'}{V_f}$ (c) $\dfrac{V_f}{V_o'}$ (d) $\dfrac{V_f}{V_o}$

24. With negative feedback, the linearity of operation of the amplifier circuit

(a)   deteriorates   (b)   improves   (c)   remian same   (d)   none of these

25. Expression for distortion D' with negative feedback, with usual notation is $D' =$

(a) $\dfrac{D}{1-\beta A}$ (b) $D(1-bA)$ (c) $\dfrac{D}{1+\beta A}$ (d) $D(1+bA)$

26. Expression for densitivity of an amplifier circuit with negative feedback is,

(a) $D = (1-\beta A)$ (b) $D = \dfrac{1}{\beta A}$ (c) $D = (1+\beta A)$ (d) $D = \dfrac{1}{(1+\beta A)}$

27. Characteristics of ideal volt age amplifier are ...

(a) $R_i = \infty, R_0 - 0$ (b) $R_i = 0, R_0 = 0$

(c) $R_i = 0, R_0 = \infty$ (d) $R_i = \infty, R_0 = \infty$

28. The product of feedback factor 'β' and amplification factor 'A' is called ...

(a)   return difference   (b)   return ratio

(c)   series ratio   (d)   shunt ratio

29. Negative series feedback voltage or current ... input resistance

(a)   decreases   (b)   no effect   (c)   can't be said   (d)   increases

30. With voltage feedback series or shunt, output ressitance R_o ....

(a)   decreases   (b)   increases   (c)   remains same   (d)   non of these

31. Johnson Noise is due to .....

    (a)   Humidity   (b)   Atmospheric conditions

    (c)   Temperature   (d)   Interference

32. The noise that arises in electronic circuits due to variation in concentrations of carriers in semiconductor devices is

    (a)   shot noise   (b)   thermal noise   (c)   Johnson noise (d) None of these

33. Expression for frequency of oscillations for the R-C phase shift oscillator circuit using JFET is, $f_0 =$

    (a) $\dfrac{1}{2\pi RC}$ (b) $\dfrac{1}{2\pi RC^2\sqrt{6}}$ (c) $\dfrac{1}{2\pi R^2 C\sqrt{6}}$ (d) $\dfrac{1}{2\pi RC\sqrt{6}}$

34. In the case of JFET R-C phase shift oscillator circuit, in order that $|\beta\,A|$ is not less than unity, μ of JFET must be at least

    (a)   16   (b)   92   (c)   29   (d)   None of these

35. Oscillator circuit in which the reactances α_1, α_2 are capacitive and α_3 is inductive is ....

    (a)   Colpitts oscillator   (b)   Hartely oscillator

    (c)   Crystal oscillator   (d)   None of these

36. In wein bridge oscillator circuits the range over which frequency of oscillations can be varied is ...

    (a)   3 : 1   (b)   10 : 1   (c)   6 : 1   (d)   None of these

37. Example for naturally occurring piezoelectric crystal material is ....

    (a) $BaTio_3$ (b)   Rochelle salt   (c)   PZT   (d)   None of these

38. Typical values for 90 KHz crystal are :

    (a) $L = 137$ H; $C = 0.0235$ μf; $R = 15$ KΩ; $Q = 5{,}500$

    (b) $L = 1$ H; $C = 0.02$ f; $R = 1\Omega$; $Q = 10$

    (c) $L = 137$ H; $C = 0.02$ μf; $R = 10$ KΩ; $Q = 1$

    (d)   None of these

39. One characteristic feature of crystals is ...

    (a)   They have low mass to elastic ratio

    (b)   They have high mass to elastic ratio

    (c)   They have low Q

    (d)   None of the above is correct

40. For Wein Bridge oscillator circuits, the minimum gain of amplifier must be

    (a)   1   (b)   3   (c)   10   (d)   None the these

CONCEPTUAL QUESTIONS (Interview Questions)

1. Why low frequency equivalent circuit is different from High frequency equivalent circuit?

Ans :   At low frequencies, capacitive reactance X_c is open circuit. Because, $X_C = \dfrac{1}{2\pi fC}$, when f is low, X_C is high or like open circuit. So junction capacitances of a transistor can be neglected. At high frequencies, when f is large, X_c is finite and can not be neglected. So the junction capacitances are to be considered. As such high frequency equivalent circuit contains capacitances, where as these are neglected (open circuited) in low frequency equivalent circuits.

Specifications of Amplifiers

Parameters		*Typical Values*
1.	Type of Amplifier	:	Audio / Typical Values / Power / RF / Video
2.	Frequency Range	:	15 Hzs - 100 KHzs
3.	Output Power	:	200 mW
4.	Voltage gain	:	20 db
5.	Current gain	:	100
6.	Power gain	:	9
7.	Input impedance	:	10 kΩ 5 pf
8.	Output impedance	:	500Ω 1 pf
9.	Band width	:	100 KHzs

2.   Give physical example for a feed back systems.

Ans : A person searching for a book in Library is one example. The book seen by eyes and signal sent to brain act as closed loop system. If the eyes are closed, it is like open loop system, as no feed back can be given to the brains.

3.   Give a practical example of a feed back system.

Ans : Controlling the temperature of a furnace is an example of feedback system. The temperature is sensed by a thermo couple and the signal is fed back to a controlling circuit. When set temperature is reached, the electric supply is turned off. When the temperature goes below the set value, electric supply is again turned on.

4.   How oscillator circuit gives output signal without input singnal ?

Ans : The circuit needs D-C input voltage supply. When the supply is switched on, the transient signal produced acts as the a.c input signal. It is amplified by the amplifier and fedback to the input. By the time the transient dies down, the feed back signal acts as the input signal. Output oscillations will be sustained for a particular frequency component for which Barkhausen criteria is satisfied. The wave shape is sinosoidal because sinosoidal function is the only function that satisfies the condition governing the exchange of energy between the energy storage elements in the circuit.

5.   Why R - C Phase shift Oscillator Circuit must have three R - C networks ? Why not two or four RC networks?

Ans : If only two R - C networks are used, each network must produce 90^0 phase shift, to get a total phase shift of 180^0 in the feed back circuit. So we must have ideal capacitor, which is not practically possible. On the otherhand we can have 4 - RC networks each producing 45° phase shift. But we will be using more number of R-C components which is not economical.

6.   Why the frequency of oscillations of a crystal oscillator circuit are stable ?

Ans : A crystal can be represented as a combination of R L and C components in its equivalent circuit form. When a crystal is connected in the circuit, it is as if R, L and C are connected in the circuit. For a crystal, these equivalent parameters of R, L and C do not change much with temperature. So frequency of oscillations of crystal oscillator circuit do not change with temperature. Thus frequency stability is high. For other oscillator circuits the R, L, C component values change significantly with temperature. So frequency stability is less.

Feedback Amplifiers

In this Chapter,

- ◆   The concept of Feedback is introduced.

- ◆   Effect of negative feedback on amplifier characteristics is explained. Necessary equations are derived.

- ◆   Voltage series and shunt amplifiers, current series and shunt amplifier circuits are given.

7.1   FEEDBACK AMPLIFIERS

Any system whether it is electrical, mechanical, hydraulic or pneumatic may be considered to have at least one input and one output. If the system is to perform smoothly, we must be able to measure or control output. If the input is 10mV, gain of the amplifier is 100, output will be 1V. If the input deviates to 9mV or 11mV, output will be 0.9 V or 1.1V. So there is no control over the output. But by introducing feedback between the output and input, there can be control over the output. If the input is increased, it can be made to increase by having a link between the output and input. By providing feedback, the input can be made to depend on output.

One example is, the temperature of a furnace. Suppose, inside the furnance, the temperature should be limited to 1000^0C. If power is supplied on, continuously, the furnace may get over heated. Therefore we must have a thermocouple, which can measure the temperature. When the output of the thermocouple, reaches a value corresponding to 1000^0C, a relay should operate which will switch off the power supply to the furnace. Then after sometime, the temperature may come down below 1000^0c. Then again another relay should operate, to switch on the power. Thus the thermocouple and relay system provides the feedback between input and output.

Another example is traffic light. If the timings of red, green and yellow lights are fixed, on one side of the road even if very few vehicles are there, the green light will be on for sometime. On the other hand, if the traffic is very heavy on the other side of the road, still if the green lamp glows for the same period, the traffic will not be cleared. So in the ideal case, the timings of the red and green lamps must be proportional to the traffic on the road. If a traffic policeman is placed, he provides the feedback.

Another example is, our human mind and eyes. If we go to a library, our eyes will search for the book which we need and indicates to the mind. We take the book, which we need. If our eyes are closed, we can't choose the book we need. So eyes will provide the feedback.

Basic definitions

Ideally an amplifier should reproduce the input signal, with change in magnitude and with or without change in phase. But some of the short comings of the amplifier circuit are

1. Change in the value of the gain due to variation in supplying voltage, temperature or due to components.
2. Distortion in wave-form due to non linearities in the operating characters of the amplifying device.
3. The amplifier may introduce noise (undesired signals)

    The above drawbacks can be minimizing if we introduce feedback

7.2   CLASSIFICATION OF AMPLIFIERS

Amplifiers can be classified broadly as,

1. Voltage amplifiers.
2. Current amplifiers.
3. Transconductance amplifiers.
4. Transresistance amplifiers.

This classification is with respect to the input and output impedances relative to the load and source impedances.

7.2.1 VOLTAGE AMPLIFIER

This circuit is a 2-port network and it represents an amplifier (see in Fig 7.1). Suppose $R_i \gg R_s$, drop across Rs is very small.

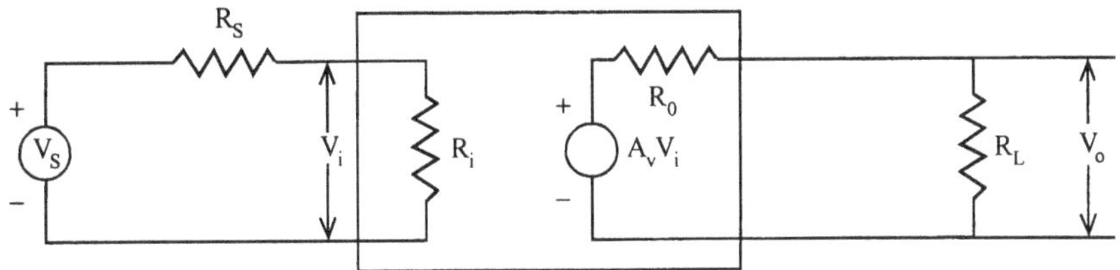

Fig 7.1 Equivalent circuit of voltage amplifiers.

$$\therefore \qquad\qquad V_i \simeq V_S.$$

Similarly, if $R_L \gg R_O$, $V_O \simeq A_V \cdot V_i$.

But $V_i \sim V_S$.

$$\therefore \qquad\qquad V_O \simeq A_V \cdot V_S.$$

$\therefore$ Output voltage is proportional to input voltage.

The constant of proportionality A_V doesn't depend on the impedances. (Source or load). Such a circuit is called as *Voltage Amplifier*.

Therefore, for ideal voltage amplifier

$$R_i = \infty.$$
$$R_O = 0.$$
$$A_V = \frac{V_o}{V_i}$$

with $\qquad\qquad R_L = \infty.$

A_V represents the open circuit voltage gain. For ideal voltage amplifier, output voltage is proportional to input voltage and the constant of proportionality is independent of R_S or R_L.

7.2.2 CURRENT AMPLIFIER

An ideal current amplifier is one which gives output current proportional to input current and the proportionality factor is independent of R_S and R_L.

The equivalent circuit of current amplifier is shown in Fig.7.2.

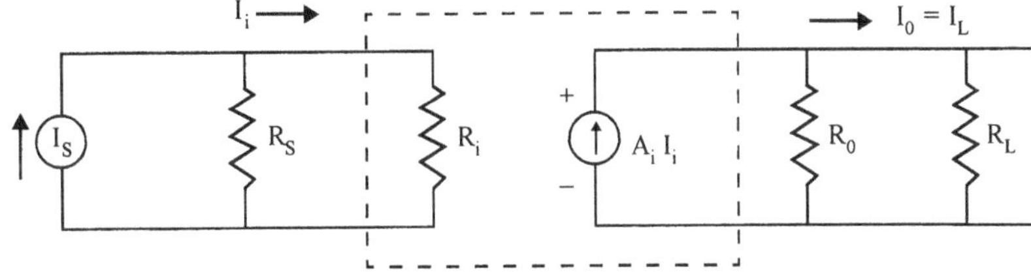

Fig 7.2 Current amplifier.

For ideal Current Amplifier,

$$R_i = 0$$
$$R_O = \infty.$$

If $$R_i = 0, I_S \sim I_i.$$

∴ $$R_O = \infty,$$
$$I_L = I_O = A_i\, I_i = A_i\, I_S.$$

∴ $$A_I = \frac{I_L}{I_i} \text{ with } R_L = 0.$$

∴ A_I represents the short circuit current amplification.

7.2.3 TRANSCONDUCTANCE AMPLIFIER

Ideal Transconductance amplifier supplies output current which is proportional to input voltage independently of the magnitude of R_S and R_L.

Ideal Transconductance amplifier will have

$$R_i = \infty.$$
$$R_O = \infty.$$

In the equivalent circuit, on the input side, it is the Thevenins' equivalent circuit. A voltage source comes in series with resistance. On the output side, it is Norton's equivalent circuit with a current source in parallel with resistance. (See Fig. 7.3).

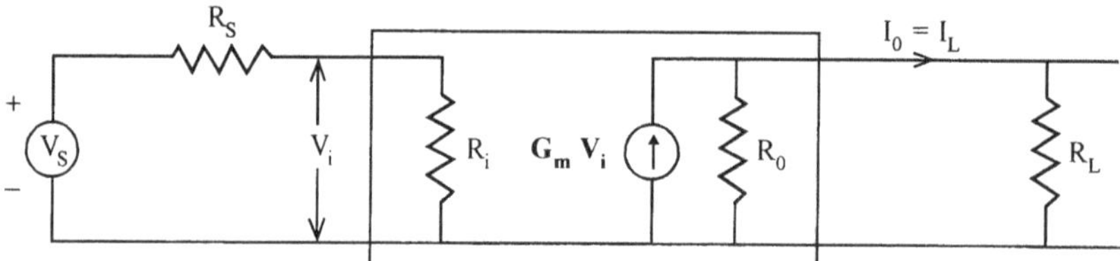

Fig 7.3 Equivalent circuit of Transconductance amplifier.

7.2.4 TRANS RESISTANCE AMPLIFIER

It gives output voltage V_0 proportional to I_s, independent of $R_s \propto R_L$. For *ideal amplifiers*

$$R_i = 0, R_0 = 0$$

Equivalent circuit

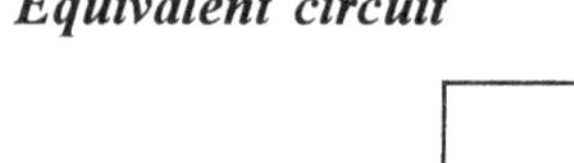

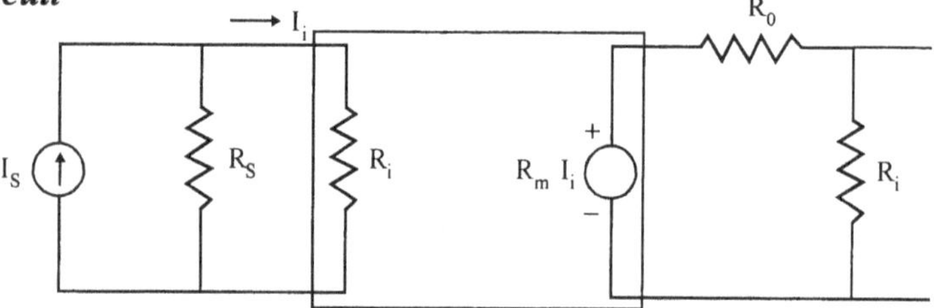

Fig 7.4 Trans resistance amplifier.

Norton equivalent circuit on the R_i input side.

Thevenins' equivalent circuit on the output side.

7.3 FEEDBACK CONCEPT

A sampling network samples the output voltage or current and this signal is applied to the input through a feedback two port network. The block diagram representation is as shown in Fig. 7.5.

GENERALIZED BLOCK SCHEMATIC

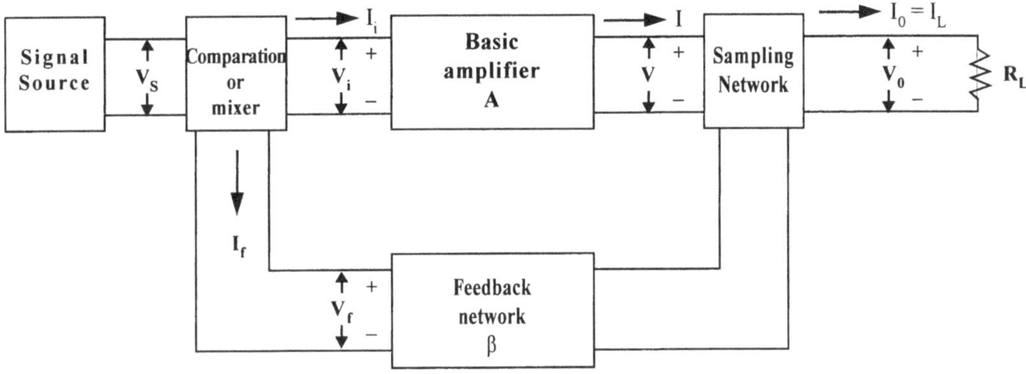

Fig 7.5 Block diagram of feedback network.

Signal Source

It can be a voltage source V_s or a current source I_s

FEEDBACK NETWORK

It is a passive two port network. It may contain resistors, capacitors or inductors. But usually a resistance is used as the feedback element. Here the output current is sampled and feedback. The feedback network is connected in series with the output. This is called as *Current Sampling or Loop Sampling*.

A voltage feedback is distinguished in this way from current feedback. For voltage feedback, the feedback element (resistor) will be in parallel with the output. For current feedback the element will be in series.

COMPARATOR OR MIXER NETWORK

This is usually a differential amplifier. It has two inputs and gives a single output which is the difference of the two inputs.

$$V = \text{Output voltage of the basic amplifier } \textit{before sampling} \text{ [see the block diagram of feedback]}$$

$$V_i = \text{Input voltage to the basic amplifier}$$

$$A_V = \text{Voltage amplification} = V/V_i$$

$$A_I = \text{Current amplification} = I/I_i$$

$$G_M = \text{Transconductance of basic amplifier} = I/V_i$$

$$R_M = \text{Transresistance} = V/I_i.$$

All these four quantities, A_v, A_I, G_M and R_M represent the transfer gains (Though G_M and R_M are not actually gains) of the basic amplifier without feedback. So the symbol 'A' is used to represent these quantities.

A_f is used to represent the ratio of the output to the input *with feedback*. This is called as the transfer gain of the amplifier with feedback.

$$\therefore \qquad A_{Vf} = \frac{V_o}{V_s}$$

$$A_{If} = \frac{I_o}{I_s}$$

$$G_{Mf} = \frac{I_o}{V_s}$$

$$R_{Mf} = \frac{V_o}{I_s}$$

Feedback amplifiers are classified as shown below.

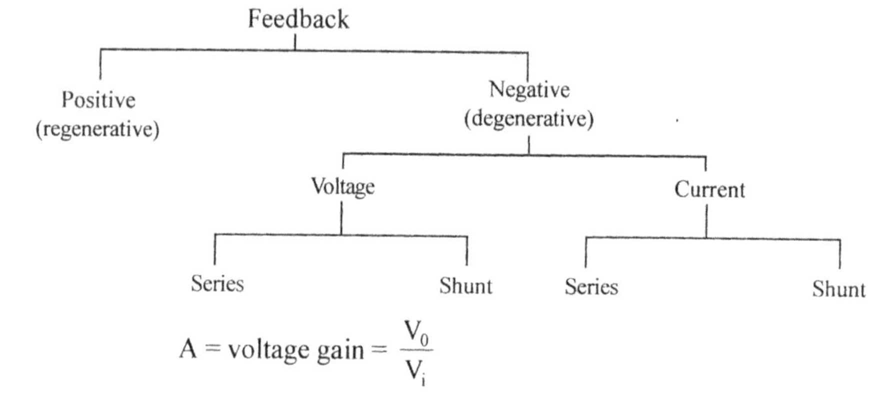

$$A = \text{voltage gain} = \frac{V_o}{V_i}$$

Feedback Factor, $\quad \beta = \dfrac{V_f}{V_0} \quad$ (This β is different from β used in BJTs)

$$(\beta)\,(A)\,(-1) = -\beta A$$

Loop gain / Return Ratio

-1 is due to phase-shift of 180^0 between input and output in Common Emitter Amplifier. Since $\text{Sin}(180^0) = -1$.

Return difference $D = 1 -$ loop gain negative sign is because it is the difference

$$D = 1 - (-\beta A)$$

$$\boxed{D = 1 + \beta.A.}$$

Types of Feedback

How to determine the type of feedback ? Whether current or voltage ? If the feedback signal is proportional to voltage, it is *Voltage Feedback.*

If the feedback signal is proportional to current, it is *Current Feedback.*

Conditions to be satisfied

1. Input signal is transmitted to the output through amplifier A and not through feedback network β.
2. The feedback signal is transmitted to the input through feedback network and not through amplifier.
3. The reverse transmission factor β is independent of R_s and R_L.

7.4 TYPES OF FEEDBACK

Feedback means a portion of the output of the amplifier circuit is sent back or given back or feedback at the input terminals. By this mechanism the characteristics of the amplifier circuit can be changed. Hence feedback is employed in circuits.

There are two types of feedback.

1. Positive Feedback
2. Negative Feedback

Negative feedback is also called as ***degenerative feedback***. Because in negative feedback, the feedback signal opposes the input signal. So it is called as degenerative feedback. But there are many advantages with negative feedback.

Advantages of Negative Feedback

1. Input impedance can be increased.
2. Output impedance can be decreased.
3. Transfer gain A_f can be stabilized against variations in ***h-parameter*** of the transistor with temperature etc.

   i.e. stability is improved.
4. Bandwidth is increased.
5. Linearity of operation is improved.
6. Distortion is reduced.
7. Noise reduces.

7.5 EFFECT OF NEGATIVE FEEDBACK ON TRANSFER GAIN

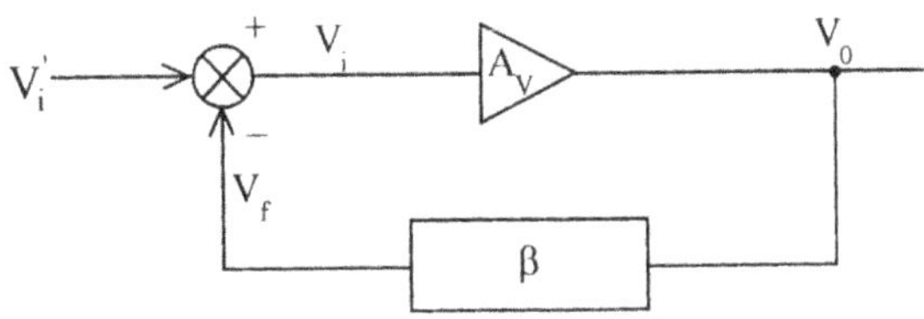

Fig 7.6 Block schematic for negative feedback.

$$A_v = \frac{V_o}{V_i}$$

$$A_v' = V_o / V_i'$$

$$V_i' = V_i - \beta V_0$$

$$V_o = A_v (V_i - \beta V_o)$$

$$V_o = A_v . V_i - A_v . \beta V_o$$

$$V_o (1 + \beta A_v) = A_v V_i$$

$$\therefore \quad A_{vf} = \frac{V_o}{V_i} = \frac{A_v}{1 + \beta A_v}$$

7.5.1 Reduction in Gain

$$A'_v = \frac{A_v}{1 - \beta A_v}$$

(for positive feedback)

A_v = Voltage gain without feedback. (Open loop gain).

If the feedback is negative, β is negative.

$$\therefore \quad A'_v = \frac{A_v}{1 - \left(- \beta . A_v \right)}$$

For negative feedback $A'_v = \dfrac{A_v}{1 + \beta A_v}$

Denominator is > 1. $\therefore$ $A'_v < A_v$

$\therefore$ There is reduction in gain.

7.5.2 Increase in Bandwidth

If, f_H is upper cutoff frequency.

f_L is lower cut off frequency.

f is any frequency.

Expression for A_v (voltage gain at any frequency f) is,

$$A_v = \frac{A_v(mid)}{1 + j . \dfrac{f}{f_H}} \qquad \qquad ..... (1)$$

A_v (mid) = Mid frequency gain

$$A_v = \frac{A_v(mid)}{1 + \beta A_v} \quad \text{for negative feedback.} \qquad ..... (2)$$

Substituting equation (1) in (2) for A_V,

$$A_v = \frac{A_v(mid) / 1 + j . \dfrac{f}{f_H}}{1 + \beta \left[\dfrac{A_v mid}{1 + jf / f_H} \right]} \qquad \text{(for negative Feedback, } \beta \text{ is } V_e)$$

Simplifying,
$$A_V = \frac{\dfrac{A_V\,(mid)}{f_H + j.f}\left[1 + j\dfrac{f}{f_H}\right]}{1 + j.f + \beta\left[A_V\ mid\right]}$$

$$= \frac{A_V\,(mid)}{f_H + j.f + \beta.A_{Vmid}\cdot f_H}$$

$$A_V{}' = \frac{A_v(mid)/1 + \beta_v A_v(mid)}{1 + \dfrac{jf}{f_H\left[1 + \beta_v A_v(mid)\right]}} \qquad\qquad ..... (3)$$

Equation (3) can be written as,

$$A_v{}' = \frac{A_V'\,(mid)}{1 + \dfrac{jf}{f_H'}}$$

Where
$$A_v{}'_{(mid)} = \frac{A_v(mid)}{1 + \beta_v A_v(mid)}$$

and
$$f_H{}' = f_H\,(1 + \beta_v\,A_{v\,(mid)})$$

$\therefore$ β is negative for negative feedback, $f_H{}' > f_H$.

$\therefore$ ***Negative feedback, increases bandwidth.***

Similarly,
$$f_L{}' = \frac{f_L}{1 + \beta_v A_{v(mid)}} \qquad A_v = \frac{A_{v(mid)}}{1 - j\left(\dfrac{f_L}{f}\right)}$$

or
$$A_v{}' = \frac{A_{V\,(mid)}/1 + \beta A_{V(mid)}}{1 + j\dfrac{f_L}{f\left(1 + \beta.A_V\right)}}$$

$$A_v{}' = \frac{A_V}{1 + j\dfrac{f_L'}{f}}$$

7.5.3 Reduction in Distortion

Suppose, the amplifier, in addition to voltage amplification is also producing distortion D.

$$V_0 = A_v \cdot V_i + D.$$

Where,
$$V_i = V_i{}' - \beta_v \cdot V_0 \quad \text{(for negative feedback)}$$
$$V_i{}' = V_i - \beta\,V_0 \quad \text{and } \beta \text{ is negative for negative feedback,}$$

$\therefore$
$$V_0{}' = A_v\left[V_i{}' - \beta_v V_0\right] + D$$

$$V_o \left[1 + A_V \ \beta_V\right] = V_i'.A_V + D$$

$$\therefore \qquad V_o = \frac{V_i'.A_v}{\left(1 + A_V.\beta_V\right)} + \frac{D}{\left(1 + \beta_V A_V\right)}$$

For negative feedback, β is negative therefore denominator is > 1

$\therefore$ The distortion in the output is reduced.

$$= \frac{D}{1 + \beta_V A_v} \text{ is} < D$$

The physical explanation is, suppose the input is pure sinusoidal wave. There is distortion in the output as shown, in Fig. 7.7.

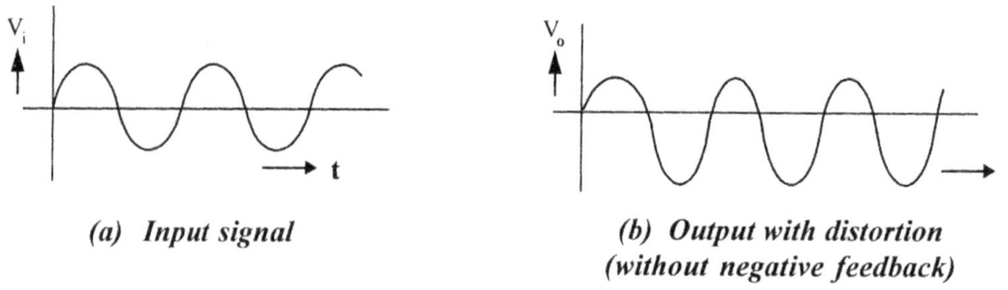

(a) Input signal   *(b) Output with distortion*
(without negative feedback)

Fig 7.7

Now if a part of the distorted output is fed back to the input, so as to oppose the input, the feedback signal and input will be out of phase.

So the new input V_i will have distortion introduced init, because of mixing of distorted V_f with pure V_i. So the distortion in the output will be reduced because, the distortion introduced in the input cancels the distortion produced by the amplifier. Because these two distortions are out of phase. The feedback signal cancels the distortion produced by the amplifier, Therefore these two are out of phase.

7.5.4 Feedback to Improve Sensitivity

Suppose an amplifier of gain A_1 is required. Build an amplifier of gain $A2 = DA_1$ in which D is large. Feedback is now introduced to divide the gain by D. Sensitivity is improved by the same factor D, because both gain and instability are divided by D. The stability will be improved by the same factor.

7.5.5   Frequency Distortion

$$A_f = \frac{A}{1 + \beta A}$$

If the feedback network does not contain reactive elements. The overall gain is not a function of frequency. So frequency duration is less. If β depends upon frequency, with negative feedback, Q factor will be high.

7.5.6 BAND WIDTH

It increases with negative Feedback (as shown in Fig.7.2)

$$f_1^1 < f_1, \; f_2^1 > f_2$$

$$BW' = (f_1' - f_2') : BW = (f_2 - f_1)$$

$$f_1' = \frac{f}{1+\beta A_m} \quad f_2' = f_2\,(1+\beta A_m) \quad BW' > BW$$

$$BW = f_2 - f_1 \simeq f_2$$

$$(BW)_f = f_2' - f_1' \sim f_2'$$

$$\therefore \quad (BW)_f = (1+\beta A_m)\,BW$$

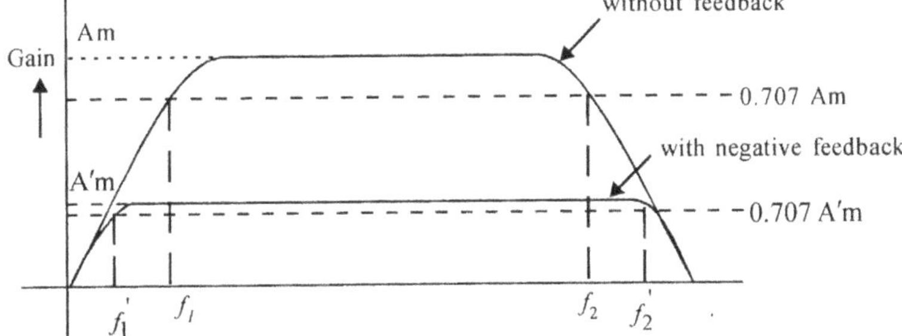

Fig 7.8 Frequency Response with and without negative feedback.

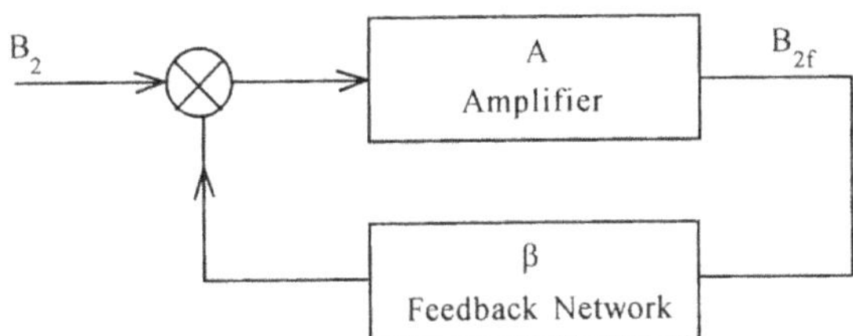

Fig 7.9 Feedback network.

7.5.7 SENSITIVITY OF TRANSISTOR GAIN

Due to aging, temperature effect etc., on circuit capacitance, transistor or FET, stability of the amplifier will be affected

Fractional change in amplification with feedback divided by the fractional change without feedback is called *Sensitivity* of Transistor.

$$= \frac{\left|\dfrac{dA_f}{A_f}\right|}{\left|\dfrac{dA}{A}\right|}$$

$$A_f = \frac{A}{1+\beta A}$$

$$\text{Differentiating,} \qquad \frac{dA_f}{A_f} = \frac{1}{|1+\beta A|}\left|\frac{dA}{A}\right|$$

$$\therefore \text{ Sensitivity,} \qquad = \frac{1}{1+\beta A}$$

Reciprocal of sensitivity is **Densitivity** $\boxed{D = (1+\beta A).}$

7.5.8 REDUCTION OF NONLINEAR DISTORTION

Suppose the input signal contains second harmonic and its value is B_2 before feedback. Because of feedback. B_{2f} appears at the output. So positive βB_{2f} is fed to the input. It is amplified to $-A\beta B_{2f}$.

$\therefore$ Output with two terms $B_2 - A\beta B_2 f$.

$$\therefore \qquad B_2 - A\beta B_2 f = B_{2f}.$$

$$\text{or} \qquad B_{2f} = \frac{B_2}{1+\beta A} \quad B_{2f} < B_2$$

So it is reduced.

7.5.9 REDUCTION OF NOISE

Let N be noise constant without feedback and N_F with feedback. N_F is fed to the input and its value is βN_F. It is amplified to $-\beta A N_F$.

$$\therefore \qquad N_F = N - \beta A N_F$$
$$N_F(1+\beta A) = N$$

$$\text{as} \qquad N_F = \frac{N}{1+\beta A}$$

$$\therefore \qquad N_F < N. \text{ Noise is reduced with negative feedback.}$$

7.6 TRANSFER GAIN WITH FEEDBACK

Consider the generalized feedback amplifier shown in Fig.7.10.

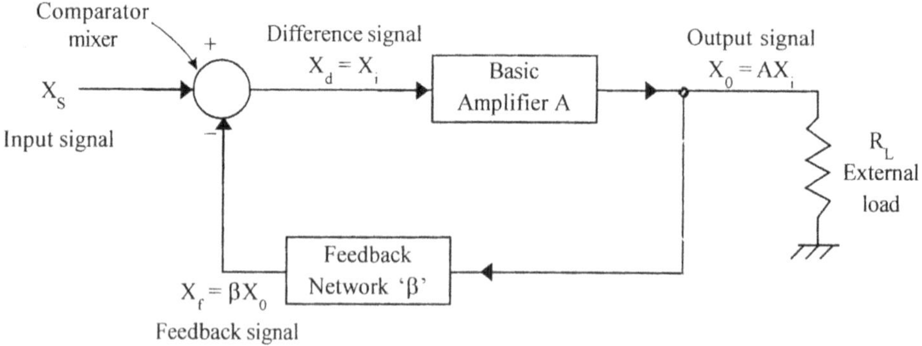

Fig 7.10

The basic amplifier shown may be a voltage amplifier or current amplifier, transconductance or transresistance amplifier.

The four different types of feedback amplifiers are,

1. Voltage series feedback.
2. Voltage shunt feedback.
3. Current series feedback.
4. Current shunt feedback.

$$X_s = \text{Input signal.}$$
$$X_0 = \text{Output signal.}$$
$$X_f = \text{feedback signal.}$$
$$X_d = \text{Difference signal. [Difference between the input signal and the feedback signal].}$$
$$\beta = \text{Reverse transmission factor or feedback factor} = X_f / X_o.$$

The feedback network can be a simple resistor. The mixer can be a difference amplifier. The output of the mixer is the difference between the input signal and the feedback signal.

$$X_d = X_s - X_f = X_i.$$

X_d is also called as error or comparison signal.

$$\beta = \frac{X_f}{X_o}$$

It is often a positive or negative real number.

This β should not be confused with the β of a transistor $\dfrac{I_C}{I_B}$

Transfer gain $\qquad A = \dfrac{X_o}{X_i} = \dfrac{X_o}{X_d}$

$$X_i = X_d \qquad\qquad\qquad\qquad\qquad .....(1)$$

Gain with feedback $A_f = \dfrac{X_o}{X_s}$

But $\qquad\qquad X_d = X_s - X_f = X_i \qquad\qquad .....(2)$

$$\beta = \frac{X_f}{X_o} \qquad\qquad\qquad\qquad .....(3)$$

From (2), $\quad X_d = X_s - X_f.$

Substitute (2) in (1).

$$\therefore \qquad\qquad A = \frac{X_0}{X_s - X_f}$$

Dividing Numerator and Denominator by X_s,

$$A = \frac{X_o/X_s}{1 - \dfrac{X_f}{X_s}} = \frac{X_o/X_s}{1 - \dfrac{X_f}{X_o} \cdot \dfrac{X_o}{X_s}}.$$

We know that $X_o / X_s = A_f$; $\dfrac{X_f}{X_o} = \beta$.

$$\therefore \quad A = \dfrac{A_f}{(1 - \beta.A_f)}$$

or $\quad A - \beta. A_f . A = A_f.$

$$A = A_f (1 + \beta A)$$

$$\therefore \quad \boxed{A_f = \dfrac{A}{1 + \beta A}}$$

A_f = gain with feedback.

A = transfer gain without feedback.

If $|A_f| < |A|$ the feedback is called as negative or degenerative, feedback

If $|A_f| > |A|$ the feedback is called as positive or regenerative, feedback

7.6.1 LOOP GAIN

In the block diagram of the feedback amplifier, the signal X_d which is the output of the comparator passes through the amplifier with gain A. So it is multiplied by A. Then it passes through a feedback network and hence gets multiplied by β, and in the comparator it gets multiplied by-1. In the process we have started from the input and after passing through the amplifier and feedback network, completed the loop. So the total product Aβ.

In order that *series feedback* is most effective, the circuit should be driven from a *constant voltage* source whose internal resistance R_s is small compared to R_i of the amplifier. If R_s is very large, compared with R_i, V_i will be modified not by V_f but because of the drop across R_s itself. So effect of V_f will not be there. Therefore for series feedback, the voltage source should have less resistance. (See Fig.7.11).

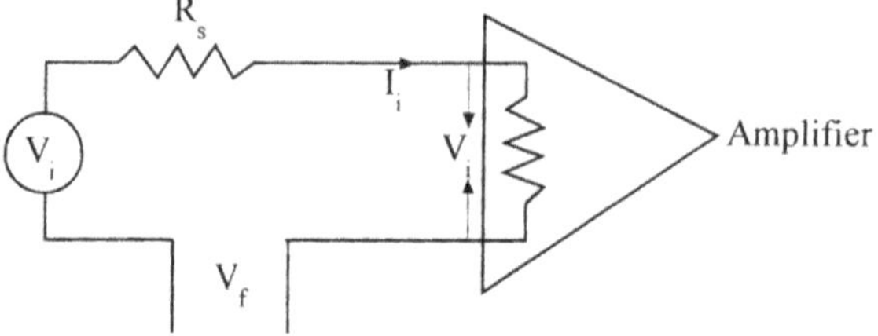

Fig 7.11   Series feedback.

In order that *shunt feedback* is most effective, the amplifier should be driven from a constant current source whose resistance R_s is very high ($R_s >> R_i$).

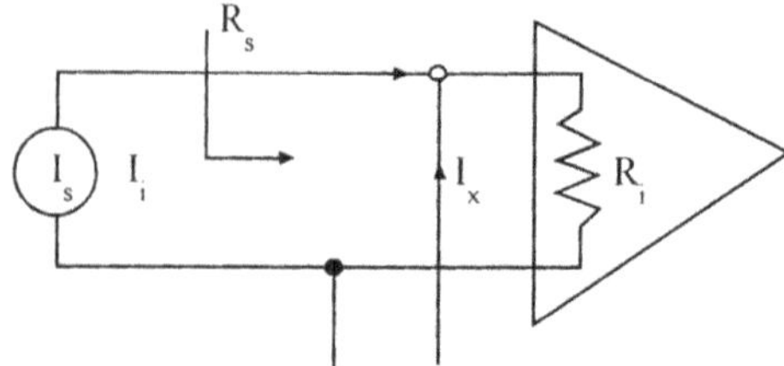

Fig 7.12 Current shunt feedback.

If the resistance of the source is very small, the feedback current will pass through the source and not through R_i. So the change in I_i will be nominal. Therefore the source resistance should be large and hence a current source should be used.

If the feedback current is same as the output current, then it is series derived feedback.

When the feedback is shunt derived, output voltage is simultaneously present across R_L and across the input to the feedback. So in this case V_f is proportional to V_o.

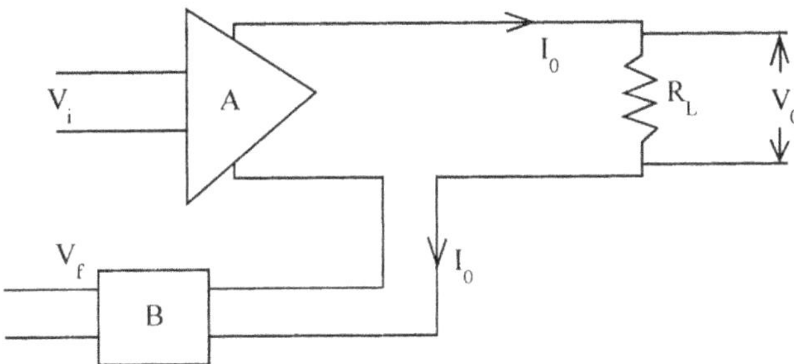

Fig 7.13 Series derived feedback.

An amplifier with shunt derived negative Feedback increases the output resistance. When the feedback is series derived, I_o remains constant, so R_o increase.

Similarly if the feedback signal is ***shunt fed*** it reduces the input resistance.

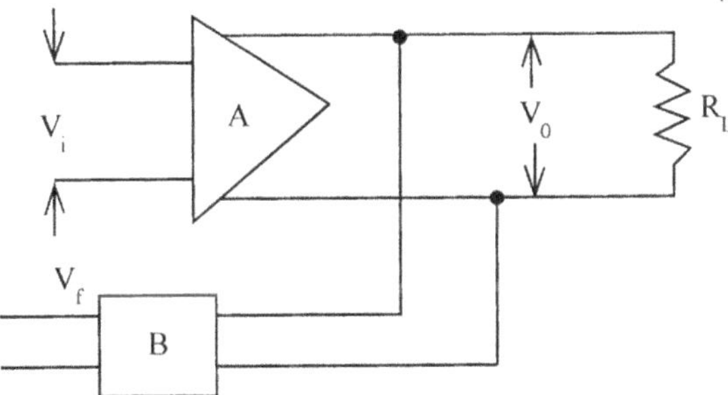

Fig. 7.14 (a) Shunt derived feedback.

If it is ***series fed***, it increases the output resistance. Therefore I_o remains constant, even through V_o increase, so R_o increases.

Return Ratio

βA = Product of feedback factor β and amplification factor A is called as ***Return Ratio***.

Return Difference (D)

The difference between unity (1) and return ratio is called as ***Return difference***.

$$D = 1 - (-\beta A) = 1 + \beta A.$$

Amount of feedback introduced is expressed in decibels

$$= N \log \left(\frac{A'}{A}\right) \quad N = 20 \log \left|\frac{A_f}{A}\right| = 20 \log \left|\frac{1}{1 + A\beta}\right|$$

If feedback is negative N will be negative because $A' < A$.

7.7 CLASSIFACTION OF FEEDBACK AMPLIFIERS

There are four types of feedback,

1. Voltage series feedback.
2. Voltage shunt feedback.
3. Current shunt feedback.
4. Current series feedback.

Voltage Series Feedback

Feedback signal is taken across R_L proportional to V_o. So it is voltage feedback. V_f is coming in series with V_i So it is Voltage series feedback.(See Fig.7.14).

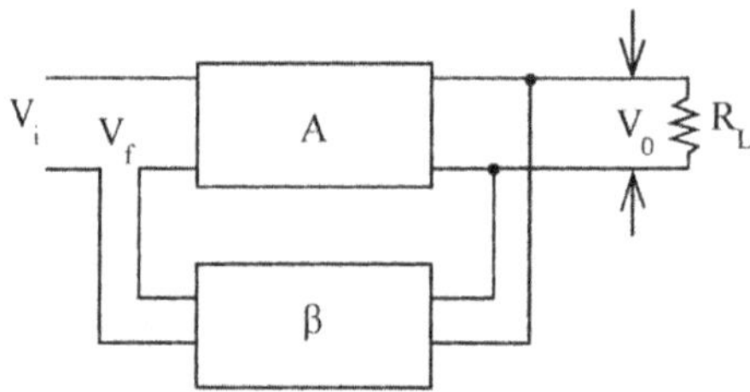

Fig 7.14   Schematic for voltage series feedback.

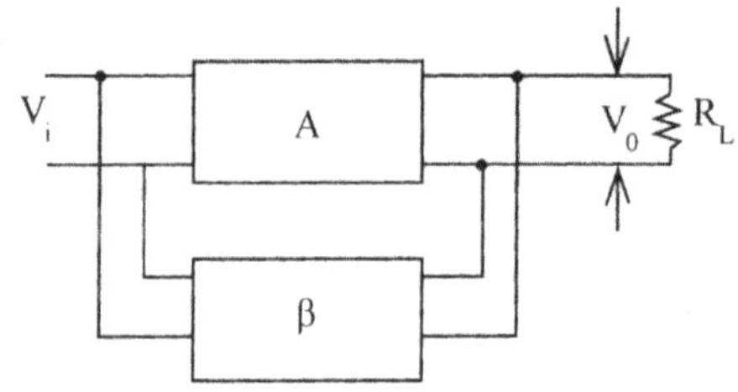

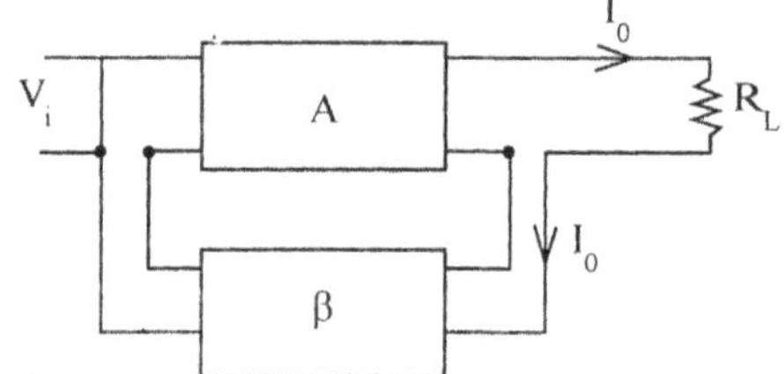

(a) Voltage shunt Feedback   *(b) Current Shunt Feedback*

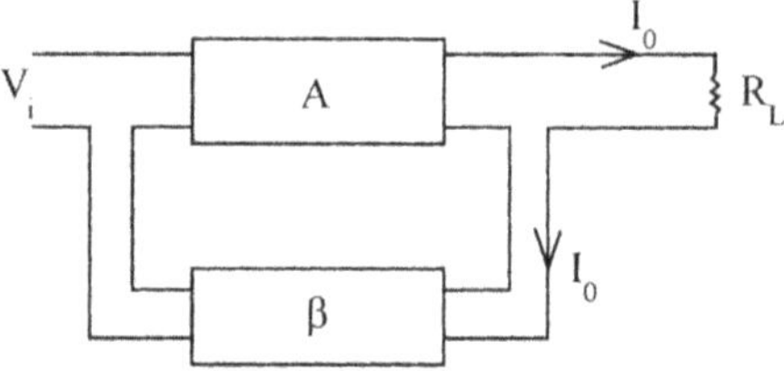

(c) Current Series Feedback

Fig 7.15

Improvement of Stability with Feedback

Stability means, the stability of the voltage gain. The voltage gain must have a stable value, with frequency. Let the change in A_v is represented by S.

$$\frac{dA_V}{A_v} = S\left(\frac{dA'_V}{A'_V}\right) = S = \frac{\left(\dfrac{d\,A'_V}{A'_V}\right)}{\left(\dfrac{d\,A_V}{A_V}\right)}$$

$$A'_V = \frac{V_0}{V_S} \quad \text{(for negative feedback)}$$

$$A'_V = \frac{A_V}{1 + A_V\,\beta_1}$$

Differentiating with respect to A_v, $\dfrac{dA_V}{A_V} = \dfrac{\left(1+\beta.A_V\right) - A_V\left(+\beta\right)}{\left(1+\beta.A_V\right)^2} = \dfrac{1}{\left(1+A_V\beta\right)^2}$

Dividing by A_v on both sides, of $\dfrac{dA'_V}{dA_V} = \dfrac{1}{\left(1+\beta A_V\right)^2}$

$$\frac{dA'_V}{A_V}\frac{1}{dA_V} = \frac{1}{\left(1+\beta A_v\right)} \cdot \frac{1}{A'_V}$$

But $\qquad A'_V = \dfrac{A'_v}{1+\beta A_v} \quad \text{and} \quad \dfrac{dA'_v}{A'_v} = S' \quad \dfrac{dA_v}{A_v} = S$

$$S = \frac{dA'_V}{dA_V} = \frac{1}{\left(1+\beta A_V\right)^2}$$

$$\boxed{S = \frac{dA_v}{\left(1+\beta A_v\right)}}$$

For negative Feedback, β is negative $\therefore$ denominator > 1.

$$\therefore \qquad\qquad S' < S$$

i.e., variation in A_v, or % change in A_v is less with -negative feedback.

$\therefore$ *Stability* is good.

7.8 EFFECT OF FEEDBACK ON INPUT RESISTANCE

7.8.1 INPUT RESISTANCE WITH SHUNT FEEDBACK

With feedback $\qquad R_{if} = \dfrac{V_i}{I_s}$

$$I_f = \beta_i\,I_o. \qquad \left[\because \beta_i = \frac{I_f}{I_o}\right]$$

$$I_s = I_i + \beta_i\,I_o$$

$$R_i' = \frac{V_i}{I_i + \beta_i I_O}$$

$$I_o = A_i . I_i$$

$$R_i' = \frac{V_i}{I_i + \beta_i A_i I_i}$$

$$R_i = \frac{V_i}{I_i(1 + \beta_i A_i)}$$

$$\boxed{R_i' = \frac{R_i}{(1 + \beta_i A_i)}}$$

∴ *Input resistance decreases with shunt feedback.*

If the feedback signal is taken across R_L, it is $\alpha \, V_o$ or so it is *Voltage feedback*.

If the feedback signal is taken in series with the output terminals, feedback signal is proportional to I_o. So it is *current feedback*.

If the feedback signal is in series with the input, it is *series feedback*.

If the feedback signal is in shunt with the input, it is *shunt feedback*.

EXPRESSION FOR R_I WITH CURRENT SHUNT FEEDBACK

A_i = Shunt circuit current gain of the BJT

A_I = practical current gain I_0 / I_i

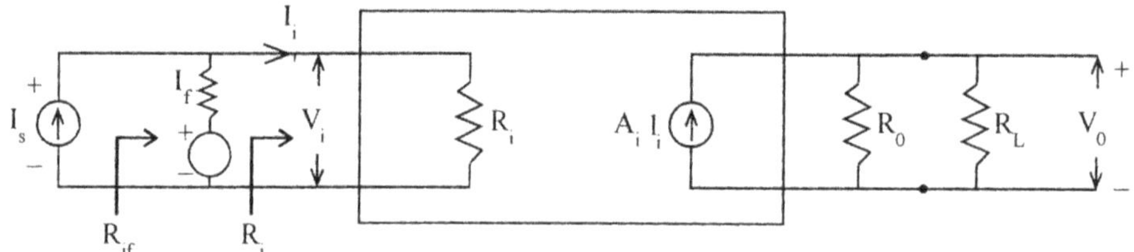

Fig. 7.16 Current shunt feedback.

A_i represents the Shunt circuit current gain taking R_s into accounts,

$$I_S = I_i + I_f = I_i + \beta \, I_0$$

and

$$I_0 = \frac{A_i R_o I_i}{R_o + R_L} = A_I I_i \qquad \therefore \quad \text{Let} \quad A_I = \frac{A_i R_o}{R_o + R_L}$$

$$A_I = \frac{I_o}{I_i} = \frac{A_i R_o}{R_o + R_i} = (I_i + I_f)$$

$$I_S = I_i + \beta . I_0 = I_i \left(1 + \beta . \frac{I_o}{I_i} \right) = I_i \, (1 + \beta . A_I)$$

$$R_{if} = V_i \, | \, I_s \qquad R_i = V_i | \, I_i$$

$$\boxed{R_{if} = \frac{V_i}{(1 + \beta A_I)I_i} = \frac{R_i}{1 + \beta . A_I}}$$

for shunt feedback the input resistance decreases.

7.8.2 INPUT IMPEDANCE WITH SERIES FEEDBACK

$$V_i' = V_i + \beta V_0 \text{ (in general case)}.$$

For negative feedback, β is negative.

$$V_i = \frac{V_i'}{(1+\beta A)}$$

$$\frac{V_i}{I_i} = \frac{V_i'}{I_i(1+\beta A)}$$

In general, R_i increases,

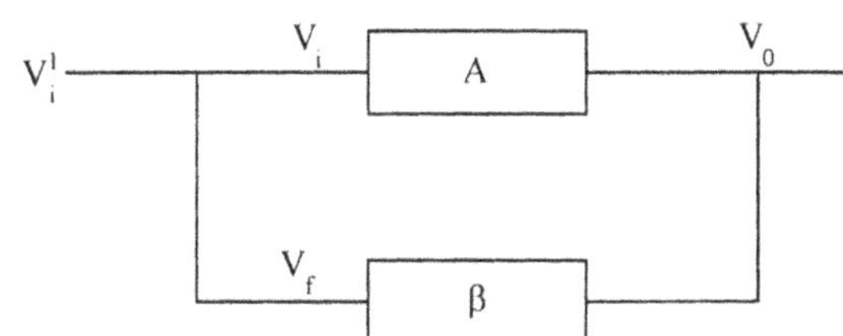

Fig 7.17 Feedback network.

$$V_i' = V_i + \beta V_0$$

$$V_i' = V_i + \beta.A.V_i \qquad \because \quad V_0 = A.V_i$$

$$V_i' = V_i (1+\beta A)$$

But $$\qquad V_i' = I_i R_i$$

$\therefore \qquad V_i' = (1+\beta A) R_i I_i$

$$\frac{V_i'}{I_i} = \text{Input Z seen by the source} = R_i (1+\beta A)$$

$\therefore \qquad R_{if} = R_i (1+\beta A)$

EXPRESSION FOR R_i WITH VOLTAGE SERIES FEEDBACK

In this circuit A_V represents the open circuit voltage gain taking R_s into account. (see Fig.7.18).

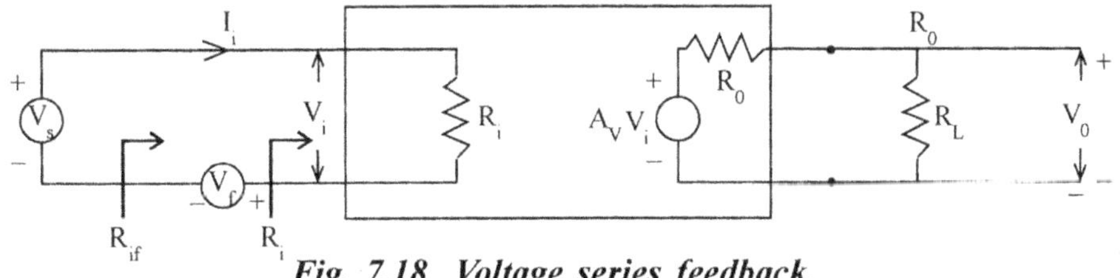

Fig. 7.18 Voltage series feedback.

$$V_i = V_S - V_f$$

or $$V_s = V_i + V_f$$

$$R_0' = R_{if} = V_S/I_i.$$

$$V_f = I_i R_i + V_f = I_i R_i + \beta V_0$$

$$V_0 = \frac{A_v V_i R_L}{R_0 + R_L} = (A_V I_i R_i)$$

Let, $$\qquad \frac{A_v.R_L}{R_0 + R_L} = A_V, \; V_i = I_i. R_i.$$

$$A_V = \frac{V_o}{V_i} = \frac{A_v R_L}{R_o + R_L}$$

$$V_S = I_i R_i + \beta . V_0$$

$$\boxed{\frac{V_s}{I_i} = R_i + \frac{\beta . V_o}{I_i} = R_i \left(1 + \frac{\beta . V_o}{I_i \times R_i} \right)}$$

$$R_{if} = R_i \left(1 + \beta . \frac{V_o}{V_i} \right) = R_i \left(1 + \beta . A_V \right)$$

A_V = voltage gain, without feedback.

for series feedback input resistance increases.

7.9   EFFECT OF NEGATIVE FEEDBACK ON R_O

Voltage feedback (series or shunt) R_o decreases.

Current feedback (series or shunt) R_o increases.

Series feedback (voltage or current) R_i increases.

Shunt feedback (voltage or current) R_i decreases.

OUTPUT RESISTANCE

Negative feedback tends to decrease the input resistances. Feeding the voltage back to the input in a degenerative manner is to cause lesser increase in V_0. Hence the output voltage tends to remain constant as R_L changes because output resistance with feedback $R_{of} \ll R_L$.

Negative feedback, which samples the output current will tend to hold the output current constant. Hence an output current source is created ($R_{0f} \gg R_L$). So this type of connection increases output resistance.

For voltage sampling $R_{of} < R_0$. For current feedback $R_{of} > R_0$.

7.9.1   VOLTAGE SERIES FEEDBACK

Expression for R_{0f} looking into output terminals with R_L disconnected,

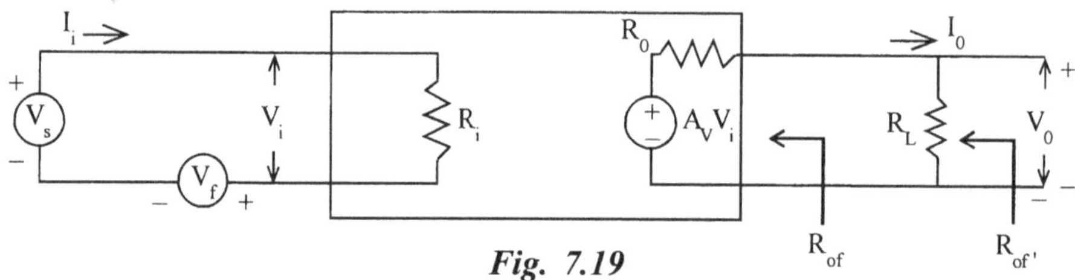

Fig. 7.19

R_0 is determined by impressing voltage 'V' at the output terminals or messing 'I', with input R_{of}' terminals shorted.

Disconnect R_L. To find R_{of}, remove external signal (set $V_s = 0$, or $I_s = 0$)

Let $R_L = \infty$.

Impress a voltage V across the output terminals and calculate the current I delivered by V. Then, $R_{0f} = V|I$.

$$I = \frac{V_0 - A_V \, V_i}{R_0} = \frac{V_0 + \beta A_v V}{R_o}$$

$\because$ V_0 = output voltage.

$$V_i = -\beta_V$$

Because with $V_S = 0, \, V_i = -V_f = -\beta_V$

Hence, $R_{of} = \dfrac{V_0}{I} = \dfrac{R_o}{1 + \beta A_v}$

This expression is excluding R_L. If we consider R_L also R_{of} is in parallel with R_L.

$$R_{of}{}' = \frac{R_{of}.R_L}{R_{of} + R_L} \qquad \text{Substitute the1 value of } R_{of}$$

$$R_{of}{}' = \frac{\dfrac{R_o}{1 + \beta A_V} \times R_L}{\dfrac{R_0}{1 + \beta A_V} + R_L} = \frac{R_o R_L}{R_o + R_L + \beta.A_v R_L}$$

7.9.2 Current Shunt Feedback

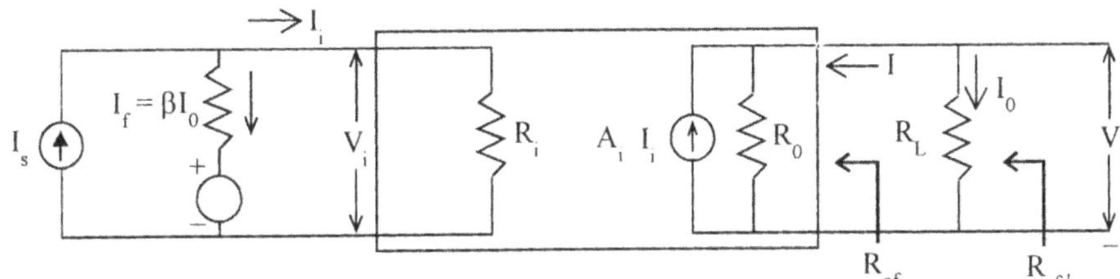

Fig. 7.20 Block schematic for current shunt feedback amplifier.

$\therefore$ $I_0 = -I$

A_i = Shunt circuit current gain

A_I = Practical current gain $\left(\dfrac{I_0}{I_i}\right)$

$$I = \frac{V}{R_o} - A_i \, I_i$$

With, $I_S = 0, \, I_i = -I_f = -\beta I = +\beta I,$

$$I = \frac{V}{R_o} - \beta \, A_i \, I \text{ or } I \, (1 + \beta \, A_i) = \frac{V}{R_o}$$

$$R_{of} = \frac{V}{I} = R_o \, (1 + \beta \, A_i)$$

A_i = short circuit current gain.

$R_{of}{}'$ includes R_L as part of the amplifier.

$$R'_{of} = \frac{R_{of}.R_L}{R_{of}+R_L}$$

$$= \frac{R_o(1+\beta A_i)R_L}{R_o(1+\beta A_i)+R_L}$$

$$= \frac{R_o R_L}{R_o+R_L} \times \frac{1+\beta A_i}{1+\beta A_i R_o}$$
$$\frac{}{R_o+R_L}$$

$$A_I = \frac{I_o}{I_i} = \frac{A_i R_o}{R_o+R_L}$$

$$R'_o = \frac{R_o.R_L}{R_o+R_L}$$

$$\therefore \qquad R'_{of} = R'_o \frac{1+\beta A_i}{1+\beta A_I}$$

If, $\qquad R_L = \infty,$

$$A_I = 0, \ R'_o = R_0$$

$$\therefore \qquad \boxed{R'_{of} = R_0 (1+\beta A_i)}$$

Problem 7.1

The following information is available for the generalized feedback network. Open loop voltage amplification $(A_v) = -100$. Input voltage to the system $(V_i') = 1mV$. Determine the closed loop voltage amplification, the output voltage, feedback voltage, input voltage to the amplifier, and type of feed back for (a) $\beta = 0.01$, (b) $\beta = -0.005$ (c) $\beta = 0$ (d) $\beta = 0.01$.

Solution

$$A'_v = \text{closed loop voltage amplification} = \frac{A_v}{1-\beta_v A_v}$$

Sign must be considered,

$$A'_V = \frac{-100}{1-0.01(-100)} = -50.$$

$\because$ It is positive feedback, A_v' is less negative $\quad \therefore$ there is increase in gain.

V_o Output voltage $= V_i A_v$

$$= -50 \times 10^{-3} \text{ mV}$$

$$= -50 \text{ mV}.$$

V_x Feedback voltage $= \beta V_o = 0.01 (-50 \times 10^{-3})$

$$= -0.5 \text{ m V}.$$

$$V_i = V_i' + \beta_v. V_o$$
$$= 10^{-3} + (-0.5 \times 10^{-3})$$
$$= 0.5 \text{ m V}.$$

This is negative feedback because, $V_i < V_i'$.

$$\therefore \qquad |A_V'| < |A_v|.$$

Problem 7.2

In the above problem determine the % variation in A_v' resulting from 100 % increase in A_v when $\beta_v = 0.01$.

When $A_v = -100$, $A_v' = -50$.

Solution

If A_v increases by 100 %, Then new value of $A_v = -200$.

$$A_V' = \frac{A_v}{1 - \beta_v A_v}$$

$$A_V' = \frac{A_v}{1 - \beta_v A_v}$$

Now, $\qquad A_V' = \dfrac{-200}{1 - 0.01(-200)} = -66.7.$

Change in A_V' is $-66.7 - 50 = 16.7$

$$\therefore \qquad \text{Variation} = \frac{16.7}{50} \times 100 = 33.3\%$$

Problem 7.3

An amplifier with open loop voltage gain $A_v = 1000 \pm 100$ is available. It is necessary to have an amplifier where voltage gain varies by not more than $\pm 0.1 \%$

(a) Find the reverse transmission factor β of the feedback network used

(b) Find the gain with feedback.

Solution

(a) $\qquad \dfrac{dA_f}{A_f} = \dfrac{1}{1 + \beta A} \cdot \dfrac{dA}{A}$

or $\qquad \dfrac{0.1}{100} = \dfrac{1}{(1 + \beta A)} \cdot \dfrac{100}{1000}$

$\therefore \qquad (1 + \beta A) = 100$ or $\beta A = (100 - 1) = 99.$

Hence, $\qquad \beta = \dfrac{99}{1000} = 0.099$

(b) $\qquad A_f = \dfrac{A}{1 + \beta A} = \dfrac{1000}{1 + 99} = 10$

Problem 7.4

A gain variation of $+10\%$ is expected for an amplifier with closed loop gain of 100. How can this variation be reduced to $+1\%$

Solution

$$S' = \frac{S}{1-\beta A_v}.$$

for positive feedback,

$$S' = \frac{S}{1-\beta A_v}$$

$$S' = 0.01, \quad S = 0.1$$

for negative Feedback, $(1+\beta A_v) = \dfrac{0.1}{0.01} = 10 = \dfrac{S}{S'}$

$$A_v' = 100 = \frac{A_v}{1+\beta A_v} = \frac{A_v}{10}$$

$\therefore \qquad A_v = 100 \times 10 = 1000.$

$\qquad 1 + \beta A_v = 10 ; \qquad \therefore \quad \beta.A_v = 10 - 1 = 9$

$\therefore \qquad \beta = \dfrac{9}{A_v} = \dfrac{9}{1000} = 0.009$

$\therefore$ By providing negative feedback, with $\beta = 0.009$, we can improve the stability to 1%.

Problem 7.5

An amplifier with $A_v = -500$, produces 5% harmonic distortion at full output. What value of β is required to reduce the distortion to 0.1% ? What is the overall gain?

Solution

$$D' = \frac{D}{1+\beta A_v} \quad \text{(for negative feedback)}$$

$$D = 5; \qquad D' = 0.1$$

$$N' = \frac{N}{(1+\beta A_v)}$$

$$0.1 = \frac{5}{1+\beta[500]}$$

or $\qquad\qquad \beta = \dfrac{4.9}{50} = 0.098.$

$$A_v' = \frac{-A_v}{1+\beta A_v} = -\frac{500}{50} = -10.$$

$$\frac{D'}{D} = \frac{A_v'}{A_v}.$$

$$\therefore \qquad A_v' = A_v \left(\frac{0.1}{5} \right) = -10$$

Problem 7.6

An amplifier has voltage gain of 10,000 with ground plate supply of 150 V and voltage gain of 8000 at reduced rate supply of 130 V. On application of negative feedback. The voltage gain at normal plate supply is reduced by a factor of 80. Calculate the 1) voltage gain of the amplifier with feedback for two values of plate supply voltage. 2) Percentage reduction in voltage gain with reduction in plate supply voltage for both condition, with and without feedback.

Solution

$$A = \frac{V_o}{V_s}. \text{ When } V_s = 150 \text{ V and } A = 10,000.$$

$$\therefore \qquad V_o = 10,000 \times 150 = 15 \times 10^5 \text{ V. (at normal supply voltages)}$$

$$\beta = \frac{V_f}{V_o}. \quad A_{fb} = \frac{A}{1 + \beta A}$$

But
$$A_{fb} = \frac{10,000}{80}$$

$\because$ it is reduced by a factor of 80 with feedback. $\left(\frac{1}{80}^{Th} \text{ of } 10,000 \right)$

$$\therefore \qquad \frac{10,000}{80} = \frac{10,000}{1 + \beta \times 10000}$$

$$1 + \beta (10,000) = 80$$

$$\therefore \qquad \beta = + \frac{79}{10^4}$$

$$= 0.0079$$

V_o when $V_s = 130$ V is gain $\times V_s = 8000 \times 130 = \mathbf{104 \times 10^4}$ **V**

Voltage gain of amplifier with feedback when $V_s = 150$V,

$$A_{fb_1} = \frac{10,000}{80} = 125.$$

$$V_s = 130 \text{ V.}$$

$$A_{fb_2} = \frac{8000}{1 + 0.0079 \times 8000} = 124.7$$

$$\% \text{ stability of gain without feedback} = \frac{10,000 - 8000}{10,000} \times 100 = 20\%$$

$$\% \text{ stability of gain with feedback} = \frac{125 - 124.7}{125} \times 100 = 0.24\%$$

$\therefore$ With negative feedback stability is improved.

7.10   ANALYSIS OF FEEDBACK AMPLIFIERS

When an amplifier circuit is given, separate the basic amplifier block and the feedback network. Determine the gain of the basic amplifier A. Determine the feedback factor β. Knowing A and β of the feedback amplifier, the characteristics of the amplifiers R_i, R_o, noise figure, A_f etc., can be determined .

Complete analysis of the feedback amplifier is done by the following steps :

Step I : First identify whether the feedback signal X_f is a voltage feedback signal or current feedback signal. If the feedback signal X_f is applied in shunt with the external signal, *it is shunt feedback*. If the feedback signal is applied in series, it is *series feedback*.

Then determine whether the sampled signal X_o is a voltage signal or current signal. If the sampled signal X_o is taken between the output node and ground, it is voltage feedback.

If the sampled signal is taken from the output loop, it is current feedback.

Step II :

2. Draw the basic amplifier without feedback

3. Replace the active device (BJT or FET) by proper model (hybrid $-\pi$ equivalent circuit or *h-parameter* model)

4. Indicate V_f and V_0 in the circuit.

5. Calculate $\beta = V_f / V_0$

6. Calculate 'A' of basic amplifier.

7. From A and β, calculate $A_f = \dfrac{A}{1+\beta A}$; R_{if} and R_{of}.

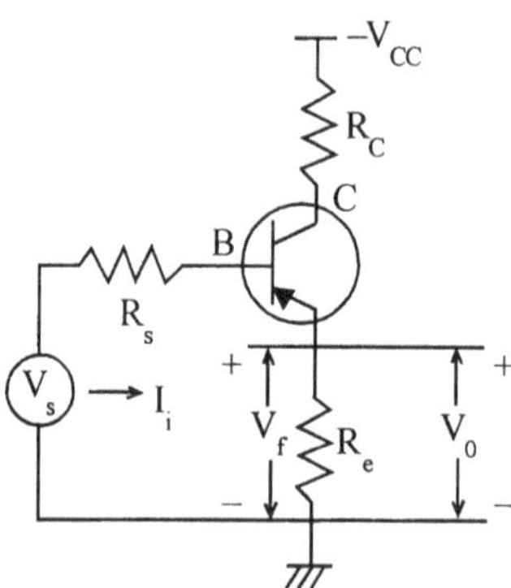

Fig. 7.21   Circuit for voltage series feedback.

7.10.1   VOLTAGE SERIES FEEDBACK

Amplifier circuit is shown in Fig. 7.22.

This is emitter follower circuit because output is taken across R_e. The feedback signal is also across R_e. So $V_f = V_o$. because feedback signal is proportional to output voltage. If V_0 increases V_f also increase and if V_0, V_f also decreases because V_f a V_o. So it is

Voltage feedback. Now the drop across R_e, i .e . V_f opposes the input voltage. It reverse biases the feedback in V_f coming in series with V_{BE} and it opposes it .So it is negative *series voltage feedback*.

In the current series feedback, V_f is the voltage across R_e but, output is taken across R_c or R_L and not Re. So in that case $V_f \propto I_o$ or I_c or not V_o . But in this case, it is emitter follower circuit. Output is taken across R_e and that itself is the feedback signal V_f.

Let us draw the base amplifier without feedback.

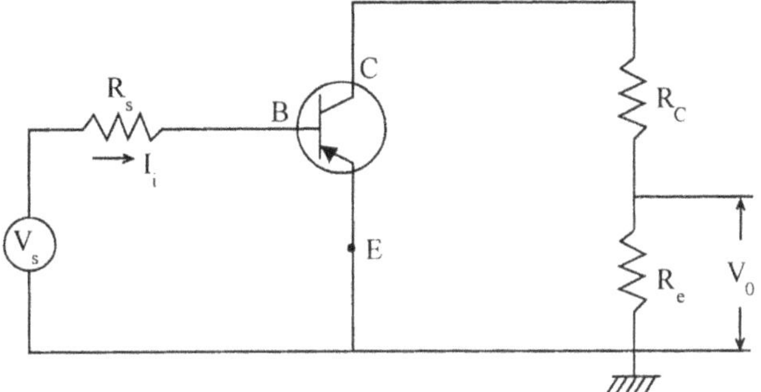

Fig. 7.22 Simplified circuit.

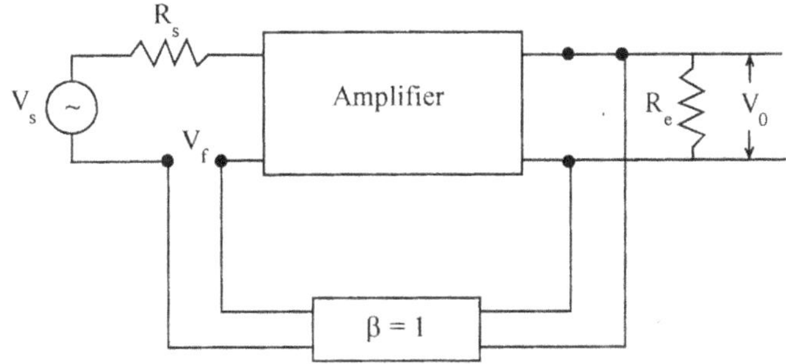

Fig. 7.23 Block schematic.

EQUIVALENT CIRCUIT

Now replace the transistor by its low frequency h-parameter equivalent circuit.

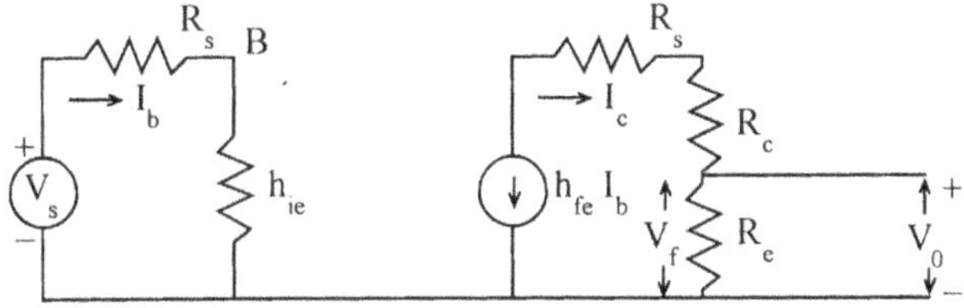

Fig. 7.24 Block Schematic

$$V_o = V_f$$

$\therefore$

$$\boxed{\beta = \frac{V_o}{V_f} = 1.}$$

V_S is considered as part of the amplifier. So $V_i = V_S$.

$$A_v = \frac{V_o}{V_i} \; . \; V_o = h_{fe}. \, I_b.R_e$$

$$I_C \simeq I_e \; ; \; I_C = h_{fe} . I_b \quad (\because \text{It is voltage source, } R_s \text{ in services with } h_{ie})$$

$$V_i = (R_s + h_{ie}) \, I_i$$

$$I_i = I_b$$

$$A_v = \frac{h_{fe} \, I_b \, R_e}{I_b (R_s + h_{ie})}$$

$\therefore \qquad A_v = \dfrac{h_{fe}.R_e}{R_s + h_{ie}} = \text{Voltage gain without feedback}$

$\mathbf{A_V'}$: Voltage gain with feedback.

$$A_V' = \frac{A_v}{1 + \beta A_v}$$

$$\beta = 1$$

$$A_V = \frac{h_{fe}.R_e}{R_s + h_{ie}}$$

$$= \frac{\dfrac{h_{fe}. R_e}{R_S + h_{ie}}}{1 + \dfrac{1.h_{fe} . R_e}{R_S + h_{ie}}}$$

$$A_V' = \frac{h_{fe}.R_e}{R_s + h_{ie} + h_{fe} R_e}$$

$$h_{fe} \, R_e >> (R_s + h_{ie})$$

$\therefore \qquad A_{Vf} \cong 1. \text{ or } <1, \text{ which is true for emitter follower.}$

R_i' = Input resistance without feedback is $R_s + h_{ie}$.

R_i' = Input resistance with feedback

$R_i' = R_i \, (1 + \beta \, A_v)$ (voltage Feedback increase input resistance)

$\beta = 1,$

$$A_V = \frac{h_{fe}.R_e}{R_s + h_{ie}}$$

$\therefore \qquad R_i' = (R_s + h_{ie}) \left(1 + \dfrac{h_{fe}.R_e.1}{R_s + h_{ie}} \right)$

$$R_i' = \frac{(R_s + h_{ie})(R_s + h_{ie} + h_{fe}R_e)}{(R_s + h_{ie})}$$

$$\boxed{R_i' = R_s + h_{ie} + h_{fe}R_e}$$

R_i' : Output resistance with feedback

Output resistance of the circuit is R_e. Because output taken across R_e and not R_c and ground. Load resistance is also R_e.

   $\therefore$ Considering load resistance,

$$R_0' = \frac{R_0'}{1 + \beta A_v}$$

$$R_0' = \frac{R_e}{1 + \dfrac{1.h_{fe}.R_e}{R_s + h_{ie}}}$$

Problem 7.7

Calculate A_{Vf}, R_{of} and R_{if} for the voltage series feedback for the circuit shown in Fig. 7.25.

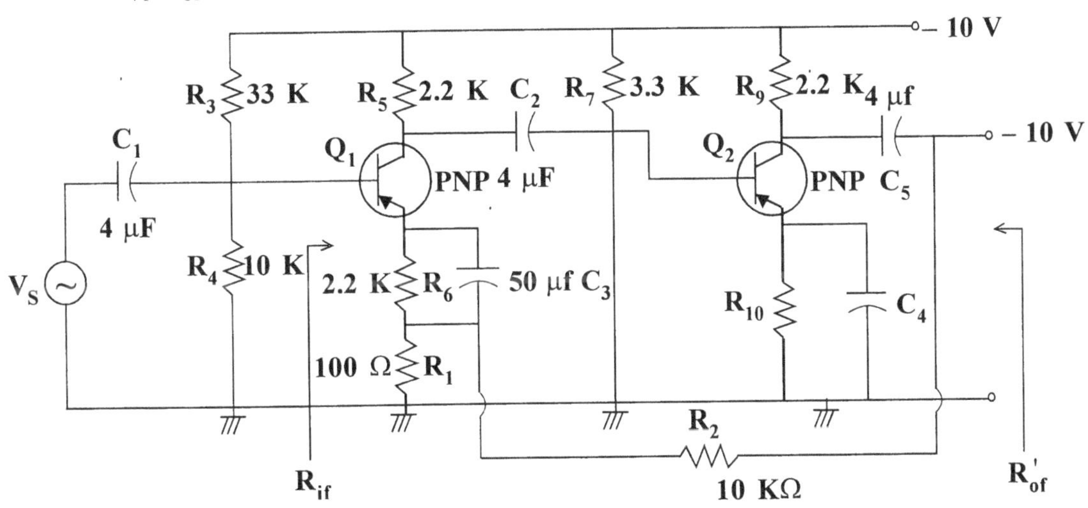

Fig. 7.25 For Problem 7.7.

Assume, $R_s = 0, h_{fe} = 50,$

        $h_{ie} = 1.1 \text{ k} h_{re}$

        $= h_{oe} = 0$ and identical transistor.

The second collector is connected to the first emitter through the voltage divider R_1 R_2. Capacitor C_1, C_2, C_5 are DC blocking capacitors. C_3 and C_4 are bypass capacitors for the emitter resistors. All these capacitances represent negligible reactance at high frequencies.

Solution

Voltage gain without feedback $A_v = A_{v_1} \times A_{v_2}$.

Load resistance, R'_{L1}

R_5 in parallel with R_7 ($\because$ at high frequency X_{C2} is negligible), in parallel with

$$R_8, \text{ and } h_{ie2} \; R'_{L1} = 2.2 \parallel 33 \parallel 4.3 \parallel 1.1 \text{ k}\Omega = 980\Omega.$$

$$R'_L = R_5 \parallel R_7 \parallel R_8 \parallel h_{ie2}$$

Effective load resistance R'_{L2} of transistor Q_2 is its collector resistance R_9 in parallel with

$$(R_1 + R_2); \; R'_{L2} = R_9 \parallel (R_1 + R_2)$$

$\therefore$ $\qquad\qquad R'_{L2} = 2.2 \text{ k}\Omega \parallel^{le} \text{ with } 10.1 \text{k}\Omega \sim 2 \text{ k}\Omega$

Effective emitter impedance R_e of Q_1 is R_1 parallel with R_6.

$$R_{e1} = R_1 \parallel^{le} R_6 \qquad \because \; C_3 \text{ is short for AC}$$
$$R_e = 0.1 \parallel^{le} \text{ with } 2.2 \text{ k}\Omega = 0.098 \text{k}\Omega = 98\Omega$$

The voltage gain A_{V_1} of Q_1 for a common emitter transistor with emitter resistance

is $\qquad\qquad A_{V_1} = \dfrac{V_1}{V_i} = \dfrac{-h_{fe} R'_L}{h_{ie} + (1 + h_{fe})R_e}$

For Q_1 emitter is not at GROUND potential. So for A_V, this formula must be used.

$$= \dfrac{-50 \times 0.980}{1.1 + (51) \times 0.098} \sim -7.78$$

Voltage gain A_{v2} of transistor Q_2,

$$A_{v2} = A_I \cdot \dfrac{R_L}{R_i} = -\dfrac{h_{fe} R_L}{h_{ie}}$$

$$A_{v2} = -\dfrac{h_{fe} \cdot R_{L2}}{h_{ie}} = -\dfrac{50 \times 2}{1.1} = -91$$

For Q_2 emitter is bypassed. So $R_e = 0$. $\therefore$ For A_V this formula is used.

Voltage gain A_v of the two stages is cascade without feedback

$$A_v = \dfrac{V_o}{V_i} = A_{v1} \times A_{v2} = 7.78 \times 91 \sim 728$$

$$\beta = \dfrac{R_1}{R_1 + R_2} = \dfrac{100}{100 + 10,000} = \dfrac{100}{10100} = \dfrac{1}{101} \sim 0.01$$

$$A_v \times \beta = 728 \times 0.01 = 7.28$$

$$D = 1 + \beta A_v = 8.28 \,.$$

$$A_{vf} = \dfrac{A}{1 + \beta A} = \dfrac{A_v}{D} = \dfrac{728}{8.28} \sim 90$$

$$A_{vf} = \dfrac{A}{1 + \beta A} = \dfrac{A_v}{D} = \dfrac{728}{8.28} \sim 90$$

Input resistance without external feedback

$$R_i = h_{ie} + \left(1 + h_{fe}\right) R_e = 1.1 + \left(1 + 50\right) 0.098 \simeq 6.1K \; \Omega$$

$$R_i' = R_{if} = R_i \left(1 + \beta A\right) = 6.1 \left(1 + 7.28\right) = 50.5 \approx 51$$

Output resistance without feedback $R' = R'_{L_2} = 2 \; k\Omega$

Output resistance without feedback $R'_{0f} = \dfrac{R_o}{1 + \beta A} = \dfrac{2}{8.28} = 0.24 \; K \; \Omega$

Equivalent Circuit

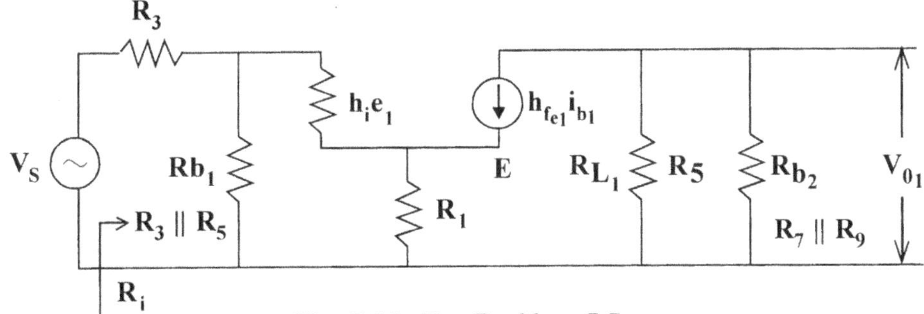

Fig. 7.26 For Problem 7.7.

BJT will not behave like a fixed resistor with h_{ie}, value . So circuit analysis is not simply parallel or series combination.

$$A_V = \frac{V_o}{V_i} = \frac{R_L}{R_s + r_{bb'}}$$

$$R_o = \frac{V_o}{R_L} = \frac{V_i}{R_s + r_{bb}'}$$

$$A_I = \frac{I_o}{I_i} = \frac{R_s}{R_s + r_{bb}'}$$

7.10.2 CURRENT SHUNT FEEDBACK

The circuit is shown in Fig. 7.27.

Amplifier circuit parameters in terms of transistor h-parameters.

$$h_{ie} = R_i$$
$$h_{fe} = \beta$$
$$h_{oe} = R_0$$

The circuit shows two transistors Q_1 and Q_2 in cascade(See Fig.7.30). Feedback is provided from the emitter of Q_2 to the base of Q_1. This is negative feedback because, V_{i2} the input voltage to Q_2 is $>> V_{i1}.V_{i2}$ is out of phase with V_{i1}. $V_{i2} >> V_{i1}$ because Q_1 is in Common emitter configuration. A_V is large. Also, V_{i2} is 180^0 out of phase with V_{i1}. Q_2 is emitter follower because emitter is not at ground potential. Voltage is taken across R_e. [This voltage follows the collector voltage. So it is emitter follower]. So $A_v < 1. \simeq 0.99$

$\therefore$ V_{e2} is slightly less than V_{i2}, and there is no phase shift. [because emitter follows action]

$\therefore$ V_{e2} is in phase with V_{i2}.

V_{i2} is out of phase with V_{i1}.

$\therefore$ V_{e2} is out of phase with V_{i1} (180^0). So it is negative feedback

$V_{i2} \gg V_{i1}$. $V_{e2} \simeq V_{i2}$.

$V_{e2} \gg V_{i1}$.

If the input signal V_s increases, I'_s the input current from source also increases. If I'_s increases, I_f also increases ($\because$ V_{e2} increases as I'_s increases)

$I_i = I'_s - I_f$. (I_i is the base current for the transistor Q_1. So it is negative feedback)

This is current shunt feedback because,

$$I_f = \frac{V_{e_2} - V_i}{R'}$$

But

$$V_{i_1} \ll V_{e_2}.$$

$$I_f \simeq + \frac{V_{e_2}}{R'}.$$

I_0 = collector current of Q_2. $\qquad\qquad$ $V_{e_2} = (I_0 - I_f)\, R_e$

$\simeq$ emitter current of Q_2.

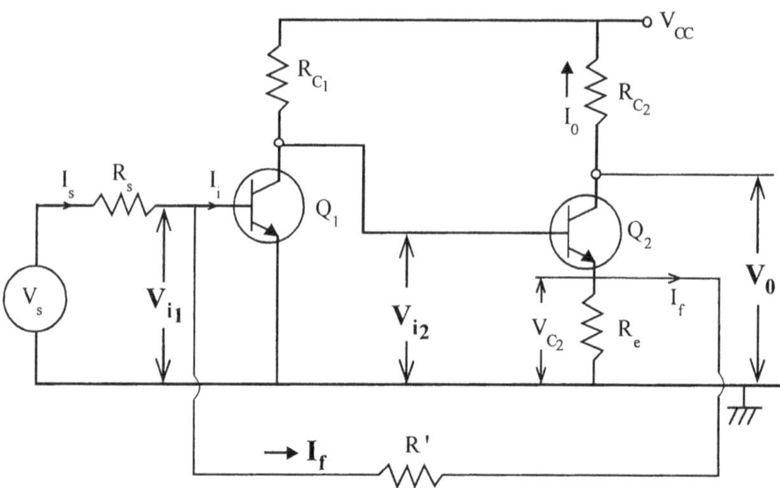

Fig. 7.27 Current shunt feedback.

Dividing by R'.

$$\frac{V_{e2}}{R'} = I_f = \frac{(I_0 - I_f)R_e}{R'}.$$

$\therefore$

$$I_f = + \frac{(I_0 - I_f)R_e}{R'};$$

$$I_f + \frac{I_f . R_e}{R'} = \frac{I_o . R_e}{R'}$$

$$I_f \left(1 + \frac{R_e}{R'}\right) = \frac{I_o . R_e}{R'}$$

$$I_f = \frac{R' . I_o . R_e}{(R' + R_e) R'} = \frac{R_e}{(R' + R_e)} I_0$$

$$\boxed{\beta = \frac{I_f}{I_o} = \frac{R_e}{R' + R_e}}$$

$$I_f \, \alpha \, I_0$$

∴ This is current feedback

A_I' : Current gain with feedback.

$$A_I' = \frac{I_o}{I_s'}$$

$$I_s' = I_i + I_f \text{ which is small } \mu A \, I_i = I_b$$

∴ $$I_s' \simeq I_f$$

∴ $$A_I' = \frac{I_o}{I_f} = \frac{1}{\beta} \qquad\qquad \because \beta = \frac{I_f}{I_o}$$

∴ $$\boxed{A_I' = \frac{R' + R_e}{R_e}}$$

A_V' : Voltage gain with feedback.

$$\frac{V_o}{V_s} \text{ (with feedback)}$$

$$V_0 = I_0 . R_{c2}$$
$$V_s = I_s . R_s.$$

$$\frac{V_o}{V_s} = \frac{I_o . R_{c2}}{I_s R_s} \qquad\qquad I_s \simeq I_f$$

$$A_V' \simeq \frac{I_o}{I_f} . \frac{R_{c2}}{R_s} \qquad\qquad \frac{I_0}{I_f} = \frac{1}{\beta}$$

$$= \frac{1}{\beta} . \frac{R_{c2}}{R_s} \qquad\qquad \beta = \frac{R_e}{R' + R_e}$$

$$\boxed{A_V' = \frac{R' + R_e}{R_e} . \frac{R_{c2}}{R_s}}$$

We shall determine R_i', R_0' taking numerical values. The actual circuit, without feedback, considering R' can be drawn as shown in Fig. 7.28 To represent R' on the input side, with output terminals open circuited ($\therefore$ it is emitter follower configuration, output is taken across emitter and ground). When E_2 is left open (To make $I_o = 0$). R' is in series with R_{e2} and thus total resistance is between base B_1, and ground. Hence on the input side, R' is in series with R_{e2}.

To find the output resistance, the base of Q_1 is shorted to ground because one terminal of R' is connected to (input shorted looking the high output terminals) B_1 is at ground. R' or R_{e2} are in parallel.

$\therefore$ The circuit is as shown below, (Fig. 7.28)

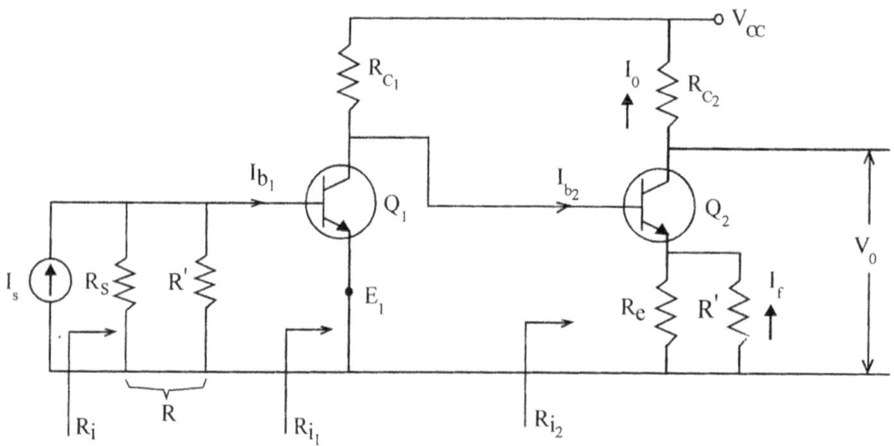

Fig. 7.28 Current shunt feedback without R'.

Because it is shunt feedback, we have shown this as a current source with R_s in parallel with I_S.

Problem 7.8

To find A_V', R_i', R_0'. A_1' for the circuit shown in Fig. 7.31.

$$R_{c_1} = 3\ k\Omega, \qquad R_{c_2} = 500\Omega; \quad R_{e_2} = 50\Omega$$
$$R' = R_s = 1.2\ k\Omega \quad h_{fe} = 50.$$
$$h_{ie} = 1.1\ k\Omega \qquad h_{re} = h_{re} = 0.$$

Solution

In this problem, output is taken at the collector 2 and not emitter 2. So the formulae derived earlier can be used directly.

$$A_1 = \frac{-I_{c_2}}{I_s} = \frac{-I_{c2}}{I_{b2}} \cdot \frac{I_{b2}}{I_{c1}} \cdot \frac{I_{c1}}{I_{b1}} \cdot \frac{I_{b1}}{I_s}$$

(Multiplying and dividing by I_{b2}, I_{c_1} and I_{b1})

$$\frac{-I_{c_2}}{I_{b_2}} = - h_{fe} = - 50; \text{ (Emitter follower)}$$

$$\frac{I_{c_1}}{I_{b_1}} = + h_{fe} = + 50 \text{ (common emitter configuration)}$$

$$I_{b_2} \# I_{c_1},$$

$\because$ The current gets divided in the ratio of the resistances, current gets divided depending upon R_{c_1} and R_{i_2} of transistor Q_2.

$$\frac{I_{b2}}{I_{c1}} = \frac{-R_{c_1}}{R_{c_1} + R_{i_2}}$$

$$= \frac{-3}{3 + 3.55} = -0.457.$$

$$R_{i2} = h_{ie} + (1 + h_{fe})(R_{e2} \| R').$$

(For emitter follower configuration this is the input resistance).

$$= 1.1 + (50 + 1)\left[\frac{0.05 \times 1.20}{1.25}\right] \quad (R_{e\,2} = 0.05 \text{ k }\Omega)$$

$$= 3.55 \text{ k }\Omega$$

Let $\qquad R = R_S$ in parallel with $(R' + R_{e2})$

$$= \frac{1.2 \times 1.25}{1.2 + 1.25} = 0.612 \text{ k }\Omega$$

$$= \frac{I_{b_1}}{I_S} = \frac{R}{R + h_{ie}}$$

$$= \frac{0.61}{0.61 + 1.1} = 0.358 \,\Omega$$

$\therefore \qquad A_I = (-50)(-0.457)(50)(0.358)$

$$= +406$$

$$\beta = \frac{R_{e_2}}{R_{e_2} + R'}$$

$$= \frac{50}{1,250} = 0.04.$$

$$(1 + \beta A_i) = 1 + (0.040)(406) = 17.2$$

$$A_i' = \frac{A_I}{1 + \beta A_i} = \frac{406}{17.2} = 23.6$$

$$A_V' = \frac{V_o}{V_s} = \frac{-I_{c_2}.R_{c_2}}{I_s.R_s} = \frac{A_I'.R_{c_2}}{R_s}$$

$$= \frac{(23.6)(0.5)}{1.2} = 9.83$$

$$[A'_V \text{ is also} = \frac{R_{c_2}}{\beta R_s} = \frac{0.5}{(0.040)(1.2)} = 10.4]$$

R_i = Input resistance without feedback

= R in parallel with h_{ie}

$$= \frac{(0.61)(1.1)}{1.71} = 0.394 \text{ k}\Omega$$

$$R'_i = \frac{R_i}{1+\beta A_i} = \frac{394}{17.2} = 23.0 \Omega$$

R_{oL} = Output resistance considering R_L,

$$= R_o \text{ in parallel with } R_{c2} = R_{c2} \quad \because R_o \text{ is large } \frac{1}{h_{ie}} = \alpha.$$

R'_L = with feedback considering R_L,

$$= \frac{R_{0L}(1+\beta A_i)}{1+\beta A_i} = R'_0 = R_{c2} = 500$$

When we represent R on the input side and output side and calculate the value of A_1, it is not the current gain with feedback. Because, R' is represented on the input side leaving E_2 terminal open and on the output side shorting B_1 to ground (to make I_o and $I_s = 0$). So thus A_1 will not be the same if R' is actually connected between E_2 and B_1. We are taking into account the effect of R' and not the feedback effect. In practice for shunt or series feedback, the signal generator will act as a current source or voltage source. Therefore it is capable of supplying current or voltage required .In theory we assume ideal voltage and current sources. Therefore for shunt feedback we must have a current sources irrespective of input voltage. The current source will supply sufficient current to drive the resistance.

$$R_i = (1+9) \ 20 \text{ k}\Omega = 200 \text{ k}\Omega$$

R_s is negligible compared to 200 kΩ.

$$R_0 = \frac{r_o}{1+\beta A_o} = \frac{20\Omega}{10} = 2\Omega$$

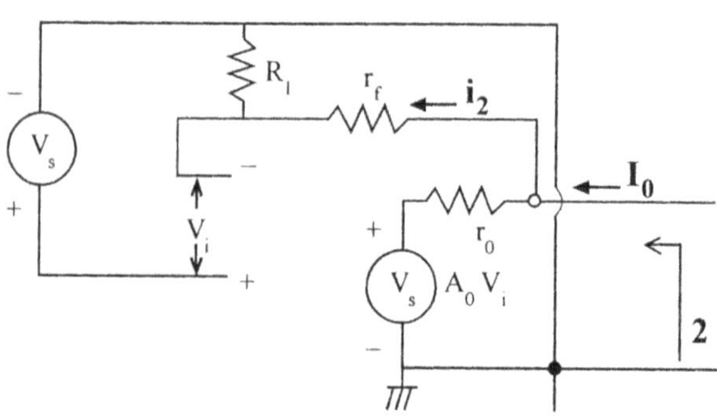

Fig. 7.29 Equivalent circuit.

7.10.3 CURRENT SERIES FEEDBACK

INPUT RESISTANCE (R_i')

$$V_i = V_i' - V_f$$

$$R_i' = \frac{V_i'}{I_i} = \frac{V_i + V_f}{I_i} = \frac{V_i}{I_i} + \frac{V_f}{I_i} \qquad \therefore \qquad V_i' = V_i + V_f.$$

$$\frac{V_i}{I_i} = R_i;$$

$$V_f = I_e R_e \; ; \; I_i = I_e - I_c \qquad [\because \text{ negative current series feedback}] \; (I_b)$$

$$\therefore \qquad R_i' = R_i + \frac{I_e R_e}{I_e - I_c}$$

$$R_i' = R_i + \frac{R_e}{1 - \dfrac{I_c}{I_e}}$$

Dividing II term by I_e.

$$\frac{I_C}{I_e} = \alpha = \frac{\beta}{1 + \beta}$$

$$\frac{I_C}{I_e} = \frac{I_C}{I_C + I_b}$$

$$= \frac{I_c \,|\, I_b}{\dfrac{I_c}{I_b} + 1} = \frac{h_{fe}}{1 + h_{fe}}.$$

$$\therefore \qquad R_i' = R_i + \frac{R_e}{1 - \dfrac{h_{fe}}{1 + h_{fe}}}$$

$$R_i' = R_i + (1 + h_{fe}).R_e$$

But $\qquad R_i = h_{ie}$ for the transistor

$$\boxed{\therefore \; R_i' = h_{ie} + (1 + h_{fe}) R_e}$$

If we consider the bias resistor also, R_1 and R_2 will come in parallel with R_i'

$$= R_i \,\|\, R_1 \,\|\, R_2$$

VOLTAGE GAIN (A_V') .

$$A_V = \frac{A_i.R_L}{R_i} \qquad [\text{This is the general expression for } A_V \text{ interms of } A_i]$$

$$\therefore \qquad A_V' = \frac{A_i.R_L}{R_i'}$$

[With feedback, A_i will not change, R_L will not change, but R_i will be R_i']

$$A_i \simeq -h_{fe}$$

$$\therefore \qquad A_V' = \frac{-h_{fe}.R_L}{h_{ie} + (1 + h_{fe})R_e}$$

$$\therefore \qquad A_V' < A_v$$

OUTPUT RESISTANCE (R_0')

$$R \text{ without feedback} \simeq \frac{1 + h_{fe}}{h_{oe}}$$

This will be very large.

Taking into effect, the feedback,

$$R_0' = R_o \parallel R_L$$

$$\frac{1}{R_o} = h_{oe} - \frac{h_{fe}h_{re}}{h_{ie} + R_s}$$

ACTUAL EXPRESSION FOR β THE FEEDBACK FACTOR :

With negative feedback, $R_i' = R_i(1 + \beta. A_v)$

$$A_v = \frac{-h_{fe}.R_L}{R_i} \; ;$$

$$\therefore \qquad R_i' = R_i\left[1 - \frac{h_{fe}.R_L}{R_i}.\beta\right]$$

$$= R_i - h_{fe}. R_L. \beta \qquad\qquad .....(1)$$

The expression we get for

$$R_i' = R_i + (1 + h_{fe}).R_e \qquad\qquad .....(2)$$

Comparing (1) and (2), we find that

$$\beta = \frac{(1 + h_{fe})}{h_{fe}} \cdot \frac{R_e}{R_L}$$

This is the actual expression for β

$$\frac{1 + h_{fe}}{h_{fe}} \simeq 1.$$

$$\therefore \qquad \beta \simeq \frac{R_e}{R_L}$$

7.10.3 CURRENT SERIES FEEDBACK (TRANS CONDUCTANCE AMPLIFIER)

Consider the circuit shown in Fig. 7.30 output is taken between collector and ground. The drop across R_e is the feedback signal V_f. The sampled signal is the load current I_o and not V_0. This is current series feedback. Because V_f is in series with V_i. It is current series feedback.

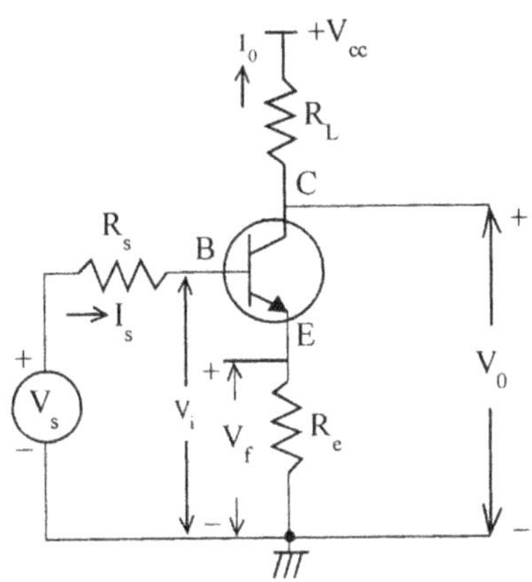

Fig. 7.30 Current series feedback.

$$V_f = I_e R_e.$$
$$I_e \simeq I_c = I_o.$$
$$\therefore \quad V_f = I_o R_e .$$

$\because$ R_e is constant,

$$V_f \, \alpha \, I_o.$$

$[\beta$ must be independent of R_s or $R_L]$

$$\beta = \frac{V_f}{V_o} = \frac{-I_o . R_e}{I_o . R_L} = -\frac{R_e}{R_L}$$

The basic amplifier circuit without feedback is shown in Fig.7.31 below,

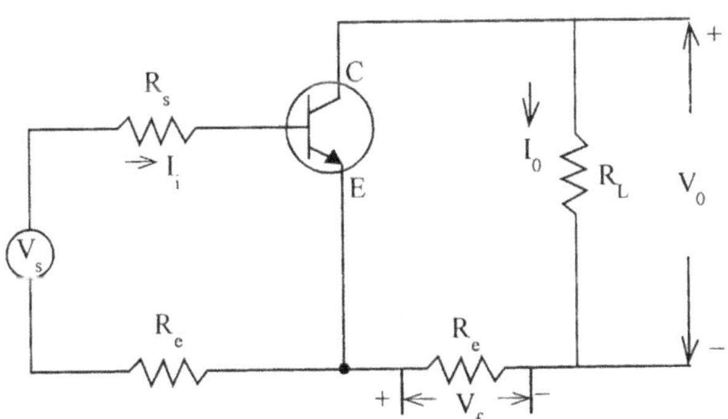

Fig.7.31 Circuit without feedback.

The equivalent circuit when the active device is replaced by *h-parameter* circuit is shown in Fig.7.32.

This has to be considered as Transconductance Amplifier, Since, β is taken as $\dfrac{V_f}{V_o}$.

It depends, on R_L, because for this circuit, G_M is considered.

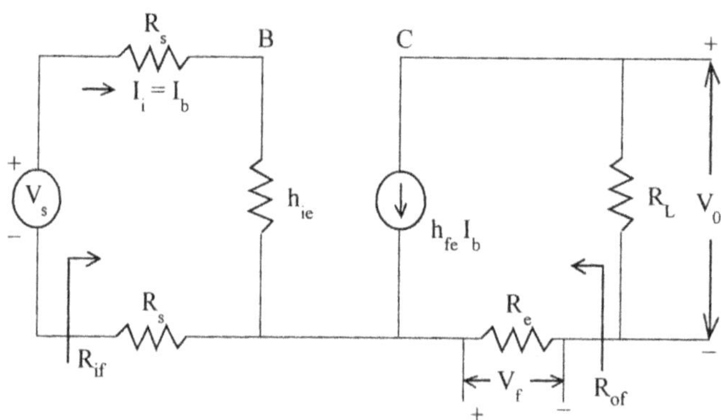

Fig. 7.32 h - parameter equivalent circuit.

$$\beta = \frac{V_f}{I_o} = \frac{-I_o . R_e}{I_o} = -R_e.$$

($\because$ the sampled signal is I_o and not V_o.)

$$G_M = \frac{I_o}{V_i} \quad I_0 = -h_{fe} . I_b \text{ (Input current gain)}$$

Since, $V_i = I_b (R_s + h_{ie} + R_e)$

$\therefore$ $$G_m = \frac{-h_{fe} . I_b}{I_b (R_s + h_{ie} + R_e)}$$

$$= \frac{-h_{fe}}{R_s + h_{ie} + R_e}$$

$$D = 1 + \beta . G_M = 1 + \frac{h_{fe} . R_e}{R_s + h_{ie} + R_e}$$

$$= \frac{R_s + h_{ie} + (1 + h_{fe}) R_e}{R_s + h_{ie} + R_e}$$

$$G_{Mf} = \frac{G_M}{D} = \frac{-h_{fe}}{R_s + h_{ie} + (1 + h_{fe}) R_e}$$

Voltage gain $$A_{vf} = \frac{I_o . R_L}{V_s} = \frac{I_o}{V_s} = G_{Mf}$$

$\therefore$ $$A_{vf} = G_{Mf} . R_L$$

$$= \frac{-h_{fe} . R_L}{R_s + h_{ie} + (1 + h_{fe}) R_e}$$

$$(1 + h_{fe}) R_e >> R_s + h_{ie}.$$

Since h_{fe} is a large quantity.

$\therefore$ Denominator $\simeq (1 + h_{fe}) R_e$.

$$h_{fe} >> 1.$$

$\therefore$ Denominator $\simeq h_{fe}. R_e$.

$\therefore$

$$A_{vf} = \frac{-h_{fe}R_L}{h_{fe}.R_e} = -\frac{R_L}{R_e}$$

$R_i = R_s + h_{ie} + R_e$ (without feedback).

$R_{if} = R_i D = R_s + h_{ie} + (1 + h_{fe}) R_e$.

7.10.4 VOLTAGE SHUNT FEEDBACK (TRANS RESISTANCE AMPLIFIER)

This is *voltage shunt feedback* because, the feedback current through R_B is proportional to the output voltage or it is in shunt with the input. So it is voltage shunt feedback. [We are not interested whether voltage is fed or current is fed. But the feedback signal is proportional to output voltage. *So it is voltage feedback*]

The circuit can be written as, shown in Fig. 7.33.

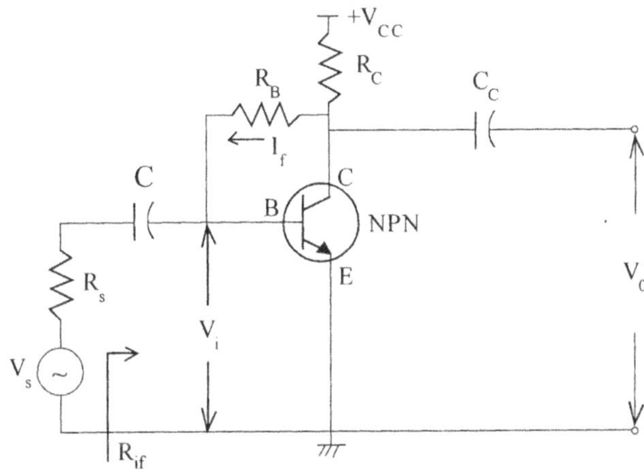

Fig. 7.33 Circuit for voltage shunt feedback.

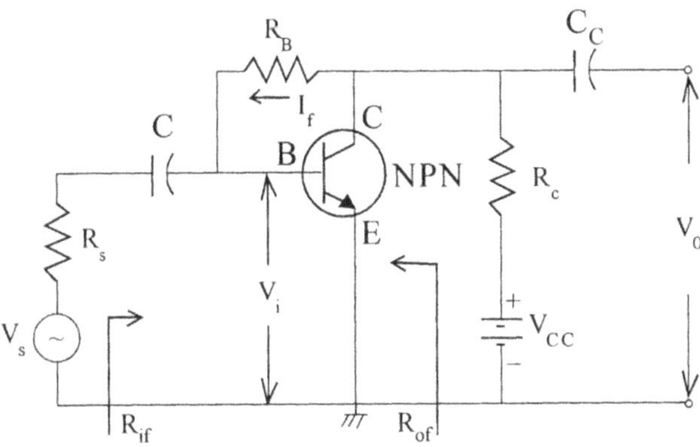

Fig. 7.34 Redrawn circuit for voltage shunt feedback.

The low frequency **h-parameter** equivalent circuit is, shown in Fig. 7.35.

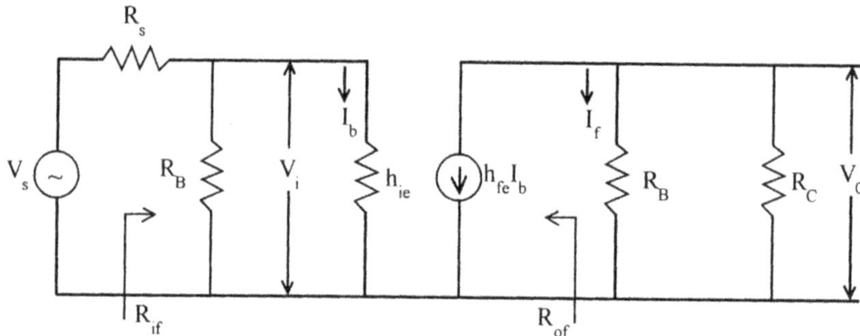

Fig. 7.35 h-parameter equivalent circuit.

$$V_0 \gg V_i \qquad \because \text{ Amplification is there.}$$

$$I_f \simeq \frac{-V_0}{R_B},$$

$$\therefore \quad I_f \simeq \frac{-V_0}{R_B};$$

$$\therefore \quad \beta = \frac{-I_f}{V_0}$$

$$\beta = \frac{-1}{R_B}$$

$$A_v = \frac{V_0}{V_i} = \frac{-h_{fe} \cdot R_L'}{h_{ie}} \text{ without feedback} \left(\frac{A_I R_L}{R_v} \right)$$

$$R_L' = R_B \parallel R_C.$$

$$A_v{}^1 = \frac{V_0}{V_s} \text{ (with feedback)}$$

$$\boxed{A_v{}^1 = A_v \cdot \frac{R_i'}{R_s + R_i'}}$$

R_i' = Input resistance with feedback

$$A_{vf} = \frac{V_0}{V_s}$$

$$= \frac{V_0}{I_s . R_s} \simeq \frac{1}{\beta R_s} = \frac{-R_B}{R_s}$$

$\mathbf{R_i'}$:

$$I_s = I_f + I_b, \; I_b \text{ is negligible}$$

$$\therefore \quad I_s \simeq I_f \quad \text{or} \quad \beta = \frac{-1}{R_B}.$$

$$R_i' = h_{ie} \text{ in parallel with } \frac{R_B}{1-A_v}$$

Transresistance
$$R_M = \frac{V_o}{I_s} = \frac{-I_o.R_c}{I_s} = \frac{-h_{fe}.I_b.R_c}{I_s}$$

$$R_C' = R_c \| R_B \; ; \; I_s = \frac{(R + h_{ie})}{R} I_b. \qquad \text{Where } R = R_s \| R_B.$$

$$\therefore \quad R_M = \frac{-h_{fe}.R_c'.R}{(R + h_{ie})};$$

$$I_b = \frac{-h_{fe}.R_e'.R}{(R + h_{ie})}$$

$$R_M' = \frac{R_M}{1 + \beta R_M}.$$

$$R_i = \frac{R \times h_{ie}}{R + h_{ie}}$$

$$R_i' = \frac{R_i}{1 + \beta R_M}$$

$$R_0' = \frac{R_B}{R_s} \times \frac{R_s + h_{ie}}{h_{fe}}$$

Problem 7.9

For the circuit shown, $R_c = 4 \text{ k}\Omega$, $R_B = 40 \text{ k}$, $R_s = 10 \text{ k}$, $h_{ie} = 1.1 \text{ k}\Omega$, $h_{fe} = 50$, $h_{re} = h_{oe} = 0$.
Find A_{vf}, R_{if}.

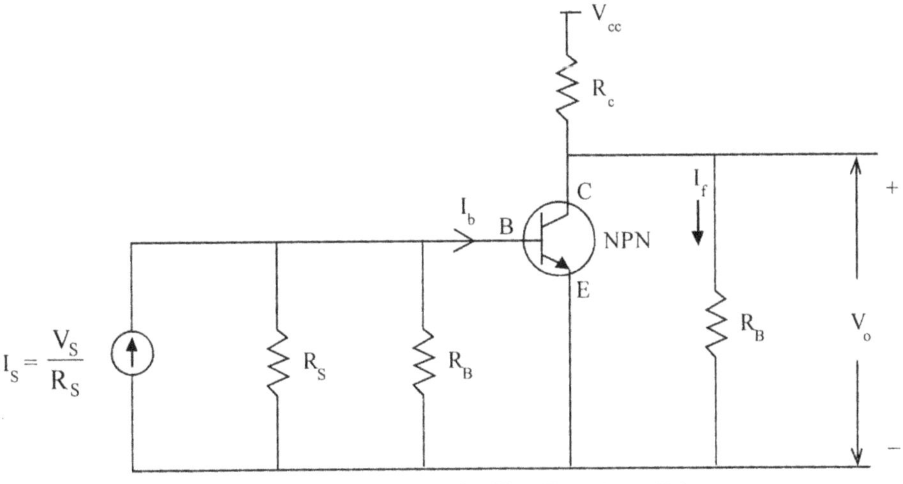

Fig. 7.36 For Problem 7.9.

Solution

$$R'_C = R_c \,||^{le}\, R_B = \frac{4 \times 40}{4 + 40} = 3.64 \text{ k}\Omega$$

$$R = R_s \,||^{le}\, R_B = \frac{10 \times 40}{10 + 40} = 8 \text{ k}\Omega$$

$$\text{Transresistance}\, R_M = \frac{V_o}{I_s} = \frac{-I_C.R_c'}{I_s} = \frac{-h_{fe}.I_b.R_c'}{I_s} \qquad \because\ I_c = h_{fe} \cdot I_b$$

$$I_s = \frac{(R + h_{ie})}{R} I_b; \qquad \therefore\ R_M = \frac{-h_{fe}.I_b\, R_C' \cdot R}{(R + h_{ie}).I_b}$$

$$R_M = -160 \text{ k}$$

$$\therefore \qquad \beta = -\frac{-1}{R_B} = -0.025 \text{ mA|V}$$

$$R'_M = \frac{R_M}{1 + \beta R_M} = \frac{-160}{5.00} = -32.0 \text{ k}\Omega$$

$$A'_v = \frac{V_o}{V_s} = \frac{V_o'}{I_s.R_s}$$

$$= \frac{V_o'}{I_s} = R'_M$$

$$\therefore \qquad A'_v = \frac{R_M'}{R_s} = -\frac{32}{10} = -3.2 \ .$$

$$R_i = \frac{R \times h_{ie}}{R + h_{ie}} = \frac{8 \times 1.1}{8 + 1.1} = 0.968 \text{ k}\Omega;$$

$$R'_i = \frac{R_i}{1 + \beta R_M} = \frac{968}{5.0} = 193\Omega$$

SUMMARY

- ♦ An Amplifier circuit is to provide voltage gain or current gain or both in the form of power gain. But the other desirable characteristics of the amplifier circuits are high Z_i, Low Z_o, Large B.W, low distortion, low noise and high stability. To achieve these characteristics a part of output signal is feedback and coupled to input, to oppose in phase with (V_i). So it is negative feedback. Though gain reduces due to negative feedback, it is employed in amplifier circuits to get the other advantages.

- Feedback factor $\beta = V_f/V_o$. The product βA is called returned ratio. $(1 + \beta A)$ is called return difference D or desensitivity factor.

- The different types of feedback are   (i) voltage series   (ii) current series   (iii) voltage shunt and   (iv) current shunt.

- With voltage feedback (series or shunt) output resistance R_0 decreases.

- With current feedback (series or shunt) R_0 increases.

- With series feedback (current or voltage) R_i increases.

- With shunt feedback (current or voltage) R_0 decreases.

- With negative feedback, distortion, noise, gain reduce by a factor $(1 + \beta A)$. Bandwidth, f_2, stability improve by $(1 + \beta A)$.

- If the feedback signal is proportional to voltage, it is voltage feedback. If the feedback signal is porportional to current, it is current feedback. If the feedback signal V_f is coming in series with input signal V_i, it is series feedback. If V_f is in parallel with V_i it is shunt feedback.

- If the feedback signal V_f is out of phase with V_i, opposing it is negative feedback. If V_f is in phase with V_i, adding to it or aiding V_i, it is positive feedback.

OBJECTIVE TYPE QUESTIONS

1.	The disadvantage of negative feedback is _______________.

2.	The expression for sensitivity of an amplifier with negative feedback is _______________.

3.	The expression for Desentivity of negative feedback amplifier is D = _______________.

4.	The relation between bandwidth of an amplifier without feedback and with negative feedback is _______________.

5.	Relation between upper cut-off frequency f_2' with negative feedback and f_2 without negative feedback is $f_2' = $ _______________.

6.	Negative feedback is also called as _______________.

7.	For Ideal transconductance amplifier $R_i = $ _______________.

8.	For practical transresistance amplifier, it is desirable that R_i is _______________ and R_0 is _______________.

9.	Voltage sampling is also known as _______________.

10.	Characteristics of ideal voltage amplifier are : _______________.

11.	Desirable characteristics of practical current amplifier are = _______________.

12.	βA in feedback amplifier circuits is called _______________.

13.	With voltage feedback, (series or shunt), output resistance R_0 of an amplifier _______________.

14. With series feedback, (voltage or current), input resistance R_i of an amplifier __________________.

15. Expression for reverse transmission factor or feedback factor β = __________________.

16. Identify the type of feedback.

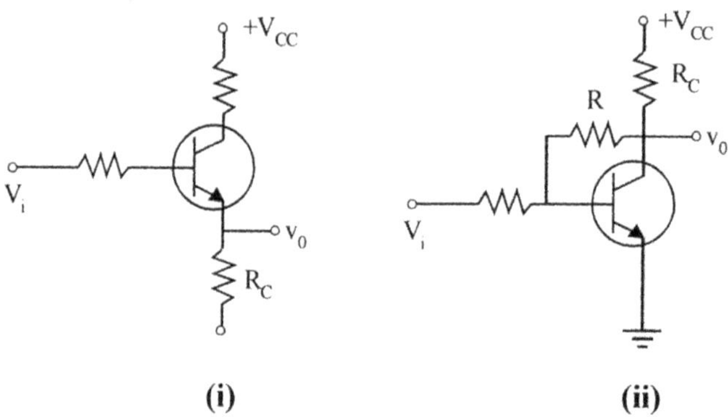

(i)   **(ii)**

17. In the case of Voltage - series feedback, expression for output impedance with feed back z_{of} is ....

18. In the case of Voltage shunt feedback amplifier, expression for input impedance with feedback is Z_{if} = ...

19. In the case of current series feedback amplifier, expression for output impedance with feedback z_{of} is ....

20. Expression for z_{if} with current shunt feedback is ....

ESSAY TYPE QUESTIONS

1. Explain the concept of feedback as applied to electronic amplifier circuits. What are the advantages and disadvantages of positive and negative feedback ?

2. With the help of a general block schematic diagram explain the term feedback.

3. What type of feedback is used in electronic amplifiers ? What are the advantages of this type of feedback ? Prove each one mathematically.

4. Define the terms Return Ratio, Return Difference feedback factor, closed loop voltage gain and open loop voltage gain. Why negative feedback is used in electronic amplifiers eventhough closed loop voltage gain decreases with this type of feedback ?

5. Give the equivalent circuits, and characteristics of ideal and practical amplifiers of the following types (i) Voltage amplifier, (ii) Current amplifiers, (iii) Transresistance amplifier, (iv) Transconductance amplifier.

6. Derive the expression for the input resistance with feedback R_{if} (or R_i') and output resistance with feedback R_{0f} (or R_i') in the case of

   (a) Voltage series feedback amplifier.

   (b) Voltage shunt feedback amplifier.

   (c) Current series feedback amplifier.

   (d) Current shunt feedback amplifier.

7. In which type of amplifier the input impedance increases and the output impedance decreases with negative impedance ? Prove the same drawing equivalent circuit.

8. Draw the circuit for Voltage series amplifier and justify the type of feedback. Derive the expressions for A_V', β, R_i' and R_0' for the circuit.

9. Draw the circuit for Current series amplifier and justify the type of feedback. Derive the expressions for A_V', β, R_i' and R_0' for the circuit.

10. Draw the circuit for Voltage shunt amplifier and justify the type of feedback. Derive the expressions for A_V', β, R_i' and R_0' for the circuit.

11. Draw the circuit for Current shunt amplifier and justify the type of feedback. Derive the expressions for A_V', β, R_i' and R_0' for the circuit.

MULTIPLE CHOICE QUESTIONS

1. Loop sampling is also known as

   (a) Voltage sampling (b) Current sampling

   (c) Power sampling (d) Node sampling

2. Positive feedback is also known as

   (a) regenerative feedback (b) degenerative feedback

   (c) Loop feedback (d) return feedback

3. Equation for feedback factor 'β' is ....

   (a) $\dfrac{V_o}{V_f}$ (b) $\dfrac{V_o'}{V_f}$ (c) $\dfrac{V_f}{V_o'}$ (d) $\dfrac{V_f}{V_o}$

4. With negative feedback, the linearity of operation of the amplifier circuit

   (a)   deteriorates   (b)   improves   (c)   remian same   (d)   none of these

5. Expression for distortion D' with negative feedback, with usual notation is $D' =$

   (a) $\dfrac{D}{1-\beta A}$ (b) $D(1-bA)$ (c) $\dfrac{D}{1+\beta A}$ (d) $D(1+bA)$

6. Expression for densitivity of an amplifier circuit with negative feedback is,

   (a) $D = (1-\beta A)$ (b) $D = \dfrac{1}{\beta A}$ (c) $D = (1+\beta A)$ (d) $D = \dfrac{1}{(1+\beta A)}$

7. Characteristics of ideal volt age amplifier are ...

   (a) $R_i = \infty,\ R_0 = 0$ (b) $R_i = 0,\ R_0 = 0$

   (c) $R_i = 0,\ R_0 = \infty$ (d) $R_i = \infty,\ R_0 = \infty$

8. The product of feedback factor 'β' and amplification factor 'A' is called ...

   (a)   return difference   (b)   return ratio

   (c)   series ratio   (d)   shunt ratio

9. Negative series feedback voltage or current ... input resistance

   (a)   decreases   (b)   no effect   (c)   can't be said   (d)   increases

10. With voltage feedback series or shunt, output ressitance R_0 ....

   (a)   decreases   (b)   increases   (c)   remains same   (d)   non of these

CONCEPTUAL QUESTIONS (Interview Questions)

1.   Give physical example for a feed back systems.

   Ans : A person searching for a book in Library is one example. The book seen by eyes and signal sent to brain act as closed loop system. If the eyes are closed, it is like open loop system, as no feed back can be given to the brains.

2.   Give a practical example of a feed back system.

   Ans : Controlling the temperature of a furnace is an example of feedback system. The temperature is sensed by a thermo couple and the signal is fed back to a controlling circuit. When set temperature is reached, the electric supply is turned off. When the temperature goes below the set value, electric supply is again turned on.

8 Oscillators

In this Chapter,

- ◆ Basic principle of oscillator circuits is explained. Generation of sinusoidal waveforms by the oscillator circuits without external A.C. input is explained.

- ◆ Barkhausen criteria to be satisfied for generation of oscillations is given.

- ◆ R-C phase shift oscillator circuit, Hartley oscillator, Colpitts oscillator, crystal oscillator, Resonant oscillator circuits are given.

8.1 OSCILLATORS

Oscillator is a source of AC voltage or current. We get A.C.output from the oscillator circuit. In alternators (AC generators) the thermal energy is converted to electric energy at 50Hz.

In the oscillator circuits that we are describing now, the electric energy in the form of DC is converted into electric energy in the form of AC. 'Invertors' in electrical engineering convert DC to AC, but there, only output power is the criterian and not the actual shape of the wave form.

An amplifier is different from oscillator in the sense that an amplifier requires some A.C. input which will be amplified. But an oscillator doesn't need any external AC signal. This is shown in Fig. 8.1 below :

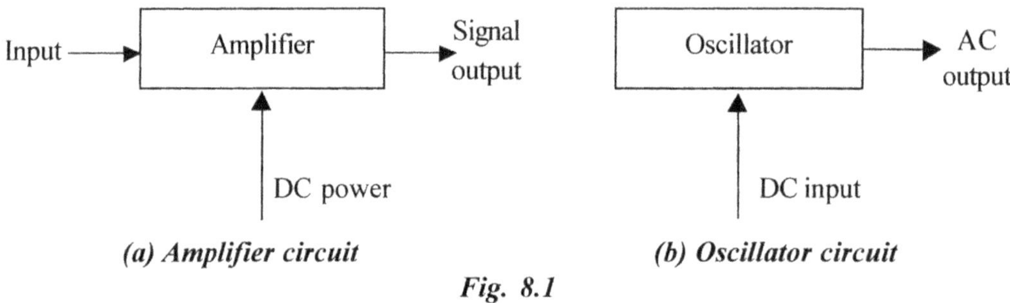

(a) Amplifier circuit   *(b) Oscillator circuit*

Fig. 8.1

For an amplifier, the additional power due to amplification is derived from the DC bias supply. So an amplifier effectively converts DC to AC. But it needs AC input. Without AC input, there is no AC output. In the oscillator circuits also DC power is converted to AC. But there is no AC input signal. So the difference between amplifier and oscillator is in amplifiers circuits, the DC power conversion to AC is controlled by the AC input signal. But in oscillators, it is not so.

There are two types of oscillators circuits :

1. Harmonic Oscillators
2. Relaxation Oscillators.

Harmonic Oscillators produce sine waves. Relaxation Oscillators produce sawtooth and square waves etc. Oscillator circuits employ both active and passive devices. Active devices convert the DC power to AC. Passive components determine the frequency of oscillators.

8.1.1 PERFORMANCE MEASURES OF OSCILLATOR CIRCUITS :

1. *Stability :* This is determined by the passive components. R,C and L determine frequency of oscillations. If R changes with T, f changes so stability is affected. Capacitors should be of high quantity with low leakage. So silver mica and ceramic capacitors are widely used.

2. *Amplitude stability :* To get large output voltage, amplification is to be done.

3. *Output Power :* Class A, B and C operations can be done. Class C gives largest output power but harmonics are more.

   Class A gives less output power but hormonics are low.

4. *Harmonics :* Undesirable frequency components are harmonics. An elementary sinusoidal oscillator circuit is shown in Fig.8.2.

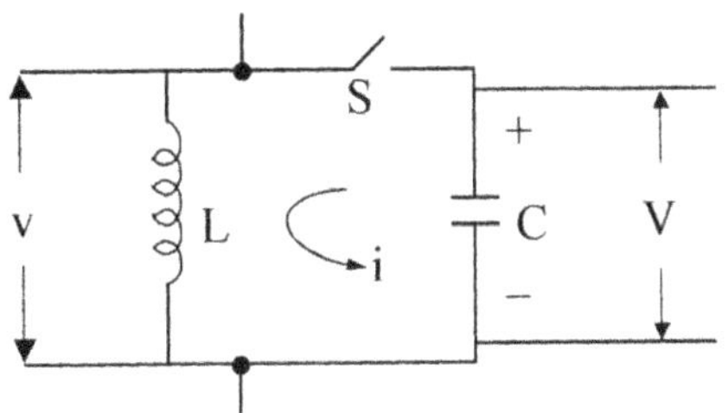

Fig. 8.2 LC Tank circuit.

L and C are reactive elements. They can store energy. The capacitor stores energy whenever there is voltage across its plates. Inductor stores energy in its magnetic field whenever current flows. Both C and L are lossless, ideal devices. So quality factor is infinity ∞. Energy is introduced into the circuit by charging capacitor to 'V' Volts. If switch is open, C cannot discharges because there is no path for discharge current to flow.

Suppose at $t = t_0$, switch 'S' is closed. Then current flows. Voltage across L will be V. At $t = t_0$, the Voltage across C is V volts. When switch is closed, current flows. So the charge across Capacitor C decreases. Voltage across C decreases, as shown in the waveform. So as the energy stored in Capacitor decreases, the energy stored in inductor L increases, because current is flowing through L Thus total energy in the circuit remains the same as before. When V across C becomes 0, current through the inductor is maximum. When the energy in C is 0, energy in L is maximum. Then the current in L starts charging C in the opposite directions. So at $t = t_1$ current in L is maximum and for $t > t_1$, the current starts charging C in the opposite direction. So V across C becomes negative as shown in the wavefrom. Thus we get sinusoidal oscillations from LC circuit.

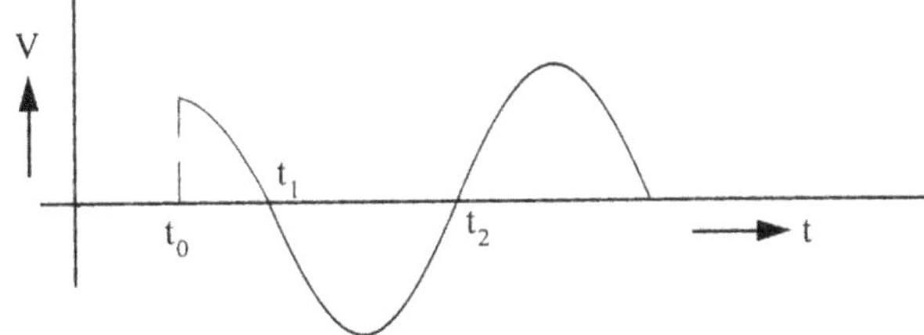

Fig. 8.3 Waveform.

Thus we are getting sinusoidal variations without giving any one input. What we have done is depositing some charge on C, so that the circuit operates on its own. But why should we get sinusoidal wave and not triangular or square wave ? Sinusoidal function is the only function that satisfies the conditions governing the exchange of the energy in the circuit.

But the above circuit is not a practical circuit. Because we have to take output from the circuit; i.e. energy has to be extracted, from the circuit. As we draw energy from the circuit, the energy stored in C and L decreases. So output also decreases or the voltage across C and L decreases so we get damped oscillation as shown below.

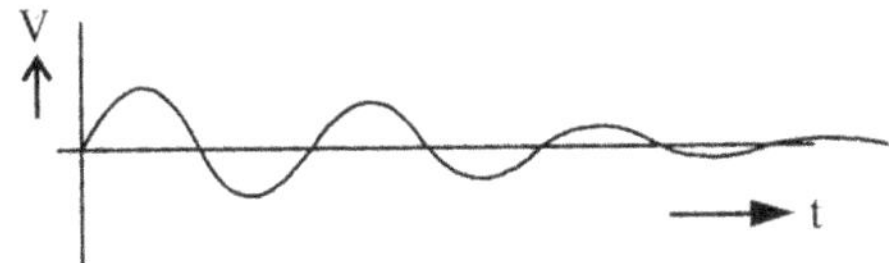

Fig. 8.4 Damped oscillations.

The energy lost by the elements must be replenished, so that oscillations are obtained continuously. If a negative resistance R is connected in the circuit, it replenishes, whatever energy that is lost in the circuit. Certain devices like tunnel diode, UJT, etc., exhibit -ve resistance. The energy supplied by the -ve resistance to the circuit actually comes from the DC bias supply.

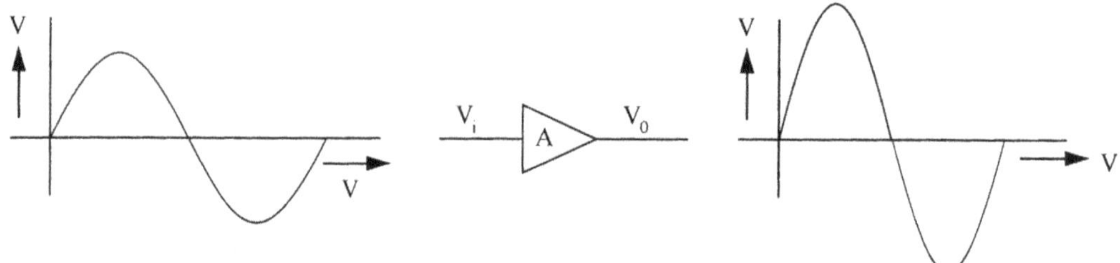

Fig. 8.5 Amplifier response.

Another method of producing sinusoidal oscillations is :

Suppose we have an amplifier with gain A_V and phase shift 180^0, $A_V \gg 1$.

If we connect V_0 through a feedback network to V_i as shown in Fig.8.6 so that after feedback, the feedback signal $V_X = V_i$, and also, the feedback network provides 180^0, phase shift, after feedback the feedback signal V_X will be removed. Thus, without any input we get sinusoidal output.

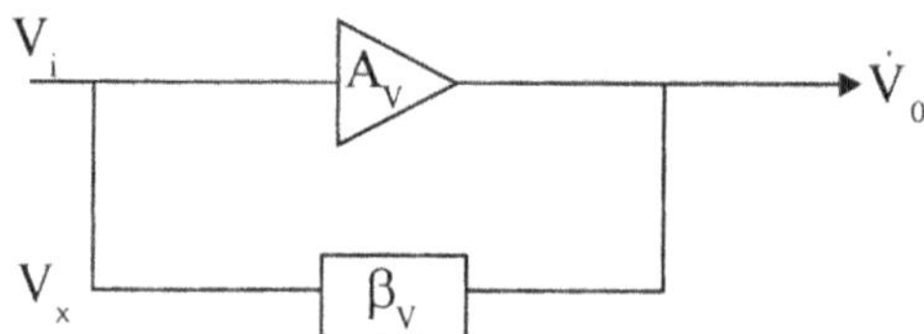

Fig. 8.6   Feedback network.

V_i is provided from the feedback signal of V_0 itself. To get initially Vo, the noise signal of transistor after switching itself is sufficient. Thus without external AC input we get sinusoidal oscillators.

$$\beta_v = \frac{V_x}{V_o}.$$

$$A_v = \frac{V_o}{V_i}.$$

$$V_x = V_i$$

$$\therefore \quad \beta . A_v = \frac{V_x}{V_o} \times \frac{V_o}{V_i}$$

$$= \frac{V_x}{V_i} \quad \text{or} \quad V_x = V_i$$

$$\therefore \quad \beta A_V = 1$$

Total phase shift = 360^0 (180 + 180). Therefore, to get sustained oscillations,

1. The loop gain must be unit 1.
2. Total Loop phase shift must be 0^0 or 360^0. (Amplifier circuit produces 180^0 phase shift and feedback network another 180^0.

8.2 SINUSOIDAL OSCILLATORS

Figure 8.7 shows an amplifier, a feedback network and input mixing circuit not yet connected to form a closed loop. The amplifier provides an output signal X_0, as a consequence of the signal X_i applied directly to the amplifier input terminal. Output of feedback network

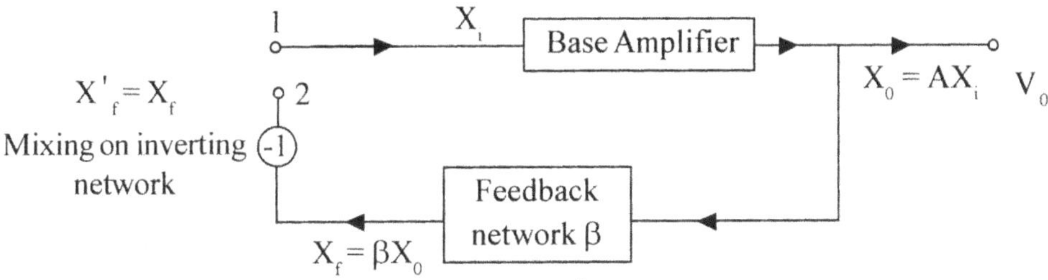

Fig. 8.7 Block schematic.

is $X_f = \beta X_0 = A\,\beta X_i$ and the output of the mixing circuit,

is $X_f' = -X_f = -A\beta X_i$

Loop gain, $\quad = \dfrac{X_f'}{X_i} = \dfrac{-A\beta X_i}{X_i} = -\beta A.$

If X_f' were to be identically equal to X_i, input signal, $-\beta A$ should be = 1, so the output will be X_0, even if the input source is removed. The condition that $-\beta A = 1$ means, the loop gain must be 1.

8.3 BARKHAUSEN CRITERION

We make an asssumption that the circuit operates only in the linear region, and the amplifier feedback network contains reactive elements. For a sinusoidal wave form, if $X_i = X_f'$, the amplitude, phase and frequency of X_i and X_f' be identical. *The frequency of a sinusoidal oscillator is determined by the condition that loop gain, Phase shift is zero at that frequency.*

For oscillator circuits positive feedback must be there i.e., V_f must be in phase with V_i to get added to V_i. When active device BJT or FET gives 180^0 phase shift, the feedback network must produce another 180^0 phase shift so that net phase shift is 0^0 or 360^0 and V_f is in phase with V_i to make it positive feedback.

Oscillations will not be sustained if, at the oscillator frequency the magnitude of the product of the transfer gain of the amplifier and of β are less than unity.

The conditions $-A\beta = 1$ is called *Barkhausen criterion*

i.e. $|\beta A| = 1$ and phase of $-A\beta = 0$.

But $\qquad A_f = \dfrac{A}{1+\beta A}$. If $\beta A = -1$, than $A_f \to \infty$.

which implies that, there exists an output voltage even in the absence of an externally applied voltage.

A 1-V signal appearing initially at the input terminals will, after a trip around the loop and back to the input terminals, appear there, with an amplitude larger than 1V. This larger voltage will then reappear as a still larger voltage and so on. So if $|\beta A|$ is larger than 1, the amplitude of oscillations will continue to increase without limit. But practically , to limit the increase of amplitude of oscillations, nonlinearity ability of the circuit will be set in. Though a circuit has been designed for $|\beta A| = 1$, since circuit components and transistor change characteristics with age, temperature, voltage etc. $|\beta A|$ will become larger or smaller than 1. If $\beta A < 1$, the oscillations will stop. If $\beta A > 1$, the amplitude practically will increase. So, to achieve a sinusoidal osciallation practically $|\beta A| > 1$ to 5%, so that with incidental variations in transistor circuit parameters, $|\beta A|$ shall not fall below unity.

The type of noise in electronic circuits and the causes are :

1. *Johnson noise or Thermal Noise :*   Due to temperature.

2. *Schottky noise or Shot noise :* Because of variation .is concentration in the Semiconductor Devices.

8.4   R – C PHASE-SHIFT OSCILLATOR (USING JFET)

This is voltage series feedback. FET amplifier is followed by three cascaded arrangements of a capacitor C, resistor R. The output of the last RC combination is returned to the gate. This forms the feedback connection. The FET amplifier shifts the phase of voltage appearing at the gate by 180^0. The RC network shifts the phase by additional amount. At some frequency, the phase-shift introduced by this network will be exactly 180^0. The total phase-shift at this frequency, from the gate around the circuit and back to the gate is $+180 - 180 = 0^0$. At this particular frequency, the circuit will oscillate. (Fig. 8.8).

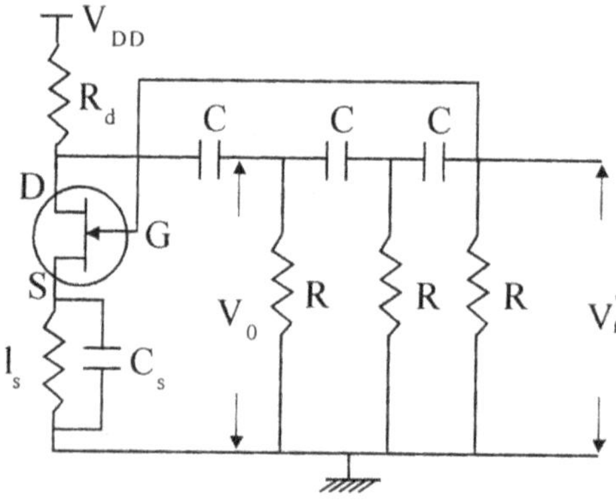

Fig. 8.8 JFET R – C phaseshift oscillator circuit.

$\therefore$ At that frequency, $V_f' = -V_f$ $\because$ 180^0 out of phase.

The equivalent circuit is, show below in Fig. 8.9.

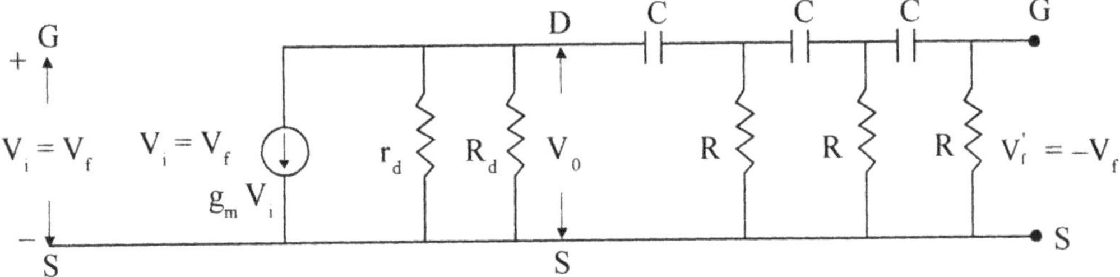

Fig. 8.9 *Equivalent circuit.*

The transformation of RC net work is $\dfrac{V_f}{V_o} = -\beta$; $\therefore$ $\beta = \dfrac{V_f}{V_o}$ and $V_f' = -V_f$

$$\therefore \quad -\beta = \frac{V_f'}{V_o} = \frac{1}{1 - 5\alpha^2 - j(6\alpha - \alpha^3)}.$$

where $\quad \alpha = \dfrac{1}{\omega RC}.$

The phase-shift of $\dfrac{V_f'}{V_o}$ is 180^0, for $\alpha^2 = 6$ or

β must be real. Therefore $j(6\alpha - \alpha^3) = 0$.

$$\therefore \quad f = \frac{1}{2\pi RC\sqrt{6}}. \quad \because \quad 6 = \frac{1}{\omega^2 R^2 C^2}.$$

$$\omega^2 = \frac{1}{6R^2 C^2}.$$

$$\therefore \quad \boxed{f = \frac{1}{2\pi RC\sqrt{6}}.}$$

when $\quad \alpha^2 = 6, -\beta = \dfrac{1}{1 - 5 \times 6 - 5(6\alpha - 6\alpha)} = \dfrac{-1}{29}$

or $\quad \boxed{\beta = \dfrac{1}{29}.}$

In order that $|\beta A|$ is not les than unity, A should be atleast $|29|$. So select F E T whose μ is atleast 29.

8.4.1   TO FIND THE β OF THE RC PHASE-SHIFT NETWORK (JFET)

Each RC network introduces a phase-shift of 60^0. Therefore, total phase-shift $= 180^0$. (See Fig.8.10).

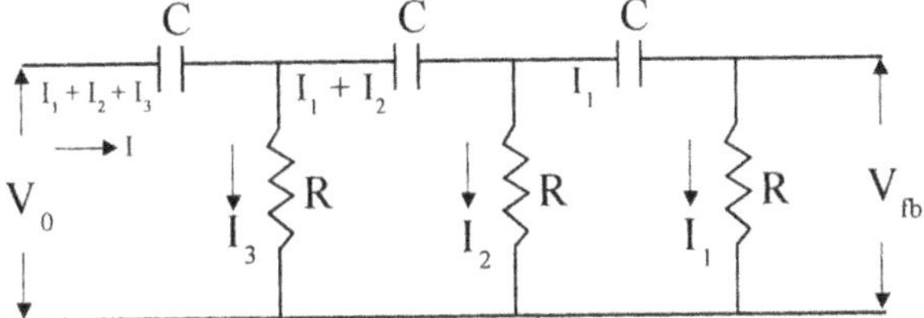

Fig. 8.10 *R C phase shift network.*

$$V_{fb} = I_L . R \qquad \qquad .....(1)$$

$$I_2 . R = \frac{I_1}{j\omega C} + I_1 . R \qquad \qquad .....(2)$$

$$I_3 . R = \frac{(I_2 + I_1)}{j\omega C} + I_2 . R \qquad \qquad .....(3)$$

$$V_o = \frac{(I_1 + I_2 + I_3)}{j\omega C} + I_3 . R \qquad \qquad .....(a)$$

But,
$$I_3 . R = \frac{(I_2 + I_1)}{j\omega C} + I_2 . R$$

$$I_2 . R = \frac{I_1}{j\omega C} + I_1 . R$$

$$I_1 R = V_{FB}$$

$\therefore$
$$I_3 . R = \frac{(I_2 + I_1)}{j\omega c} + \frac{I_1}{j\omega c} + V_{fb}.$$

By substituting 1, 2, and 3 in (a) we get,

$$V_o = \frac{(I_1 + I_2 + I_3)}{j\omega c} + \frac{(I_2 + I_1)}{j\omega c} + \frac{I_1}{j\omega c} + V_{fb}.$$

$$V_o = \frac{3I_1 + 2I_2 + I_3}{j\omega c} + V_{fb}. \qquad \qquad .....(b)$$

To eliminate I_1, I_2, and I_3,

$$I_1 = \frac{V_{fb}}{R} \qquad \qquad .....(c)$$

$$I_2 = \frac{I_1}{j\omega cR} + I_1 = \frac{V_{fb}}{j\omega cR^2} + \frac{V_{fb}}{R}$$

$$I_2 = V_{fb} \left(\frac{1}{R} + \frac{1}{J\omega CR^2} \right) \qquad \qquad .....(d)$$

$$I_3 = \frac{(I_2 + I_1)}{j\omega CR} + I_2$$

$$I_3 = \frac{V_{fb}}{j\omega cR} \left(\frac{2}{R} + \frac{1}{j\omega cR^2} \right) + V_{fb} \left(\frac{1}{R} + \frac{1}{j\omega CR^2} \right) \qquad \qquad .....(e)$$

Substitute c, d, and e equation in (b) and simplify

$$V_o = \frac{3V_{fb}}{R(j\omega c)} + \frac{2V_{fb}}{j\omega c} \left(\frac{1}{R} + \frac{1}{j\omega cR^2} \right) + \frac{V_{fb}}{(j\omega c)^2 R} \left(\frac{2}{R} + \frac{1}{j\omega cR^2} \right)$$

$$+ \frac{V_{fb}}{j\omega c} \left(\frac{1}{R} + \frac{1}{j\omega cR^2} \right) + V_{fb}$$

$$\frac{V_o}{V_{fb}} = \frac{1}{\beta} = 1 - \frac{5}{\omega^2 c^2 R^2} + j\left(\frac{1}{\omega^3 C^3 R^3} - \frac{6}{\omega CR}\right)$$

Let,
$$\alpha = \frac{1}{\omega CR}$$

$\therefore$
$$\frac{1}{\beta} = 1 - 5\alpha^2 + j(\alpha^3 - 6\alpha)$$

or
$$\beta = \frac{1}{1 - 5\alpha^2 - j(6\alpha - \alpha^3)}$$

$$\beta A = 1$$

But β is a complex number. Therefore equating imaginary part to zero.

$$6\alpha - \alpha^3 = 0.$$

$$6\alpha = \alpha^3. \quad \text{or} \quad \alpha = \sqrt{6}.$$

But
$$\alpha = \frac{1}{\omega CR}.$$

$\therefore$
$$\frac{1}{\omega CR} = \sqrt{6}.$$

or
$$\omega = \frac{1}{\sqrt{6}CR}$$

or
$$\boxed{f = \frac{1}{2\pi CR\sqrt{6}}}$$

The gain A, corresponding to this frequency will be 29.

8.5 TRANSISTOR RC PHASE-SHIFT OSCILLATOR

For transistor circuit, voltage shunt feedback is employed, because the input impedance of transistor is small. If voltage series feedback is employed, the resistance of feedback network will be shunted by the low 'R' of the transistor. (Fig. 8.11(a) and Fig.8.11(b)).

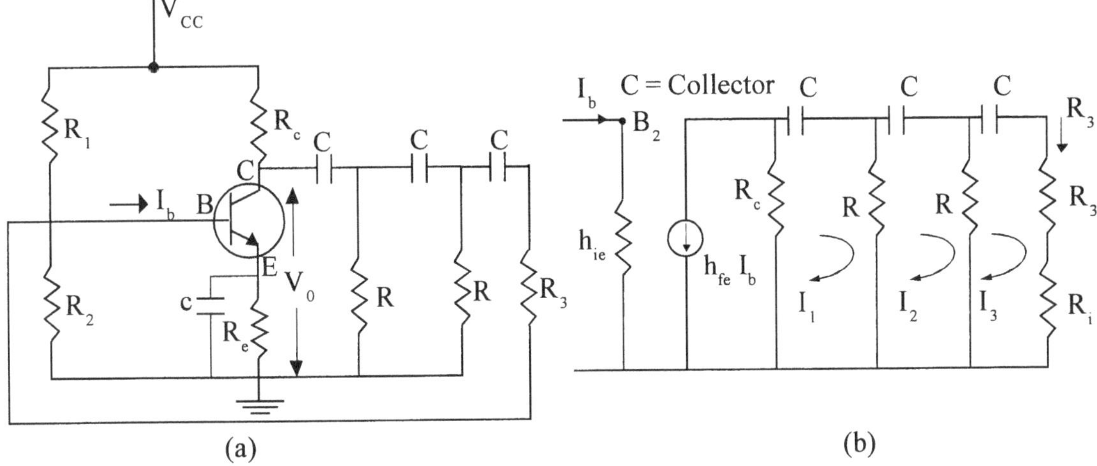

(a)   (b)

Fig. 8.11 Transistor phase shift oscillator.

The value of $R_3 = R - R_i$, where $R_i \simeq h_{ie}$, the input resistance of transistor. $\therefore$ The three RC sections of the phase-shifting network are identical R_1 R_2 and R_e are biasing resistors.

The feedback current $x_f = - I_3$ Input current $x_i = I_b$ (Negative sign is because it is negative feedback)

$\therefore$ Loop current gain $= \dfrac{-x_f}{x_i} = \dfrac{I_3}{I_b}$

By writing K V L for the three nodes, $\dfrac{I_3}{I_b}$ can be found out.

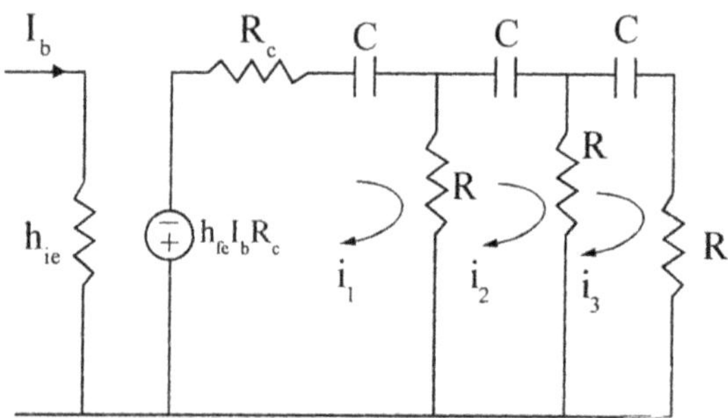

Fig. 8.12 R - C Equivalent circuit.

Loop 1 : $(R_C + R - \dfrac{j}{\omega c}) I_1 - R I_2 = - h_{fe} I_b R_C$ .....(1)

Loop 2 : $-RI_1 + (2R - \dfrac{j}{\omega c}) I_2 - RI_3 = 0$ .....(2)

Loop 3 : $-RI_2 + (2R - \dfrac{j}{\omega c}) I_3 = 0$ .....(3)

By taking 'R' as common, the modified given by equations are

$$\left(\dfrac{R_C}{R} + 1 - \dfrac{j}{\omega_{RC}} \right) I_1 - I_2 = -h_{fe} I_b \dfrac{R_C}{R}$$ .....(4)

$$-I_1 + \left(2 - \dfrac{j}{\omega RC} \right) I_2 - I_3 = 0$$ .....(5)

$$-I_2 + \left(2 - \dfrac{j}{\omega RC} \right) I_3 = 0$$ .....(6)

Let $\dfrac{R_C}{R} = K$ and $\alpha = \dfrac{1}{\omega RC}$, the equations are

$$(1 + K - j\alpha) I_1 - I_2 = - h_{fe} I_b = 0$$ .....(7)
$$-I_1 + (2 - j\alpha) I_2 - I_3 = 0$$ .....(8)
$$-I_2 + (2 - j\alpha) I_3 = 0$$ .....(9)

$$\begin{bmatrix} 1+k-j\alpha & -1 & 0 \\ -1 & 2-j\alpha & -1 \\ 0 & -1 & 2-j\alpha \end{bmatrix} \begin{bmatrix} I_1 \\ I_2 \\ I_3 \end{bmatrix} = \begin{bmatrix} -h_{fe}I_b k \\ 0 \\ 0 \end{bmatrix}$$

$$I_3 = \frac{1}{\Delta} \begin{vmatrix} 1-k-j\alpha & -1 & -h_{fe}I_b K \\ -1 & 2-j\alpha & 0 \\ 0 & -1 & 0 \end{vmatrix} = \frac{1}{\Delta}\left[-h_{fe}I_b K\right]$$

$$\Delta = [1 + K - j\alpha]\,[(2 - j\alpha)^2 - 1] + [-(2 - j\alpha)]$$
$$= [1 + K - j\alpha]\,[4 - \alpha^2 - j4\alpha - 1] - 2 + j\alpha$$
$$= -j\alpha^3 - j6\alpha - j4K\alpha - 5\alpha^2 - \alpha^2 K + 3K + 1$$
$$\Delta = -(5 + K)\,\alpha^2 + 3K + 1 + j\,[\alpha^3 - (6 + 4K)\,\alpha]$$

$$\therefore\ I_3 / I_b = \frac{-h_{fe}K}{-(5+K)\alpha^2 + 3K + 1 + j\,[\alpha^3 - (6+4K)\alpha]} \qquad \text{.....(10)}$$

The Barkhausen condition that the loop gain $\dfrac{I_3}{I_b}$ phase shift must equal zero. The phase shift equals zero provided imaginary part is zero.

Hence
$$\alpha^3 = (6 + 4K)\,\alpha$$
$$\alpha^2 = 6 + 4K$$

$$\alpha = \sqrt{6 + 4K}$$

Since
$$\alpha = \frac{1}{\omega RC}, \qquad \omega = \frac{1}{RC\sqrt{6 + 4K}}$$

$\therefore$ The frequency of oscillation f is given by

$$f = \frac{1}{2\pi RC\sqrt{6 + 4K}} \qquad \text{.....(11)}$$

For mantaining the oscillations at the above frequency, $|I_3 / I_b| > 1$

$$\left|\frac{I_3}{I_b}\right| = \left|\frac{-h_{fe}K}{-(5+K)(6+4K) + 3K + 1}\right| > 1$$

$$\left|-h_{fe}K\right| > \left|-(5+K)(6+4K) + 3K + 1\right|$$

$$\left|-h_{fe}K\right| > \left|-4K^2 - 23K - 29\right|$$

$$h_{fe}\,K > 4K^2 + 23K + 29$$

$$h_{fe} > 4K + 23 + \frac{29}{K} \qquad \text{.....(12)}$$

To determine the minimum value of h_{fe}, the optimum value of K should be determined by differentiating h_{fe} w.r.t K and equate it to zero.

$$\frac{dh_{fe}}{dk} > \frac{d}{dk}\left(4K + 23 + \frac{29}{K}\right)$$

$$0 > 4 - \frac{29}{K^2}$$

$$\therefore K^2 < \frac{29}{4}, \qquad \therefore \mathbf{K < 2.7} \qquad\qquad .....(13)$$

Substitute the optimum $K = 2.7$ in eq. (12)

$$h_{fe} > 4 \times 2.7 + 23 + \frac{29}{2.7}$$

$$\therefore h_{fe} > 44.5 \qquad\qquad .....(14)$$

The BJT with a small-signal common-emitter short-circuit gain h_{fe} lessthan 44.5 cannot be used in this phase shift oscillator.

In R–C phase shift oscillator circuit, each RC network produces 60^0 phase shift. Thus 3 sections produce the required 180^0 phase shift. If there are 4-sections, each sections must produce 45^0 phase shift. But more number of components are to be used. If only 2 sections are these, 90^0 phase shift must be produced, which is not possible for practical R-C network.

8.6   A GENERAL FORM OF LC OSCILLATOR CIRCUIT

Many Oscillator Circuits fall in to the general form as shown in Fig.8.13 (a) and (b).

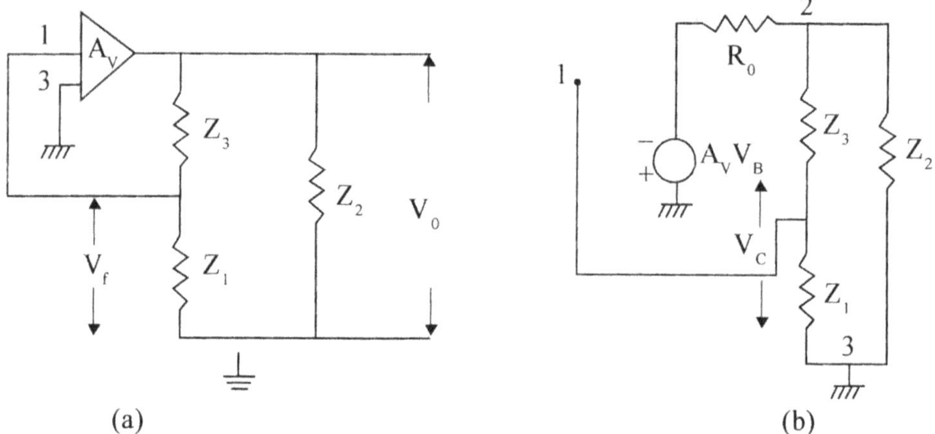

(a)   (b)

Fig. 8.13   General form of oscillator circuit

The active device can be FET, transistor or operational amplifier, Fig.8.13(b) shows the equivalent circuit using an amplifier with negative gain A_v and output resistance R_o. This is Voltage Series Feedback.

8.7   LOOP GAIN

$$\beta = -\frac{V_f}{V_o}$$

$$\therefore \quad \beta = \frac{-V_f}{V_0} = -\frac{Z_1}{Z_1 + Z_3}.$$

The load impedance, $Z_L = (Z_3$ in series with $Z_1)$ parallel with Z_L.

R_0 is the output resistance.(See Fig. 8.14).

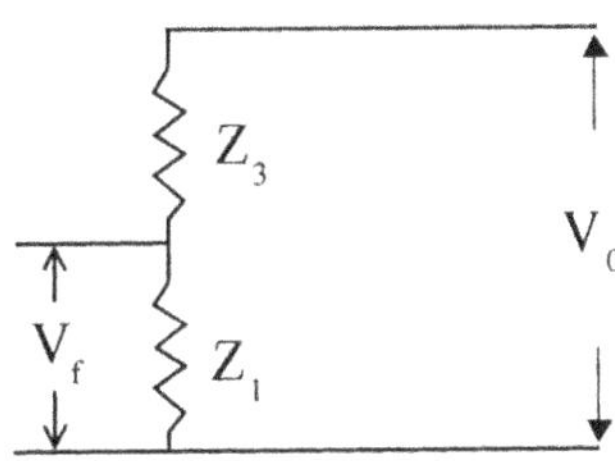

Fig. 8.14   Potential divider network.

Gain without feedback $A = \dfrac{-A_V . Z_L}{Z_L + R_O}$

$$\therefore \quad -\beta A = \frac{-A_V . Z_1 Z_L}{(Z_L + R_O)(Z_1 + Z_3)} ; \quad Z_L = \frac{Z_2(Z_1 + Z_3)}{Z_2 + Z_1 + Z_3}$$

$$= \frac{-A_V . Z_2(Z_1 + Z_3)Z_1}{(Z_1 + Z_2 + Z_3)\dfrac{(Z_2(Z_1 + Z_3))}{(Z_1 + Z_2 + Z_3)} + R_O)(Z_1 + Z_3)}$$

$$-\beta A = \frac{-A_v . Z_1 \, Z_2}{R_O(Z_1 + Z_2 + Z_3) + Z_2(Z_1 + Z_3)}$$

If the impedances are pure reactances, then $Z_1 = j\,x_1, \quad Z_2 = jx_2; \quad Z_3 = jx_3$

$$\therefore \quad -A\beta = \frac{A_v X_1 \, X_2 (j)^2}{jR_o(X_1 + X_2 + X_3) - X_2(X_1 + X_3)}$$

If $\beta A = 1,$ or for zero phase-shift, imaginary part must be zero.

$$jR_o(X_1 + X_2 + X_3) = 0.$$

or $\qquad X_1 + X_2 + X_3 = 0$

$$-A\beta = \frac{A_v X_1 \, X_2}{-X_2(X_1 + X_3)} = -\frac{A_v \, X_1}{X_1 + X_3}$$

But $\quad X_1 + X_2 + X_3 = 0 \quad \therefore \quad X_1 + X_3 = -X_2$

$$\therefore \quad -A\beta = \frac{A_v X_1}{X_2}.$$

$\therefore -A\beta$ must be positive, and at least unity in magnitude. Than X_1 and X_2 must have the same sign.

So if X_1 and X_2 are capacitive, X_3 should be inductive and vice versa.

If X_1 and X_2 are capacitors, the circuit is called *Colpitts Oscillator (Fig. 8.15(a))*

If X_1 and X_2 are inductors, the circuit is called *Hartley Oscillators (Fig. 8.15(b))*

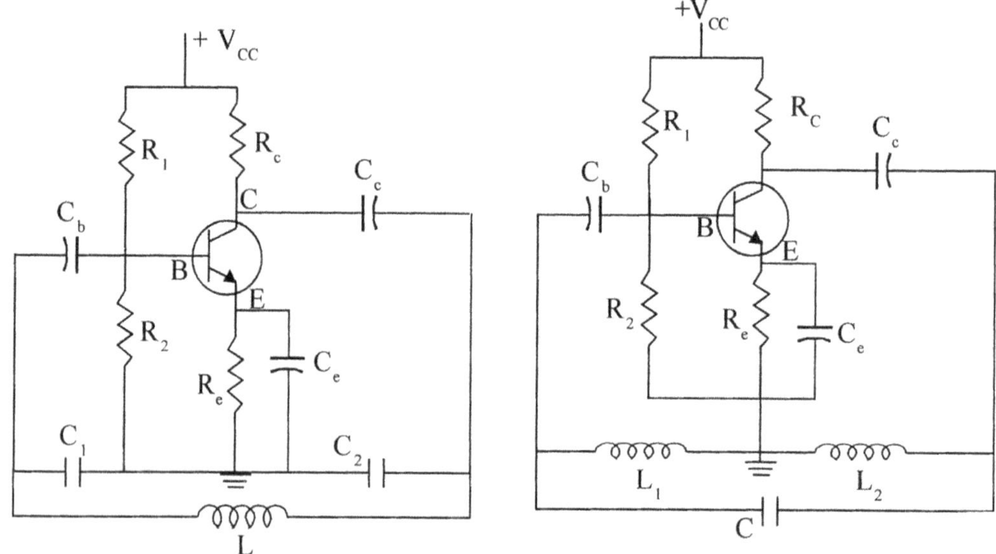

    (a) Colpitts oscillator   *(b) Hartley oscillator circuit*

Fig. 8.15

8.7.1   For Hartley Oscillator

Condition for oscillations, $X_1 + X_2 + X_3 = 0$.

where
$$X_1 = j\omega L_1 \quad X_2 = j\omega L_2; \quad X_3 = + \frac{1}{j\omega C}$$

$$\therefore \quad j\omega L_1 + j\omega L_2 + \frac{1}{j\omega C} = 0 \quad \text{or} \quad \omega L_1 + \omega L_2 = \frac{1}{\omega C}$$

$$\omega^2(L_1 + L_2) = \frac{1}{C}; \quad \frac{1}{\sqrt{(L_1 + L_2)C}} = \omega$$

$$\boxed{f = \frac{1}{2\pi\sqrt{(L_1 + L_2)C_3}}}$$

8.7.2 For Colpitts Oscillator

$$X_1 = -\frac{j}{\omega C_1}; \quad X_2 = \frac{-j}{\omega C_2}; \quad X_3 = j\omega L.$$

$$X_1 + X_2 + X_3 = 0;$$

$$\frac{-j}{\omega C_1} - \frac{-j}{\omega C_2} + j\omega L = 0 \quad \text{or}$$

$$\omega L = \frac{1}{\omega C_1} + \frac{1}{\omega C_2}$$

$$\omega L = \frac{\omega C_2 + \omega C_1}{\omega C_1 C_2} = \frac{C_2 + C_1}{\omega C_1 C_2} \quad \text{or} \quad \omega = \left(\frac{1}{\sqrt{L_3}} \sqrt{\frac{C_2 + C_1}{C_1 C_2}} \right)$$

$$\boxed{f = \frac{1}{2\pi \sqrt{L\, C_T}}}$$

where

$$C_T = \frac{C_1 C_2}{C_1 + C_2}$$

8.8   WIEN BRIDGE OSCILLATOR

In this circuit, a balanced bridge is used as the feedback network. The active element is an operational amplifier. It employs lead-lag Network. Frequency f_0 can be varied in the ratio of 10 : 1 compared to 3 : 1 in other oscillator circuits.

External voltage V_0' is applied betweeen 3 and 4, as shown in Fig.8.16.

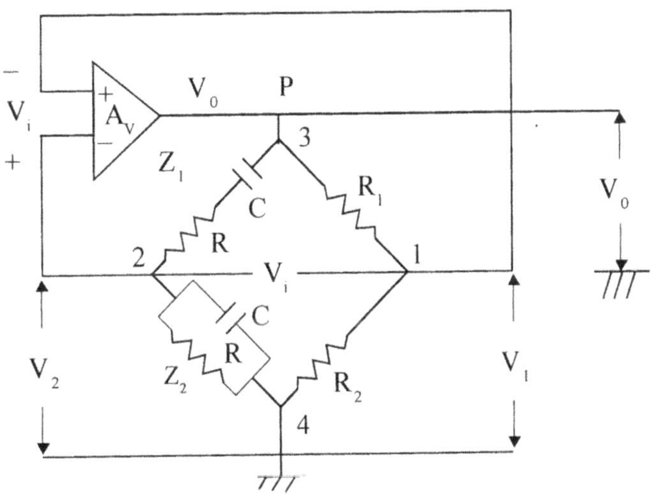

Fig. 8.16 Wien bridge oscillator circuit.

To find loop gain, $-\beta A$, (–sign because phase-shift feedback)

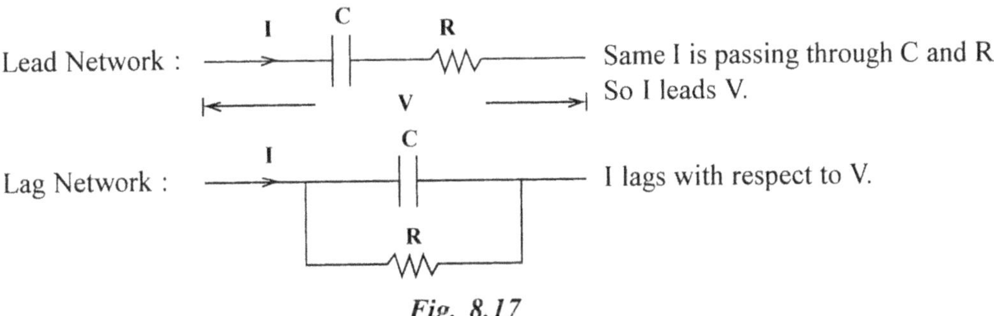

Lead Network :   Same I is passing through C and R. So I leads V.

Lag Network :   I lags with respect to V.

Fig. 8.17

Frequency variation of $10:1$ is possible in Wein Bridge compared to $3:1$ in other oscillator circuits.

$$V_0 = A_v\,V_i; \quad V_i \text{ is } (V_2 - V_1); \quad V_i = V_f.$$

Loop gain
$$= \frac{V_0}{V_0{}'} = \frac{A_v.V_i}{V_0{}'} = -\beta A. \qquad .....(1)$$

V_1 and V_2 are auxillary voltages. $V_i = V_2 - V_1$.

$$-\beta = \frac{V_i}{V_0{}'} \quad \because\ V_f = V_i \quad \therefore\ A = A_v$$

$$-\beta = \frac{V_i}{V_0{}'} = \frac{V_2 - V_1}{V_0{}'} = \left[\frac{Z_2}{Z_1 + Z_2} - \frac{R_L}{R_1 + R_2}\right]$$

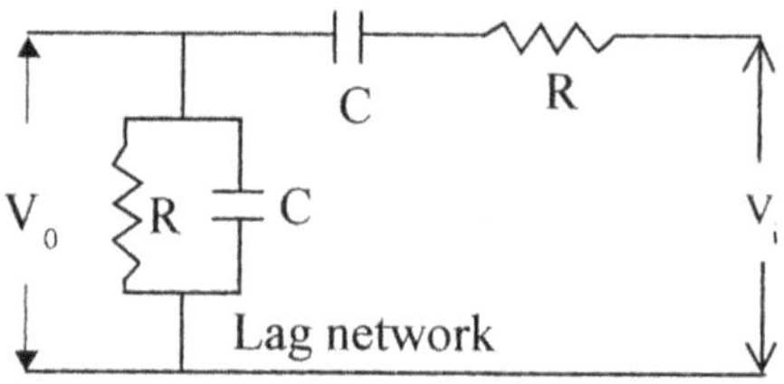

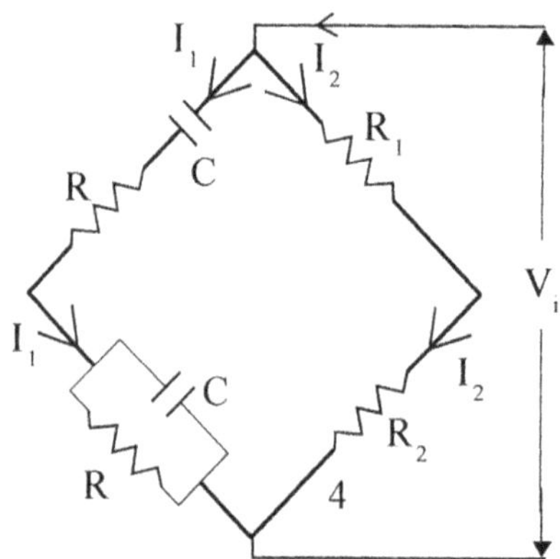

Fig. 8.18

8.9 EXPRESSION FOR f

Fig. 8.19 Wien Bridge oscillator circuit.

$$I_1 = \frac{\left(R + \dfrac{1}{j\omega c}\right)}{I_1\left(\dfrac{R}{1 + j\omega cR}\right)} = \frac{I_2 R_1}{I_2 R_L}$$

$$R_1 = \frac{\left(R_2 R + \dfrac{R_2}{j\omega C}\right)(1 + j\omega CR)}{R} \ ;$$

$$R_1 R = \left(R_2 R + \frac{R_2}{jwC} \right) (1 + jwCR)$$

$$R_2 \cdot R + \frac{R_2}{j\omega C} = \left(\frac{R_1 R}{1 + j\omega CR} \right)$$

$$[(R\,R_2)\,(j\omega C) + R_2]\,R_2\,j\omega CR - \omega^2\,C^2\,R^2\,R_2 = R_1\,R\,j\omega C$$

$$(R_1\,R_2)\,(j\omega C) + R_2 + R_2\,j\omega CR - \omega^2 c^2 R^2 R_2 = R_1\,R\,j\omega C$$

Equating imaginary parts,

$$R_1\,R_2\,\omega C + R_2\,\omega CR = R_1\,R\omega C$$

Equating real parts,

$$A = 1 + \frac{R_1}{R_2} + \frac{C_1}{C_2}, \quad R_2 = \omega^2\,C^2\,R^2\,R_2$$

$$\therefore \qquad \omega^2 = \frac{1}{C^2 R^2} \quad \text{or} \quad \boxed{f = \frac{1}{2\pi RC}}$$

$$R_1 = \frac{R_2 \cdot R}{R} + \frac{R_2}{j\omega CR} + j\omega CRR_2 + R_2$$

$$2R_2 = R_1 \quad \text{or} \quad \boxed{\frac{R_2}{R_1 + R_2} = \frac{1}{3}} \quad \text{So the minimum gain of the amplifier must be 3.}$$

$$\frac{-R_2}{\omega CR} = \omega CRR_2$$

$$\omega^2 = \frac{1}{(RC)^2} \; ; \quad \text{or} \quad \boxed{f = \frac{1}{2\pi RC}}$$

This is the frequency at which the circuit oscillates. Continuous variation of frequency is accomplished by using the capacitors 'C'.

8.10 THERMISTOR

Conductivity of Germanium and Si increases with temperature. A semiconductor when used in this way, taking advantage of this property is called a ***Thermistor***. Ex : NiO, Mn_2O_3 etc. These are used for temperature compensation in oscillator circuits.

8.11 SENSISTOR

A heavily doped semiconductor can exhibit a positive temperature coefficient of resistance because under heavy doping, semiconductor acquires the properties of a metal. So R increases because mobility decreases. Such a device is called ***Sensistor***. These are also used for temperature compensation like thermistors.

8.12 AMPLITUDE STABILIZATION

The amplitude of oscillations can be stabilized by replacing R_2 with a senistor. If β is fixed, as A increases, amplitude of oscillations increases. If a senistor is introduced, as its 'R' changes with temperature, it changes β, so that βA is always constant, (when A changes).

By equating the imaginary part to zero we get $f = \dfrac{1}{2\pi\,RC\sqrt{6+4K}}$ where $K = \dfrac{R_C}{R}$

Forward Current Gain $\boxed{h_{fe} = 4k + 23 + \dfrac{29}{K}.}$

Practically RC phase-shift oscillators can be used from several hertz to several hundred kilohertzs. In the mega hertz range, tuned LC circuits are more advantageous. Frequency of oscillators can be changed by changing R and C. Amplitude of oscillations will not vary if any C is varied, because X_C varies, but the imaginary part will be zero. Phase-shift oscillator is operated in class A in order to keep distortion minimum.

8.13 APPLICATIONS

Elevation levelling systems, Burglar detection: frequency of oscillators change as a result of local disturbance. The change in frequency causes further electronic action or alarm signal.

8.14 RESONANT CIRCUIT OSCILLATORS

Initial transient due to switching will initiate electrical signals. These are feedback to the transistor as input. The input is amplified and obtained as output. Oscillations are sustained. Output is coupled to the base of transistor through L_1. LL_1 is a transformer. R represents the resistance in series with the winding in order to account for the losses in the transformer shown in Fig. 8.20. If

'r' is very small, then at $\omega = \dfrac{1}{\sqrt{LC}}$ the Z is purely resistive. Then voltage drop across inductor is

180^0 out of phase with the applied input voltage to the FET. If the direction of winding of the secondary connected to the gate is such that it introduces another 180^0 phase-shift, the total phase-shift is zero.

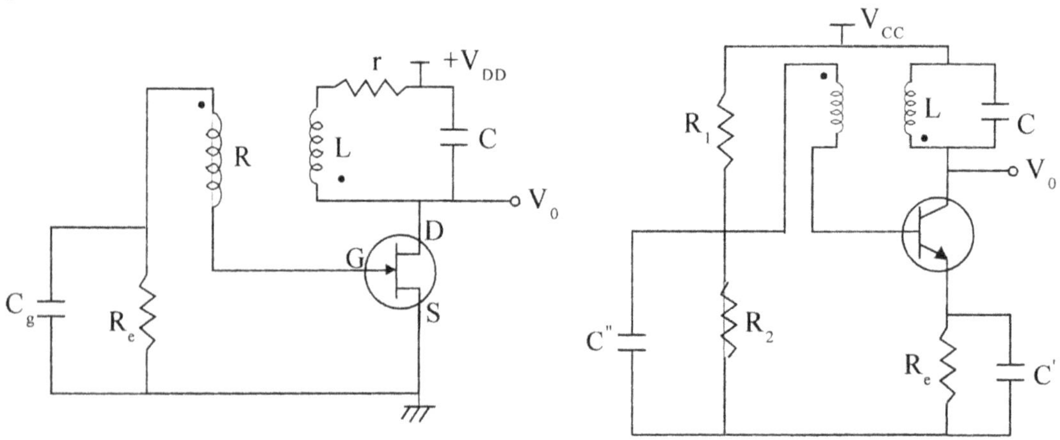

Fig. 8.20 Resonant circuit oscillator.

The ratio of the amplitude of secondary to the primary voltage is M/L where M is the inductance.

$$\text{i.e.,}\quad \frac{V_f}{V_o} = \frac{M}{L} = \beta$$

Voltage gain $A_v = -\mu$

$$\beta A_v = -1$$

$\therefore \quad -\mu\beta = -1;$

$$\mu = \frac{1}{\beta}$$

$\therefore \quad \mu = \dfrac{L}{M}$

$\therefore \quad \mu = L/M$

If we consider resistance 'r' also, $\omega^2 = \dfrac{1}{LC}\left(1+\dfrac{r}{r_d}\right)$

$$\boxed{g_m = \frac{\mu r C}{\mu M - L}.}$$

8.15 CRYSTAL OSCILLATORS

When certain solid materials are deformed, they generate within them, an electric charge. This effect is reversible in that, if a charge is applied, the material will mechanically deform in response. This is called *Piezoelectric effect*.

Naturally available materials : 1. Quartz   2. Rochelle salt.

Synthetic materials : 1. Lithium sulphate 2. Ammonium-di-hydrogen phosphate, PZT (Lead Zirconate Titanate), $BaTiO_3$ (Barium Titanate).

If the crystal is properly mounted, deformations take place within the crystal, and an electro mechanical system is formed which will vibrate when properly excited. The resonant frequency and Q depend upon crystal dimensions etc. With these, frequencies from few KHz to MHz and Q in the range from 1000s to 100,000 can be obtained. Since Q is high, and for Quartz, the characteristics are extremely stable, with respect to time, temperature etc., very stable oscillators can be designed. The frequency stability will be $\pm$ 0.001%. It is same as $\pm$ 10 parts per million (10 ppm).

The electrical equivalent circuit of a crystal is as shown in Fig.8.21. L,C,R are analogous to mass, spring constant, and viscous damping factor of the mechanical system.

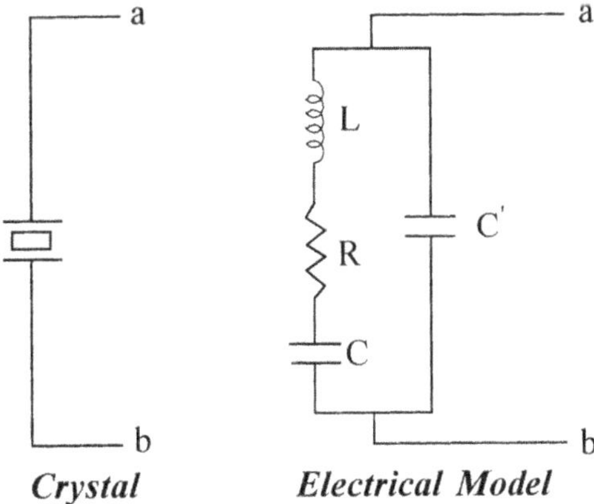

Crystal   *Electrical Model*

Fig. 8.21

Values for a 90 kHz crystal are L = 137 H, C = 0.023pF; R = 15kΩ corresponding to Q = 5,500. The dimension of a crystal will be 30 × 4 ×1.5 mm. C' is the electostatic capacitance between electrodes with the crystal as a dielectric. $C' = 3.5$ pF and is larger than C.

When the crystal slabs are cut in proper directions, with regard to the crystal axis, a potential difference exists between the faces of the crystal slab when pressure is brought to bear on them. And if the slab is placed in an electrostatic field, the slab undergoes deformation. (Fig.8.22). If the electric field is an alternating one, with a frequency which sets the slab into mechanical resonance, the slab will physically vibrate vigorously. Such a crystal can be employed to maintain an oscillation of great frequency stability. When the L.C. circuit in the plate circuit is tuned close to the crystal resonant frequency, steady oscillations will be established. These are maintained by C whose oscillations value is small. By placing crystal between the gate and source, of the FET amplifier, and feeding back a small A.C. voltage from the output, to keep crystal vibrating, the circuit becomes an oscillator with precise stability. Accuracy $\simeq 0.01\%$ 'f range is 0.1 to 20 MHz. By keeping crystal in an oven, accuracy can be improved to 0.001%.

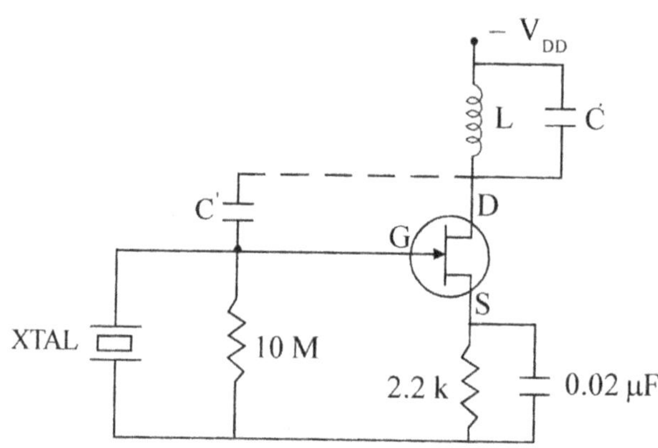

Fig. 8.22 Crystal oscillator circuit.

8.16   FREQUENCY STABILITY

It is a measure of the ability of the circuit to maintain exactly the same frequency for which it is designed over a long time interval. But actually in many circuits the 'f will not remain fixed but it drifts from the designed frequency continuously. This is because of variations of number of parameters, circuit components, transistor parameters, supply voltages temperature, stray capacitances etc. In order to increase the stability the factors which effect the 'f largely should be taken care of. If 'f depends only on R and C high precision R and C should be employed. Also temperature compensating elements are to be employed.

Effect of temperature on inductors and capacitors amounts to more than 10 parts per million per degree change.

Crystal will have mass, elasticity and damping. Crystal will have very high mass to elastic ratio $\left(\dfrac{L}{C}\right)$ and to a high ratio of mass to damping (high Q). Its value is of the order of 10,000 to 30,000.

The crystal is coupled with external C, and the LC circuit oscillates. In the oscillator circuits, instead of external L and C, crystal is connected.i.e. crystal L and C are being used. So 'f' is fixed by crystal itself. 'f' will not vary with temperature. To change 'f' another crystal is used.

8.17 FREQUENCY OF OSCILLATIONS FOR PARALLEL RESONANCE CIRCUIT

$$Z = \frac{(R - j\omega C) \times \dfrac{1}{j\omega C}}{(R + j\omega C) + \dfrac{1}{j\omega C}} = \frac{R + j\omega C}{(1 - \omega^2 LC) + j\omega RC}$$

$$Y_1 = \frac{1}{R + j\omega C} = \frac{R - j\omega C}{R^2 + \omega^2 C^2} \qquad Y_2 = j\omega C$$

Total admittance $\quad Y = Y_1 + Y_2 \quad = \dfrac{R}{R^2 + \omega^2 C^2} - j\left(\dfrac{\omega C}{R^2 + \omega^2 C^2} - \omega C\right)$

At resonance, imaginary part is zero $\qquad \therefore \quad R \to \infty$.

$$\therefore \qquad = \frac{\omega_2}{R^2 + \omega^2 C^2} = \omega C \quad \text{or} \quad \frac{L}{C} = R^2 + \omega^2 C^2$$

$$\omega^2 C^2 = \frac{L}{C} - R^2 \quad \text{or} \quad \omega^2 = \frac{1}{LC} - \frac{R^2}{L^2}$$

or $$f = \frac{1}{2\pi} \sqrt{\frac{1}{LC} - \frac{R^2}{L^2}}$$

This is the expression for frequency of oscillations.

Amplitude of oscillator will be very small (v is Less) for series resonance circuit. So parallel tuned circuits are employed.

8.18 1-MHz FET CRYSTAL OSCILLATOR CIRCUIT

In the basic configuration of oscillator circuit,

$$Z_1 + Z_2 + Z_3 = 0.$$

Z_1 is crystal Z_2 is L and C combination.

Z_3 is C_{gd}. The frequency of oscillator essentially depends up on crystal only as shown in Fig. 8.23. Z_2 and Z_3 values are insignificant compared to Z_1 of the crystal. Therefore the stability is high.

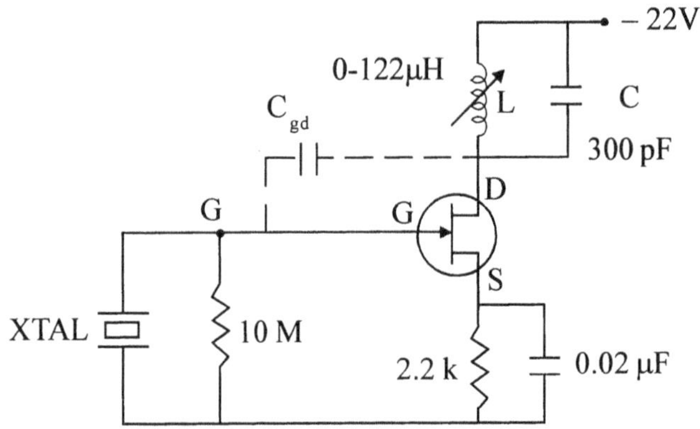

Fig. 8.23 Crystal oscillator circuit.

When electrical input is given, mechanical vibrations are set in the crystal, due to piezoelectric effect. But the crystal is being considered as an inductor capacitor combination only. So it acts like a $|Z|$ R_o L, C combination. Frequency of oscillations depend only on crystal. Other L and Cs are insignificant, as their values are less compared to L, C and R of the crystal.

SUMMARY

- Oscillators generate A.C. output without external A.C input by using Noise A.C signal generated in switching. D.C power from V_{CC} or V_{DD} is converted to A.C. power. The range of frequency signals generated by the circuit can be high and output power is small.

- For sustained oscillations, the conditions known as Barkhausen criterion to be satisfied are (i) $|\beta A| \geq 1$ (ii) Total loop phase shift must be 0^0 or 360^0 (or phase shift of feedback network must be 180^0 when the amplifying device produces another 180^0).

- In the case of JFET, R - C phase shift oscillator circuit, $f_0 = \dfrac{1}{2\pi RC\sqrt{6}}$

- In the case BJT, R - C phase shift oscillator circuit, $f_0 = \dfrac{1}{2\pi RC\sqrt{6+4k}}$ where $K = \dfrac{R_C}{R}$

- For Hartley oscillator circuit which employs two inductors and one capacitor in the feedback network is, $f_0 = \dfrac{1}{2\pi\sqrt{(L_1+L_2)C_3}}$

- For Colpitts oscillator circuit with two capacitors and one inductor in the feedback network, $f_0 = \dfrac{1}{2\pi} \dfrac{1}{\sqrt{L_3}} \sqrt{\dfrac{C_2 + C_1}{C_1 C_2}}$

- For Wein Bridge oscillator circuit, the minimum gain of amplifier must be 3. It can produce variations in f_0 in the ratio of 10 : 1 compared to a variation of 3 : 1 in other types of oscillator circuits.

- Thermistors with (NTCR) and sensistor with positive temperature coefficient of resistance (PTCR) are used for frequency stability in oscillator circuits.

- The specification parameters of oscillator circuits are (i) Amplitude stability (ii) Frequency stability (iii) Frequency range (iv) Distortion in output waveform etc.

- Crystal oscillators produce highly stable output waveform in the high frequency range of MHz also.

OBJECTIVE TYPE QUESTIONS

1. The difference between an amplifier circuit and oscillator circuit is ________________.

2. A.C. output signal is generated by the oscillator circuits without external A.C input, by amplifying ______________ signal.

3. Oscillator circuits employ ______________ type of feedback.

4. One condition for sustained oscillations is total loop phase shift must be ______________ degrees.

5. As per Berkauhsen criteria for sustained oscillations, $|\beta A|$ must be ______________.

6. The range of frequencies over which R - C phase shift oscillator circuit is used is ______________.

7. In the Mega Hertzs frequency range, the type of oscillator circuit used is ______________.

8. The range of frequency over which Wein Bridge oscillator used is ______________.

9. In the feedback network if two inductors and one capacitor elements are used, the oscillator circuit is ______________.

10. The oscillator circuit which employs two capacitors and one inductor in the feedback network, is ______________ oscillator circuit.

11. Naturally occuring materials used for in crystal oscillator circuits, exhibiting piezoelectric effect are 1. ______________ 2. ______________.

12.   Synthetic materials which exhibit piezoelectric effect are,

      1. _________________   2.   _________________

      3. _________________   4.   _________________

13.   The ratio of frequency variation possible with Wein Bridge Oscillator circuit is _________________ is compared to the ratio of _________________ with others oscillator circuits.

14.   Thermistors have temperature coefficient and the materials used are _________________.

15.   Typical values of L, C, R and Q of a crystal used in oscillator circuits are
      L = _________________   C = _________________   R = _________________
      Q = _________________.

16.   Expression for frequency of Osillations in the case of Phase-shaft oscillator circuit is f_0 = ... and the value of B must be at least ....

17.   Expression for frequency of oscillations in the case of Wien Bridge Oscillator is f_0 = ...

18.   In the case of Colpitts oscillator f_0 = ...

19.   For Hartley oscillator, f_0 = ...

20.   Frequency of oscillations, in the case of uJT oscillator circuit is fo$\sim$ ....

ESSAY TYPE QUESTIONS

1.   Explain the basic principle of generation of oscillations in LC tank circuits. What are the considerations to be made in the case of practical L.C. Oscillator Circuits ?

2.   Deduce the Barkausen Criterion for the generation of sustained oscillations. How are the oscillations initiated?

3.   Draw the circuit and explain the princple of operaiton of R.C.phase-shift oscillator circuit. What is the frequency range of generation of oscillations ? Derive the expression for the frequency of oscillations.

4.   Derive the expression for the frequency of Hartely oscillators.

5.   Derive the expression for the frequency of Colpitt Oscillators.

6.   Derive the expression for the frequency of Wein Bridge Oscillators.

7.   Derive the expression for the frequency of Crystal Oscillators.

8.   Explain how better frequency stability is obtained in crystal oscillator ?

9.   Draw the equivalent circuit for a crystal and explain how oscillations can be generated in electronic circuits, using crystals.

10.   Why three identical R-C sections are used in R-C phase-shift oscillator circuits? Consider the other possible combinations and limitations.

MULTIPLE CHOICE QUESTIONS

1. **Johnson Noise is due to .....**
   - (a) Humidity
   - (b) Atmospheric conditions
   - (c) Temperature
   - (d) Interference

2. **The noise that arises in electronic circuits due to variation in concentrations of carriers in semiconductor devices is**
   - (a) shot noise
   - (b) thermal noise
   - (c) Johnson noise
   - (d) None of these

3. **Expression for frequency of oscillations for the R-C phase shift oscillator circuit using JFET is, $f_0 =$**
   - (a) $\dfrac{1}{2\pi RC}$
   - (b) $\dfrac{1}{2\pi RC^2\sqrt{6}}$
   - (c) $\dfrac{1}{2\pi R^2 C\sqrt{6}}$
   - (d) $\dfrac{1}{2\pi RC\sqrt{6}}$

4. In the case of JFET R-C phase shift oscillator circuit, in order that $|\beta\,A|$ is not less than unity, μ **of JFET must be at least**
   - (a) 16
   - (b) 92
   - (c) 29
   - (d) None of these

5. **Oscillator circuit in which the reactances α_1, α_2 are capacitive and α_3 is inductive is ....**
   - (a) Colpitts oscillator
   - (b) Hartely oscillator
   - (c) Crystal oscillator
   - (d) None of these

6. **In wein bridge oscillator circuits the range over which frequency of oscillations can be varied is ...**
   - (a) $3:1$
   - (b) $10:1$
   - (c) $6:1$
   - (d) None of these

7. **Example for naturally occurring piezoelectric crystal material is ....**
   - (a) $BaTio_3$
   - (b) Rochelle salt
   - (c) PZT
   - (d) None of these

8. **Typical values for 90 KHz crystal are :**
   - (a) $L = 137$ H; $C = 0.0235$ μf; $R = 15$ KΩ; $Q = 5,500$
   - (b) $L = 1$ H; $C = 0.02$ f; $R = 1Ω$; $Q = 10$
   - (c) $L = 137$ H; $C = 0.02$ μf; $R = 10$ KΩ; $Q = 1$
   - (d) None of these

9. **One characteristic feature of crystals is ...**
   - (a) They have low mass to elastic ratio
   - (b) They have high mass to elastic ratio
   - (c) They have low Q
   - (d) None of the above is correct

10. **For Wein Bridge oscillator circuits, the minimum gain of amplifier must be**
    - (a) 1
    - (b) 3
    - (c) 10
    - (d) None the these

CONCEPTUAL QUESTIONS (Interview Questions)

1. How oscillator circuit gives output signal without input singnal ?

Ans : The circuit needs D-C input voltage supply. When the supply is switched on, the transient signal produced acts as the a.c input signal. It is amplified by the amplifier and fedback to the input. By the time the transient dies down, the feed back signal acts as the input signal. Output oscillations will be sustained for a particular frequency component for which Barkhausen criteria is satisfied. The wave shape is sinosoidal because sinosoidal function is the only function that satisfies the condition governing the exchange of energy between the energy storage elements in the circuit.

2. Why R - C Phase shift Oscillator Circuit must have three R - C networks ? Why not two or four RC networks?

Ans : If only two R - C networks are used, each network must produce 90^0 phase shift, to get a total phase shift of 180^0 in the feed back circuit. So we must have ideal capacitor, which is not practically possible. On the otherhand we can have 4 - RC networks each producing 45° phase shift. But we will be using more number of R-C components which is not economical.

3. Why the frequency of oscillations of a crystal oscillator circuit are stable ?

Ans : A crystal can be represented as a combination of R L and C components in its equivalent circuit form. When a crystal is connected in the circuit, it is as if R, L and C are connected in the circuit. For a crystal, these equivalent parameters of R, L and C do not change much with temperature. So frequency of oscillations of crystal oscillator circuit do not change with temperature. Thus frequency stability is high. For other oscillator circuits the R, L, C component values change significantly with temperature. So frequency stability is less.

Additional Objective Type Questions (Chapter 1-8)

1.   In a BJT the arrow on the emitter lead specifies the direction of _______________ where the emitter-base junction is _______________ biased.

2.   The emitter efficiency of a BJT is defined as the ratio of current of injected carriers at _______________ junction to total _______________ current.

3.   _______________ type of npn junction transistor is made by drawing a single crystal from a melt of silicon whose _______________ concentration is changed during the crystal drawing operation by adding n or p type atoms as requires.

4.   In an _______________ type of pnp transistor two small dots of indium are attached to opposite sides of a semiconductor.

5.   In the active region of common base output characteristics, the collector junction is _______________ biased and the emitter junction is _______________ biased.

6.   If both emitter and collector junctions are (a) forward biased, the transistor is said to be operating the _______________ region. (b) reverse biased, the transistor is said to be operating in the _______________ region.

7.   The operation of a FET depends upon the flow of _______________ carriers only. It is therefore a _______________ device.

8.   The maximum voltage that can be applied between any two terminals of the FET is the _______________ voltage that will cause avalanche breakdown across the _______________.

9.   A MOSFET of the depletion type may also be operated in an _______________ mode. Some times the symbol for the JFET is also is used for the MOSFET with the understanding that _______________ is internally connected to source.

10.   The emitter diode of a UJT drives the junction of R_{B1} and R_{B2} and $R_{BB} = R_{B1} + R_{B2}$. The intrinsic stand off ratio is defined as $\eta =$ _______________ and its value usually lies, between _______________.

> ### *Answers to Additional Objective Type Questions*

1. Holes or conventional current, forward

2. Emitter base, emitter

3. Grown junction, doping

4. Alloy junction

5. Reverse, Forward

6. Saturation, Cut-off

7. Majority, Unipolar

8. Breakdown, drain-source junction

9. Enhancement, the gate

10. $\dfrac{R_{B_1} + R_{B_2}}{R_{BB}}$, 0.5 & 0.82

Select Bibliography

1. Electronic Devices and Circuits

     - Millman and Halkias TMH

2. Micro-electronics

     - Millman and Grabel TMH

3. Electronic Devices and Circuits

     - R.L Boylestad and Loni Nashelsky, Pearson Ed. Asia, PH

4. Engineering Electronics

     - John. D. Ryder, Mc-Graw Hill

5. Electronic Circuits: Discrete and Integrated

     - David L. Schilling and Charles Belove McGraw-Hill

6. Basic Electronics for Scientists

     - James J. Brophy, Tata McGraw-Hill